Geospatial Technologies for Land Degradation Assessment and Management

Geospatial Technologies for Land Degradation Assessment and Management

R. S. Dwivedi

CRC Press
Taylor & Francis Group
Boca Raton London New York

CRC Press is an imprint of the
Taylor & Francis Group, an **informa** business

CRC Press
Taylor & Francis Group
6000 Broken Sound Parkway NW, Suite 300
Boca Raton, FL 33487-2742

First issued in paperback 2020

ISBN 13: 978-0-367-57089-7 (pbk)
ISBN 13: 978-1-4987-4960-2 (hbk)

Visit the Taylor & Francis Web site at
http://www.taylorandfrancis.com

and the CRC Press Web site at
http://www.crcpress.com

Dedicated to

my beloved wife

Asha

Contents

Foreword

Land and soil degradation are increasingly important issues because of the ever-expanding demands of the growing and increasingly affluent world population. The world population (billion people) was 2.5 in 1950, 7.6 in 2017, and is projected to be 8.6 in 2030, 9.8 in 2050, and 11.2 in 2100. Similarly, the population of India (million people) was 376 in 1950, 1339 in 2017, and is projected to be 1513 in 2013, 1659 in 2050, and 1517 in 2100. Thus, it is the decline in per capita availability of land/soil resources which is the major cause of concern. Further, the concern is aggravated by the decrease in productivity and delivery of essential ecosystems services because of the decline of soil quality and health by a range of degradation processes. Important among these are physical (decline in soil structure leading to slaking, crusting, compaction, hard setting, erosion, runoff, inundation, drought), chemical (elemental imbalance leading to salinization, acidification, nutrient depletion, toxicity of certain heavy metals, decline in cation exchange capacity), biological (depletion of soil organic carbon concentration ad stock, decline in activity and species diversity of soil biota, methanogenesis, nitrification and denitrification, reduction in enzyme activity), and ecological (disruption in biogeochemical and biogeophysical cycling, perturbation of hydrological and energy balance, change in climax vegetation and the surface cover, decline in gross, net ecosystem and biome productivity).

Despite the widespread severity of the problem, the credible estimates of land area affected by diverse soil degradation processes are not available. Global estimates of the total land area affected by diverse degradation processes range from 1 to over 6 billion hectares, with a large range in estimates of spatial/geographical distribution; the annual rate of degradation; economical and ecological impacts; and predominant processes, causes and factors affecting the extent and severity. Similarly, the recent estimates of soil degradation in India range from 114 to 147 Mha.

It is widely believed that what cannot be precisely measured cannot be improved or restored. Peter Drucker stated that "You can't manage what you can't measure." To know whether one is successful in restoration of degraded soils, it is important to know precisely what is being changed and by what criteria. It was Edward Deming who stated that "There are many things that cannot be measured and still must be managed." Soil degradation is one such issue which, with the present state-of-the-knowledge, cannot be precisely measured. In addition to the lack of credible data on the extent and severity of degradation by different processes, there is also an insufficient knowledge regarding the threshold levels and critical limits of key soil properties in relation to specific land use (e.g., arable, pastoral, plantations, horticultural crops). Key soil properties, which can also differ among soil types, land uses, and climates, include bulk density soil organic carbon concentration, plant available water capacity, infiltration rate, air porosity at field moisture capacity, pH, EC, CEC, nutrient reserves (macro and micro), enzyme activity, microbial biomass C, respiration quotient, etc. These critical properties are important to "soil functionality."

Soil functionality refers to the capacity of soil to perform numerous functions. Important among these, and adversely affected by soil degradation, are (1) production of biomass, (2) moderation of climate, (3) cycling of elements, (4) decomposition of waste, (5) renewal and purification of water, (6) providing habitat for biodiversity, (7) creating media for plant growth, (8) being foundation of civil structures, (9) source of archive of human and planetary history, and (10) providing spiritual, aesthetical, cultural, and recreational opportunities. These functions are difficult to measure directly and are estimated through indices of soil quality and soil health. Soil degradation, its extent and severity, can also be estimated by assessing indices of soil quality and health.

Land/soil degradation is also relevant to several global issues. It affects food and nutritional security by reducing agronomic productivity and jeopardizing nutritional quality. Degradation aggravates the emission of greenhouse gases from soil into the atmosphere especially of CO_2, CH_4, and N_2O. Gaseous emission is exacerbated by accelerated erosion, and inundation and anaerobiosis which increase methanogenesis, nitrification, denitrification, and mineralization. Water quality and renewability are adversely affected because of the increase in non-point source pollution and transport of dissolved and suspended loads. Algal bloom is aggravated by surface runoff from watersheds prone to soil degradation. Depletion of organic carbon and vegetation cover in ecosystem and soils reduce energy supply and habitat and adversely impact above-and below-ground biodiversity. Degraded landscapes are mirror images of the people who depend on it, and their well-being is adversely impacted by reduction in soil functionality.

The 13-chapter volume addresses pertinent themes related to the specific title such as in chapters: (1–3) Remote Sensing, (4) Digital Image Processing, (5–12) Soil Degradation Processes, and (13) Land Degradation Information Systems. These chapters address pertinent information on commonly observed and widely distributed degradation types. In addition to the processes addressed in these 13 chapters, soils of India and elsewhere in developing countries are also prone to degradation by depletion of plant nutrients and soil organic carbon, removal of topsoil or scalping to ~1 m depth for brickmaking, pollution and contamination by industrial effluent and radioactive wastes, encroachment by urbanization and surface sealing, etc. These, rather uncommon but important, degradation processes are adversely impacting soil health and reducing the per capita availability of scarce but precious land resources. Therefore, land/soil degradation by these unconventional processes must also be assessed, restored, and managed. Yet, this volume is complimentary to several other books such as Characterization and Classification of Salt-Affected Soils by J.K. Jena and P. Parichita, and Remote Sensing for Soil Erosion Prediction by A.S. Budiharso.

Satellite imagery and remote sensing techniques, used to measure landscape parameters and terrain attributes, can be important tools in assessing the extent and severity of land/soil degradation, temporal changes, and geospatial distribution. It is precisely in this context that this volume "Geospatial Technology for Land Degradation Assessment and Management" by Dr. R. S. Dwivedi is timely and highly pertinent. The state-of-the-knowledge presented in the book is of interest to researchers but also to land managers and policymakers. The information presented is also relevant to advancing the Sustainable Development Goals (SDGs) of the United Nations and in implementing programs adopted by COP 21 (4 per Thousand) and COP 22 (Adapting African Agriculture). I commend Dr. R. S. Dwivedi for undertaking this timely initiative.

Rattan Lal,
Distinguished University Professor of Soil Science, SENR
Director, Carbon Management and Sequestration Center
President, International Union of Soil Sciences

Preface

In order to ensure food security, bringing additional land under agriculture and enhancing the productivity of available agricultural land are prerequisites. Globally, an estimated 1.9 billion hectares land is subject to land degradation—defined as a decline in land quality caused by human activities (Oldeman, 1994). Land degradation covers various forms of soil degradation, adverse human impacts on water resources, deforestation, and lowering of the productive capacity of rangelands and loss of biodiversity. The consequences of land degradation are reduced land productivity, socioeconomic problems, including uncertainty in food security, migration, limited development, and damage to ecosystems. Information on nature, magnitude, spatial extent, and dynamics of land degradation is a prerequisite for developing strategies to combat these processes and mitigate their effects at the land-management and policy level. Remote sensing and allied technologies like Geographic Information System (GIS), Personal Digital Assistant (PDA), Global Positioning System (GPS), and internet technology—collectively known as geospatial technologies—provide such information in timely and cost-effective manner.

The book initially introduces the geospatial technologies—basics of remote sensing, GIS and GPS remote sensors and systems, digital data processing, and data analysis/interpretation techniques. It is followed by an overview of various land degradation processes. The subsequent chapters are dedicated to applications of geospatial technologies to major individual land degradation processes, namely soil erosion by water and wind, soil salinization and/or alkalinization, and waterlogging. Other land degradation processes like soil acidification, mining, aquaculture and shifting cultivation are also addressed appropriately in other chapters. The drought which accelerates the land degradation process is dealt with in the next chapter. The last chapter addresses the development of the information system on degraded lands that would enable updation and dissemination of digital data on degraded lands to the end users.

It is shown how remote sensing data may be utilized for inventorying, assessing, and monitoring affected ecosystems and how this information may be assimilated into integrated interpretation and modeling concepts. Additionally, case studies demonstrating the utility of geospatial technologies in generating information on degraded lands required for their rehabilitation are embedded in the frame of different local settings. Lastly, the future challenges in applications of geospatial technologies in land degradation studies are also enumerated.

R. S. Dwivedi
Hyderabad, India

Reference

Oldeman, L.R., 1994. The global extent of land degradation. In: Greenland, D.J., and Szabolcs, I., (eds.) Land Resilience and Sustainable Land Use. CABI, Wallingford, pp. 99–118.

Acknowledgments

At the outset, I express my heartfelt thanks to Krish, my grandson, for inspiration that I received for the stupendous task of preparing the manuscript of the book. In fact, I conceived the idea of this task when I was in Fremont, California, USA with him. My sincere thanks also go to Dr. Ram Shankar Dwivedi, brother, and Ms. Suman Dwivedi, sister-in-law for their blessings in this endeavor. Furthermore, the incessant encouragement and moral support received from Ms. Dolly Dwivedi, daughter, and Shri Karik Vittal, son-in-law, is beyond the words of expression. The concepts and illustrations are the soul of any literary work. In this context, the technical support received from Shri Samik Saha, Team Lead, Ms. M.S. Mahathi, Associate Analyst, Content Engineering Global Logic Technologies Pvt. Ltd., Hyderabad in generating the illustration from digital images has been apt and timely. I express my sincere gratitude to Prof. Rattan Lal, Distinguished University Professor of Soil Science, Ohio State University, USA for sparing his valuable in giving foreword of the book.

I am indebted to Dr. Y.V.N. Krishnamurthy, Distinguished Scientist and Director, National Remote Sensing Centre (NRSC), Indian Space Research Organization, Hyderabad for providing satellite data, and thematic maps and library facility that have added strength to the technical content of the book. Special thanks are due to Dr. T. Ravisankar, Group Director Land Resources, Land Use Mapping and Monitoring Group, NRSC for providing illustrations on potential of Resourcesesat-2 LISS-III data for mapping land degradation. Besides, timely technical and logistics support provided by Shri P. Ravinder, In-charge Library, NRSC, and his colleagues. Ms. Seema Kukarni and Ms. Suman S. Paul have been instrumental in preparing the manuscript.

I am extremely thankful to Dr. Ch V. Rao, Group Head, Special Products, NRSC for offering necessary suggestions and comments on the chapter on Digital Image Processing. Discussions with Dr. M.V.R. Sesha Sai, Deputy Director, Earth and Atmospheric Sciences Area, Dr. T. Ravisankar, and Dr. K. Srinivas, NRSC have been of immense help in finalizing the manuscript.

I also express my sincere thanks to Dr. N. Aparna, Group Head, NDC and her Associate Ms. K. Swarupa Rani for sharing promotional satellite data products, and to Shri G.P. Swamy, IRS Optical Data Processing Group, and Shri R. Anjaneyulu, Shri E. Venkateswarulu and Ms. Asra Majid, Senior Research Fellow, NRSC for providing sample images of the Indian Earth observation mission data and necessary technical support.

I am also thankful to the Food and Agriculture Organization (FAO) and International Soil Reference and Information Centre (ISRIC) for generously granting the permission to use and reproduce the material related to maps on land degradation. Thanks are also due to Dr. Reich Paul, USDA-NRCS, Soil Science Division, World Soil Resources for permitting to reproduce soil pH map of the world. The technical support received from the CRC Press publisher-deserves special thanks.

The encouragement and moral support provided by Shri J. Venkatesh, Associate Professor, Prof. K. Ramamohan Reddy, Prof. C. Sarala and Prof. M. Vishwanadham, Centre for Spatial Information Technology (CSIT), JNTUH is duly acknowledged. Thanks are also due to Shri L. Ravi, Doctoral scholar, Shri Ballu Harish, Faculty, CSIT for providing necessary technical support, and to Shri R. Naresh for secretarial support in finalizing the manuscript. Last but not the least, I am also thankful to all the authors of articles and the books that I have referred to and to those who have directly or indirectly contributed to the success of this mission.

Author

Dr. R. S. Dwivedi earned his master's and PhD degrees in Agricultural Chemistry from University of Allahabad, India, in 1973 and 1977, respectively, and Advanced Diploma in Remote Sensing from the University of Berlin, Germany, in 1979. He had joined the National Remote Sensing Agency (now National Remote Sensing Centre), Department of Space, Government of India, in December 1977. Dr. Dwivedi pursued his research career at National Remote Sensing Agency in applications of remote sensing technology for inventory and mapping of soil resources including land degradation. The author played a key role in mapping of salt-affected soils, wastelands, and land use/land cover mapping and monitoring for the Indian subcontinent. He is endowed with around 40 years' experience in applications of remote sensing. The testimony of such an achievement is reflected in terms of research articles (84) in national and international journals, awards (from Indian Society of Remote Sensing, and Doreen Mashler Team Award from International Crops Research Institute for Semi-arid Tropics, Hyderabad), and fellowships of the National Academy of Agricultural Sciences, India, Indian Geophysical Union and A.P. Academy of Sciences. He has contributed 13 book chapters, co-edited two books—*Remote Sensing Applications* and *Geospatial Technology for Integrated Natural Resources Management*. Also he authored a book titled *"Remote Sensing of Soils"* with Springer-Verlag, Germany. Furthermore, he had also lead a team of soil scientists, as Head Land Degradation Division, and superannuated as group director, Land Resources in 2011. He has been offered assignments from the Government of Ethiopia as a professor, Soil Science for a period of 2 years (2013 and 2014) at Hamaraya University, and at Mekelle University from October 2015 to September 2017. He has been associated with Jawaharlal Nehru Technological University, Hyderabad (JNTUH), India, since 2012 as a guest faculty member. Currently, he is academic adviser at, Geospatial Information Technology, JNTUH.

1

An Introduction to Geospatial Technology

1.1 Introduction

Derived from two Latin words—*remotus,* meaning far away or distant in time or place, and *sensus* meaning to detect a stimulus by means of any of the five senses—remote sensing refers to detecting an object/feature/phenomenon with an observation device that is not in intimate physical contact with it. According to Colwell (1966), the term *remote sensing* in its broadest sense merely means "reconnaissance at a distance." The process of making measurements or taking observations in the laboratory or field is termed *proximal sensing.* Remote sensing thus differs from proximal sensing in the way information is gathered about an object/feature or phenomenon. While the instruments are immersed in, or physically touch, the objects of measurement in proximal sensing or *in situ* measurements, the sensing device is invariably not in physical contact with the objects in the case of remote sensing. In the context of Earth observations and environmental management practice, remote sensing refers, in general sense, to the instrumentation, techniques, and methods used to observe, or sense, the surface of the Earth usually by the formation of an image in a position, stationary or mobile, at a certain distance from that surface (Buiten and Clevers, 1993).

During the pre-1960s period, the concept of remote sensing was confined to the development of different components of photography, i.e., various types of films, camera, film processing and printing systems, interpretation techniques, and photogrammetry. By the early 1960s, many new types of sensing devices were being introduced that could detect electromagnetic radiation (EMR) in spectral regions far beyond the range of visible spectrum or human vision and photographic films. In order to accommodate these developments, Evelyn L. Pruitt, a geographer formerly with the US Office of Naval Research, coined the term *remote sensing* to replace the more limiting term *aerial photographs.*

1.1.1 Geospatial Technology

With the development of computer technology in 1960s and realization of the potential of integrating the information on natural resources to derive more comprehensive and meaningful information and arriving at decisions for planning and management of natural resources and environment, the concept of a new technology called Geographical Information System (GIS) was developed. In this context, precise location of features assumes greater significance. For precise location of observations and information on natural resources and environment, satellite-based navigation system was developed in 1960.

Developments of Internet technology and its integration with the computer technology ultimately culminated into the development of personal digital assistant (PDA) devices that enabled transmitting the *in situ* observations on natural disasters and other highly dynamic phenomenon requiring real-or near-real-time field data for analysis and/or interpretation of remote sensing data. In an attempt to accommodate a host of these newly developed technologies, i.e., remote sensing, GIS, satellite-based navigation, and PDAs, a new term, *geospatial technology* or *geomatics* or *geoinformatics*, was coined.

Geospatial technology refers to equipment used in visualization, measurement, and analysis of Earth's features, typically involving such systems as remote sensing, GIS, and global positioning system (GPS). In other words, the term *geospatial technology* is used to define the collective data and associated technology that has a geographic or locational component. It has been defined as" An integrated science and technology that deals with acquisition and manipulation of geographical data, transforming it into useful information using geoscientific, analytical, and visualization techniques for making better decisions" (Jaganathan, 2011). *Geoinformatics*—another term used sometimes as synonym to *geospatial technology*—has been defined as "a descriptive which integrates the acquisition, modeling, analysis and management of spatially referenced data" (Lein, 2012).

In remote sensing systems, the information from the object/target is carried through EMR. That is to say, the EMR acts as a carrier of information from the object/target to the sensing instrument(s). The radiant energy thus captured by the sensing instruments is used by the image interpreter/analysts to derive information on the object/target. In the case of spectral measurements in the laboratory or *in situ* (field), the radiation reflected/scattered or emitted is captured as it is without any interference by the medium between the objects/target and sensing instruments. This is also true to some extent in case of airborne sensors operating in the optical region of the electromagnetic spectrum (EMS). When the spectral measurements are made from the space, the intervening atmosphere modulates both radiant energy emanating from the source to the object/target and that reflected/scattered from the target reaching the sensing instruments. The sensors capture reflected/emitted/backscattered EMR from the object/feature. These sensors/instruments are of two types, i.e., *passive sensors* and *active sensors*. *Passive sensors* detect natural energy (radiation) that is reflected or emitted by the object. Reflected sunlight is the most common source of radiation measured by passive sensors. Examples of passive remote sensors include film photography, charge-coupled devices, and radiometers. *Active sensors*, on the other hand, have their own source of energy to illuminate the object or terrain and record the backscattered energy from there. Radar and LiDAR are examples of active sensors where the time delay between disseminated and return radiation is measured and is used for establishing the location, height, speed, and direction of an object/feature.

1.2 History of Remote Sensing

Historically, the airborne remote sensing was primarily focused on surveying, reconnaissance, strategic land use mapping, and military surveillance during the First and Second World Wars. The focus was shifted subsequently to *early space-borne systems* dominated by the launch of Sputnik-1 and Explorer-1 erstwhile USSR and USA in 1957 and 1958, respectively. The launch of advanced meteorological satellites, namely Television Infrared Observation Satellite (TIROS) series in 1960, marked the beginning of space-borne Earth observation systems. In fact, the major breakthrough in Earth observation from space was achieved with the launch of Landsat-1 in 1972. Several satellites, namely Landsat series, Landsat-2 through -8, le Système Pour Observation de la Terre (SPOT), and Indian Remote

Sensing series of satellites with the state-of-the-art sensors, were subsequently launched. And the process is continuing with improved fervor and spirit to cater to newer and challenging applications. For further details on historical sketch of and the major milestones in remote sensing with respect to the development of sensors, platform, and launch vehicles, the readers may refer to Campbell and Wynne (2011) and Jensen (2007).

1.3 Electromagnetic Radiation

Electromagnetic (*EM*) *energy* refers to all energy that moves with the velocity of light in a harmonic wave pattern. A harmonic wave pattern consists of waves that occur at equal interval in time. There are two models of the EMR: *wave model* and *particle model*. The wave model explains how EM energy propagates (moves). However, this energy can only be detected when it interacts with the matter. In this interaction, EM energy behaves as though it consists of many individual bodies/particles called *photons*, which have such particle-like properties as energy and momentum. The reviews on the nature of EM radiation and physical principles are available in Silva (1978) and Suits (1983).

1.3.1 Particle Model

The EMR was primarily thought of as a smooth and continuous wave. Albert Einstein observed that when light interacts with electrons, it has a different character. He concluded that when EMR interacts with matter, it behaves as though it is composed of many individual bodies called *photons*, which carry such particle-like properties as energy and momentum which confirms its duality.

1.3.2 Wave Model

The EMR has been conceptualized as waves that travel through space at the speed of light (3×10^8 ms^{-1}). It consists of two fluctuating fields—one electric (E) and the other magnetic (M) (Figure 1.1). The two vectors are orthogonal to one another, and both are perpendicular to the direction of travel. Important parameters characterizing any EMR under study are as follows: wavelength/frequency/amplitude/phase/the direction of propagation and polarization. The wavelength and frequency of EMR are related as follows:

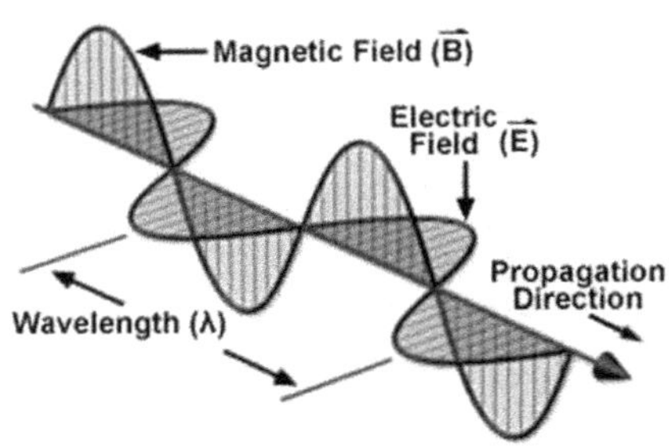

FIGURE 1.1
The electromagnetic radiation.

$$c = \lambda \upsilon \tag{1.1}$$

where c is the velocity of light (3×10^8 m/s)

$$\upsilon = \frac{c}{\lambda} \tag{1.2}$$

$$\lambda = \frac{c}{\upsilon} \tag{1.3}$$

When EMR passes from one substance to another, the speed of the light and its wavelength change, while the frequency remains the same.

1.3.3 Amplitude

The amplitude of EM waves refers to its intensity or brightness. For the visible light, the brightness is usually measured in lumens (Figure 1.2). In the case of other wavelengths, the intensity of the radiation, which is power per unit area or watts per square meter, is used. The square of the amplitude of a wave is the intensity of EMR.

1.3.4 Phase

Phase denotes a particular point in the cycle of a waveform, measured as an angle in degrees. Phase is a number describing the position of the wave within its repeating cycle at any instant in time. It is represented in degrees. The phase ranges from 0° to 360° before repeating (Figure 1.3). The phase of a waveform specifies the extent to which the peaks of one waveform align with those of another.

1.3.5 Polarization

The polarization of EMR refers to the orientation of the oscillation within the electric field of the EM energy. The process of transforming unpolarized light into polarized light is known as polarization. Typically, radar signals are transmitted in a plane of polarization that is either parallel to the antenna axis (*parallel polarization, H*) or perpendicular to that axis (*vertical*

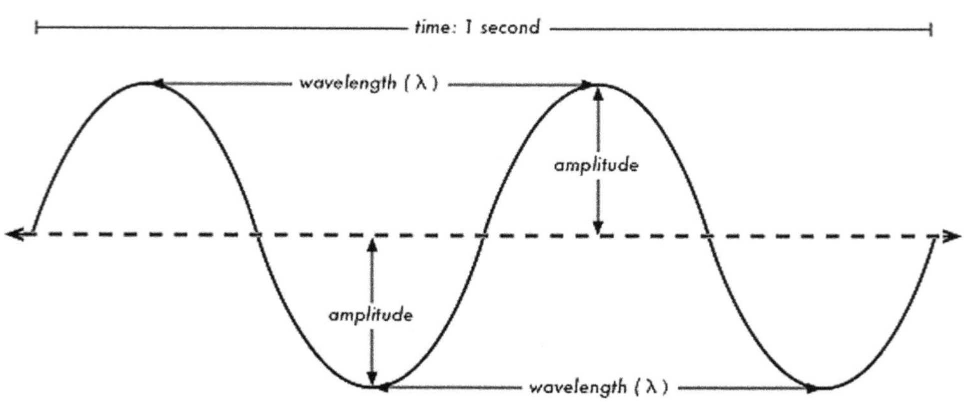

FIGURE 1.2
Wavelength and amplitude of the electromagnetic radiation.

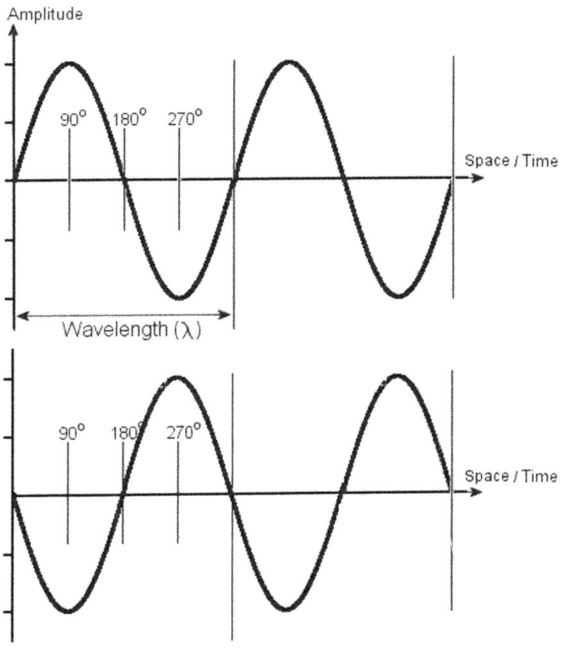

FIGURE 1.3
The concept of the phase of electromagnetic radiation.

polarization, V), as shown in Figure 1.4a and b. Thus, there is a possibility of having four different combinations of signal transmission and reception (HH, VV, HV, and VH), where the first letter indicates the transmitted polarization and the second indicates the received polarization. The HH and VV are referred to as *like polarized* or *co-polarized* signals, while HV and VH are referred to as *cross-polarized* signals. Various objects modify the polarization of the energy they reflect to varying degrees. The mode of signal polarization influences the manifestation of objects/features on the resulting imagery (Lillesand et al., 2004).

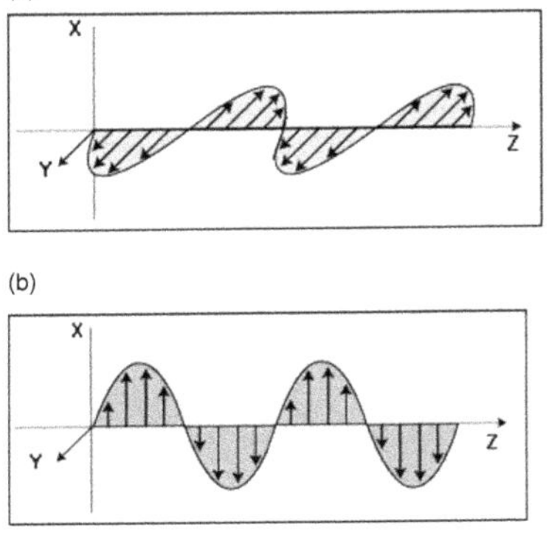

FIGURE 1.4
(a) Horizontally polarized wave is one for which the electric field lies only in the y–z plane. (b) Vertically polarized wave is one for which the electric field lies only in the x–z plane.

1.4 Electromagnetic Spectrum

The EMS is the continuum of energy that ranges from meters to nanometers in wavelength, travels at the speed of light (3×10^8 ms^{-1}), and propagates through a vacuum such as outer space. The EMS spans from 10^{-10} µm (cosmic rays) to 10^{10} µm (radio waves), the broadcast wavelengths (Figure 1.5). The EMS has been divided broadly into ultraviolet, visible, infrared, and microwave regions. However, these divisions are arbitrarily defined. There is no clear-cut dividing line between one nominal spectral region and the next.

The term *optical wavelengths*, extending from 0.30 to 15 µm, is used to denote the region of the EMS where optical techniques of refraction and reflection can be used to focus and redirect radiation. At these wavelengths, EM energy can be reflected and refracted with solid materials. The region between 0.38 and 3.0 µm is frequently referred to as the reflective portion of the spectrum. Energy sensed in these wavelengths is primarily radiation originating from the sun and reflected by objects on the Earth.

1.4.1 The Ultraviolet Spectrum

The ultraviolet literally means "beyond violet," a region of short-wavelength radiation that lies between the X-ray region and the visible region (0.40–0.70 µm) of the EMS. The ultraviolet region is often subdivided into the *near ultraviolet* (0.32–0.40 µm), the far ultraviolet (0.32–0.28 µm), and the extreme ultraviolet (below 0.28 µm), sometimes known as UV-A, UA-B, and UA-C, respectively.

1.4.2 The Visible Spectrum

The term *visible spectrum* is derived from the fact that the human eyes respond to these wavelengths which span from 0.40 to 0.70 µm. The visible spectrum can be divided into three segments: 0.40–0.50 µm (blue), 0.50–0.60 µm (green), and 0.60–0.70 µm (red), the three primary colors.

1.4.3 The Infrared Spectrum

The infrared region extends from 0.72 to 15 µm and has been divided into three broad categories: (i) near infrared (NIR) (0.72–1.30 µm), (ii) middle infrared/shortwave infrared (1.30–3.0 µm), and (iii) far infrared (7.0–15.0 µm). Radiation in the NIR region behaves in a manner analogous to radiation in the visible spectrum. Therefore, remote sensing in the NIR can use films, filters, and cameras with designs similar to those intended for use with the visible light. The far-infrared region (7.0–15 µm) comprises wavelengths well beyond

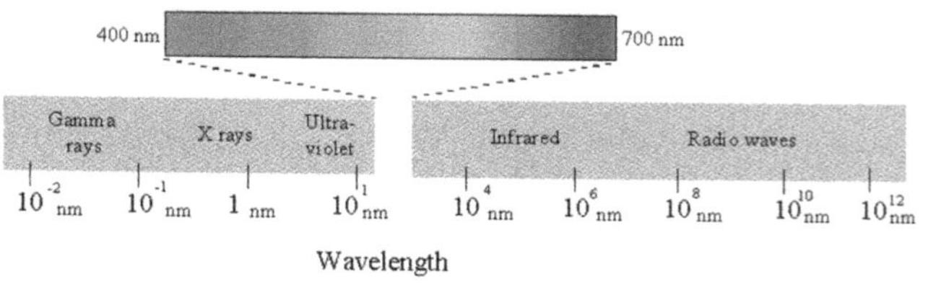

FIGURE 1.5
The electromagnetic spectrum.

TABLE 1.1
Some of the Commonly Used Microwave Frequencies

Band Frequency (GHz)	Wavelength (cm)
P band: 0.3–1	30–100
L band: 1–2	15–30
S band: 2–4	7.5–15
C band: 4–8	3.8–7.5
X band: 8–12.5	2.4–3.8
Ku band: 12.5–18	1.7–2.4
K band: 18–26.5	1.1–1.7
Ka band: 26.5–40	0.75–1.1

the visible extending into regions that border the microwave region. The far-infrared radiation is basically emitted by the Earth. There is no specific term usually applied to the wavelength region from 3.0 to 7.0 μm.

1.4.4 The Microwave Spectrum

The microwave region extends from 1 mm to 1 m. Microwaves are the longest wavelengths commonly used in remote sensing. The shortest wavelengths in this range have much in common with the thermal energy of the far-infrared region. It is further divided into different frequency bands which are commonly used in remote sensing (1 GHz = 10^9 Hz) (Table 1.1). Unlike in the optical region of EMS where wavelength is used to define the spectral bands, frequencies are commonly used for defining spectral bands within the microwave region.

1.5 Energy–Matter Interactions in the Atmosphere

The EM energy that encounters the matter, whether solid, liquid, or gas, is called *incident* radiation. Interaction with the matter can change the intensity, direction, wavelength, polarization, and phase of the EM energy. These changes are recorded, and then the resulting data/images are interpreted to determine the characteristics of the matter that interacted with incident EM energy. The EMR from the sun is propagated through the Earth's atmosphere almost at the speed of light in a vacuum. If the sensor is carried by a low flying aircraft, effects of the atmosphere on image quality may be negligible. In contrast, energy that reaches sensors on board Earth observation satellites must pass through entire depth of the Earth's atmosphere.

Unlike a vacuum where nothing happens, however, atmosphere may affect not only the speed of light but also its wavelength, intensity, and spectral distribution. Besides, in the atmosphere, the EM energy is subject to modification by several physical processes, namely scattering, absorption, and emission. Moreover, a considerable amount of incident radiant flux from the sun is reflected from the top of clouds and other materials in the atmosphere. A substantial amount of this energy is reradiated back to space. It is this reflected energy that is captured by sensors aboard satellites.

1.5.1 Scattering

Scattering of radiation by atmospheric particles has pronounced effects on the reflected EM energy captured by the sensors on board Earth observation satellites. The amount of scattering that occurs depends on sizes of these particles, their abundance, the wavelength of radiation, and the depth of the atmosphere through which the EM energy is traveling. Based on the size of atmospheric particles the EMR is interacting, scattering can be grouped into three types: Rayleigh, Mie, and nonselective scattering.

1.5.1.1 *Rayleigh Scattering*

Rayleigh scattering—sometimes also referred to as *molecular scattering*—dominates when radiation interacts with the atmospheric molecules such as oxygen and nitrogen and other tiny particles which have much smaller diameter (usually <0.1) than the wavelength of the incident EM radiation. Most Rayleigh scattering takes place in the upper 4.5 km of the atmosphere. Rayleigh scattering is inversely proportional to the fourth power of the EM wavelength. Shorter wavelengths, therefore, are scattered much more than their longer counterparts. The blue color of the sky is due to Rayleigh scattering, since it scatters blue light more than the radiation with longer wavelengths such as green and red. Rayleigh scattering is one of the major causes of haze in images.

1.5.1.2 *Mie Scattering*

Mie scattering, sometimes also referred to as nonmolecular scattering, takes place in the lower 4.5 km atmosphere, where there may be many essentially spherical particles present with diameters approximately equal to the size of the wavelength of incident energy. Water vapor, dust, and various aerosols are primarily responsible for Mie scattering. The actual size of the particles may range from 0.1 to 10 times the wavelength of incident energy present in the atmosphere. Mie scattering can influence a broad range of wavelengths in and near the visible spectrum.

1.5.1.3 *Nonselective Scattering*

Nonselective scattering takes place in the lowest portions of the atmosphere where the size of the particles is much larger (>10 times) than the wavelength of the incident EM radiation. Thus, water droplets, having diameters from 50 to 1,000 nm, scatter visible, NIR, and shortwave-infrared wavelengths nearly equally. This type of scattering is nonselective in the sense that light of all wavelengths is scattered.

1.5.2 Absorption

Absorption is a process by which radiant energy is absorbed and converted into another form of energy. The absorption of the incident energy may take place in the atmosphere or on the Earth's surface. Absorption of radiation occurs when the atmosphere prevents, or strongly attenuates, transmission of radiation through it. Energy acquired by the atmosphere is subsequently reradiated at longer wavelengths. Three gases—ozone (O_3), carbon dioxide (CO_2), and water vapor (H_2O)—are responsible for most absorption of solar radiation in the atmosphere.

Wavelengths shorter than 0.30 μm are completely absorbed by the ozone (O_3). Absorption of the high-energy, short-wavelength portion of the ultraviolet spectrum (mainly λ less than 0.24 μm) prevents transmission of this radiation to the lower atmosphere. Hence, it precludes the usage of these wavelength regions' remote sensing.

CO_2 is important in remote sensing because it effectively absorbs radiation in the mid- and far-infrared regions of the spectrum. Its strongest absorption occurs in the region from about 13 to 17.5 μm. Water vapor is several times as effective in absorbing radiation as are all other gases combined. Two of the most important regions of absorption are in several bands between 5.5 and 7.0 μm, and above 27 μm. Absorption in these regions can exceed 80% if atmosphere contains an appreciable amount of water vapor.

1.5.3 Emission

Like Earth, the atmosphere also emits EM radiation due to its thermal state. Owing to its gaseous nature, only discrete bands of radiation are emitted by the atmosphere. The atmospheric emission would tend to increase the path radiance, which would act as a background noise, superposed over the ground signal. However, as spectral emissivity equals spectral absorptivity, *atmospheric windows* are marked by low atmospheric emission. Therefore, for terrestrial sensing, the effects of self-emission by the atmosphere can be significantly reduced by restricting remote sensing observations to well-defined atmospheric windows.

1.6 Atmospheric Windows

After striking the Earth's surface, the reflected component of the incident radiation again travels back to space through the atmosphere. As mentioned in previous sections, the atmosphere attenuates (scatters and absorbs) the incident/outgoing EMR in certain wavelength regions and allows it to pass through the radiation of selective wavelengths. Consequently, radiation in certain wavelength regions only can pass through the atmosphere well. These regions are called *atmospheric windows*. These are the regions of the EMR which are useful in Earth observation. The dominant wavelengths within atmospheric windows are in the visible and radio-frequency regions, while X-rays and UV seem to be very strongly absorbed and gamma rays and infrared are somewhat less strongly absorbed.

1.6.1 Atmospheric Windows in Optical Region

Extending from X-rays (0.02 μm wavelength) through visible and far-infrared (1 mm wavelength), the optical range refers to that of the EMS in which optical phenomena of reflection and refraction can be used to focus the radiation. Atmospheric windows in the optical infrared region include (i) 0.3–1.3 μm, (ii) 1.5–1.8 μm, and (iii) 2.0–2.6 μm (Figure 1.6).

CO_2 exhibits its strongest absorption in the region from about 13 to 17.5 μm. Water vapor is several times as effective in absorbing radiation as are all other gases combined. Two of the most important regions of absorption are in several bands between 5.5 and 7.0 μm, and above 27 μm. Absorption in these regions can exceed 80% if atmosphere contains an appreciable amount of water vapor. In addition, nitrous oxide (N_2O) present in the atmosphere absorbs the radiant energy in certain portions of the EMS. The cumulative effect of the absorption by the various constituents can cause atmosphere to be opaque in certain regions of the spectrum. Consequently, in these regions, practically no energy is available for remote sensing.

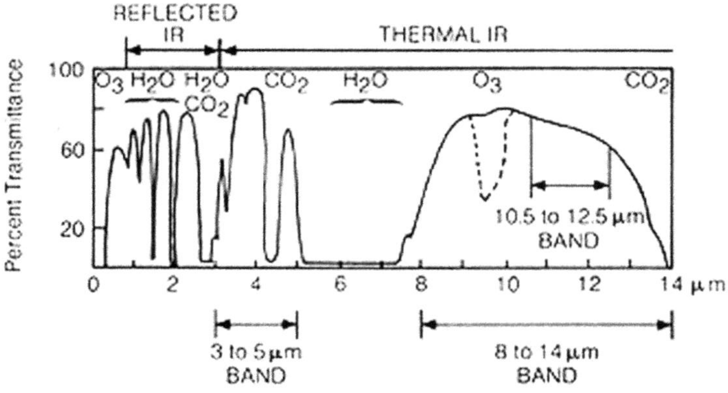

FIGURE 1.6
The atmospheric windows in visible and infrared regions.

1.6.2 Atmospheric Windows in Microwave Region

The atmosphere is opaque in the range from 22 μm to 1 mm, and hence, this part of the EM spectrum is not used for remote sensing. Molecular oxygen and water vapor are the major absorbing constituents in the microwave region. Microwaves are generally less affected by atmosphere; even in this region, there are preferred windows for observations especially for passive sensing. As evident from Figure 1.7, at 1–40 GHz, the atmosphere is fairly

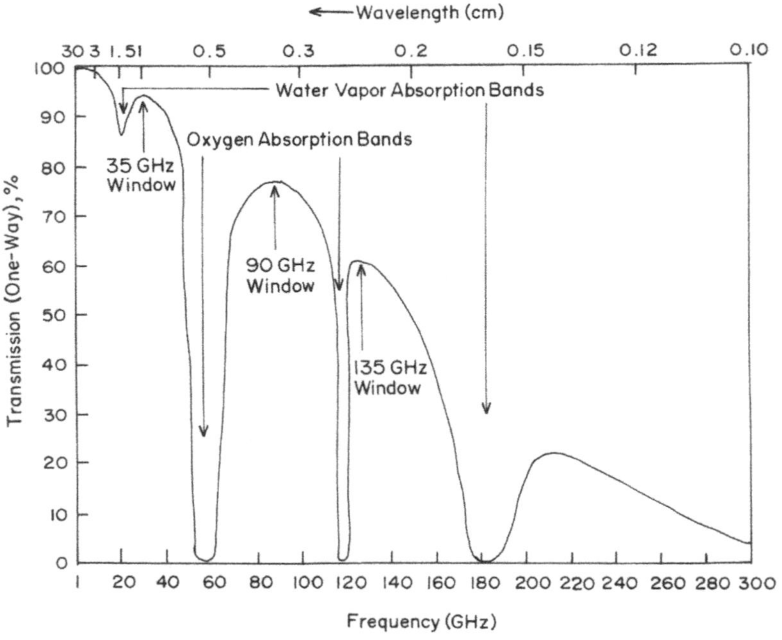

FIGURE 1.7
Absorption bands in microwave region.

transparent under clear sky conditions. In addition, other possible windows include 90 and 135 GHz. Thus, for remote sensing, generally frequencies below 40 GHz are chosen. Even measurements at window frequencies are affected to some extent by clouds, water vapor, etc., especially at higher frequencies. Hence, an accurate estimate of surface radiance needs correction for these absorptions and emissions from the atmosphere. Most surface sensing radiometers include frequency channels also sensitive to water vapor and liquid water mainly to correct for their effects. For observation of the atmospheric parameters, frequencies are selected in the vicinities of the absorption peaks of atmospheric gases, which generally correspond to the frequencies above 50 GHz.

1.7 Energy–Matter Interactions with the Terrain

When radiant energy from the Sun strikes the Earth's surface, a portion of it is reflected back to the atmosphere, and the rest is transmitted and absorbed into the terrain. Following the law of conservation of energy, the radiation budget equation states that the total amount of radiant flux in specific wavelengths (λ) incident to the terrain ($\Phi_{i\lambda}$) must be accounted for by evaluating the amount of energy reflected from the surface (r_λ), the amount of energy absorbed by the surface (α_λ), and the amount of radiant energy transmitted through the surface (τ_λ):

$$\Phi_{i\lambda} = r_\lambda + \alpha_\lambda + \tau_\lambda \tag{1.4}$$

It is important to note that these radiometric quantities are based on the amount of radiant energy incident to a surface from any angle in a hemisphere. And the proportions of energy reflected, absorbed, and transmitted will vary for different Earth features, depending upon their material type and conditions (Lillesand et al., 2008), which permit us to distinguish different features on image. Furthermore, the wavelength dependency means that even within a given feature type, the proportion of energy reflected, absorbed, and transmitted will vary at different wavelengths.

1.7.1 Reflection Mechanism

Reflection is the process whereby radiation bounces off an object like Earth's surface, cloud top, etc. In fact, the process is more complicated, involving reradiation of photons in unison by atoms or molecules in a layer of approximately one-half wavelength deep. There are different types of reflecting surfaces. *Specular reflection* occurs when the surface from which the radiation is reflected is essentially smooth (Figure 1.8). That is, the average surface profile height is several times smaller than the wavelength of the radiation striking the surface. For example, calm water bodies behave as near-perfect specular reflector.

If the terrain feature has a large surface height relative to the size of the wavelength of the incident energy, the reflected rays go in many directions, depending upon the orientation of the small reflecting surfaces. This kind of *diffuse reflection* produces diffuse radiation. Lambert defined a perfect diffuse surface (Figure 1.9). *Lambertian surface* is one for which the radiant flux (light) leaving the surface is constant for any angle of reflectance to the surface.

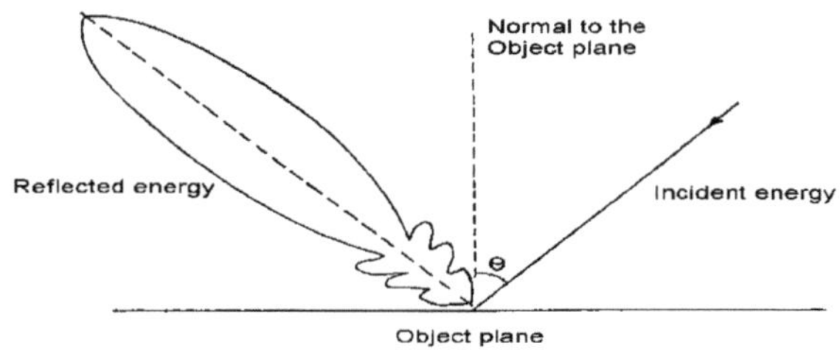

FIGURE 1.8
Schematic of a reflection from a specular reflector.

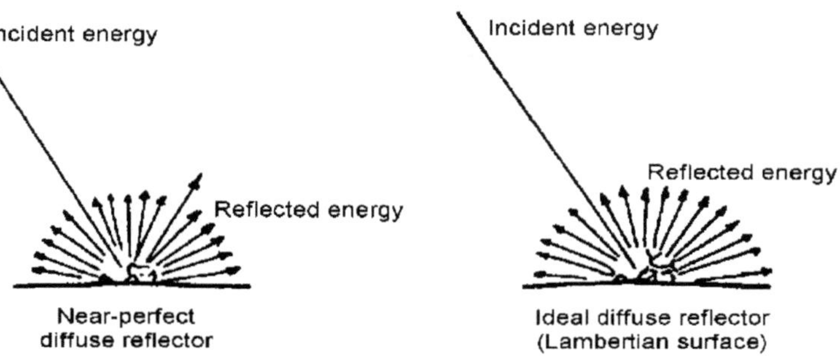

FIGURE 1.9
Near-perfect diffuse reflector and *Lambertian* surface.

1.7.2 Transmission Mechanism

When a beam of EM energy is incident on a boundary, for example, Earth's surface, one part of the energy gets scattered from the surface (surface scattering) and the other part may get transmitted into the medium. If the material is homogenous, then this wave is simply transmitted (Figure 1.10). If, on the other hand, the material is heterogeneous, the transmitted rays get further scattered, leading to volume scattering in the medium. In nature, both surface and volume scattering happen concurrently, and both the processes contribute to the total signal received at the sensor. The depth of penetration is considered as that depth below the surface at which the magnitude of the power of the transmitted wave is equal to 36.8% (1/e) of the power transmitted, at a point just beneath the surface (Ulaby and Goetz, 1987).

1.7.3 Absorption Mechanism

Interaction of incident energy with matter on the atomic-molecular scale leads to selective absorption of the EM radiation. An atomic-molecular system is characterized by a set of energy inherent states (i.e., rotational, vibrational, and electronic). A different amount of energy is required for transition from one energy level to another. An object absorbs radiation of a particular wavelength if the corresponding photon energy is just sufficient to cause a set of permissible transitions in the atomic-molecular energy levels of the object. The wavelengths absorbed are related to many factors, such as dominant cations and anions present, impurities, trace elements, and crystal lattice.

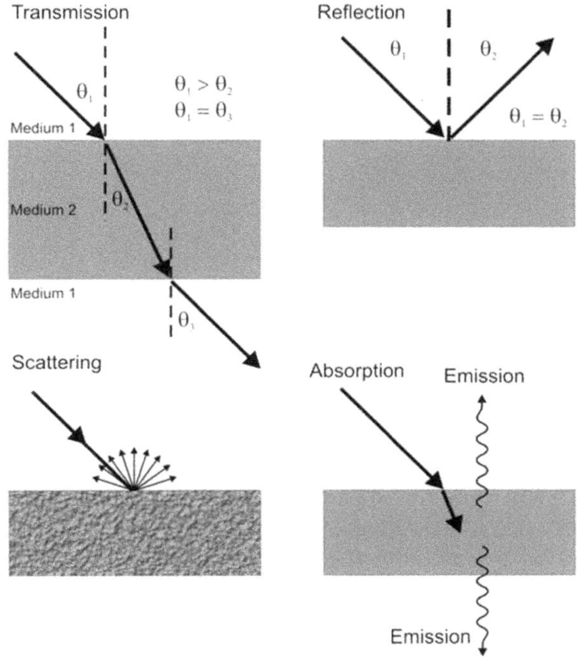

FIGURE 1.10
Reflection/scattering, absorption, transmission, and emission.

1.7.4 Emission Mechanism

The Earth, owing to its ambient temperature, is a source of blackbody radiation, which constitutes the predominant energy available for terrestrial sensing at wavelengths >3.5 μm (Figure 1.10). The emitted radiation depends upon temperature and emissivity of the materials.

1.8 EMR Laws

The propagation of EM energy follows certain physical laws. Some of these are briefly outlined here.

1.8.1 Planck's Law

Planck observed that the EM energy is absorbed and emitted in discrete units called *quanta* or *photons*. The size of each unit is directly proportional to the frequency of the energy's radiation. Planck defined a constant (h) to relate frequency (υ) to radiant energy (Q):

$$Q = h \cdot \upsilon \tag{1.5}$$

where Q is the energy of quantum (J) and h is Planck's constant (6.626×10^{-34} Js).

By substituting Q for hυ, we can express the wavelength associated with a quantum of energy as

$$\lambda = \frac{hc}{Q} \tag{1.6}$$

or

$$Q = \frac{hc}{\lambda} \tag{1.7}$$

where c is the speed of light (3×10^8 ms^{-1}).

The above equation implies that the longer the wavelength involved, the lower is its energy content. It has relevance in remote sensing in that when we are measuring the reflected/emitted EMR from the objects or features at longer wavelengths, in order to have detectable signals, the reflected or emitted energy from a larger area needs to be integrated. It implies that while other conditions remain constant, the spatial resolutions of the sensor become coarser with increasing wavelengths.

Planck's law allows us to calculate the total energy radiated in all directions from a blackbody (radiator) for a particular temperature and wavelength:

$$M_\lambda = \frac{\varepsilon c_1}{\lambda^5 (e^{c_2/\lambda T} - 1)} \; W/(m^2\,\mu m) \tag{1.8}$$

where M_λ = spectral radiant exitance, $W/(m^2\,\mu m)$

ε = emittance (emissivity), dimensionless

C_1 = first radiation constant, 3.74×10^{-8} W $(\mu m)^4/m^2$

C_2 = second radiation constant, 1.43884×10^4 μm K

λ = radiation wavelength, μm

T = absolute temperature, K

The Earth radiates roughly like the 300 K curve and the Sun like the 5,800 K curve (Figure 1.11). It may be noted that the maximum of the Sun's radiation is at the wavelengths that are visible to human eyes.

1.8.2 Stefan–Boltzmann Law

The Stefan–Boltzmann law states that the total emitted radiation from a blackbody (M) measured in Watts per square meter is proportional to the fourth power of its absolute temperature (T) measured in kelvin. This is expressed as

$$M_b = \sigma T^4 \tag{1.9}$$

where σ is the Stefan–Boltzmann constant (5.6697×10^{-8} Wm^{-2} K^{-4}) and T is the absolute temperature in kelvin.

A blackbody is a theoretical construct that absorbs all the radiant that falls on it and radiates at the maximum possible rate per unit area at each wavelength for any given temperature (Mulligan, 1980). The emissivity of blackbody, also known as Planckian radiator, is

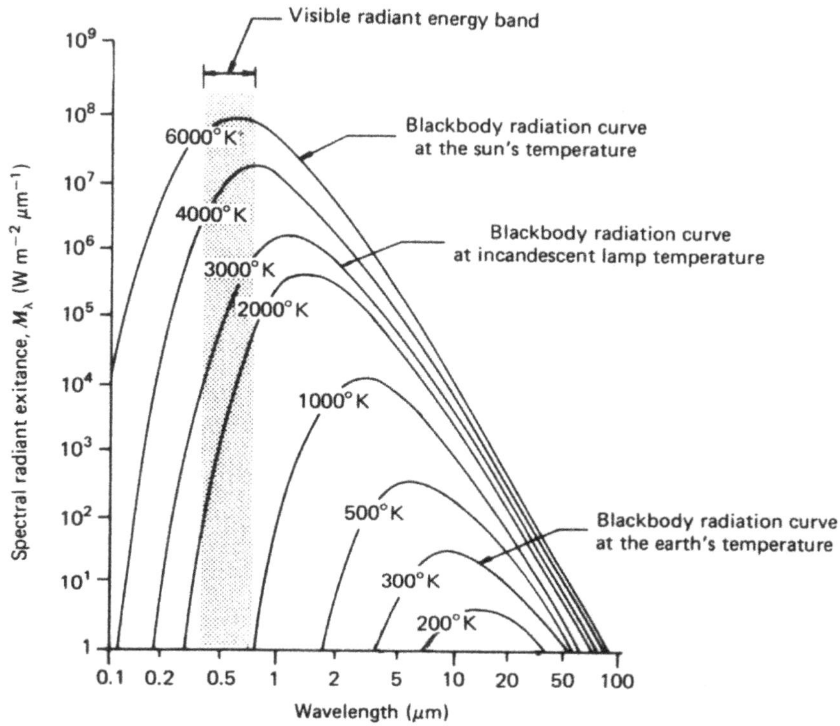

FIGURE 1.11
Spectral distribution of energy radiated by blackbodies at various temperatures.

equal to 1. In other words, it radiates the entire energy whatever it absorbed. In nature, all objects reflect at least a fraction of the radiation that strikes them and do not act as perfect reradiators of absorbed energy. A gray body, on the other hand, is one for which emissivity value is constant at all wavelengths but less than unity. A selective radiator is one for which emissivity value varies with wavelength.

The total radiant exitance is the integration of all the area under the blackbody radiation curve (Figure 1.11). In essence, the Stefan–Boltzmann law states that hot blackbodies emit more energy per unit area than do cool blackbodies.

1.8.3 Wein's Radiation Law

Wien's law may be used when the product of wavelength and temperature is less than 3×10^3 μm K. The dominant wavelength, or wavelength at which a blackbody radiation curve reaches a maximum, is related to its temperature:

$$M_\lambda = \frac{K_2}{\lambda^5 (e^{hc/\lambda T} - 1)} \; W/(m^2 \, \mu m) \tag{1.10}$$

where K_2 is the unit-fitting constant and h is Planck's constant.

$$\lambda_{max} = (k/T) \tag{1.11}$$

where λ_{max} is the wavelength of maximum spectral radiant exitance (μm),

k = 2,897.8 μm K,
T is the absolute temperature in K.

Three observations may be made from this diagram. As temperature increases, (i) the emissive power increases at each wavelength, (ii) relatively more energy is emitted at shorter wavelengths, and (iii) the position of maximum emissive power shifts toward shorter wavelengths. The Wien displacement law helps us determining the dominant wavelength at which an Earth feature will radiate the maximum radiant energy. This, in turn, helps us designing a sensor that is sensitive to that dominant wavelength. For example, the dominant wavelength for glacier at 20°C (~253 K) can be computed as 11.45 μm (2,898/253). Similarly, for Earth at 300 K, forest fire at 800 K, volcanoes at 1,200 K, and the Sun at around 6,000 K, the values of λ_{max} could be worked out as 9.66, 3.66, 1.97, and 0.48 μm, respectively.

1.8.4 Rayleigh–Jeans Law

This law explains blackbody emission at longer wavelengths:

$$M_\lambda = \frac{k_1 T}{\lambda^4}\ \mathrm{W/(m^2\ \mu m)} \tag{1.12}$$

where M_λ is the spectral radiant exitance [W/(m² μm)], K_1 is the unit-fitting constant, and T is the absolute radiant temperature. The Rayleigh–Jeans law may be used when the product of wavelength and temperature exceeds approximately 10^5 μm K.

1.8.5 Kirchhoff's Law

In the infrared portion of the EMS, the Russian physicist Kirchhoff had observed that the spectral emissivity of an object generally equals its spectral absorptance, i.e., $\alpha_\lambda = \varepsilon_\lambda$. The observation is often phrased as "*good absorbers are good emitters and good reflectors are poor emitters.*" Kirchhoff's law states that the ratio of emitted radiation to absorbed radiation flux is the same for all blackbodies at the same temperature. This law forms the basis for the definition of emissivity (ϵ), the ratio between the emittance of a given object (M) and that of a blackbody at the same temperature (M_b):

$$\varepsilon = M/M_b \tag{1.13}$$

1.9 Spectral Response Pattern

The Earth's land surface reflects about 4% of all incoming solar radiation back to space. The rest is either reflected by the atmosphere or absorbed and reradiated as infrared energy. The various objects that make up the Earth's surface absorb and reflect different amounts of energy at different wavelengths. The magnitude of energy that an object reflects or emits across a range of wavelengths is called its *spectral response pattern*. Because spectral responses measured by remote sensors over various features often permit an assessment

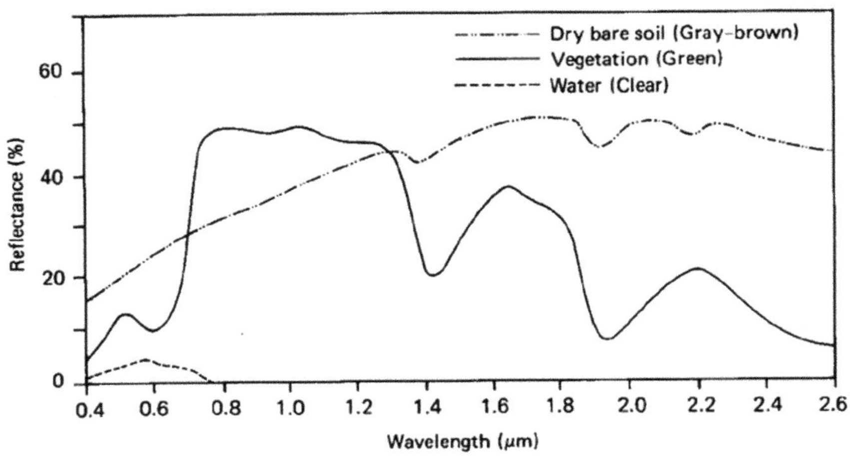

FIGURE 1.12
Spectral reflectance pattern of water and other major terrain features.

of the type and/or condition of the features, these responses have often been referred to as *spectral signature*. Although it is true that many Earth surface features manifest very distinctive spectral reflectance and/or emittance characteristics, these characteristics result in spectral "response patterns" rather than in spectral "signatures." The reason for this is that the term *signature* tends to imply a pattern that is absolute and unique. This is not the case with the spectral patterns observed in the natural world.

Shown in Figure 1.12 is the spectral reflectance pattern of major terrain features, namely soils, vegetation, and water (shallow/deep). The absorption of incident radiation in blue (0.4–0.50 μm), red (0.60–0.70 μm), and shortwave-infrared region (at 1.4, 1.9, and 2.6 μm) by vegetation is very conspicuous. Whereas absorptions in blue and red regions is due to the presence of chlorophyll (green color) in plant leaves, the absorption of incident radiation in the shortwave-infrared region of the spectrum is attributed to the presence of water in plant leaves.

A comparison of the spectral response pattern of vegetation with that of water (Figure 1.12) reveals the fact that contrary to vegetation, water reflects maximum in the blue region (0.4–0.5 μm) and absorbs maximum in the NIR region (0.7–1.3 μm). Importantly, vegetation reflects maximum in the NIR region. This contrasting feature enables detection of vegetation from water bodies using air/space borne multispectral images. Soils, on the other hand, exhibit an increasing trend in the spectral reflectance pattern with increasing wavelengths except for two absorption bands centered around 1.4 and 1.9 μm.

1.10 Hyperspectral Remote Sensing

The "hyper" in hyperspectral means "over" as in "too many" and refers to the large number of measured wavelength bands. Hyperspectral images are spectrally over determined, which means that they provide ample spectral information to identify and distinguish spectrally unique materials. Hyperspectral imagery provides the potential for more accurate and detailed information extraction than possible with any other type of remotely sensed data. Although most hyperspectral sensors measure hundreds of wavelengths, it is not the number of measured wavelengths that defines a sensor as hyperspectral. Rather, it is the narrowness and contiguous nature of the measurements. For example, a sensor that measured only 20 bands could be considered hyperspectral if those bands were contiguous and, say, 10 nm wide. If a sensor measured 20 wavelength bands that were, say, 100 nm wide, or that were separated by non-measured wavelength ranges, the sensor would no longer be considered hyperspectral.

Every object—both living and nonliving—has a distinctive spectral signature embedded in the spectra of the light reflected or emitted by it. These spectral characteristics of the object are unique and are determined by the electronic and vibrational energy states of the constituent substances. In turn, these spectral characteristics allow that object or substance to be identified through various spectral analyses techniques.

Most multispectral imagers (e.g., Landsat-Operational Land Imager, SPOT-HRV, NOAA-AVHRR) measure radiation reflected from a surface at a few wide, separated wavelength bands. As a result, subtle spectral features associated with certain material characteristics are not available in multispectral spectra (Figure 1.13).

In an example given here, the strength of the measurements of spectral response patterns in several narrow and contiguous spectral bands is demonstrated. The spectral bands of Landsat -TM are indicated with numerals 1 to 7 along with their band width in discrete blue lines. The spectral response of green vegetation at 7 points within 350 to 2,500 nm region has been joined and assigned blue color. Hyperspectral response pattern of green vegetation is portrayed in continuous green color. Except for the region between 1,800 to 2,000 nm, spectral response pattern of green vegetation is continuous.

Owing to discrete and broad spectral bands Thematic Mapper could not capture the water absorptions band centered around 1,400 nm, and two absorption features at around 1,000 nm and 1,200 nm within near infrared plateau.

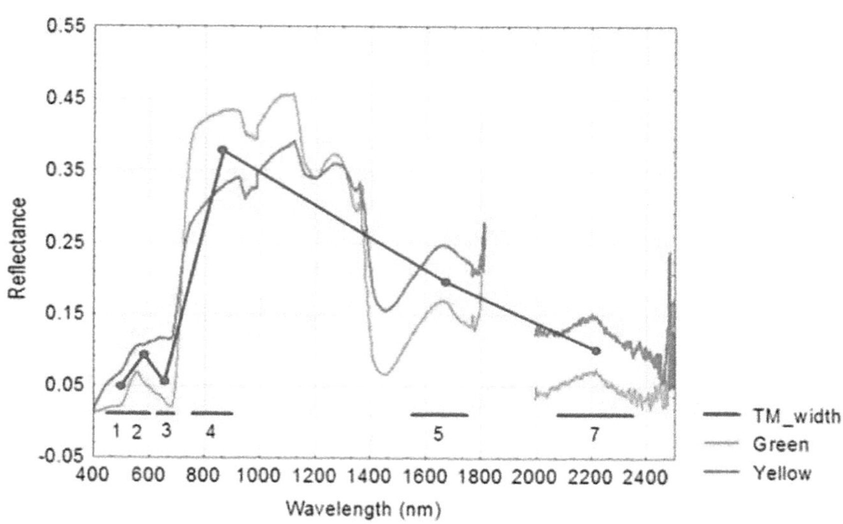

FIGURE 1.13
A Comparison of multispectral and hyperspectral response patterns of vegetation (Source: http://www.iasri.res.in/ebook/GIS_TA/M2_4_HYSRS.pdf. Accessed on 03-06-2018).

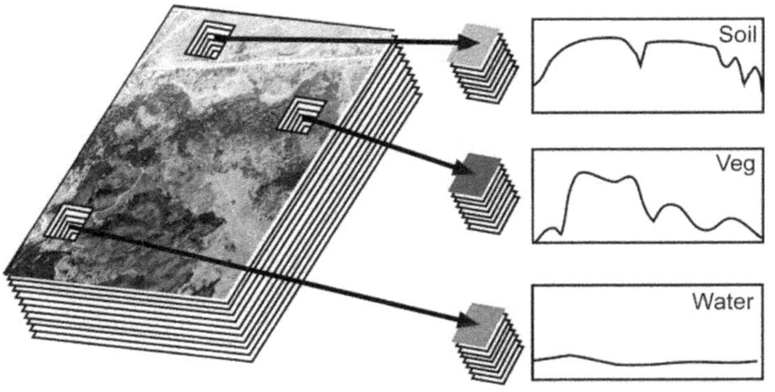

FIGURE 1.14
The concept of hyperspectral imagery. Image measurements are made at many narrow contiguous wavelength bands, resulting in a complete spectrum for each pixel.

Similarly, the spectra for various terrain features, namely soil, water, and vegetation, for instance, can be generated (Figure 1.14).

Owing to several spectral bands, the hyperspectral data are difficult to visualize all at once. Because each ground scene can be made up of hundreds of images (bands), one way of understanding the patterns in the data is to create an image cube (Figure 1.15). The x and y axes are the spatial dimensions showing the ground surface of terrain. The z axis is made up of all the other bands as if they were stacked like a ream of paper and placed on its side. The top image is a three-band composite made from any three of the bands for presentation purposes (generally R, G, and B). The colors streaming away along the edges represent the edge pixel values in the z axis colored from blue to red as a rainbow. Thus, following an edge pixel in this cube along the z axis, one can see how the spectra vary and that there is an enormous amount of information contained in spectra.

For further details on hyperspectral remote sensing, readers may refer to Chapter 2 and van der Meer et al. (2012) and Ben Dor et al. (2012).

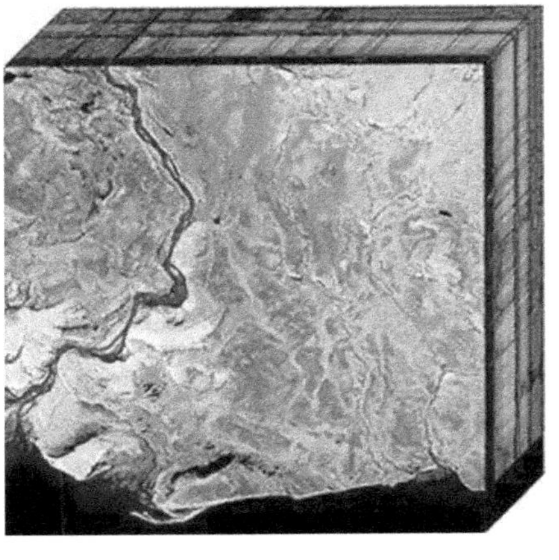

FIGURE 1.15
Two-dimensional projection of a hyperspectral cube.

1.11 Remote Sensing Process

A remote sensing system consists of platform, instrumentation (sensor), data reception, processing, and analysis designed to measure, monitor, and predict the physical, chemical, and biological aspects of the Earth's system (Figure 1.16).

1.11.1 The Source of Illumination

The sensors or instruments record reflected/emitted/scattered radiation from the object or feature. The source of illumination is, generally, the sun that radiates its EM energy in the visible region (0.4–0.7 μm). In the case of sensors with their own source of illumination (active sensors) like radar and LiDAR, there is freedom with respect to direction and angle of illumination and time of observation.

1.11.2 The Sensor

Beginning with the simple photographic cameras sensitive to the visible (0.4–0.7 μm) region of the EMS, there has been a phenomenal development in sensor technology. Films with the sensitivity in NIR have been developed. With the development of rocket technology that facilitated acquisition of images from satellites, the major challenge was to retrieve the measurements from the satellite platform. Electro-optical sensors with the capability of conversion of light energy (photons) to digital signals (electrons) were subsequently developed. As a follow-up, sensors with the capability to capture the microwave energy, namely microwave radiometers and radars, were developed. The development of LiDAR and imaging spectrometers is the other major milestone in the field of sensor technology.

1.11.3 Platforms

The reflected/emitted/scattered EMR could be recorded *in situ/in-place/in laboratory* by either manually holding the sensor or mounting it onto stable platform like tripod stand or hydraulic platform. Such measurements are useful for sensor calibration and serve as ground truth for interpretation/analysis of remote sensing data. Although balloons,

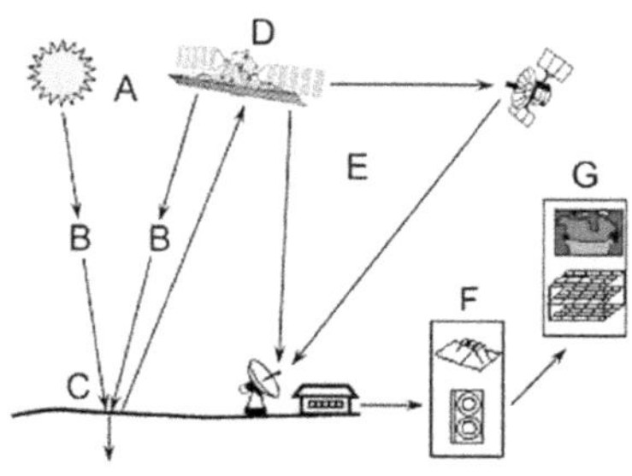

FIGURE 1.16
The remote sensing system. A = energy source/illumination; B = radiation and the atmosphere; C = interaction with the object; D = recording of energy by the sensor; E = transmission, reception, and processing; F = interpretation/analysis; and G = applications.

aircrafts, and rockets have been used as platforms, aircrafts and satellites have been the most widely used platforms. Aircrafts are used for covering smaller area or limited region of interest, while satellites cover larger areas and provide synoptic views at regular intervals. The ability to support the sensor system in terms of weight, velocity, power, etc. and the stability of the platform are the major considerations while selecting the platform for remote sensing surveys.

1.11.4 Data Reception

As mentioned in Section 1.13.3, the sensors aboard air or space borne platforms measure reflected/emitted/backscattered radiation. In the case of aerial platform, remote sensing data (films/digital data) can be retrieved immediately after completion of the flying. In the case of space platforms, the mechanism needs to be in place to retrieve the measurements made by the sensors. It is achieved by transmitting the data recorded by space borne platforms to the ground receiving stations (GRSs). There are three main options for transmitting data acquired by space borne sensors:

i. The data can be directly transmitted to the Earth if a GRS is in the line of sight of the satellite.
ii. The data can be recorded on board satellite for transmission to GRS at a later time when the satellite is in the visibility range of GRSs.
iii. The data can also be relayed to GRSs through the Tracking and Data Relay Satellite System (TDRSS), which consists of communication satellites in geostationary orbit. The data are transmitted from one satellite to another until they reach the appropriate GRS (Liang et al., 2012). The transmission frequencies of some of the Earth observation satellites are provided in Table 1.2.

1.11.5 Data Product Generation

The data acquired with the sensor aboard aircraft/satellite have a number of errors due to (i) instability and orbital characteristics of the platform, (ii) imaging characteristics of the sensor, (iii) scene/surface characteristics, (iv) Earth's motion, and (v) and atmospheric effects in the case of space borne sensors. For deriving information on Earth resources, appropriate corrections, therefore, need to be applied while generating the data products.

TABLE 1.2

Data Reception Frequencies of Some Earth Observation Mission

Terra	8.2125 GHz (X band)
Aqua	8.160 GHz (X band)
Resourcesat-1/-2	8.025–8.4 GHz (X band)
	2.2–2.3 GHz (S band)
NOAA-17 and-18	1.70705 MHz (L band)
ERS-2 (high rate)	8,140.0 MHz (X band)
SPOT-4 and-5	8,253.0 MHz (X band)
ERS-2 (high rate)	8,140.0 MHz (X band)
EROS-A1	8,150 and 8,250 MHz
Landsat-5 and -7	8,212.5 MHz

Source: Modified after Cracknell and Hayes (2007).

The data products are of two types, namely analog or photographic product, and digital. The analog product consists of black-and-white (B&W) prints of individual spectral bands in the case of panchromatic data or color prints developed from three spectral bands' of multispectral data by exposing them through three primary colors.

When blue, green, and red spectral bands are exposed through corresponding primary colors, namely blue, green, and red, the resultant color composite is called *true or natural color composite*. On the contrary, when primary colors—blue, green, and red—are assigned to three different spectral bands data, namely green, red, and NIR, the resultant color composite is referred to as *false color composite* (*FCC*). After necessary geometric and radiometric corrections, the *digital data products* are stored in Digital Linear Tapes (DLTs) and CDs. The data in these media are stored in well-defined standard format for easy retrieval across the globe, which can be analyzed for generating the information on natural resources and environment.

1.11.6 Data Analysis/Interpretation

Depending upon the type of remote sensing data, two approaches, namely visual interpretation and digital analysis, are employed. Visual interpretation could be employed to both hard-copy photo products of remote sensing data and digital data as well. For digital data, on-screen or heads-up visual interpretation is employed to derive information from digital remote sensing data. Stereoscopic images enable deriving information on the third dimension of terrain feature, namely height or depth. Stereoscopic images are interpreted using stereoscope. Nowadays, digital photogrammetric techniques are employed to derive precise measurements on terrain features.

Computer-assisted digital analysis, in general, is solely based on the spectral response pattern of terrain features in different spectral bands. However, algorithms mimicking the human capability of integrating image elements, context, and association for deriving information from digital panchromatic/multispectral data have been recently developed.

1.11.7 Data/Information Storage

Aircraft data are usually acquired in a campaign that is commissioned by, or on behalf of, a particular user and is carried out in a predetermined area. Aerial data are also acquired for a particular purpose, such as making maps or monitoring some given natural resources. The instruments, wavelengths, and spatial resolution used are chosen to suit the purpose. Such data are, generally, not available in public domain. In the case of Earth observation space borne missions, the output signal from an instrument, or a number of instruments, on board a spacecraft is superimposed on a carrier wave, and this carrier wave, at radio frequency, is transmitted back to GRS. The data transmitted from a remote sensing satellite can, in principle, be received not only by the owner of the spacecraft but also by anyone who has the appropriate receiving equipment and necessary technical information. The data transmitted by a civilian remote sensing satellite are not encrypted, and the technical information on transmission frequencies and signal formats is usually available.

1.11.8 Archival and Distribution

In early days, the philosophy of archiving distribution was to store the data in a raw state at the ground station where they were received, and produce quick look (B&W) image in one of the spectral bands. Increased computing power enabled applying various levels of processing of raw data. The processed data or information extracted from those data can then be supplied to users. Thus, all the data may be geometrically rectified, i.e., presented in a standard map projection, or any of several geophysical quantities (such as sea surface temperature and vegetation indices) may be calculated routinely on pixel-by-pixel basis from that data.

Most of the users want processed data instead of raw data. The NOAA's National Environmental Satellite, Data, and Information Service (NESDIS) has taken a lead in this endeavor. NESDIS operates the Comprehensive Large Array-Data Stewardship System (CLASS) which is an electronic library of NOAA environmental data. It enables a user, in principle from anywhere in the world, to view the soft copy quick looks, and then access or order the data online. India too has made substantial progress in this direction. The Department of Space, Government of India, has developed an electronic library of satellite data called BHUVAN for the benefit of the users worldwide.

1.12 Geographical Information System

Historically, the development of geographical information system (GIS) could be traced back to the outbreak of cholerain London in the year 1854 when Dr. John Snow was able to trace the source of pollutant (contaminated water that caused the cholera). Through the study that Snow carried out, officials from the government were able to determine the cause of the disease. It was contaminated water from one of the major pumps that caused the disease. Photozincography - a type of photoengraving using a sensitized zinc plate, capable of dividing into layers was developed in the earlier years of the 1900s. In the initial stages, the process of drawing these maps was lengthy since it involved free hand, but this changed later on with the introduction of the computer. The first GIS was designed and developed by Dr. Roger Tomlinson and then introduced in the early 1960s in Canada. During its inception, this system was mainly meant for collecting, storing, and then analyzing the capability and potential which the land in the rural areas had. Prior to this, mapping by the use of computers was being done for such cases, but this is a method that had numerous limitations associated with it.

By the end of the 1980s, the use of GIS had already become popular in other related fields, which is why it led to a spur in the growth of the industrial sector. Recently, designers came up with open-source software for GIS so that the brilliant technology could be enhanced in a much simpler manner while being made available to all. As evident from the above-mentioned text, the basic needs to access, organize, update, and analyze the geographic information, and to utilize it in an optimal way led to the concept of the GIS. Geographic or geographical information system or geospatial information system is the science and technology dealing with the structure and character of spatial information, its capture, its classification and qualification, its storage, processing, portrayal, and dissemination, including the infrastructure necessary to secure optimal use of this information (Groot, 1989). ArcGIS (Esri), Geomedia (Hexagon Geospatial), MapInfo Professional (Pitney Bowes), Global Mapper (Blue Marble), Manifold GIS (Manifold), Smallworld (General Electric), MapViewer and Surfer (Golden Software), and Bentley Map are some of the commonly used GIS packages. For further details, readers may refer to Burrough and Rachael (2005) and Longley et al. (2011).

1.12.1 Components of GIS

The components a GIS comprise are hardware, software, and data (Figure 1.17).

1.12.1.1 Hardware

Hardware comprises the equipment needed to support the many activities needed for geospatial analysis ranging from data collection to data analysis. The central piece of equipment is the workstation, which runs the GIS software and is the attachment point for ancillary equipment. Data collection efforts can also require a digitizer for conversion of hard copy data to digital data and a GPS data logger to collect data in the field. The use of handheld field technology, i.e., PDA device, is also becoming an important data collection tool in GIS. With the advent of web mapping, web servers have also become an important piece of equipment.

1.12.1.2 Software

The GIS application package is the core software package. Such software is essential for creating, editing, and analyzing spatial and attribute data; therefore, these packages contain a myriad of geospatial functions inherent to them. Extensions or add-ons are software packages that extend the capabilities of the GIS software package. Component GIS seeks to build software applications that meet a specific purpose and thus are limited in their spatial analysis capabilities. Utilities are stand-alone programs that perform a specific function, for example, a file format utility that converts from on type of GIS file to another. There is also web GIS software that helps serve data and interactive maps through Internet browsers.

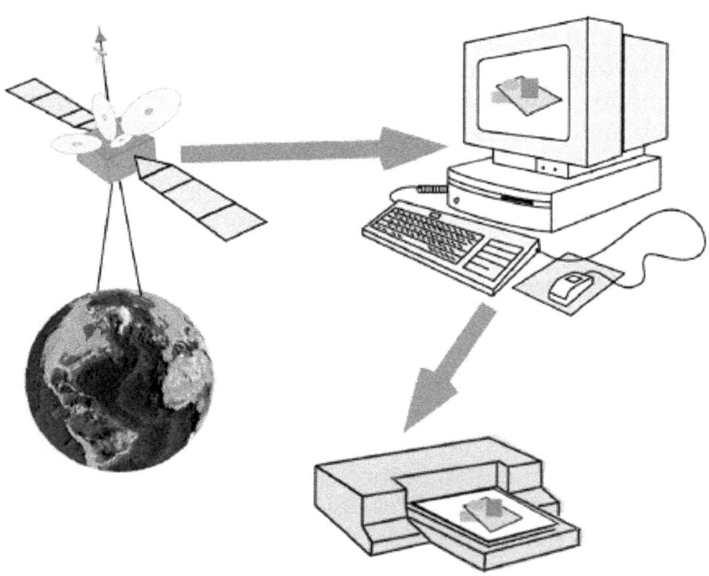

FIGURE 1.17
Three major components of a Geographic Information System. These components consist of input, computer hardware and software, and output subsystems.

FIGURE 1.18
GIS data–thematic data layers.

1.12.1.3 Data

Data are the core of any GIS (Figure 1.18). There are two primary types of data that are used in GIS: vector and raster data. A geodatabase is a database that is in some way referenced to locations on the Earth. Geodatabases are grouped into two different types: vector and raster. Vector data are spatial data represented as points, lines, and polygons. Raster data are cell-based data such as aerial imagery and digital elevation models. Coupled with these data are usually data known as attribute data. Attribute data are generally defined as additional information about each spatial feature housed in tabular format. Documentation of GIS datasets is known as metadata. Metadata contain such information as the coordinate system, when the data were created, when they were last updated, who created them, and how to contact them and definitions for any of the code attribute data.

1.13 Global Navigation Satellite Systems

Inventory and monitoring of Earth resources and environment is carried out using an appropriate database. Additionally, precise location of the site/s where observation/s about a feature/features or phenomenon/phenomena has been made is very crucial. The space-based navigation technology provides a viable solution. Satellite-based positioning or navigation is the determination or observation of sites on land or at sea, in the air and in space by means of artificial satellites. The term *global navigation satellite system* (GNSS) covers each individual global satellite-based positioning system as well as the combination or augmentation of these systems. The GNSSs comprise four systems, namely the United States' NAVSTAR (NAVigation System with Time And Ranging; informally the "navigation star")-Global Positioning System, the Russian Federation's Global'naya Navigatsionnaya Sputnikovaya Sistema (GLObal NAvigation Satellite System) (GLONASS), Europe's Galileo, and China'sBeiDou (formerly known as COMPASS). The former two of them are fully operational, while the latter two are in the process of development. For review on the development of satellite-based positioning system, the readers may refer to Guier and Weiffenbach (1997) and Hoffmann-Waellenhof et al. (2008).

1.13.1 GPS Segments

GPS is a US space-based radio navigation system that helps pinpoint a three-dimensional position to about a meter of accuracy (e.g., latitude, longitude, and altitude) and provide nanosecond precise time anywhere on Earth. GPS comprises three different parts.

1.13.1.1 Space Segment

A constellation of at least 24 US government satellites distributed in six orbital planes inclined 55° from the equator in a medium Earth orbit (MEO) at about 20,200 km (12,550 miles) and circling the Earth every 12 hours (Figure 1.19).

1.13.1.2 Control Segment

The control segment has a master control station in Colorado Springs, with three antennas and five monitor stations located throughout the world. The monitor stations keep track of all GPS satellites and collect information from the satellite broadcasts. The monitor stations send the collected information to the master control station that computes precise satellite orbits. They serve as uplink installations, capable of transmitting data to the satellites, including new ephemerides (satellite positions as a function of time), clock corrections, and other broadcast message data, while the Colorado Springs serves as the master control station. This processing involves the computation of satellite ephemerides

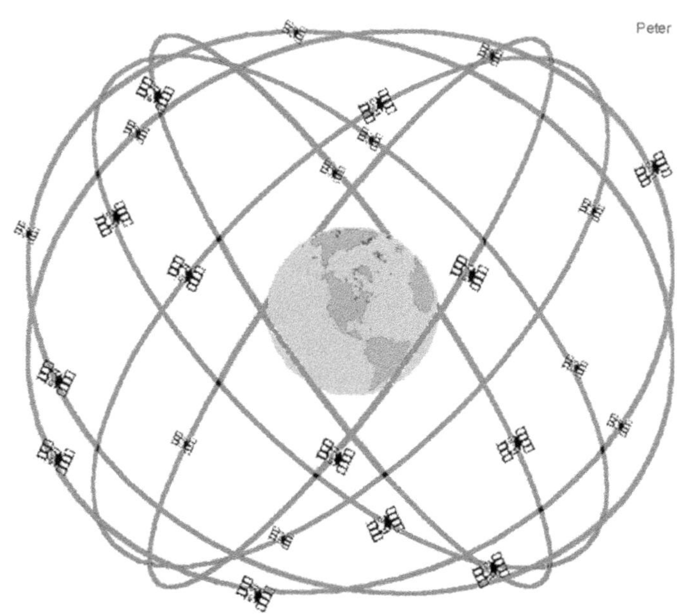

FIGURE 1.19
GPS nominal satellite constellation.

and satellite clock corrections. The master station controls orbital corrections, when any satellite strays too far from its assigned position, and necessary repositioning to compensate for unhealthy (not fully functioning) satellites.

1.13.1.3 The User Segment

The user segment is a total user and supplier community, both civilian and military. The user segment consists of all Earth-based GPS receivers and antennas that permit land, airborne, and sea operators to receive the GPS satellite broadcasts and make a precise calculation of their velocity, position, and time. Receivers vary greatly in size and complexity, though the basic design is rather simple. The typical receiver is composed of an antenna and preamplifier, radio signal microprocessor, control and display device, data recording unit, and power supply (Figure 1.20). The GPS receiver decodes the timing signals from the "visible" satellites (four or more) and, having calculated their distances, computes its own latitude, longitude, elevation, and time. This is a continuous process and generally the position is updated on a second-by-second basis, displayed onto display panel of the receiver, if it has the display device. In the case of receiver having data capture capabilities, it is stored by the receiver-logging unit.

1.13.2 Operating Principle of GPS

A GPS receiver calculates its position by precisely timing the signals sent by GPS satellites high above the Earth. Each satellite continually transmits messages that include (a) the time the message was transmitted, (b) precise orbital information (the ephemeris), and (c) the general system health and rough orbits of all GPS satellites (the almanac). The receiver uses the messages it receives to determine the transit time of each message and computes the distance to each satellite. These distances along with the satellites' locations are used with the possible aid of trilateration, depending on which algorithm is used to

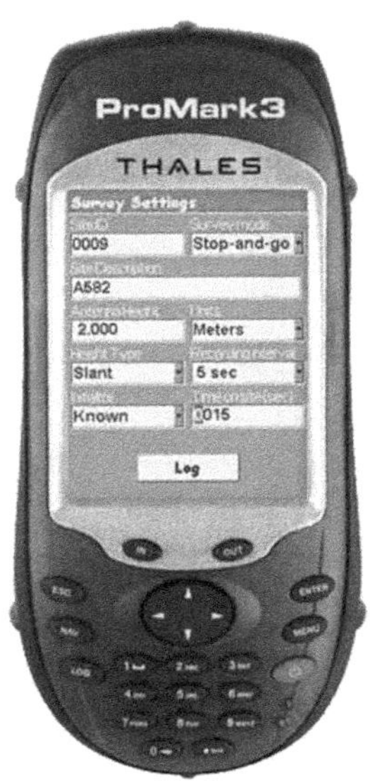

FIGURE 1.20
A GPS receiver.

compute the position of the receiver. This position is then displayed, perhaps with a moving map display or latitude and longitude; elevation information may be included. Many GPS units show derived information such as direction and speed, calculated from position changes. For determining the location of a point on the ground, three satellites might seem enough, since space has three dimensions and a position near the Earth's surface can be assumed. However, even a very small clock error multiplied by the very large speed of light, the speed at which satellite signals propagate, results in a large positional error. Therefore, receivers use four or more satellites to find the receiver's location and time (Figure 1.21).

A satellite's position and pseudo range define a sphere, centered on the satellite with radius equal to the pseudo range. The position of the receiver is somewhere on the surface of this sphere. Thus, with four satellites, the indicated position of the GPS receiver is at or near the intersection of the surfaces of four spheres. In the ideal case of no errors, the GPS receiver would be at a precise intersection of the four surfaces.

Each signal consists of a carrier wave at a frequency near 1.6 GHz, modulated by a stream of digital bits at a rate of about 1 million bits per second (Mbps). The digital bits are generated in a way that is actually systematic but which appears random, and are called pseudorandom noise code or PRN code. Each satellite has its own specific PRN code. The PRN code is itself modulated by digital navigation data at a slow rate (typically 50 bits per second). The frequency of each satellite's signal and the bit rate of its PRN code are controlled by an extremely precise clock (an atomic clock) on board the satellite. The satellite signal is designed such that a receiver which "hears" the signal can read the satellite's exact time at the instant the signal was transmitted, with an error of a few nanoseconds.

1.13.3 Navigation

It is possible to navigate with GNSS. The signals, thus received, could be used for navigation using a variety of configurations. The most common setups are as follows.

1.13.3.1 Stand-Alone Satellite Navigation

This is the basic method of GNSS navigation where only the received signals from a GNSS constellation, such as the publicly available GPS standard positioning service, are used. The performance of stand-alone GNSS is sufficient only for a limited number of applications.

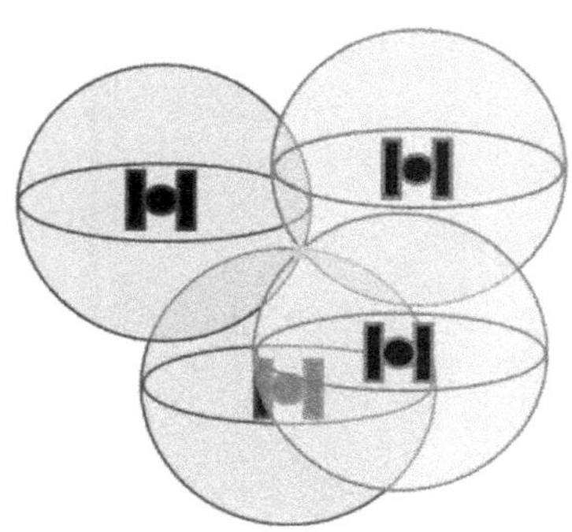

FIGURE 1.21
Third dimension positioning using GPS.

1.13.3.2 Differential GNSS Navigation

Relative or differential GPS carries the triangulation principles one step further, with a second receiver at a known reference point. The differential GNSS (DGNSS) navigation facilitates determination of a point's position, relative to the known Earth surface point, which demands collection of an error-correcting message from the reference receiver. Differential corrections may be used in real time or later, with post-processing techniques. The reference station is placed on the control point, a triangulated position, the control point coordinate. This allows for a correction factor to be calculated and applied to other roving GPS units used in the same area and in the same time series.

1.13.3.3 Network-Assisted GNSS Navigation

Whenever any communication network is used to relay information to a GNSS receiver, it can be said to be receiving assistance. This is called network-assisted GNSS (A-GNSS). The DGNSS described above can be thought of as a subset of A-GNSS. This assistance is often a correction to raw measurements calculated elsewhere and sent over a radio link to remote receivers. However, unlike DGNSS, in A-GNSS this assistance can often include more basic information used to assist the receiver in performing an accelerated position fix or to extend the validity of the satellite information used during positioning.

1.13.3.4 Carrier-Phase Differential (Kinematic) GPS

The differential correction technique described above applies to code-phase GPS receivers, which use the transmitted GNSS code information to compute pseudo ranges (distances) from the Earth to the GPS satellites in space. When a receiver operates in carrier-phase mode, it is measuring a different GNSS observable, namely the GNSS carrier wave. In order to obtain high accuracy with carrier-phase measurements, it is necessary for a roving GPS receiver to use information from a base receiver to compute the integer number of GPS wavelengths between the roving GPS receiver's antenna and the satellite(s). This technique yields accuracies in the cm range and can yield mm-level accuracies in static environments. In dynamic environments (called "real-time kinematic," or RTK), GNSS is capable of providing accuracies in the 1–5 cm range.

1.14 Organization of This Book

Beginning with the introduction to geospatial technology, namely remote sensing, GIS and GNSS, the author intends to take the readers to different kind of remote sensing sensors operating in different portions of the EMS; data processing and analysis/interpretation techniques to derive information on various land degradation processes. It is followed logically by an introduction to land degradation, and the applications of geospatial technologies to the assessment and management of soil erosion by water and wind, soil salinization and/or alkalization, waterlogging, mining, aquaculture and shifting cultivation, soil acidification and drought in subsequent chapters. The book concludes with the development of information systems for degraded lands.

References

Ben Dor, E., Malthus, T., Plaza, A., and Schläpfer, D., 2012. Hyperspectral remote sensing. In: Wendisch, M., and Brenguier, J.L., (eds.) *Airborne Measurements for Environmental Research,* Wiley, Hobcon, New Jersey, USA, pp. 419–691.

Buiten, H.J., and Clevers, J.G.P.W., 1993. Gordon and Breach Science Publishers. 642 pp.

Burrough, P.A., and McDonnell, R.A., 2005. *Principles of Geographic Information Systems.* Oxford University Press, Oxford.

Campbell, J.B., and Wynne, R.H., 2011. *Introduction to Remote Sensing,* 5th edn. The Guilford Press, London, NY, 622 pp.

Colwell, R.N., 1966. *Manual of Photographic Interpretation.* American Society of Photogrammetry, Falls Church, VA.

Cracknell, A.P., and Hayes, 2007. *Introduction to Remote Sensing.* CRC Press, Taylor & Francis Group, Boca Raton, FL.

Groot, R., 1989. Meeting educational requirements in Geomatics. *ITC Journal,* 1:1–4.

Guier, W.H., and Weiffenbach, G.C., 1997. Genesis of satellite navigation. *John Hopkins APL Technical Digest,* 18(2):178–181.

Hoffmann-Waellenhof, B., Listenegger, H., and Wasle, E., 2008. *GNSS-Global Navigation Satellite Systems; GPS, GLONASS, Galileo, and More.* Springer Wien, New York.

Jaganathan, C., 2011. Geoinformatics: An overview and recent trends. In: Anbazhgan, S., Subramanian, S.K., and Yang, X., (eds.), *Geoinformatics in Applied Geomorphology.* CRC Press, Taylor & Francis Group, Boca Raton, FL.

Jensen, J.R., 2007. *Remote Sensing of the Environment-An Earth Resources Perspective.* Prentice Hall, Upper Saddle River, NJ.

Lein, J.K., 2012. *Environmental Sensing: Analytical Techniques for Earth Observations.* Springer-Verlag, Berlin.

Liang, S., Li, X., and Wang, J., 2012. *Advanced Remote Sensing: Terrestrial Information Extraction and Applications.* Elsevier. Amsterdam, The Netherlands 800 pp.

Lillesand, T.M., Kiefer, R.W., and Chipman, J.W., 2004. *Remote Sensing and Image Interpretation,* 5th edn. John Willey and Sons, New York, 724 pp.

Lillesand, T.M., Kiefer, R.W., and Chipman, J.W., 2008. *Remote Sensing and Image Interpretation,* 6th edn. John Willey and Sons, New York, 756 pp.

Longley, P.A., Goodchild, M.F., Maguire, D.J., and Rhind, D.W., 2011. *Geographic Information Systems and Science.* Wiley, Hoboken, NJ.

Mulligan, J.F., 1980. *Practical Physics: The production and Conservation of Energy.* McGraw Hill, New York, 526 pp.

Silva, L.F., 1978. Radiation and instrumentation in remote sensing. In: Swain, P.S., and Davis, S.M., (eds.) *Remote Sensing: The Quantitative Approach.* McGraw Hill, New York, pp. 21–135.

Suits, G.H., 1983. The nature of electromagnetic radiation. In: Colwell, R.N., (ed.) *Manual of remote Sensing.* American Society of Photogrammetry, Falls Church, VA, pp. 37–60.

Ulaby, F.T., and Goetz, A.F.H., 1987. *Remote sensing techniques. Encyclopedia of physical science and technology.* Vol.12. Academic Press, New York, pp. 164–196.

Van der Meer, F.D., van der Werff, H.M.A., Ruitenbeek, F.J.A., Hecker, C., Bakker, W.H., Noomen, M.F., van der Meijde, M., Carranza, E.J.M., and Boudewijn de Smeth, J., 2012. Multi- and hyperspectral geologic remote sensing: A review. *International Journal of Applied Earth Observation and Geo-information* 14 (1):112–128.

2

Passive Remote Sensing

2.1 Introduction

An introduction to the basic concept of remote sensing including electromagnetic radiation, its behavior, and interactions with the atmospheric constituents and with the matter has been dealt with in Chapter 1. In fact, the interactions of the electromagnetic radiation with the matter lead to reflection, absorption, transmission, and emission of the incident radiation. Various objects reflect/emit the incident radiation in varying degrees in different portions of the electromagnetic spectrum, which is referred to as spectral response pattern and helps us in detection/identification of the particular target/feature. The instruments that are used to measure the properties of electromagnetic radiation leaving a surface due to reflection/scattering/emission are referred to as sensors. While making the measurement of reflected/scattered/backscattered radiation, the sensor needs to be either handheld or clamped to a stand or mounted onto a cherry picker or aircraft or spacecraft. The device which provides the mechanical support to the sensors is called remote sensing platform. It is quite evident that for reflection/emission to take place the object/target needs to be illuminated using either natural or artificial radiation source. Based on the source of radiation used for illuminating the object/target, remote sensing sensors have been categorized into two primary types—*active* and *passive*. *Passive sensors* capture natural radiation, either reflected or emitted, from the target. Reflected sunlight is the most common source of radiation measured by passive sensors. Passive remote sensors include photographic camera, digital camera, video camera, radiometers, and imaging spectrometers. It is also possible to generate electromagnetic radiation of a specific wavelength or a band of wavelengths as a part of the sensor system. The interaction of the electromagnetic radiation with the target could then be studied by sensing the scattered radiation. Such sensors which produce their own electromagnetic radiation are called *active sensors*. Radar and LiDAR are examples of active sensors.

The passive sensors whose response covers a wavelength region extending from about 0.4 to 20 μm are considered optical infrared (OIR) sensors. The OIR sensors could be further classified into two categories—photographic and electro-optical. Passive sensors are the most common sensor type used for natural resources and environmental management. This is not only because passive sensor systems are generally simpler in design built only to receive energy, but also because portions of the solar spectrum provide very useful information for inventory and monitoring the natural resources and environment. The major limitation of passive systems is that in most cases they require sunlight in order for valid and useful data to be acquired. Consequently, deployment of or data acquisition by passive sensors is dependent on lighting (time of day, time of year, latitude) and weather conditions, as cloud cover can interfere with the path of solar radiation from the sun to the earth surface and then to the sensor.

Generally, radiance is the property that is measured by the sensor as a function of wavelength, but could include other parameters such as state of polarization. Most remote sensing instruments (sensors) are designed to measure photons. The fundamental principle underlying sensor operation centers on what happens in the detector. In fact, when a negatively charged plate of some appropriate light-sensitive material is subjected to a beam of photons, there is an emission of negative particles (electrons). The electrons can then be made to flow from the plate, collected, and counted as a signal. Interestingly, the magnitude of the electric current produced (number of photoelectrons per unit time) is directly proportional to the light intensity. Thus, changes in the electric current can be used to measure changes in the photons (numbers; intensity) that strike the plate (detector) during a given time interval. The kinetic energy of the released photoelectrons varies with frequency (or wavelength) of the impinging radiation. But, different materials undergo photoelectric effect (e.g., release of electrons) over different wavelength intervals; each has a threshold wavelength at which the phenomenon begins and a longer wavelength at which it ceases.

2.2 Remote Sensing Platforms

As discussed in Section 1.13, remote sensing platform comprises one of the components of the system. The reflected/emitted/scattered electromagnetic radiation leaving the target surface could be recorded using an appropriate instrument (sensor) either *in situ* (in-place) or in laboratory by manually holding it or mounting onto a stable platform like tripod stand or a hydraulic platform. The remote sensing platform refers to any system onto which remote sensing instruments (sensors) are mounted. The purpose of platform is to position the sensor over area of interest. Therefore, the type of platform is determined by the requirements of the measurements to be made. Tripod stand, cherry pickers, towers, crane, tall buildings or scaffolding, balloons, aircrafts, rockets, and satellites are generally used as platforms. Remote sensing data for Earth resources surveys are generally collected from air/space borne platforms. Although balloons, aircrafts, and rockets have been used as platforms, aircrafts and satellites covering larger geographical area have been by the most widely used platforms. Aircrafts are used for covering smaller area or limited region of interest over a given timeframe, while satellites cover larger areas and provide synoptic view at regular intervals (Figure 2.1).

2.2.1 Airborne Platforms

A great deal of remote sensing studies in the past have been conducted, and still being carried out using aircrafts of various types as a platform. However, balloons are also sometimes used for studies related to atmospheric applications. Besides, unmanned aerial vehicles (UAVs) are being increasingly used for various applications involving smaller aerial extent. Recently, there has been a phenomenal increase in the use of small UAVs. The acronym UAV or UAS (unmanned aircraft systems) refers to a power-driven, reusable airplane operated without a human pilot on board. With this definition, remote-controlled aircrafts also fall into this category. Actually, most UAVs have remote control abilities to avoid some severe failures that may cause crashes.

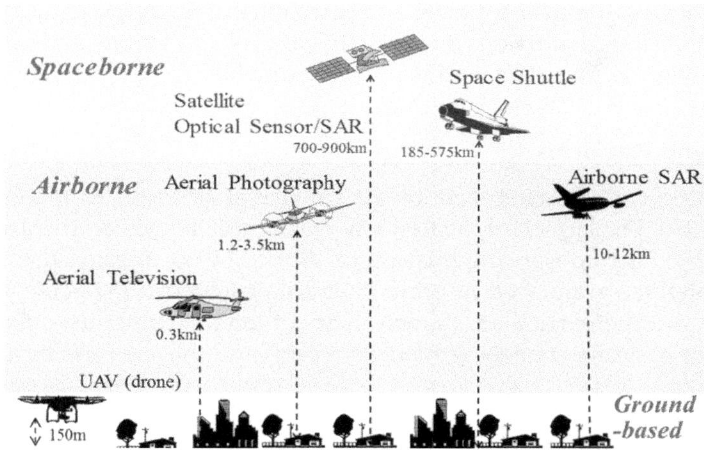

FIGURE 2.1
Remote sensing platforms.

UAVs can be remote-controlled aircraft (e.g., flown by a pilot at a ground control station) or can fly autonomously based on preprogrammed flight plans or more complex dynamic automation systems. UAVs are currently used for a number of missions, including reconnaissance and attack roles. A UAV is defined as being capable of controlled, sustained level flight and powered by a jet or reciprocating engine. In addition, a cruise missile can be considered to be a UAV, but is treated separately on the basis that the vehicle is the weapon. The acronym UAV has been expanded in some cases to UAVS (unmanned aircraft vehicle system). The FAA has adopted the acronym UAS to reflect the fact that these complex systems include ground stations and other elements besides the actual air vehicles. The term *unmanned aerial vehicle* was changed to "unmanned aircraft system" to reflect the fact that these complex systems include ground stations and other elements besides the actual air vehicles. The term *UAS*, however, is not widely used as the term *UAV* has become part of the modern lexicon.

With the emergence of high-power-density batteries, long-range and low-power micro radio devices, cheap airframes, and powerful microprocessors and motors, small/micro UAVs have been used in civilian circumstances like remote sensing, mapping, traffic monitoring, search and rescue, etc. Small UAVs have a relatively short wingspan and light weight. They are expendable, easy to be built and operated. Most of them can be operated by one to two people, or even be hand-carried and hand-launched (Wu et al., 2004; Cambone et al., 2005). As a matter of fact, small UAVs are designed to fly at low altitude (normally less than 1,000 m) to provide a close observation of the ground objects. This low-altitude flight makes the UAVs easy to crash. A robust and accurate autopilot system is indispensable for small UAVs to successfully perform tasks like low-altitude surveillance (Chao et al., 2010).

The major advantage of aircrafts is their versatility. Aircrafts can be flown at short notice where and when required, of course subject to weather conditions. The flying height can be altered to adjust the scale of the photo or image or to fly under cloud cover. Besides, their flight lines can be arranged for specific purposes so as to cover a specific area, to observe that area from a particular angle, or to produce overlapping images for stereoscopy. Disadvantages include higher cost of flying operation, seeking permission from competent authorities in some countries, and instability of the aerial platforms, and images may suffer from various distortions due to drift, yaw, roll, and pitch, and positioning of the aircraft may be slightly uncertain and is not reproducible.

2.2.2 Spaceborne Platforms

With the advancement of rocket technology, the use of satellites as remote sensing platform began in 1960. The launch of the first low-earth orbit Television Infrared Observation Satellite (TIROS-1) Earth observing mission on April 1, 1960, marked the beginning such an initiative. Satellites are placed in orbits that are designed for specific purposes of the mission, and to suit the particular characteristics of the instrument(s) onboard. There are two main classes of orbits, namely *geostationary* orbit and the *low-earth* or *near-polar* orbit.

2.2.2.1 Geosynchronous Satellites

A geostationary orbit, geostationary Earth orbit, or geosynchronous equatorial orbit (GEO) is a circular orbit 35,786 km above the Earth's equator and following the direction of the Earth's rotation (e.g., west to east). In fact, a satellite placed at this altitude above the earth takes approximately 23 hours 56 minutes and 4 seconds, matching the Earth's sidereal rotation period around its axis. As a result, the satellite appears to remain in position over a point on the equator. Such an orbit is sometimes referred to as *Clarke orbit* and preferred for communication satellites, as they remain in view of small fixed antennae and give continuous reception. There are several meteorological satellites, namely Meteosat, GOES, GOMS, INSAT, etc., distributed around the equator, each viewing nearly 40% of the earth's surface and providing almost continuous coverage of the global weather patterns. Figure 2.2 is the schematic representation of a geostationary orbit.

2.2.2.2 Polar Orbiting Satellites

Most earth observation satellites are in near-polar orbits that range from 600 to 2,000 km above the ground. With the Earth's radius of about 6,300 km, such satellites take about 100 minutes to complete one orbit. If such an orbit passes over the Earth's poles, the subsatellite track would always pass through the same points on the Earth's surface. As a result only that particular track would be observed by the sensor onboard. If, however, orbits are slightly inclined away from the poles, then the orbit would *process* with respect to the Earth and subsequent tracks (paths) would be displaced by an amount that depends on the angle made by the plane of the satellite orbit and the Earth's rotational axis. The rate of

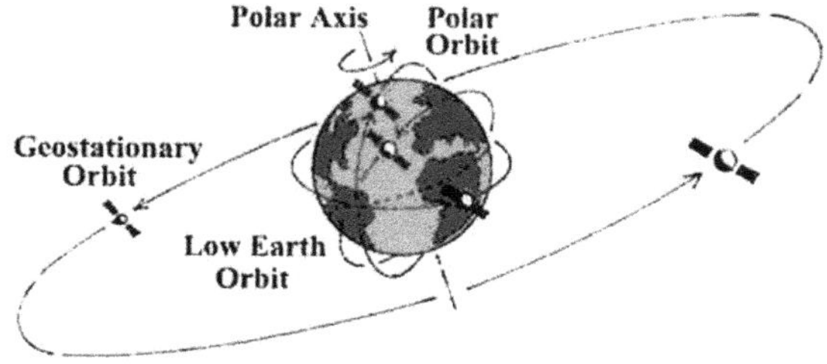

FIGURE 2.2
Schematic of a geostationary orbits. Polar as well as low-earth orbits are also shown.

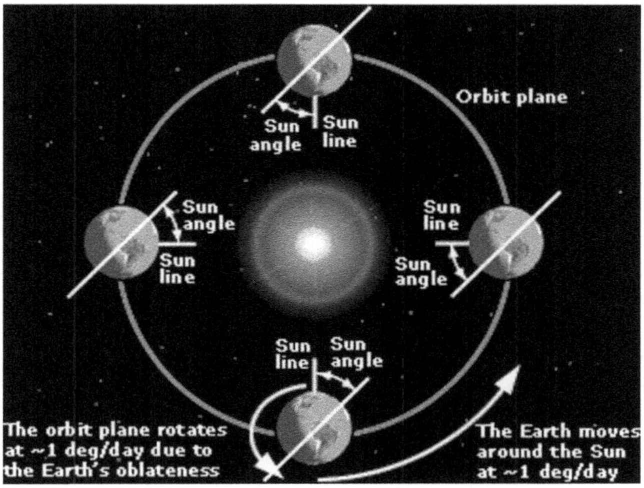

FIGURE 2.3
Schematic representation of a sun-synchronous orbit.

precession will determine the number of orbits required before the satellite track repeats itself, and the time taken to accomplish it will be the satellite revisit time, i.e., the time between successive observations of a particular point on the earth surface.

Such orbits are called *polar orbits* or more correctly *near-polar orbits.* A special case of a polar orbit is one where the orbital precession is exactly equal to the earth's solar precession, so that the satellite crosses the equator at exactly the same local solar time on each orbit. Such an orbit is referred to as *sun-synchronous orbit* (Figure 2.3). The advantage of this type of orbit is that the solar angle is approximately the same each time a point is imaged, and hence variability of illumination and shadow angles will be minimized. This makes them particularly suitable for monitoring highly dynamic features like vegetation.

2.3 Passive Sensors

As mentioned in Section 2.1 passive sensors operate both in optical and in microwave regions of electromagnetic spectrum. The passive optical sensors generally consist of two components: optics and the detector. Optics includes radiation-collecting optics and radiation-sorting optics.

2.3.1 The Optics

The optics for radiation collection primarily comprise mirrors, lenses, and a telescopic setup to collect the radiation from the ground and focus it onto radiation-sorting optics. A calibration source is often provided on board, and a chopper enables radiation from the calibration source to be viewed by the detectors at regular intervals. *Radiation-sorting optics* use devices such as gratings, prisms, and interferometer to separate radiation of different wavelength ranges. In some cases, such as vidicons and push-broom scanners, spectral separation may be achieved by using appropriate band-pass filters which cover the lens or detectors. The radiation—normally visible infrared (VIR) and/or near infrared (NIR) and shortwave infrared (SWIR), and/or thermal infrared (TIR) in nature—must then be broken into its spectral elements/into narrow bands. The width in wavelength units of a band or channel is defined by the instrument's spectral resolution. Prisms and diffraction gratings can be used to break selected parts of the electromagnetic spectrum into spectral bands. Filters are also used to segregate a portion of the electromagnetic radiation into spectral bands of varying widths.

Absorption filters pass only a limited range of radiation wavelengths, absorbing radiation outside this range. They may be either broad or narrow band-pass filters. These filters may be high band-pass (selectively removes shorter wavelengths) or low band-pass (absorbs longer wavelengths). Interference filters work by reflecting unwanted wavelengths and transmitting others in a specific interval. A common type of filter used in general optics and in many scanning spectroradiometers is the dichroic filter. A *dichroic filter*, thin-film filter, or interference filter is a very accurate color filter used to selectively pass light of a small range of colors while reflecting other colors. This uses an optical glass substrate over which are deposited (in a vacuum setup) from 20 to 50 thin (typically, 0.001 mm thick) layers of a special refractive index dielectric material (or materials in certain combinations) that selectively transmits a specific range or band of wavelengths. Absorption is nearly zero. These filters can be either additive or subtractive color filters when operating in the visible range. After sorting, the radiation of selected wavelength ranges is directed to the detectors.

The passive sensors generally consist of two components—optics and the detector. *Optics* includes radiation-collecting optics and radiation-sorting optics. The optics for radiation collection primarily comprise mirrors, lenses, and a telescopic setup to collect the radiation from the ground and focus it onto radiation-sorting optics. A calibration source is often provided on board and a chopper enables radiation from the calibration source to be viewed by the detectors at regular intervals. *Radiation-sorting optics* use devices such as gratings, prisms, and interferometer to separate radiation of different wavelength ranges. In some cases, such as vidicons and push-broom scanners, spectral separation may be achieved by using appropriate band-pass filters which cover the lens or detectors. The radiation—normally VIR and/or NIR and SWIR, and/or TIR in nature—must then be broken into its spectral elements/into narrow bands. The width in wavelength units of a band or channel is defined by the instrument's spectral resolution.

Prisms and diffraction gratings can be used to break selected parts of the electromagnetic spectrum into spectral bands. Filters are also used to segregate a portion of the electromagnetic radiation into spectral bands of varying widths. *Absorption filters* pass only a limited range of radiation wavelengths, absorbing radiation outside this range. They may be either broad or narrow band-pass filters. These filters may be high band-pass (selectively removes shorter wavelengths) or low band-pass (absorbs longer wavelengths). Interference filters work by reflecting unwanted wavelengths and transmitting others in a specific interval. A common type of filter used in general optics and in many scanning spectroradiometers is the dichroic filter. A *dichroic filter*, thin-film filter, or interference filter is a very accurate color filter used to selectively pass light of a small range of colors while reflecting other colors. This uses an optical glass substrate over which are deposited (in a vacuum setup) from 20 to 50 thin (typically, 0.001 mm thick) layers of a special refractive index dielectric material (or materials in certain combinations) that selectively transmits a specific range or band of wavelengths. Absorption is nearly zero. These filters can be either additive or subtractive color filters when operating in the visible range. After sorting, the radiation of selected wavelength ranges is directed to the detectors.

The *detector* primarily includes devices which transform optical energy into electrical energy (e.g., photons into electrons). The core of device is a quantum or photodetector unit. A photodetector is a device which absorbs light and converts the optical energy to measurable electric current. The incident photons interact with electronic energy level of detector material, and electrons or charge material is released (photoelectric effect). The response in photodetector is very quick and, as mentioned in Section 2.1, the intensity of the electrical signal output is proportional to the intensity of photons incident in a specific energy range. Major limitations of photodetectors include (1) their response varies quickly with wavelength and (2) photoconductors operating at longer wavelengths have to be operated at very low temperature (195 K, 77 K or sometimes 5 K) to avoid noise. This is achieved by placing the detector within a double-walled vessel called Devar, filled with liquid helium or nitrogen for cryogenic cooling.

2.3.2 Detectors

As mentioned in the previous section, *detector* is a device that produces an output signal, which depends upon radiation striking the active area of the detector. In remote sensing several types of detectors are used to capture the reflected/emitted electromagnetic radiation from the object. *Photographic film* is one of the typical examples of detectors. The film, in general, consists of a photosensitive photographic emulsion coated on to a base for support. The emulsion consists of silver halide crystals, generally referred to as grains, of different sizes embedded in a gelatin matrix. Upon striking the light on the emulsion, it undergoes a photochemical reaction forming a latent image. The chemicals used during processing the film reduce (reduction—a chemical reaction) the exposed silver salts to silver grains that appear black. The areas in the film which were not exposed to light—silver halide—are dissolved during the processing. Consequently, that portion appears transparent. Thus, a "negative image"—those areas not exposed to light—appears bright whereas the rest of the areas exposed to light appear dark. Positive images are subsequently produced on a paper or on a transparent positive (diapositive) (Joseph, 2003).

With the development of detector technology photomultipliers were used as detectors. Most detectors today are made of solid-state semiconductor metals or alloys. A semiconductor has a conductivity intermediate between a metal and an insulator. Under certain conditions, such as the interaction with photons, electrons in the semiconductor are excited and moved from a filled energy level to another level called the conduction band which is deficient in electrons in the unexcited state. The resistance to flow varies inversely with the number of incident photons. Different materials respond to different wavelengths—to photon energy levels and are thus spectrally separable.

Photodiodes developed from silicon metal and lead oxide (PbO) are commonly used as detector materials for capturing the reflected radiation in the visible region of the electromagnetic radiation. For detection reflected radiation in the NIR lead sulfide (PbS) and indium arsenide (InAs) are used whereas indium stibnium/antimonide (InSb) is responsive in the mid-infrared (3–6 μm). Mercury cadmium telluride (Hg-Cd-Te) is the most common detector material for the 8–14 μm range. When operating it is necessary to cool these detectors to near 0 K using *Dewars* coolers to optimize the efficiency of electron release. Other detector materials are also used and performed under specific conditions. Other detectors used in remote sensing include photoemissive, photodiode, photovoltage, and thermal (absorption of radiation) detectors. Charge-coupled device (CCD) is the most commonly used detector in the VIR, NIR, and SWIR regions. A brief overview of detectors is presented hereunder.

2.3.2.1 Quantum Detectors

Quantum detectors, also referred to as photon detectors, respond to the rate at which the photon is absorbed. In a photon detector, since the response is dependent on the number of photons per second, the response decreases as wavelength becomes shorter. On the other hand, since thermal detectors respond to the total energy absorbed, for a constant energy input, response is independent of wavelength. There are three basic modes in which the quantum detectors perform: *photoemissive*, *photoconductive*, and *photovoltaic* modes.

2.3.2.2 Photoemissive Detectors

Photoemissive detectors are based on the photoelectric effect, in which incident photons release electrons from the surface of the detector material. The free electrons are then collected in an external circuit. If a photon of energy greater than a certain critical energy (i.e., wavelength less than a critical value) is incident on certain materials, it can liberate electrons with sufficient energy to escape from the surface. This property is called photoelectric or photoemissive effect.

2.3.2.3 Semiconductor Detectors

Semiconductor detectors could be categorized into *extrinsic* and *intrinsic*. The intrinsic semiconductor is one made of single type of atoms like silicon (Si) or germanium (Ge). It is possible to influence the properties of a semiconductor in a controlled manner by adding impurities, known as *doping*, and these semiconductors are called 'extrinsic" semiconductors. Two types of extrinsic semiconductors can be created, depending upon the parent material and impurity doped. In the first type the impurity is such that their valence electron is within the band gap, normally close to the conduction band of the parent semiconductor. These electrons are then more easily exited into the conduction band and constitute the *n*-type semiconductors. In the second case, the dopant is such that it accepts electrons from the valence band of the parent semiconductor, and thus creates holes. The resulting material is called *p*-type semiconductor. Thus, in the *n*-type of semiconductor, conduction is mainly due to the negative charge or electrons, with holes as minority carriers. In a *p*-type of semiconductor, conduction is mainly due to positive charges or holes, with electrons as minority carriers.

2.3.2.4 Photoconductive Detectors

The incoming light produces free electrons which can carry electrical current so that the electrical conductivity of the detector material changes as a function of the intensity of the incident light. Photoconductive detectors are fabricated from semiconductor materials such as silicon. When electrons are excited to the conduction band, the conductivity of the detector increases, and can be measured. Thus, photoconductive detectors can be considered to be light-sensitive variable resistors, whose conductivity varies with the number of photons absorbed. The band gaps and hence the cut-off wavelength can be altered by suitable doping. Some of the commonly used photoconductive detectors for remote sensing include indium antimonite, indium gallium arsenide, mercury, cadmium telluride. By adjusting the relative concentration of cadmium telluride and mercury telluride, it is possible to cover a range of wavelengths extending from NIR to TIR.

2.3.2.5 Photovoltaic Detectors

Photovoltaic detector contains a junction in a semiconductor material between a region where the conductivity is due to electrons and a region where the conductivity is due to holes (a so-called *p*–*n* junction). A voltage is generated when optical energy strikes the device. When *p* and *n* extrinsic semiconductors are brought together, we have a *p*–*n* junction. Since the *n*-type region has excess electrons, the concentration gradient causes electrons to diffuse from the *n*-type region to the *p*-type region and holes from *p*-type to the *n*-type region. This diffusion generates a net positive charge in the vicinity of the junction on the *n* side and a negative potential on the *p* side, thus creating an electric field. However, electron-hole movement across the junction stops when the electric field generated is equal to the concentration gradient. The potential difference created across the junction is called the *contact potential.* A device operating in this mode to detect radiation is called a *photodiode.*

When a photodiode is exposed to radiation with energy greater than band gap energy, additional charge carriers are generated. These additional charge carriers increase the potential across the junction. The voltage generated across the junction (open circuit) increases logarithmically with the radiation intensity. However, if the photodiode is operated in the current mode, the current generated increases linearly (within limits) with intensity of radiation. This mode of operation in the zero bias mode is called the photovoltaic mode. Here the photodiode is actually generating an electrical potential.

2.3.2.6 Thermal Detectors

The detectors based on the thermal and quantum effects, in general, convert optical energy into an electrical signal which is then measured suitably. These types of detectors are called "electro-optical" detectors. Opto-mechanical scanners use one or other types of electro-optical detectors. Thermal detectors make use of the heating effects of the electromagnetic radiation. The consequent rise in temperature causes a change in some physical properties of the detector. The response of the thermal detector is only dependent upon the radiant power which they absorb and hence independent of wavelength. Some of the common thermal detectors include bolometers, thermocouples, and pyroelectric detectors.

In bolometers, a change in temperature caused by radiation changes the electrical resistance, which is suitably measured. A thermocouple is formed by joining two dissimilar materials having different thermoelectric powers. Two of the commonly used materials are bismuth and antimony. One of the junctions is blackened to absorb the radiation. The difference in temperature between the junctions produces a thermoelectric electromagnetic field (emf), which is a measure of the incident radiant power. In order to increase the sensitivity, a number of thermocouples are connoted in s series, called a thermopile.

When the temperature is below the Curie temperature, certain ferroelectric crystals exhibit spontaneous polarization with change in temperature. This is called pyroelectric effect. A pyroelectric detector is basically a ferroelectric capacitor that is thermally isolated and exposed to incident radiation. When the incident radiation changes the temperature of the detector element, it exhibits spontaneous polarization, i.e., opposite faces of the crystallographic orientation exhibit opposite electrical charges in response to temperature changes. This can cause a displacement to flow in an external circuit, which can be detected. Pyroelectric detectors respond only to temperature.

2.4 Optical Sensors

By combining a number of detectors or radiometers into detector arrays, it is possible to create a sensor that can acquire a 2D image of an area. There are three basic designs for imaging sensors: frame, push-broom, and mechanical scanner. The first two designs are similar. The frame sensor is a 2D array of detectors that acquires an entire image in one exposure similar to the way a camera captures an image on film. A push broom sensor is a 1D array that obtains an image one line at a time. Each new data line is added as the platform moves forward, building up an image over time. In a mechanical scanner system the sensor acquires only one or several pixels in any given instant, but since the scanner physically sweeps or rotates the sensor (a radiometer) or a mirror back and forth, an image is produced.

This category of sensor (passive visible, infrared, and thermal imaging systems) contains numerous instruments that have been deployed on a wide variety of platforms and used for many applications. Most modern imaging systems are multispectral (acquiring data for more than one limited spectral area). The recording of each discrete spectral sampling is referred to as an image band or channel. Using image processing techniques, multiple (usually three) bands selected from a multispectral image database can be combined to make a single color composite image.

2.4.1 Conventional Photographic Cameras

Photographic cameras are the oldest sensors used even today. The photographic camera consists of a lens assembly, and the film magazine. The lens cone assembly includes the lens, filter, shutter, and diaphragm (Figure 2.4). The lens is usually a multielement lens assembly. The filter limits the wavelength region of the scene radiance reaching the film. Filters are transparent (glass or gelatin) material placed in front of the lens, the most common being the absorption films, which absorb certain wavelengths. The diaphragm (aperture stop) is located in between the lens elements. The diaphragm controls the aperture of the lens, which decides the amount of light passing through the lens. The diaphragm diameter can be adjusted to suit lighting conditions and film sensitivity. The shutter controls the duration of the exposure. The shutter is incorporated at the focal plane or within the lens assembly.

The camera magazine, which holds the film supply and takes up reels, can be invariably detached from the camera. During exposure, the film is held stationary and flat at the focal plane. For cameras designed for precision measurement, a vacuum system usually ensures this. A camera body to which the lens cone and the film magazine are attached

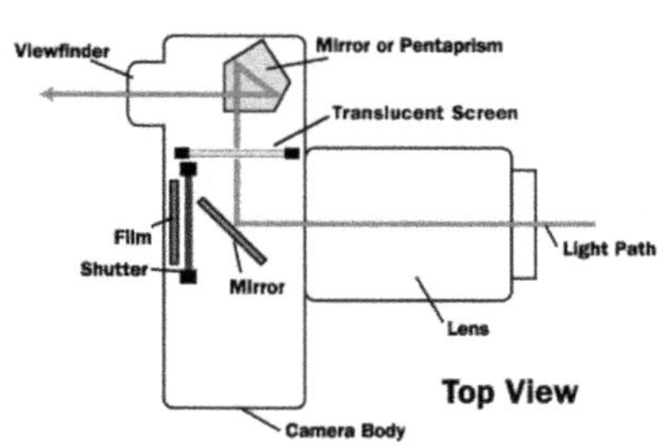

FIGURE 2.4
Schematic of a conventional photographic camera.

also contains the film drive mechanism. The most commonly used camera is the mapping camera—generally referred to as metric camera or cartographic camera. The important feature of the mapping camera is its high degree of distortion correction and provision for fixed marks (fiducials) to be recorded on the film. The lens has a fixed focal length and is rigidly fixed relative to the film plane. The fiducial markers are exposed on film simultaneously with the exposure of ground scene. The camera can be operated with varying overlaps to produce stereoscopic pairs to generate height information.

2.4.2 Digital Aerial Cameras

Digital aerial cameras are the advanced version of photographic cameras wherein the (CCD) arrays are used in place of films to produce the image of the terrain features. Digital photography is capable of delivering photogrammetric accuracy and coverage as well as multispectral data at any user-defined resolution down to 0.1 m ground sampling distance. Digital aerial cameras typically have CCD arrays that produce images containing about 3,000 × 2,000 to 7,000 × 5,000 lines/pixel. Most record an 8-bit to 12-bit black-and-white (B&W) image (256–4,096 gray levels). The shutter can be mechanical or electronic. The exposure time and aperture setting are adjusted before the over flight, depending on conditions and brightness of the mapped features. Each frame is instantaneously recorded so, unlike multispectral line scanners, there is a minimum distortion due to aircraft motion during acquisition. The aircraft altitude and the focal length of the lens system determine the ground resolution or ground sampling distance (GSD). Typical GSD values range from 15 cm to 3 m.

2.4.3 Video Cameras

In the video/vidicon/television camera, an optical system is made to focus the ground scene on to a photoconductive surface. The incident photons vary the conductivity of the surface locally according to the intensity of light. An electron beam is made to scan the photoconductive surface from the rear side. The resulting target current will be proportional to the conductivity of the photoconductive surface (and hence the intensity of light) and the signal is further amplified and recorded or transmitted. In the case of Return Beam Vidicon (RBV), the signal is derived from the depleted electron beam which is reflected from the photo-conducting surface. This is further amplified by a multi-stage electron multiplier. The RBV used in the Landsat series and the television camera system used in the Indian experimental remote sensing satellites, Bhaskara-I and -II are examples of video cameras.

2.4.4 Radiometers

A radiometer is the basic element of all electro-optical and microwave sensors. Simply, it is a device for measuring the intensity of electromagnetic radiation falling on its detector with in a defined spectral range. The technical detail depends upon the particular part of the spectrum in which it is used, but all radiometers comprise three elements: an optical system to focus the radiation, and to select the wavelength, detectors that produce an electrical signal, and a signal processor to provide an output.

2.4.4.1 Radiometers Operating in Optical Region

The simplest radiometers are non-imaging detectors that integrate the radiation that arrives from within a field of view and within a specific wave band. Handheld spectroradiometers with high spectral resolutions are often used in the field to measure the reflectance spectrum of particular targets or vegetation species, often as part of calibration process whereas radiometers mounted on aero planes or satellites will measure the radiation from individual large areas of ground along the flight line or subsatellite track.

2.4.4.1.1 Electro-Optical Scanners

The opto-mechanical scanners operate in the VIR (400–700 nm), the reflected infrared (700–3,000 nm) and the TIR (3,000–14,000 nm) region of the electromagnetic spectrum. After being focused by a mirror system, the radiation from each image element or pixel is segregated into spectral bands, with one image being produced in each spectral band. Based on the mode of scanning electro-optical scanner could be categorized into the *cross-track or whisk-broom* and the *along track or push broom* types. The whisk-broom or spotlight or across track scanners use a mirror to reflect light onto a single detector. A rotating mirror moves back and forth, to collect the spectral measurements from one pixel in the image at a time (Figure 2.5). The scanner scans the Earth in a series of lines. The lines are oriented perpendicular to the direction of motion of the sensor platform, i.e., across the swath. Each line is scanned from one side of the sensor to the other, using a rotating mirror. As the platform moves forward over the Earth, successive scans build up a 2D image of the Earth's surface. The incoming reflected or emitted radiation is segregated into several spectral components that are detected independently.

The instantaneous field of view (IFOV) and the altitude of the platform determine the ground resolution cell viewed, and thus the sensor's spatial resolution. The angular field of view of the sensor is the sweep of the mirror, measured in degrees, used to record a scan line, and determines the width of the imaged swath. Also, the length of time when the IFOV views or "sees" a ground resolution cell as the rotating mirror scans is called the dwell time or integration time. It is generally quite short (a few micro- or milliseconds), and influences the design of the spatial, spectral, and radiometric resolution of the sensor.

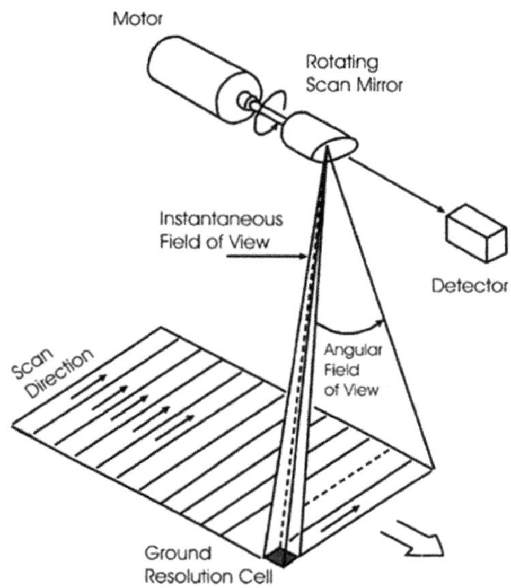

FIGURE 2.5
Sketch of an opto-mechanical scanner.

Push broom scanners, also sometimes referred to as along-track scanners, use a line of detectors arranged perpendicular to the flight direction of the aircraft or spacecraft (Figure 2.6). As the aircraft or spacecraft moves forward, the image is collected one line at a time, with all of the pixels in a line being imaged simultaneously. Each individual detector measures the energy for a single ground resolution cell and thus the size, and IFOV of the detectors determines the spatial resolution of the system. A separate linear array is required to measure each spectral band or channel. For each scan line, the energy detected by each detector of each linear array is sampled electronically and digitally recorded.

2.4.4.2 Radiometers Operating in Microwave Region

A microwave radiometer is a passive sensor that detects radiation emitted or reflected by an object in the microwave range at frequencies of about 1–100 GHz. Passive microwave systems are based on a type of radiometer that detects wavelengths in the microwave region of the spectrum. Because of their nature of microwave radiation, optical systems cannot be used for the detection of this range of wavelengths. The microwave power has to be detected to find some measure of its mean. Two straightforward detector types can be made, using microwave semiconductor diodes: the linear detector and the square-law detector. In the present case, it is very reasonable to use the square-law detector. Then the output voltage will be proportional to the input power and hence the input temperature. Finally, we indicate where the integration takes place: The signal from the detector is smoothed by the integrator to reduce fluctuations in the output, and the longer the integration time.

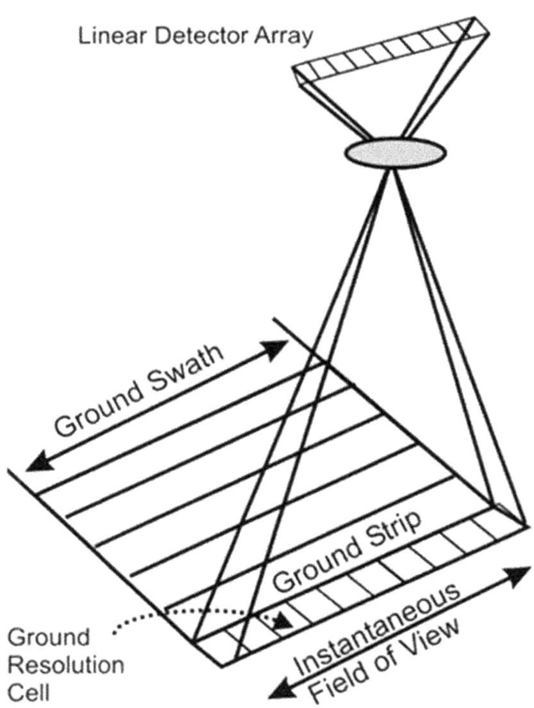

FIGURE 2.6
Sketch of a push-broom scanner.

Solids, liquids (e.g., the Earth's surface, ocean, sea ice, snow, vegetation), but also gases emit and absorb microwave radiation. Traditionally, the amount of radiation a microwave radiometer receives is expressed as the equivalent blackbody temperature also called brightness temperature. In the microwave range several atmospheric gases exhibit rotational lines.

A microwave radiometer consists of an antenna system, microwave radiofrequency components (front-end), and a back-end for signal processing at intermediate frequencies. The atmospheric signal is very weak and the signal needs to be amplified by around 80 dB. Therefore, often heterodyne techniques are used to convert the signal down to lower frequencies that allow the use of commercial amplifiers and signal processing. Increasingly low-noise amplifiers become available at higher frequencies, i.e., up to 100 GHz, making heterodyne techniques obsolete. Thermal stabilization is highly important to avoid receiver drifts.

Usually ground-based radiometers are also equipped with environmental sensors (rain, temperature, humidity) and GPS receivers (time and location reference). The antenna itself often measures through a window made of foam which is transparent in the microwave spectrum in order to keep the antenna clean of dust, liquid water, and ice. Often, also a heated blower system is attached to the radiometer which helps to keep the window free of liquid drops or dew (strong emitters in the MW) but also free of ice and snow.

As is seen from Figure 2.7, attached after being received at the antenna the radio-frequency (RF) signal is down converted to the intermediate frequency with the help of a stable local oscillator signal. After amplification with a low-noise amplifier and band-pass filtering the signal can be detected in full power mode, by splitting or splitting it into multiple frequency bands with a spectrometer. For high-frequency calibrations a Dicke switch is used here.

As with optical systems though, both non-imaging and imaging systems are available. The components of a microwave radiometer are an antenna, receiver, and recording device. Microwave energy emitted from Earth's surface is collected by the antenna, converted by a receiver into a signal, and recorded (Figure 2.8). The features of electromagnetic energy measured by microwave radiometers are polarity, wavelength, and intensity. These properties provide useful information about the structure and composition of an object. Most of the applications of passive microwave radiometers have been in the fields of atmospheric and oceanographic research. It has also proven to be an effective tool for the measurement of soil moisture, an important parameter in studying vegetation.

2.4.4.2.1 Aperture Synthesis

One of the major limitations of the passive microwave sensor is its coarser spatial resolution. In order to improve spatial resolution aperture synthesis has been tried out. Aperture

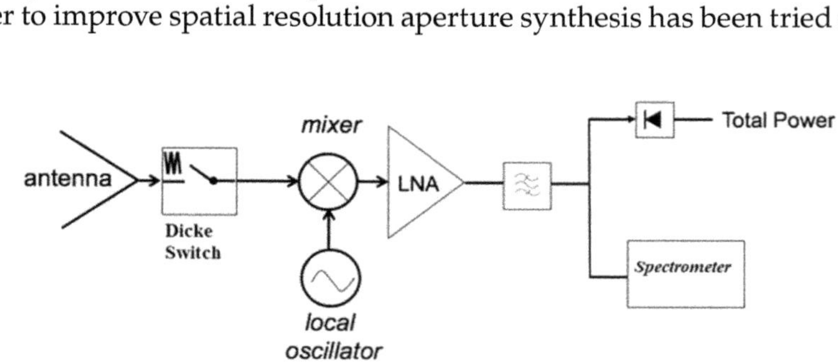

FIGURE 2.7
Schematic diagram of a microwave radiometer using heterodyne principle.

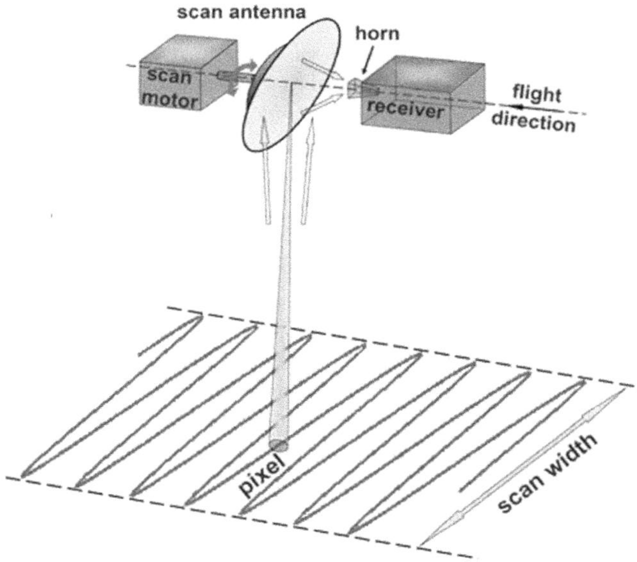

FIGURE 2.8
Schematic diagram showing working principle of a microwave radiometer.

synthesis is an interferometric technique in which the complex correlation of the output voltage from pairs of antennas is measured at different antenna spacings or baselines. The correlation measurement is proportional to the spatial Fourier transform of the intensity of a distant scene at a frequency that depends upon the antenna spacing. Each baseline measurement produces a sample point in the 2D Fourier transform of the scene.

By making measurements at many different spacings and selectively distributing the antenna elements so as to obtain optimum sampling in the Fourier domain, a set of Fourier samples suitable for inverting the transform may be obtained. High-resolution maps of the source may be retrieved using a set of relatively small antennas without the need for scanning the antenna aperture. As in a conventional antenna array, resolution is determined by the maximum spacing (baseline) and the minimum spacing determines the location of grating lobes. However, in contrast to conventional arrays, each spacing needs to appear only once, and no mechanical scanning is necessary—it is done in software as part of the image reconstruction.

The Advanced Microwave Sounding Unit (AMSU) is a multichannel microwave radiometer installed on meteorological satellites. The instrument examines several bands of microwave radiation from the atmosphere to perform atmospheric sounding of temperature and moisture levels.

There are two types of microwave radiometers: total power radiometer and Dicke radiometers are in vogue. Total power radiometer, an instrument that measures the total noise power from the antenna (and from the receiver), can be considered a total power radiometer. It essentially consists of a linear receiver with a well-defined band width (pre-detection band width) stable gain characteristics, connected to the antenna. This is followed by a square-law detector, a video amplifier, and an integrator. Dicke radiometers reduce, to some extent, the gain instability problem, by measuring the difference between the antenna temperature and a known reference temperature, instead of measuring only the antenna temperature.

2.4.4.2.2 *Spaceborne Microwave Radiometers*

An imaging microwave radiometer consists of an antenna which receives the incoming radiation, a scanning mechanism (mechanical or electrical), a receiver and associated electronics, which detects and amplifies the received radiation and produces a voltage output, in-flight calibration systems, hot body, sky horn, etc., auxiliary logic system providing signals for timing, multiplexing data, formatting, etc., and house-keeping systems which monitor various temperatures, voltage, etc.

2.4.4.2.3 *Push Broom and Synthetic Aperture Radiometer*

To overcome the issue of poor spatial resolution, the major constraints of currently operating microwave radiometers, the push broom and synthetic aperture radiometers have been developed. An alternate solution to the problem of obtaining high spatial resolution without any mechanical scanning is a push broom radiometer. Here a suitably designed array of radiators (as feed elements) is placed at the focus of an offset paraboloid reflector so as to produce a number of simultaneous circular footprints to cover the desired swath, in a circular arc such that the same incident angle is maintained. It is essentially a single reflector with multiple feeds.

The interferometric aperture synthesis is another technique to generate high-resolution microwave radiometer data. A pair of small antennas measures the scene to be imaged and outputs of the radiometers are cross-correlated. The image can be reconstructed through signal processing. Thus, instead of a full aperture, there are a number of small apertures distributed within the same aperture area to produce a high-resolution image.

Microwave Imaging Radiometer with Aperture Synthesis (MIRAS) is the major instrument on the Soil Moisture and Ocean Salinity satellite (SMOS). MIRAS employs a planar antenna composed of a central body (the so-called hub) and three telescoping, deployable arms, in total 69 receivers on the unit. Each receiver is composed of one Lightweight and Cost-Effective Front-end (LICEF) module, which detects radiation in the microwave L-band, in both horizontal and vertical polarizations. The apertures on the LICEF detectors, planar in arrangement on MIRAS, point directly toward the Earth's surface as the satellite orbits.

2.4.4.3 Imaging Spectrometer

Hyperspectral remote sensing involves the convergence of two related but distinct technologies: *spectroscopy* and the *remote imaging* of Earth and planetary surfaces. Spectroscopy refers to the study of light that is emitted by or reflected from materials and its variation in energy with wavelength. As applied to the field of optical remote sensing, spectroscopy deals with the spectrum of sunlight that is diffusely reflected (scattered) by materials at the Earth's surface. Spectrometers or spectroradiometers are used to make ground-based or laboratory measurements of the light reflected from a test material. By introducing imaging capability in spectroradiometer a new technology has been developed. The concept has been termed *imaging spectroscopy/imaging spectrometry/hyperspectral imaging* (Goetz et al., 1985). The imaging spectrometry is defined as "*The simultaneous acquisition of images in many relatively narrow, contiguous spectral bands throughout the ultraviolet, visible and infrared portions of the spectrum.*" (Jensen, 2000).

Imaging spectrometry or hyperspectral remote sensing involves the acquisition of image data in many contiguous spectral bands with an ultimate goal of producing laboratory quality reflectance spectra for each pixel in an image. An optical dispersing element such as a grating or prism in the spectrometer splits this light into many narrow, adjacent wavelength bands and the energy in each band is measured by a separate detector. By using

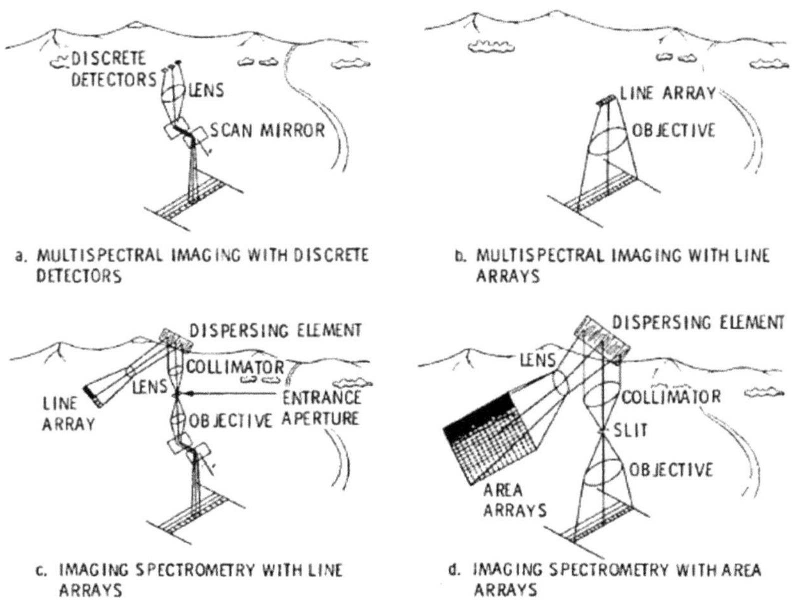

FIGURE 2.9
Various types of scanning mechanisms. In imaging spectrometer both line array detectors and area array detectors are used. Area array detector is, however, very common (Source: Adapted from Goetz et al. (1985)).

hundreds or even thousands of detectors, spectrometers can make spectral measurements of bands as narrow as 10 nm over a wide wavelength range, typically at least 400–2,500 nm (visible through middle infrared wavelength ranges). As evident from Figure 2.9, discrete detectors are used in opto-mechanical scanner or whiskbroom scanners. Linear array detectors are used in push broom scanners. Although both line array and area array detectors are used for imaging in imaging spectrometer, the use of area array detector is most common.

Since conceptualization of hyperspectral remote sensing or imaging spectrometry in mid-1980s several airborne and a few spaceborne sensors have been developed (Table 2.1). As evident from Table 2.1 most of the instruments except for hyperion aboard EO-1 and FTHSI on Mighty Sat II are used in airborne platforms only.

2.5 Resolution of a Sensor

With the wide variety of remote sensing systems available choosing the proper data source fora particular applications, namely inventory, and monitoring degraded land, soil resources mapping and land use/land cover mapping, observing oceans and atmosphere can be challenging. A thorough understanding of different types of resolution of remote sensing data is a prerequisite for its judicious utilization. Spatial, spectral, radiometric, temporal, and angular resolutions are generally used to compare remote sensing analogue and digital data. However, in the context of remote sensing, resolution normally refers to

TABLE 2.1
Main Air and Spaceborne Hyperspectral Sensors

Types of Sensors	Number of Spectral Bands	Spectral Range (μm)
FTHSI on MightySat II	256	0.35–1.05
Hyperion on EO-1	242	0.40–2.50
AVIRIS	224	0.40–2.50
HYDICE	210	0.40–2.50
PROBE-1	128	0.40–2.50
CASI	Over 228	0.40–1.00
HyMap	100 to 200	Visible to TIR
EPS-H	VIS/NIR (76)	VIS/NIR (0.43–1.05)
	SWIR1 (32)	SWIR1 (1.50–1.80)
	SWIR2 (32)	SWIR2 (2.00–2.50)
	TIR (12)	TIR (8–12.50)
DAIS 7915	VIS/NIR (32)	VIS/NIR (0.43–1.05)
	SWIR1 (8)	SWIR1 (1.50–1.80)
	SWIR2 (32)	SWIR2 (2.00–2.50)
	MIR (1)	MIR (3.00–5.00)
	TIR (12)	TIR (8.70–12.30)
DAIS 21115	VIS/NIR (76)	VIS/NIR (0.40–1.00)
	SWIR1 (64)	SWIR1 (1.00–1.80)
	SWIR2 (64)	SWIR2 (2.00–2.50)
	MIR (1)	MIR (3.00–5.00)
	TIR (6)	TIR (8.00–12.00)
AISA	Over 288	0.43–1.00

Source: Vorovenci, (2009).
Note: *AVIRIS,* Airborne Visible Infrared Imaging Spectrometer; *HYDICE,* Hyperspectral Digital Imagery Collection Experiment; *CASI,* Compact Airborne Spectrographic Imager; *EPS-H,* Environmental Protection System; *DAIS 7915,* Digital Airborne Imaging Spectrometer; *AISA,* Airborne Imaging Spectrometer.

the ability of a sensor to separate and distinguish adjacent objects or items in a scene, be it in a photo, an image, or real life.

2.5.1 Spatial Resolution

Spatial resolution refers to the size of the smallest object that can be resolved on the ground. In a digital image, the resolution is limited by the pixel size. The intrinsic resolution of an imaging system is determined primarily by the IFOV of the sensor, which is a measure of the ground area viewed by a single detector element in a given instant in time. However, this intrinsic resolution can often be degraded by other factors which introduce blurring of the image, such as improper focusing, atmospheric scattering and target motion. Figure 2.10 illustrates the effect of spatial resolution on detectability of terrain features. Shown in the figure are the parts of Vizag city, Andhra Pradesh, southern India captured by Cartosat-2 panchromatic camera (PAN) image with 1 m spatial resolution.

The image has been resampled to 2.5 m, 5 m, 10 m, and 20 m using nearest neighbor algorithm. The images vividly demonstrate the effects of deteriorating spatial resolution on detectability of objects. In this case the crude oil storage tanks (circular white/dark

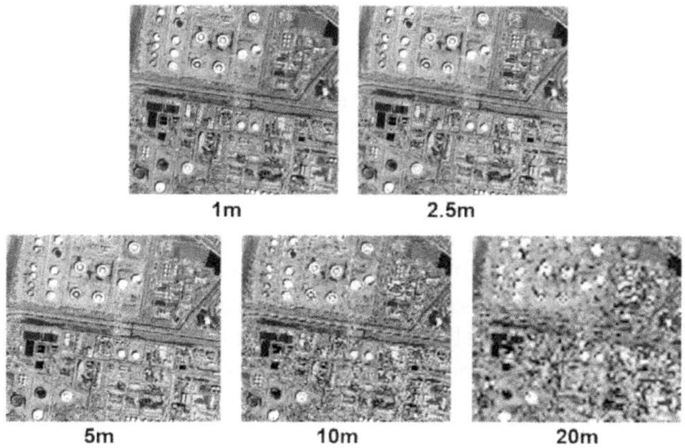

FIGURE 2.10
An illustration of the effects of spatial resolution on detectability of terrain features (Courtesy: National Remote Sensing Centre, Indian Space Research Organization, Department of Space, Government of India).

features) are clearly detectable in images up to 5 m resampled image. There after these features are discernible only to limited extent in 10 m resampled image while the features are totally blurred in 20 m resampled image.

There is another term *super-resolution* (SR) that is often used in remote sensing. SR are techniques that construct high-resolution images from several observed low-resolution images, thereby increasing the high-frequency components and removing the degradation caused by the imaging processes of the low-resolution camera. The basic idea behind SR is to combine the nonredundant information contained in multiple low-resolution frames to generate a high-resolution image (Yang and Huang, 2011).

2.5.2 Spectral Resolution

Spectral resolution is the sensitivity of an air/spaceborne sensor to respond to a specific wavelength range. For digital images spectral resolution corresponds to the number and location of spectral bands, their width, and the range of sensitivity within each band (Jensen, 2007).

2.5.3 Radiometric Resolution

Radiometric resolution describes the ability of the sensor to measure the signal strength/brightness of objects. It is a measure of a sensor's ability to distinguish between two objects of similar reflectance. The more sensitive a sensor is to the reflectance/emittance/back-scattered radiation of an object compared to its surroundings, the smaller an object that can be detected and identified. For example, while the Resourcesat-2 Linear Imaging Self–scanning Sensor (LISS-III) has a radiometric resolution of 1,024 (10 bit = 2 raise to the power of 10), the Moderate Resolution Imaging Spectrometer (MODIS) has a radiometric resolution of 4,096 (12 bit = 2 raise to the power of 12). It implies that LISS-III can identify 128 levels of reflectance in each band while MODIS can differentiate 4,096. Thus, MODIS imagery can potentially show more and finer distinctions between objects of similar reflectance.

2.5.4 Temporal Resolution

Temporal resolution refers to the length of time it takes for a satellite to complete one entire orbit cycle. It denotes the observation frequency (revisit period) provided by the sensor. It is a measure of how often same area is visited/imaged by the sensor. Unlike spatial, spectral, and radiometric resolutions, temporal resolution does not describe a single image, but rather a series of images that are captured by the same sensor over time. The revisit period of a satellite sensor is usually several days. Therefore, the absolute temporal resolution of a remote sensing system to image the exact same area at the same viewing angle a second time is equal to this period. However, because of some degree of overlap in the imaging swaths of adjacent orbits for most satellites and the increase in this overlap with increasing latitude, some areas of the Earth tend to be reimaged more frequently. Also, some satellite systems by virtue of off-nadir viewing capability (e.g., SPOT series of satellites) are able to point their sensors to image the same area between different satellite passes separated by periods from 1 to 5 days. Furthermore, increasing the number of satellite operating in tandem brings down the revisit time. For example, the Rapid Eye constellation of five satellites can afford daily coverage of the same area. Thus, the actual temporal resolution of a sensor depends on a variety of factors, including the satellite/sensor capabilities, the swath overlap, and latitude.

2.5.5 Angular Resolution

The concept of angular resolution is very recent and refers to the sensor's capacity to make observation of the same area from different viewing angles (Diner et al., 1999). It is commonly assumed that terrestrial surfaces exhibit Lambertial reflection, and therefore have similar reflectance independent of the observation angle. In real-world scenario, this is not the case, especially in those surfaces with strong bidirectional reflectivity effects. One way to model these effects is to observe the surfaces from different directions, thus facilitating a better characterization. Polarization and directionality of the Earth's Reflectance (POLDER) aboard the Japanese satellite ADEOS in 1997 and MISR (multi-angle imaging spectroradiometer) onboard Terra platform launched in 1999 providing nine observation angles almost simultaneously from the same zone and at different wavelengths are some examples of multi-angular observations (Chuvieco and Huete, 2010).

2.6 Spaceborne Missions with Passive Sensors

Some of the currently operating major spaceborne missions with passive sensors are briefly described hereunder.

2.6.1 The Landsat Mission

Landsat-6, 7 and 8 missions represent the last generation of Landsat satellites. The Landsat-6 was lost just after its launch on October 3, 1993. Landsat-7 was launched on April 15, 1999 and is equipped with a multispectral sensor known as the Enhanced Thematic Mapper Plus (ETM+). The observation bands are essentially the same seven bands as TM, and a panchromatic band 8 (0.5–0.90 μm) with 15 m spatial resolution has been added. The unique feature of this sensor is the scan line corrector (SLC) which removes the zig-zag effect caused by the combined along-track and cross-track motions (Figure 2.11). However, due to malfunctioning of SLC on May 31, 2003 all Landsat-7 scenes acquired since July 14, 2003 have been collected in "SLC-off" mode.

Landsat-8 was launched on February 11, 2013 as a backup of Landsat-7 to ensure the continuity of data well beyond the duration of the current Landsat-7 mission. Initially named as Landsat Data Continuity Mission (LDCM), it was renamed later as Landsat-8. It has two sensors, namely operational land imager (OLI) and thermal infrared sensor (TIS) onboard.

Operational Land Imager (OLI): The OLI is a push-broom sensor collecting image data for nine shortwave spectral bands with 12-bit quantization level over a 185 km swath with a 30 m spatial resolution for all bands spanning from 0.43 to 2.29 µm, except a 15 m panchromatic band. The OLI also collects data for two new bands, a coastal band (0.43–0.45 µm) and a cirrus band (1.36–1.38 µm) (Table 2.2).

Thermal Infrared Sensor (TIRS): The TIRS was added to the Landsat-8 payload to continue thermal imaging and to support emerging applications such as evapotranspiration rate measurements for water management. The 100 m TIRS data could be registered to the OLI data to create radiometrically, geometrically, and terrain-corrected 12-bit LDCM data products.

2.6.2 The SPOT Mission

SPOT-6 and-7 constitute the currently operating SPOT mission. SPOT-6 was launched on September 9, 2012 and SPOT-7 on June 30, 2014 from Satish Dhawan Space Centre, Sriharikota, Andhra Pradesh, India. Both satellites offer 2 m resolution data in a 60 km by 60 km swath. The satellites will be co-orbital with the high-resolution Pléiades-HR satellites. The 1.5-m-resolution natural-color products, ortho-rectified as standard product and

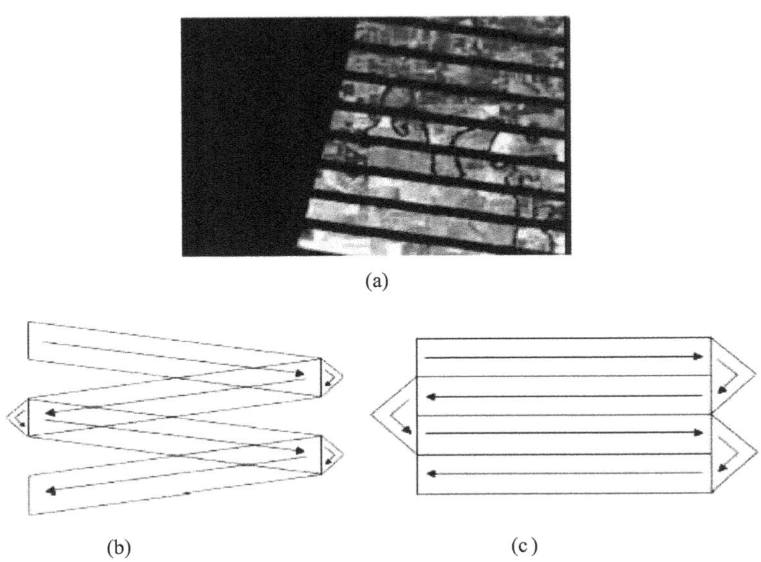

FIGURE 2.11
Showing the effect of malfunctioning of scan line corrector (SLC) (a), sketch of part of the uncorrected image (b), and after correction (c) Data gaps produced from the SLC-off mode have alternating wedges with the widest parts occurring at the scene edge.

TABLE 2.2
Salient Features of Landsat-8 Sensors

Spectral Channel	Wavelength (µm)	Spatial Resolution (m)
Operational land imager (OLI) spectral channels		
Band 1	0.43–0.45	30
Band 2	0.45–0.51	30
Band 3	0.53–0.59	30
Band 4	0.64–0.67	30
Band 5	0.85–0.88	30
Band 6	1.57–1.65	30
Band 7	2.11–2.29	30
Band 8 (PAN)	0.50–0.68	15
Band 9 (cirrus)	1.36–1.38	30
Thermal infrared sensor (TIRS)		
Band 10	10.6–11.19	100
Band 11	11.5–12.51	100

Source: http://landsat.gsfc.nasa.gov/about/ldcm.html.

daily revisits to any point on the globe, are the significant improvements in the SPOT-6 and SPOT-7 missions. With location accuracy better than 10 m (CE90) and a resolution of 1.5 m, SPOT-6 and SPOT-7 are the ideal solution for national 1:25,000 scale map series. Operating in both panchromatic (0.450–745 nm) with 1.5 m spatial resolution, and multispectral (450–525 nm, 530–590 nm, 625–695 nm, and 760–890 nm) mode simultaneously, the sensor provides 1.5 and 6 m spatial resolution, respectively.

Automatic ortho-image with a location accuracy of 10 m CE90 using Reference3D 120 km × 120 km bi-strip or 60 km × 180 km tri-strip mapping in a single pass and delivery of mosaic product stereo and tri-stereo acquisition of 60 km × 60 km scenes for production of DEM 6 tasking plans per day. Two imaging systems aboard the spacecraft, the New AstroSat Optical Modular Instruments (NAOMI), are capable of producing panchromatic images at a resolution of 1.5–2.2 m, and multispectral images at a resolution of 6.0–8.8 m. These instruments can cover a swath of 60 km.

2.6.3 Pleiades Mission

Pleiades-HR is a two-spacecraft constellation of CNES (Space Agency of France), representing a long-term engagement with the introduction of advanced technologies in Earth observation capabilities. Starting with the first launch in 2009, the Pleiades program follows the SPOT satellite series services. The identical twin satellites deliver very high optical resolution (0.5 m resolution) and offer a daily revisit capability to any point on the globe. Other salient features of the mission are as follows:

- Provision of an optical high-resolution panchromatic (0.7 m) and multispectral (2.8 m) imagery with high-quality product standards in terms of resolution, MTF (0.2 at system level), and a high image location accuracy.
- Global coverage and a daily observation accessibility to any point on Earth.

- Service provision of so-called "level-2 products" to customers consisting of a panchromatic image with a merged multispectral image ortho-rectified on a DTM (Digital Terrain Matrix).
- Provision of stereo imagery (up to 350 km × 20 km or 150 km × 40 km) and mosaic imagery of size up to 120 km × 120 km.

2.6.4 The Indian Earth Observation Mission

Resourcesat-1, -2, and -2A are three major earth observation missions currently operating.

2.6.4.1 Resourcesat-1

Resourcesat-1 was launched in October 17, 2003 with three unique sensors, namely Advanced Wide Field Sensor (AWiFS), LISS-III, and LISS-IV which offer immense potential in deriving regional, macro- and micro-level information on natural resources and environment.

Advanced Wide Field Sensor (AWiFS): AWiFS operates in four spectral bands identical to WiFS with 10-bit radiometry and a spatial resolution of 56 m covering a swath of 740 km. It has a 5-day revisit capability for 80% of the area covered. It is extremely useful in species-level vegetation mapping, sub district-level agricultural drought assessment, and integrated land and water resources-related applications.

Linear Imaging Self-scanning Sensor (LISS-III) operates in four spectral bands, namely B2 (0.52–0.59 μm), B3 (0.62–0.68 μm), B4 (0.77–0.86 μm), and SWIR band (B5 1.55–1.75 μm) with 23.5 m spatial resolution 145 km swath.

The Linear Imaging Self-scanning Sensor (LISS-IV) is a high-resolution multispectral sensor operating in three spectral bands, namely B2 (0.52–0.59 μm), B3 (0.62–0.68 μm), and B4 (0.77–0.86 μm). LISS-IV provides a spatial resolution of 5.86 m (at nadir) and can be operated in two modes: multispectral and mono mode. In multispectral mode it covers a swath of 23 km selectable from 70 km of total swath in three bands. In mono mode (panchromatic mode), the full swath of 70 km is covered in one single band which is selectable by ground command. LISS-IV can be tilted up to ±26° in the across track direction thereby providing a revisit period of 5 days. The oblique viewing (off-nadir viewing) capability can be used to acquire stereo pairs in mono mode only.

2.6.4.2 Resourcesat-2

To maintain the continuity of remote sensing data services to global users provided by Resourcesat-1, and to provide data with enhanced multispectral and spatial coverage, Resourcesat-2 was launched on April 20, 2011. Important improvements in Resourcesat-2 compared to Resourcesat-1 are as follows: enhancement of LISS-IV multispectral swath from 23 to 70 km and improved radiometric accuracy from 7 to 10 bits for LISS-III and LISS-IV, and 10 to 12 bits for AWiFS (Table 2.3). Resourcesat-2 carries two solid-state recorders with a capacity of 200 GB each to store the images taken by its cameras which can be read out later to ground stations.

2.6.4.3 Resourcesat-2A

Launched on December 6, 2016, Resourcesat-2A is a follow-on mission to its predecessors Resourcesat-1 and Resourcesat-2, launched in 2003 and 2011, respectively. It is intended to continue the remote sensing data services to global users provided by its two predecessors.

TABLE 2.3
Salient Features of Resourcesat-2 Sensors

Specifications	AWiFS	LISS-III	LISS-IV
No. of bands	4	4	1 (mono), 3 (multispectral)
Spectral bands (μm)	B2 0.52–0.59 B3 0.62–0.68 B4 0.77–0.86 B5 1.55–1.70	B2 0.52–0.59 B3 0.62–0.68 B4 0.77–0.86 B5 1.55–1.70	B2 0.52–0.59 B3 0.62–0.68 B4 0.77–0.86 B5 1.55–1.70 B3-default band for mono
Spatial resolution (m)	56	23.5	5.8
Swath (km)	740	140	70/23
Revisit (days)	5	24	5
Data rate (Mbs per stream)	105	105	105
Quantization	12-bit	10-bit	10-bit

2.6.5 The Earth Observing System Mission

The Terra and Aqua platforms are part of NASA's Earth Observing Systems.

2.6.5.1 Terra (EO-AM)

Terra was launched on December 18, 1999. With the equatorial crossing time of 10:30 a.m. of Terra and 1:30 p.m. for Aqua, they are also known as EOS-AM (Terra) and EOS-PM (Aqua). The principal instruments among others on Terra and Aqua are as follows: MOderate Resolution Imaging Spectrometer (MODIS) and Advanced Spaceborne Thermal Emission and Reflection Spectrometer (ASTER). Salient features of these sensors are given hereunder:

MOderate Resolution Imaging Spectrometer (MODIS): MODIS is a 36-band spectrometer providing a global dataset every 1–2 days with a 16-day repeat cycle. The spatial resolution of MODIS (pixel size at nadir) is 250 m for channels 1 and 2 (0.6–0.9 μm), 500 m for channels 3–7 (0.4–2.1 μm), and 1,000 m for channels 8–36 (0.4–14.4 μm). The MODIS instrument consists of a cross-track scan mirror, collecting optics and individual detector elements. The swath dimensions of MODIS are 2,330 km (across track) by 10 km (along track at nadir). The along track swath dimension is due to the optical setup as well as the scanning mechanism of MODIS. In contrast to other scanning sensors like, e.g., AVHRR, during one scan MODIS images ten lines of 1km spatial resolution (40 lines of 250 m resolution and 20 lines of 500 m resolution, respectively), and 12-bit radiometry.

Advanced Space borne Thermal Emission and Reflection Radiometer (ASTER): ASTER operates in the VIR through TIR portions of the electromagnetic spectrum. Of its 14 bands, three are in the visible and near-infrared (VNIR) between 0.5 and 0.9 μm, six are in the SWIR between 1.6 and 2.43 μm, and five are in the TIR between 8 and 12 μm. VNIR channels have 15-m resolution, SWIR have 30-m resolution, and TIR channels have 90-m resolution. ASTER has a 60-km swath width, with a cross-track adjustable swath center. A special feature of ASTER is an aft pointing additional VNIR telescope for creating stereo views as in Table 2.4. The stereo images have a base-to-height ratio of 0.6. ASTER's repeat cycle is 16 days.

2.6.5.2 Aqua (EOS PM-1)

Aqua is a multinational NASA scientific research satellite in orbit around the Earth. It was launched on May 4, 2002 in a sun-synchronous polar orbit. Aqua carries six

TABLE 2.4
Salient Features of ASTER Sensor

Instrument's Parameters	VNIR	SWIR	TIR
Bands	1–3	4–9	10–14
Spatial resolution	15 m	30 m	90 m
Swath width	60 km	60 km	60 km
Crosstrack pointing	±318 km (±24°)	±116 km (±8.55°)	±116 km (±8.55°)
Quantization (bits)	8	8	12

instruments, namely Advanced Microwave Scanning Radiometer(AMSR-E), MODIS, AMSU, Atmospheric Infrared Sounder(AIRS), Humidity Sounder for Brazil (HSB), and Clouds and the Earth's Radiant Energy System(CERES) for studies of water on the Earth's surface and in the atmosphere.

Moderate Resolution Imaging Spectroradiometer (MODIS): The MODIS aboard Aqua mission is similar to the one that is aboard Terra satellite. Since Terra passes over equator in the forenoon (around 10:30 hours) and Aqua in the afternoon around 14:30 hours, they provide daily two coverage on an area of interest that is very useful in studying the dynamic phenomenon.

Multi-Angle Imaging Spectroradiometer (MISR): The MISR instrument measures the Earth's brightness in four spectral bands, at each of nine look angles spread out in the forward and aft directions along the flight line. Spatial samples are acquired every 275 m. Over a period of 7 minutes, a 360 km wide swath of earth comes into view at all nine angles. Each MISR camera sees instantaneously a single row of pixels at right angles to the ground track in a push broom format. It records data in four bands: blue, green, red, and NIR. Each camera has four independent linear CCD arrays (one per filter), with 1,504 active pixels per linear array, and spectral bands (0.4–2.5 μm) with a 30 m resolution. The instrument can image a 7.5 km by 100 km land area per image. Hyperion has a single telescope and two spectrometers, one VNIR (380–1,000 nm) spectrometer and one SWIR (900–2,500 nm) spectrometer.

Advanced Land Imager (ALI): The Advanced Land Imager (ALI) employs novel wide-angle optics and a highly integrated multispectral and panchromatic spectrometer. Operating in a push broom fashion at an altitude of 705 km, the ALI provides Landsat-type panchromatic and multispectral bands. These bands have been designed to mimic six Landsat bands with three additional bands covering 0.433–0.453 μm, 0.845–0.890 μm, and 1.20–1.30 μm (Table 2.5). The ALI also contains wide-angle optics designed to provide a continuous 15° × 1.625° field of view for a fully populated focal plane with 30-m resolution for the multispectral pixels and 10 m resolution for the panchromatic pixels.

Atmospheric Corrector (AC): The images of the Earth acquired by satellites are degraded by atmospheric absorption and scattering. Sensors—Linear Etalon Imaging Spectrometer Array (LEISA) Atmospheric Corrector (LAC). Earth imagery is degraded by atmospheric absorption and scattering. The New Millennium Program's EO-1 mission provided the first space-based test of an AC for increasing the accuracy of surface reflectance estimates. The LAC provided the following capabilities via a compact and simple bolt on design for future Earth Science, land imaging missions:

- High spectral, moderate spatial resolution hyperspectral imager using a wedge filter technology.

TABLE 2.5

Salient Features of EO-1 Advanced Land Imager

Band	Wavelength (μm)	Ground Sample Distance (m)
PAN	0.48–0.69	10
MS-1	0.433–0.453	30
MS-1′	0.45–0.515	30
MS-2	0.525–0.605	30
MS-3	0.63–0.69	30
MS-4	0.775–0.805	30
MS-4′	0.845–0.89	30
MS-5′	1.2–1.3	30
MS-5	1.55–1.75	30
MS-7	2.08–2.35	30

- Spectral coverage of 85–1.5 μm—bands are selected for optimal correction of high spatial resolution images.
- Correction of surface imagery for atmospheric variability (primarily water vapor).

These data are no longer being acquired as part of the EO-1 Extended Mission.

2.6.6 Earth Observing-1 Mission (EO-1)

Launched on November 21, 2000 EO-1 houses three sensors, namely hyperion, advanced land imager (ALI), and AC (Table 2.6). Hyperspectral imager is capable of resolving 220 spectral bands (0.4–2.5 μm) with a 30 m resolution. The instrument can image a 7.5 km by 100 km land area per image. Hyperion has a single telescope and two spectrometers, one VNIR (380–1,000 nm) spectrometer and one SWIR (900–2,500 nm) spectrometer.

2.6.7 RapidEye

The RapidEye satellite constellation is a dedicated system, consisting of five mini-satellites, which carry a CCD-based imaging system. The CCD-based earth imaging system (six spectral bands including VIR, NIR, and panchromatic) consists of two cameras allowing the generation of images of up to 150 km × 1,000 km at a resolution of 6.5 m. The RapidEye mission was launched in 2008 (Table 2.7). The five satellites were launched on one vehicle and placed in a common sun-synchronous orbit of 620 km, with the satellites equally spaced about 19 minutes apart in their orbit, ensuring frequent imaging of particular areas of interest. Each

TABLE 2.6

Salient Features of EO-1 Sensors

Parameter	ALI	Hyperion	AC
Spectral range	0.4–2.4 μm	0.4–2.4 μm	0.9–1.6 μm
Spatial resolution	30 m	30 m	250 m
Swath width	36 km	7.6 km	185 km
Spectral resolution	Variable	10 nm	6 nm
Spectral coverage	Discrete	Contiguous	Contiguous
PAN band resolution	10 m	N/A	N/A
Total number of bands	10	220	256

TABLE 2.7
RapidEye Satellite Sensor Specifications

Number of satellites	5
Orbit altitude	630 km in sun-synchronous orbit
Equator crossing time	11:00 a.m. local time (approximately)
Sensor type	Multi-spectral push broom imager
Spectral bands	Wavelength (nm)
	440–510
	520–590
	630–685
	690–730
	760–850
Ground sampling distance (nadir)	6.5 m
Pixel size (ortho-rectified)	5 m
Swath width	77 km
On board data storage	1,500 km of image data per orbit
Revisit time	Daily (off-nadir)/5.5 days (at nadir)
Dynamic range	12 bit

satellite has the capability of performing an off-track rotation. The camera's imaging swath of approx. 150 km combined with an off-track angle of ±22° ensures daily global accessibility. On November 6, 2013 RapidEye officially changed its name to BlackBridge.

2.6.8 Hyperspatial Resolution Earth Missions

Some of the major earth observation missions with high spatial resolution are described hereunder.

2.6.8.1 WorldView Mission

WorldView-1: Launchedon September 18, 2007, WorldView-1 is operating at an altitude of 496 km. WorldView-1 has an average revisit time of 1.7 days and captures panchromatic images of 17.6 km wide strip of the earth with 0.50 m spatial resolution and 11-bit radiometry. The satellite is also equipped with state-of-the-art geo-location capabilities and exhibits stunning agility with rapid targeting and efficient in-track stereo collection.

WorldView-2: Launched in October 8, 2009, WorldView-2 satellite provides 0.46 m panchromatic (B&W) mono and stereo satellite image data. With its improved agility, WorldView-2 is able to act like a paintbrush, sweeping back and forth to collect very large areas of multispectral imagery in a single pass. And the combination of WorldView-2's increased agility and high altitude enables it to typically revisit any place on earth in 1.1 days. When added to the satellite constellation, revisit time drops below 1 day and never exceeds 2 days, providing the most same-day passes of any commercial high-resolution constellation. The WorldView-2 sensor provides a high-resolution panchromatic band and eight multispectral bands; four standard colors (red, green, blue, and NIR 1) and four new bands (coastal 400–450 nm, yellow 585–625 nm, red edge 705–745 nm, and NIR 2,860–1,040 nm), full-color images for various applications.

WorldView-3: WorldView-3 was launched on August 13, 2014 at an altitude of 617 km. It provides 31 cm panchromatic resolution, 1.24 m multispectral resolution, and 3.7 m SWIR resolution. WorldView-3 has an average revisit time of less than 1 day (Table 2.8).

TABLE 2.8
Salient Features of World View-3 Mission

Orbit	Altitude: 617 km Type: Sun sync, 1:30 p.m. descending node Period: 97 minutes.
Sensor bands	Panchromatic: 450–800 nm 8 Multispectral: (red, red edge, coastal, blue, green, yellow, near-IR1, and near-IR2) 400 nm–1,040 nm, SWIR: 1,195 nm–2,365 nm 12 CAVIS bands: (desert clouds, aerosol-1, aerosol-2, aerosol-3, green, water-1, water-2, water-3, NDVI-SWIR, cirrus, snow) 405 nm–2,245 nm
Sensor resolution (or GSD, ground sample distance; off-nadir is geometric mean)	Panchromatic nadir: 0.31 m GSD at nadir 0.34 m at 20° off-nadir Multispectral nadir: 1.24 m at nadir, 1.38 m At 20° off-nadir SWIR nadir: 3.70 m at nadir, 4.10 m At 20° off-nadir CAVIS nadir: 30.00 m
Dynamic range	11-bits per pixel PAN and MS; 14-bits per pixel SWIR
Swath width	At nadir: 13.1 km
Onboard storage	2,199 GB solid state with EDAC
Revisit frequency (at 40°N Latitude)	1 m GSD: <1.0 day 4.5 days at 20° off-nadir or less
Geolocation accuracy (CE90)	Predicted performance: <3.5 m CE90 without ground control

WorldView-4: WorldView-4 satellite was successfully launched on Friday, November 11, 2016 at 10:30 a.m. PST from Vandenberg Air Force Base, California. DigitalGlobe's WorldView-4 is capable of discerning objects on the Earth's surface as small as 31 cm in the panchromatic and will collect 4-band multispectral at 1.24-m resolution (Table 2.9). WorldView-4 will have similar resolution as the WorldView-3 satellite sensor. This advanced resolution will offer customers unprecedented, precise views for 2D or 3D mapping, change detection, and image analysis.

2.6.8.2 Cartosat Mission

The Cartosat mission aimed at designing and developing an advanced 3-axis body stabilized remote sensing satellite for providing enhanced spatial resolution with stereo imaging capability for cartographic applications.

Cartosat-1 was launched by the Indian Space Research Organization (ISRO) on May 5, 2005. Cartosat-1 carries two PANs that take B&W stereoscopic images in the visible region of the electromagnetic spectrum. The satellite images have a spatial resolution of 2.5 m and cover a swath of 30 km. The cameras are mounted on the satellite in such a way that near-simultaneous imaging of the same area from two different angles is possible. This facilitates the generation of accurate 3D maps. The cameras maneuver across the direction of the satellite's movement also facilitates the imaging of an area more frequently. Cartosat-1 also carries a solid-state recorder with a capacity of 120 GB to store the images taken by its cameras.

Cartosat-2 is an advanced remote sensing satellite capable of providing scene-specific spot imagery. The PAN on board the satellite can provide imagery with a spatial resolution

TABLE 2.9
WorldView-4 Satellite Sensor Specifications

Orbit altitude	617 km
Type	SunSync, 10:30 a.m. descending node
Sensor bands	Panchromatic: 450–800 nm
Multispectral	655–690 nm
	510–580 nm
	450–510 nm
	780–920 nm
Sensor resolution	Panchromatic nadir: 0.31 m
	Multispectral nadir: 1.24 m
Dynamic range	11 bits per pixel
Swath width: at nadir	13.1 km
Mono	66.5 km × 112 km (5 strips)
Stereo	26.6 km × 112 km (2 pairs)
Revisit frequency (at 40°N latitude) 1 m GSD	<1.0 day
Total constellation >4.5 accesses/day	
Geo-location accuracy (CE90) Predicted <4 m CE90 without ground control	

of better than 1 m and a swath of 9.6 km. The satellite can be steered up to 45° along as well as across the track. The satellite has been placed in sun-synchronous polar orbit at an altitude of 630 km. It has a revisit period of 4 days which can be improved to one day with suitable orbit maneuvers.

Cartosat-2A is the 13th satellite in the Indian Remote Sensing Satellite series (IRS). It is a sophisticated and rugged remote sensing satellite that can provide scene specific spot imagery. The satellite carries a PAN. The spatial resolution of this camera is better than 1 m and swath of 9.6 km.

Cartost-2B was launched on July 12, 2010. It carries a PAN similar to those of its predecessors—Cartosat-2 and 2A. It is capable of imaging a swath (geographical strip) of 9.6 km with a resolution of better than 1 m. The highly agile Cartosat 2B is steerable up to ±26° along as well as across track to obtain stereoscopic imagery and achieve a 4–5 day revisit capability. Beginning with Cartosat 2C, a multi-spectral camera was added and a lower 505 km orbit for better ground resolution was introduced. Cartosat 2C was launched in June 2016, Cartosat 2D on February15, 2016, and Cartosat 2E on June 23, 2017, and Cartosat 2F on January 12, 2018.

In the focal plane of Cartosat 2F resides a panchromatic and multispectral detection system using CCDs with 7 µm pixels. The detector system covers a spectral range of 450–900nm and reaches a ground resolution of 65 cm when employing apparent velocity reduction for along-track imaging. Four MX detectors with band-pass filters between 450 and 860 nm deliver imagery at a 2-m ground resolution along a 10-km swath.

2.6.8.3 GeoEye-1

The GeoEye-1 was launched on September 6, 2008, and is capable of acquiring image data at 0.41 m panchromatic (B&W) and 1.65 m multispectral resolution. It also features a revisit time of less than 3 days, as well as the ability to locate an object within just 3 m of its physical location. It was renamed to WorldView-4 in July 2014 which was launched on November11, 2016. With a maximum resolution of 31 cm, WorldView-4 provides similar imagery as WorldView-3, the highest resolution commercially available at the time of its launch.

2.6.9 Passive Microwave Missions

Some of the important earth observation missions with passive microwave sensors are described hereunder.

2.6.9.1 National Oceanic and Atmospheric Administration AMSU-A

AMSU-A, onboard National Oceanic and Atmospheric Administration (NOAA) mission—part of the National Polar-orbiting Operational Environmental Satellite System (NPOESS), is a cross-track, line-scanning instrument designed to measure scene radiances in 15 discrete frequency channels which permit the calculation of the vertical temperature and moisture profile from about 3 millibars (~45 km) to the Earth's surface. Each scan has 30 cells, with a resolution of 3.3° (50 km at nadir). The swath width is 2,343 km (96.66°). The AMSU-A instrument is made up of two separate modules, A1 and A2. AMSU-A1 contains the 13 highest frequencies (50.3–89 GHz) of various polarizations. AMSU-A2 contains the two lowest frequency channels (23.8 and 31.4 GHz), both vertically polarized.

NOAA-15 was the first spacecraft to fly AMSU. Launched on May 13, 1998, NOAA-15 is in a sun-synchronous near-polar orbit. Flying at approximately 833 k min altitude at an orbital inclination of 98.7°, NOAA-15 completes an orbit every 101.2 minutes. The ascending equator crossing time is at approximately 1,920 local time.

The Advanced Technology Microwave Sounder (ATMS), a cross-track scanner with 22 channels, provides sounding observations needed to retrieve profiles of atmospheric temperature and moisture for civilian operational weather forecasting as well as continuity of these measurements for climate monitoring purposes. Like the long heritage of its predecessors, ATMS combines all the channels of the preceding AMSU-A1, AMSU-A2, and AMSU-B sensors into a single package with considerable savings in mass, power, and volume.

ATMS is a key instrument that collects microwave radiation from the Earth's atmosphere and surface all day and all night, even through clouds. ATMS currently flies on the Suomi NPP satellite mission and will fly on the JPSS-1, or NOAA-20 as it will be known once operational, JPSS-2, JPSS-3, and JPSS-4 satellite missions.

2.6.9.2 Defense Meteorological Satellite Program

The Special Sensor Microwave Imager (SSM/I) aboard Defense Meteorological Satellite Program (DMSP) is a seven-channel, four-frequency, linearly polarized, passive microwave radiometer which measures atmospheric, ocean, and terrain microwave brightness temperatures which are converted into environmental parameters such as sea surface winds, rain rates, cloud water, precipitation, soil moisture, ice edge, and ice age. SSM/I consists of an offset parabolic reflector (61 cm × 66 cm) that is fed by a corrugated broadband seven-port horn antenna. The reflector and feed are mounted on a drum which contains the radiometers, digital data subsystem, mechanical scanning subsystem, and power subsystem. The entire reflector, feed horn, and drum assembly rotate about the axis of the drum by a coaxially mounted BAPTA (Bearing and Power Transfer Assembly). SSM/I consumes 45 W of power and has a mass of 48.5 kg.

The SSM/I sensor executes a 45° conical scan of the Earth's surface from nadir. This gives a nominal incidence (zenith) angle of 53.1° to the Earth's surface from the nominal orbit. Only part of the possible 360° scan in azimuth is used to collect data. The active azimuth scan angle is 102.4° is ahead of the S/C for an afternoon ascending orbit and behind the S/C for a morning ascending orbit. The sensor electronics perform an integration and hold sequence on each channel, timed so that the odd 85 GHz reading is centered with the 37 GHz reading. The sensor conically scans the Earth and atmosphere at a scan rate of 31.9 scans/min (or 1.88 s/scan). The sampling scheme results in 128 samples (called scene stations) per scan for the 85 GHz channels, and 64 scene stations per two scans for the other channels. The footprints are sampled every 25 km for the 19, 22, and 37 GHz channels, and every 12.5 km for the 85 GHz channel.

2.6.9.3 Aqua (EO: PM-1)

Aqua with Advanced Microwave Scanning Radiometer–Earth Observing System (AMSR-E) along with other sensors was launched on May 4, 2002. AMSR-E is a 12-channel, six-frequency, passive-microwave radiometer system. It measures horizontally and vertically polarized brightness temperatures at 6.9 GHz, 10.7 GHz, 18.7 GHz, 23.8 GHz, 36.5 GHz, and 89.0 GHz. Spatial resolution of the individual measurements varies from 5.4 km at 89 GHz to 56 km at 6.9 GHz. AMSR-E uses an offset parabolic reflector, 1.6 m in diameter, to focus Earth-emitted microwave radiation into an array of six feed horns, which then feed the radiation to the detectors. It measures cloud properties, sea surface temperature, near-surface wind speed, radiative energy flux, surface water, ice, and snow. The AMSR-E data has been found quite useful in studying the regional-level soil moisture status.

2.6.9.4 Soil Moisture and Ocean Salinity Mission

Launched on November 2, 2009 the Soil Moisture and Ocean Salinity (SMOS) carries a L-band (1.4 GHz) radiometer known as microwave imaging radiometer with aperture synthesis (MIRAS), which provides the best sensitivity to variations of moisture in the soil and changes in the salinity of the ocean, coupled with minimal disturbance from weather, atmosphere, and vegetation cover. For capturing emitted microwave radiation with improved spatial resolution in L-band, a huge antenna would have been required—too big for a satellite to carry. To overcome this challenge, the antenna needed for MIRAS has been simulated through 69 small antennas, distributed over the three arms and central hub of the instrument. During the launch the three deployable arms are folded up, but once SMOS is in orbit, each of the arms fold out into an unusual three-pointed star shape. Hence, with a diameter of 8 m, MIRAS is often dubbed a "star in the sky." The 69 antenna elements, called LICEFs, are antenna-receiver integrated units, each measure radiation emitted from Earth's surface at L-band (Figure 2.12).

2.6.9.4.1 Measurement Principle

SMOS measures microwave radiation emitted from Earth's surface within the L-band (1.4 GHz) using an interferometric radiometer. The radiometer exploits the interferometry principle, which by way of 69 small receivers measures the phase difference of incident radiation (Table 2.10). The technique is based on cross-correlation of observations from all

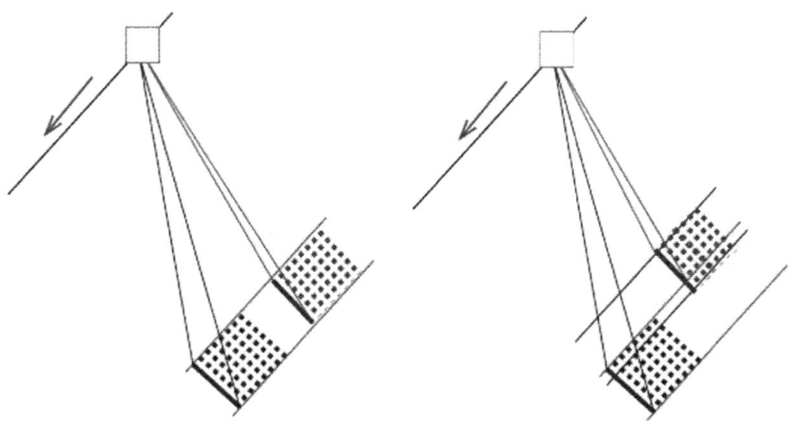

FIGURE 2.12
Cartosat 1 stereo and wide swath imaging (Credit: Indian Space Research Organization, Department of Space, Government of India).

TABLE 2.10
Salient Features of SMOS Mission

Instrument	Microwave imaging radiometer using aperture synthesis—MIRAS
Frequency	L-band (21 cm–1.4 GHz)
Number of receivers	69
Receiver spacing	0.875 lambda = 18.37 cm
Polarization	H & V (polarimetric mode optional)
Spatial resolution	35 km at center of field of view
Tilt angle	32.5°
Radiometric resolution	0.8–2.2 K
Angular range	0–55°
Temporal resolution	3 days revisit at Equator
Instrument data rate	89 kbps H & V pol.

Source: Kerr et al. (2001).

possible combinations of receiver pairs. A 2D "measurement image" is taken every 1.2 seconds. As the satellite moves along its orbital path each observed area is seen under various viewing angles. From an altitude of around 758 km, the antenna views an area of almost 3,000 km in diameter. However, due to the interferometry principle and the Y-shaped antenna, the field of view is limited to a hexagon-like shape about 1,000 km across called the "alias-free zone." This area corresponds to observations where there is no ambiguity in the phase difference. The MIRAS instrument has three main operational modes: (i) dual-polarization mode, in which all receivers are switched synchronously to either H or V polarization, (ii) full polarimetric mode, in which segments of the array are switched according to a predefined sequence between H and V, and (iii) calibration mode, in which measurements of the internal load, the noise diodes, or the so-called "fringe washing function" are determined (Figure 2.13).

FIGURE 2.13
The microwave imaging radiometer with aperture synthesis (MIRAS) (Credit: European Space Agency).

2.6.9.5 Soil Moisture Active Passive Mission

Launched on January 13, 2015 into a 680-km near-polar, sun-synchronous orbit, with equator crossings at 6 a.m. and 6 p.m. local time Soil Moisture Active Passive Mission (SMAP) provides global measurements of soil moisture and its freeze/thaw state. The SMAP payload includes a radiometer and synthetic aperture radar operating at L-band (1.20–1.41 GHz). The radiometer is a very sensitive receiver that accurately measures the naturally occurring RF energy given off by the Earth's surface. This frequency has been set aside by international agreement for applications which involve only a receiver (and no transmitting is allowed in that band). This allows SMAP's radiometer (and, in other frequencies, ground-based radio telescopes) to operate without interference. This frequency band also allows the radiometer to be not much affected by weather or by a moderate amount of vegetation that may cover the soil. Within this frequency band (L-band), water appears relatively "cold" (about 100 K) and dry soil appears relatively "warm" (about 300 K) to the radiometer. With this great difference between wet and dry soils, the radiometer allows SMAP to produce very high soil moisture accuracy (4%) by simply measuring the microwave "temperature" of the land surface. The salient features of radiometer are as follows:

- Frequency: 1.41 GHz
- Polarizations: H, V, 3rd and 4th Stokes

- Relative accuracy (40 km grid): 1.3 K
- Data collection:
  - High-rate (16-sub-band) data acquired over land
  - Integration time for high-rate, full-band data with averaging is typically ~40 ms
  - Low-rate data acquired globally

2.7 Conclusion

A thorough understanding of the fundamental principles and functioning of sensors, one of the crucial components of any remote sensing systems, is essential for developing any viable remote sensing applications program. Of the two types of sensors: *passive* and *active*, the chapter deals with the basic elements of passive sensors—those sensors capturing natural reflected/emitted radiation. As evident from the body of the text, the sensor technology is a fast-evolving and developing field—with conventional photographic camera at initial stages of its developments to state-of-the-art technology sensors available now. With the developments in detector technology newer sensors with improved capabilities in terms of spectral, spatial, radiometric, and temporal resolutions operating in various parts of the electromagnetic spectrum are likely to be available that may help improve assessment, monitoring, and modeling of our natural resources and environment.

References

Cambone, S.A., Krieg, K.J., Pace, P., and Wells, L., 2005. 'USA's Unmanned Aircraft Roadmap, 2005-2030, National Defense.

Chao, H., Cao, Y.C., and Chen, Y.Q., 2010. Autopilots for small unmanned aerial vehicles: A survey, *International Journal of Control, Automation, and Systems* 8(1):36–44.

Chuvieco, E., and Huete, A., 2010. *Fundamentals of Satellite Remote Sensing*. CRC Press , Taylor & Francis Group, Boca Raton, FL.

Diner, D.J., Asner, G.P., Davies, R., Knyazikhin, Y., Muller, J.P., Nolin, A.W., Pinty, B., and Stroeve, J., 1999. New directions in earth observing: Scientific applications of multiangle remote sensing. *Bulletin of the American Meteorological Society* 80:2209–2228.

Goetz, A.F.H., Vane, G., Solomo, J.E., and Rock, B.N., 1985. Imaging spectrometry for earth remote sensing. *Science* 228:1147–1153.

Jensen, J.R., 2000. *Remote Sensing of the Environment-An Earth Resources Perspective*. Prentice Hall, Upper Saddle River, NJ.

Jensen, J.R., 2007. *Remote Sensing of the Environment-An Earth Resources Perspective*. Prentice Hall, Upper Saddle River, NJ.

Joseph, G., 2003. *Fundamentals of Remote Sensing*. University Press, Hyderabad, 433 pp.

Kerr, Y.H., Waldteufel, P., Wigneron, J.P., Martinuzzi, J., Font, J., and Berger, M., 2001. Soil moisture retrieval from space: The Soil Moisture and Ocean Salinity (SMOS) mission. *IEEE Transactions on Geoscience and Remote Sensing* 39(8):1729–1735.

Vorovenci, I., 2009. The hyperspectral sensors used in satellite and aerial remote sensing. *Bulletin of the Transilvania University of Braşov* 2:51–56.

Wu, H., Sun, D., and Zhou, Z.,2004. Micro air vehicle: Configuration, analysis, fabrication, and test. *IEEE/ASME Transactions on Mechatronics* 9(1):108–117.

Yang, J., and Huang, T., 2011. Image super- resolution: Historical overview and future challenges. In: Milanfar, P., (ed.) *Super Resolution Imaging*. CRC Press, Taylor & Francis Group, Boca Raton, FL.

3

Active Remote Sensing

3.1 Introduction

As mentioned in Section 1.1, remote sensing instruments (sensors) record reflected/backscattered/emitted electromagnetic radiation from the objects/features. In order to make it happen the objects/features need to be illuminated. Once it is illuminated part of the incident radiation striking the object/feature is reflected/backscattered and the rest is absorbed, which is eventually converted into heat energy. All the objects/features with a temperature above absolute zero emit electromagnetic radiation. Sensors which depend on external source of radiation for illumination of the objects/features are termed *passive sensors*, whereas those having their own source of illumination are called *active sensors*. The active remote sensing refers to a process where in the instruments with a capability to illuminate object/target and to collect the reflected or backscattered radiation therefrom are used for deriving information on the objects/targets. Remote sensing using radar, LiDAR, and sonar represents this category of remote sensing.

3.2 Active Microwave Sensors

Active microwave sensors could be grouped into imaging sensors and non-imaging sensors. Radar—both real aperture and synthetic aperture—belongs to imaging sensors category, whereas scatterometer and altimeter represent non-imaging sensors group.

3.2.1 Imaging Sensors

The microwave imaging sensors are called RADAR—an acronym for *RAdio Detection And Ranging*. Radars measure range—distance of the target/object from the sensor—by measuring the two-way travel time. Besides, radars also measure amplitude and the phase of backscattered radiation. As the speed of the light is known, by multiplying it with the travel time of the pulse, the distance of the target from the sensor can be computed. Additionally, microwave sensors also record polarization both at the time of transmission toward the object/target and from the object/target to the sensor (antenna). Unlike in optical sensors wherein sensors, generally, record the reflected or emitted radiation from the sensor platform by looking vertically down, microwave imaging radars are side looking. The rationale for side-looking nature of microwave imaging radars is given below:

Radars used for remote sensing of Earth are almost always side-looking systems. This difference exists because of the fact that the optical instruments are able to distinguish among targets based upon their angular distance from the sensor. However, a radar can only distinguish the returns from various targets based upon the arrival time of the received signals. A nadir-looking radar could not distinguish between two scatterers that are equidistant from the sensor because a single incident wave front illuminates both points at the same instant. So the scattered returns from both points arrive at the receiving antenna simultaneously. This leads to an ambiguity in range for any right/left-symmetric, equidistant points. If the radar illumination is restricted to one side of the platform, the wave front illuminates the same two points at different times. Their backscattered returns arrive at the sensor that are separated in time and are thereby distinguishable from each other.

Most imaging radars used for remote sensing are side-looking radars (SLRs). They are divided into two categories: real aperture and synthetic aperture systems.

3.2.1.1 Real Aperture Radar

Real aperture radars (RARs) are also called side-looking radar/side-looking airborne radar (SLAR). In the real aperture system, resolution is determined by the actual beam width determined by the antenna size. A typical microwave SLAR consists of a pulse-generating device, a transmitter, a duplexer that carefully coordinates when the microwave energy is transmitted and received, an antenna, a receiver, a recording device such as hard disk, and typically cathode ray tube (CRT) monitor (Figure 3.1). The radar can be mounted onboard an aircraft or a satellite. The antenna beam produces a narrow beam width in the along-track (azimuth) direction, while the beam width in the cross-track (range) direction is broad and defines the swath. Generally, a short pulse is transmitted by the radar, which when strikes the ground, a part is scattered back to the antenna. The time delay associated with the transmitted and received signal gives the distance of the target from the antenna and hence the location. The backscattered intensity or amplitude which depends on the sensor and terrain characteristics is also measured by the radar.

When the sensor craft has moved one beam width forward, the backscattered signals come from a different strip on the ground. Thus, the radar beam produces an image

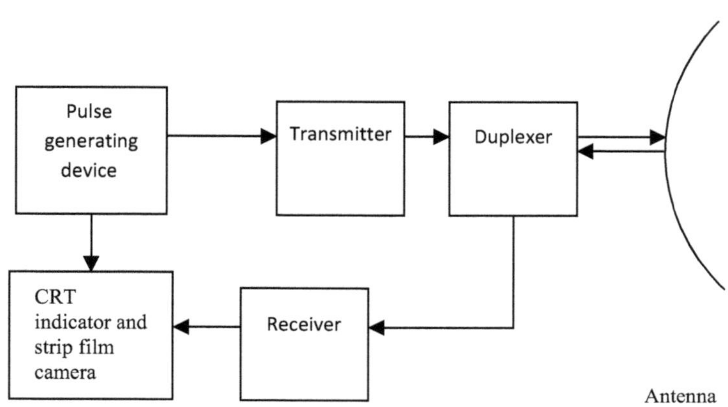

FIGURE 3.1
Schematic of a typical active microwave system components.

TABLE 3.1

Some of the Commonly Used Microwave Frequencies

Band	Frequency (GHz)	Wavelength (cm)
P band	0.3–1	30–100
L band	1–2	15–30
S band	2–4	7.5–15
C band	4–8	3.8–7.5
X band	8–12.5	2.4–3.8
Ku band	12.5–18	1.7–2.4
K band	18–26.5	1.1–1.7
Ka band	26.5–40	0.75–1.1

of the strip across-track and the sensor craft motion produces a successive image strip along-track, thus building up a contiguous 2D image. The microwave beam sent out by the antenna illuminates an area on the ground, which is known as the antenna's "footprint." In radar imaging, the recorded signal strength depends on the microwave energy backscattered from the ground targets within the footprint. Increasing the length of the antenna will decrease the width of the footprint. Hence, it will improve the spatial resolution along range (across-track) direction.

The pulse of electromagnetic radiation sent out by the transmitter through the antenna is of a specific wavelength and duration—pulse length is measured in microseconds. Most commonly used wavelengths of microwave energy are listed in Table 3.1. Since microwave energy used in remote sensing is much longer than optical wavelengths, it is usually measured in centimeters rather than in micrometers. Furthermore, it is the frequency of operating microwave energy that is commonly used for a sensor rather than its wavelength.

The alphabets associated with microwave wavelengths (e.g., Ka, K, Ku, X, C, L, S, and P) are artifacts of original secret work on radar remote sensing when it was customary to use the alphabet descriptor instead of actual wavelength or frequencies. These descriptors are continued to be used even today in microwave literature. Another noteworthy point in Table 3.1 is the use of band designation in units of frequency measured in billions of cycles per second (Gigahertz-GHz or 10^9 cycles s^{-1}).

3.2.1.1.1 Imaging Geometry

As mentioned in the previous section, most of imaging radars are side looking. In other words, the geometry of illumination of SLRs is different from optical sensors geometry (Figure 3.2). Since the geometry of illumination has a direct bearing on spatial resolution its clear understanding is required for interpretation/analysis of radar data. Unlike in optical regions wherein the spatial resolution of the sensor is generally same along and across the image swath, being side looking, in imaging radar systems, the spatial resolution along the azimuth direction (along the flight line in the case of airborne sensors or along the swath for spaceborne systems), and range resolution across the flight line or swath are different.

3.2.1.1.2 Spatial Resolution

Knowledge of variables that determine resolution and the range of resolutions for SLRs are important considerations in determining the feasibility of using a given sensor for

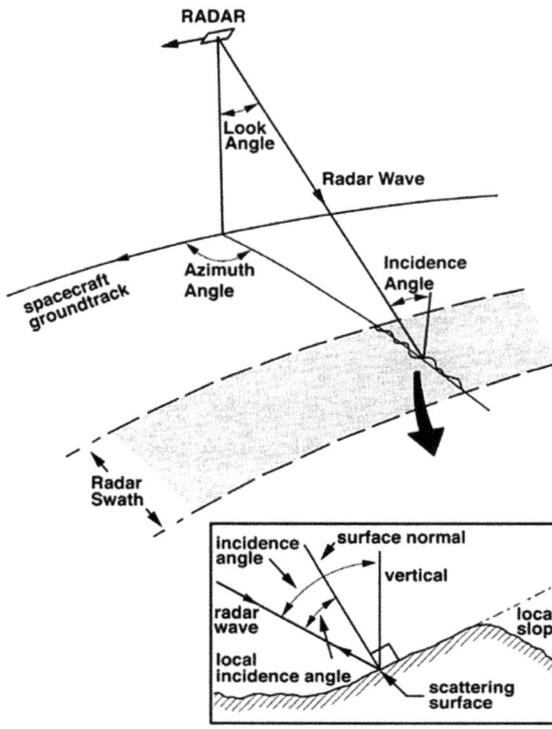

FIGURE 3.2
Imaging geometry of a side-looking airborne real aperture radar.

any application. The spatial resolution of a radar image is defined in two directions: parallel and perpendicular to flight line/satellite ground track. Radar resolution parallel to the flight line/satellite ground track is termed *azimuth resolution*. Resolution perpendicular to the radar flight line/satellite ground track is termed *range resolution*. The parameters that control radar resolution depend on the defined direction. In addition, azimuth resolution varies with the type of imaging radar system: real aperture or synthetic aperture.

Range Resolution: Range resolution in real aperture/SLR is directly related to the pulse length of the transmitted radar signal. The shorter the pulse length, the finer the range resolution. Pulse length (τ), the physical length of the microwave signal, is a product of the speed of light (c) multiplied by the pulse duration of the transmission (τ), usually given in microseconds. The value of τ typically ranges from 0.4 to 1.0 microseconds (10^{-6} seconds), which translates into a pulse length ranging from approximately 8 to 210 m.

Slant range resolution (S_r) is constant across the entire range (near to far) of the image or dataset. The ground range resolution varies across the range and is more complex to calculate. To accommodate the different geometries of slant and ground range, the pulse length is divided by either the cosine of the depression angle (γ) or the sine of the look angle (ϕ). The resultant equation for range resolution (R_r) is as follows:

$$R_r = \frac{c \cdot \tau}{2 \cdot \cos\gamma} \tag{3.1}$$

Azimuth Resolution: In the azimuth direction (parallel to the flight line or satellite ground track), the resolution is determined by the type of imaging system and operational parameters. Azimuth resolution (R_a) with a real aperture (brute force) imaging radar system is most simply defined in the equation (3.2) by the slant-range distance (S) to the point of interest, operating wavelength (λ), and antenna length (L).

$$R_a = \frac{S \times \lambda}{L} \tag{3.2}$$

As evident from equation 3.2, by increasing antenna length (L), we can improve the azimuth resolution. However, there is practical limitation on the length of the antenna whether it is airborne or spaceborne system. Hence advanced signal processing is used to achieve finer spatial resolution from a short-length physical antenna.

3.2.1.2 Synthetic Aperture Radar

As we have seen in Section 3.2.1.1, in a RAR system the antenna length (L) is inversely proportional to the angular width, which defines the azimuth resolution. By increasing the length of the antenna, substantial improvement in azimuth resolution could be achieved. Like a RAR, synthetic aperture radar (SAR) also uses a relatively small antenna (e.g., 1 m) that sends out a relatively broad beam perpendicular to the sensor platform. The major difference is that a greater number of additional beams are sent toward the object. Doppler principle is subsequently used to record and monitor the return signals from all these microwave pulses to achieve finer azimuth resolution.

The Doppler principle is related to change in the frequency of sound waves as the source approaches or moves away from the listener. It states that the frequency or pitch of a sound changes if listener and/or source are in motion relative to one another. For example, an approaching train whistle will have increasingly higher pitch as it approaches the listener. The pitch will be highest when it is directly perpendicular to the listener. This is called the point of zero Doppler. In the case of a SAR, as the sensor platform moves forward over the object/target, it remains illuminated until it becomes out of sight from the sensor (Figure 3.3). When the sensor is approaching the object, the Doppler shift is positive and becomes zero, as mentioned earlier, when the sensor is overhead or orthogonal to the object. As the sensor moves away from the object the Doppler shift becomes negative. The sensor records the Doppler history of radar pulse.

Over time, individual transmit/receive cycles are completed with the data from each cycle being stored electronically. The signal processing uses magnitude and phase of the received signals over the successive pulses from elements of a synthetic aperture. After a given number of cycles, the stored data are recombined (taking into account the Doppler effects inherent in the different transmitter to target geometry in each succeeding cycle) to create a high-resolution image of the terrain being over flown/imaged by a spaceborne sensor.

Azimuth resolution in SAR system is radically different than in RAR. The equation for the maximum attainable azimuth or along-track resolution for a point target in a SAR system is given as

$$A_s = l/2 \tag{3.3}$$

where "l" denotes the antenna length.

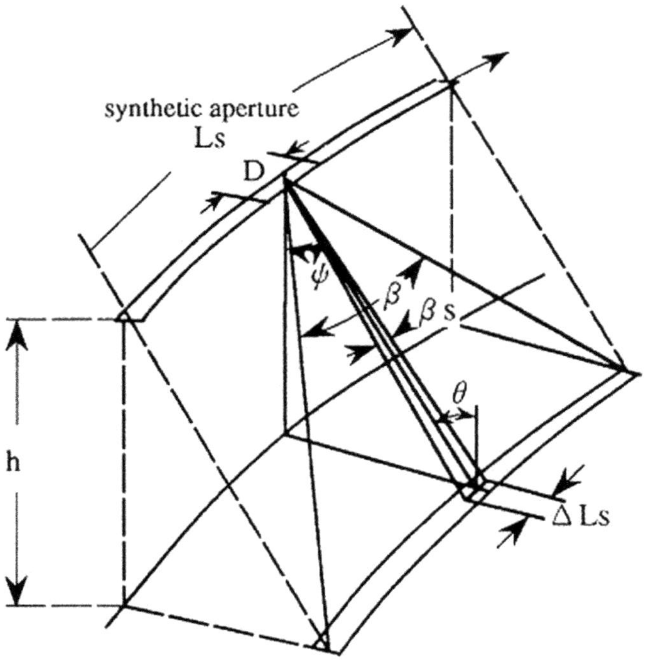

FIGURE 3.3
Relationship between real aperture and synthetic aperture radar. Where D is real aperture; β is real beam width, β_s is synthetic aperture beam, h is height, ΔLs is azimuth resolution, ψ is off-nadir angle (Source: http://wtlab.iis.u-tokyo.ac.jp/wataru/lecture/rsgis/rsnote/cp4/4-3-1.gif, accessed on 04-06-2018).

A comparison of the azimuth resolution equations (equations 3.2 and 3.3) reveals several important differences. The most notable is the omission of R (range distance) or its equivalent H/cos ϕ from the equation for defining azimuth resolution. It means that the azimuth resolution of a SAR system is independent of range distance or sensor altitude. Theoretically, the spatial resolution in the azimuth direction from a SAR imaging system will be the same from an aircraft platform as it is from a spacecraft. However, there is a limitation in equation 3.3. The coherent nature of the SAR signal results in speckle or fading. To improve the image quality, speckle is minimized but at the expense of azimuth resolution. The image is divided into sub-beams. Each sub-beam provides an independent "look." Each look (N) is incoherently averaged to produce an image with reduced speckle and reduced azimuth resolution. The approximate resolution of the incoherently averaged image is given as

$$A_s = (N)\,1/2 \tag{3.4}$$

An interesting characteristic of equation 3.4 is that the antenna length (l) is now in numerator, whereas in equation 3.2, of SLR, antenna length is part of denominator. This reversal means that, technically, the smaller the antenna, the better the resolution (the opposite of SLR). The wavelength of the system (λ) in equation 3.2 of SLR is also omitted from equation 3.4, indicating that azimuth resolution in a SAR system is also independent of wavelength (λ).

3.2.1.3 The Operating Modes of SAR

Initially, only a fixed angle for SLR beam was feasible. With the development in sensor technology, it is now possible to image the terrain with varying spatial and the coverage across the swath or flight line to cater to users' requirements. Some of the commonly used modes of SAR system include (i) strip map mode, (ii) spotlight mode, and (iii) scan mode.

3.2.1.3.1 SAR Strip Map Mode

The most popular is probably the strip map mode. The radar antenna points along a fixed pointing direction with respect to the sensor platform path, and the antenna footprint covers a strip on the illuminated surface as the platform moves and the system operates (Figure 3.4). The strip map SAR image dimension is limited in the across-track (range) but not in the along-track (azimuth) direction.

The strip map mode involves two imaging geometries. The most conventional one is referred to as boresight, with the antenna beam pointing in the plane perpendicular to the flight/orbital direction in the case of spaceborne SAR (Figure 3.5).

The squinted mode of operation is another imaging mode of SAR operation (Figure 3.6).

In this mode of operation the antenna makes a pointing angle, referred to as (forward or backward) the squint angle, with respect to the boresight direction. The squint angle can be a desired system characteristic or can be due to undesired motions of the platform. The squint angle can reach some tens of degrees and is useful for several applications. The analysis of the backscattering properties of the illuminated surfaces with respect to the azimuth angle is one such application.

3.2.1.3.2 SAR Spotlight Mode

Spotlight SAR is capable of extending the high-resolution SAR imaging capability significantly. The radar antenna is steered during the overall acquisition time to illuminate the same area (Figure 3.7). As more pulses are used, the available synthetic antenna length can

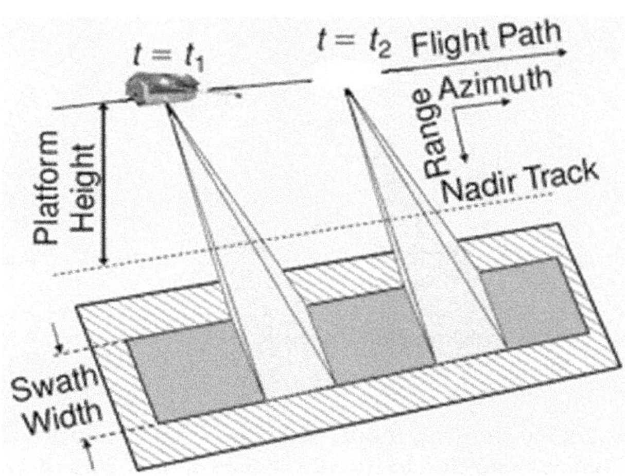

FIGURE 3.4
The strip map SAR operation mode (Source: https://www.researchgate.net/figure/Illustration-of-different-SAR-operation-modes-which-are-usedto-increase-the-swath-width_fig3_257355346?_sg=OKCVq-svq7oYD9__G0Ag4FnXH0y8BQc51kia9aa39NvHlZKjY76DT8-Tfy0m0NDu3JdRF-8dkQMI5p_Y8JREJw).

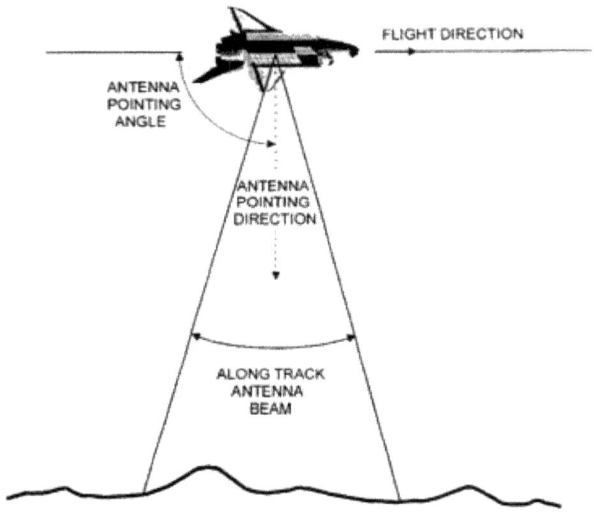

FIGURE 3.5
Bore sight imaging geometry: the antenna pointing angle is equal to 90° (Source: Adapted from Franceschetti and Lanari (1999)).

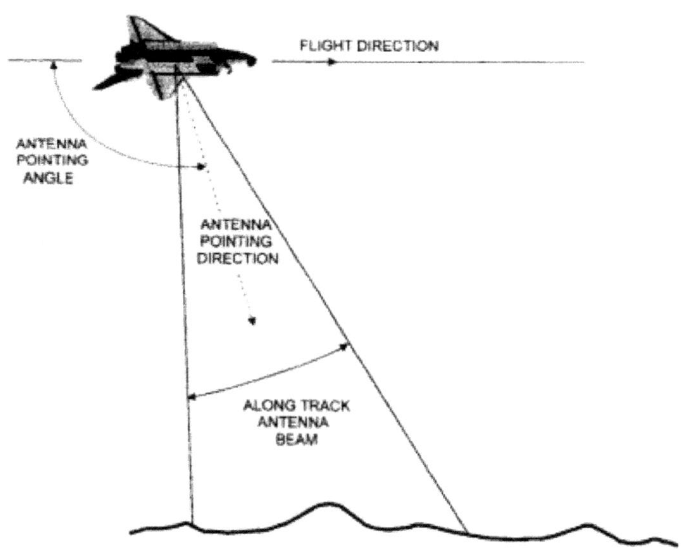

FIGURE 3.6
Squinted imaging geometry: the antenna pointing angle is different from 90° (Source: Adapted from Franceschetti and Lanari (1999)).

be increased with respect to the strip mode, thus improving azimuth resolution. This gain is traded off by loss of coverage due to the illumination of a limited area along the sensor flight path/imaging swath. The spotlight SAR mode of operation is usually at the expense of spatial coverage, as other areas within a given accessibility swath of the SAR cannot be illuminated while the radar beam is spotlighting over a particular target area. Again, extension to the interferometric SAR (InSAR) configuration is possible.

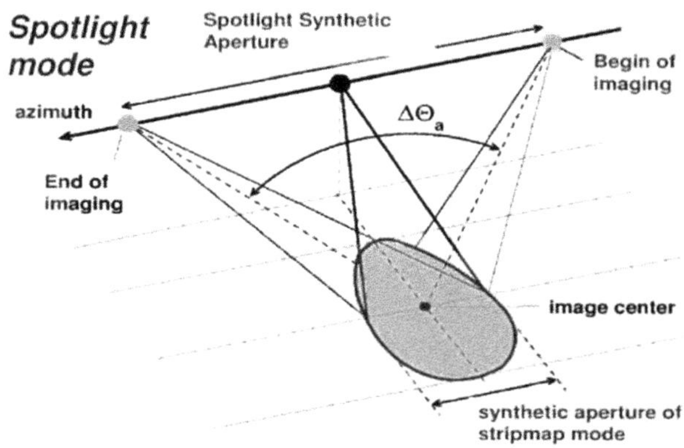

FIGURE 3.7
Spotlight SAR operation mode.

3.2.1.3.3 SAR Scan Mode

The scan mode allows a drastic increase of the range swath dimension. This is achieved by periodically stepping the antenna beam to neighboring sub-swaths (in the range direction), as shown in Figure 3.8. In this case the radar is continuously on, but only portions of the full synthetic antenna length are available for each target in a sub-swath. This causes a degradation of the achievable azimuth resolution with respect to the strip map mode. In other words, the range swath dimension increases at the expense of azimuth resolution.

Compared to real operations, azimuth swath dimension is significantly reduced to highlight stepping function. The scan mode allows an interferometric extension, too, but a careful synchronization between the acquisitions is required for two-pass operations. This is not the case for single-pass operations.

3.2.2 Non-imaging Microwave Sensors

Non-imaging remote sensing radars are either scatterometers or altimeters. Any calibrated radar that measures the scattering properties of a surface is called scatterometer. Thus, a scatterometer may be a radar specifically designed for backscatter measurements. An

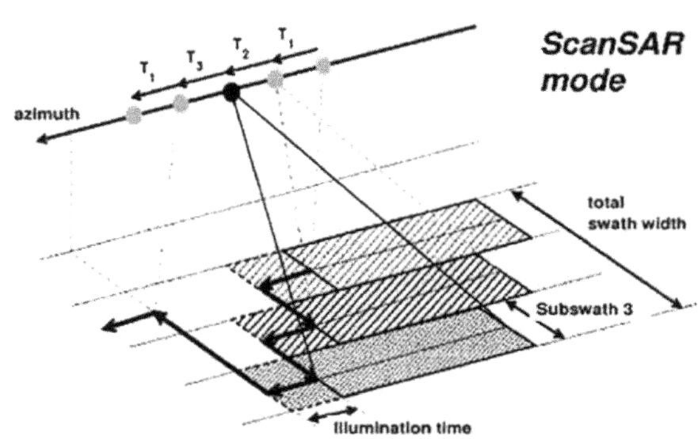

FIGURE 3.8
Scan SAR operation mode; two-sub swath case.

altimeter is a sensor which is able to measure the height with reference to a prescribed reference level. The most classical altimeter is based on the Earth atmospheric pressure which changes with height.

3.2.2.1 Scatterometers

A scatterometer is an incoherent surface-based radar that measures reflectivity over a set of different incident angles. Scatterometers average the detected returns from a sequence of pulses, a process known as post-detection integration (Henderson and Lewis, 1998). Averaging ensures more accurate measurements of the backscattering coefficient, since single return pulses are typically noisy. Because pulses from multiple ground targets interfere with one another, the return signal is distorted, a process referred to as fading. Averaging together a sufficient number of return pulses has the effect of canceling out the noise caused by fading, often achieving ± 0.10–0.15 dB accuracy. The downside to averaging together multiple return pulses is a significant reduction in spatial resolution: 25–50 km, compared to 1–10 km with SAR.

The intensity of the backscattered signal depends on the roughness and the dielectric properties of the target. Changes in wind velocity cause changes in ocean surface roughness, modifying the radar cross section of the ocean and the magnitude of the backscattered power. Multiple collated measurements acquired from several directions can be used to solve wind speed and direction simultaneously (Kramer, 1994).

For ice and snow, the backscatter is influenced by surface roughness (including orientation of the surface scatterers), liquid water content, snow grain size, brine concentration in sea ice, and density. Scatterometers can be calibrated to less than a few tenths of a decibel (dB); thus, seasonal and interannual differences of only 1–2 dB can be accurately monitored using scatterometry data.

Scatterometers use different techniques to collect measurements at multiple incident and azimuth angles. The Seasat-A Scatterometer System (SASS), AMI ERS-1/2 satellites in scatterometer mode (ESCAT), and NASA Scatteromter (NSCAT) instruments used a fan-beam approach, where the scatterometer fans out, or points, several fixed-angle beams of radar pulses at the ground simultaneously (Figure 3.9). Another approach called the pencil-beam method, which is used by SeaWinds, is to rotate a single beam of pulses at multiple angles (Figure 3.10).

Scatterometers are designed to measure radar backscatter very precisely, but typically have lower spatial resolution than SARs. They tend to be less complicated than SARs.

To date, four spaceborne scatterometers have been deployed. All were originally designed to measure ocean winds.

3.2.2.1.1 Satellite Scatterometer System

The Satellite Scatterometer System (SASS) was a payload aboard NASA's Seasat-A satellite, which was launched on June 27, 1978. The mission lasted until October 10, 1978, when a satellite power failure terminated the mission. However, the mission did provide a baseline for studies of global change.

3.2.2.1.2 ESCAT: ERS Scatterometer

ESCAT is the name given to the Active Microwave Instrument (AMI) on board the European Space Agency's (ESA) Earth Remote Sensing (ERS)-1 and -2 satellites when it is in scatterometer mode. ERS-1 was launched on July 17, 1991, and ERS-2 was launched on April 21, 1995.

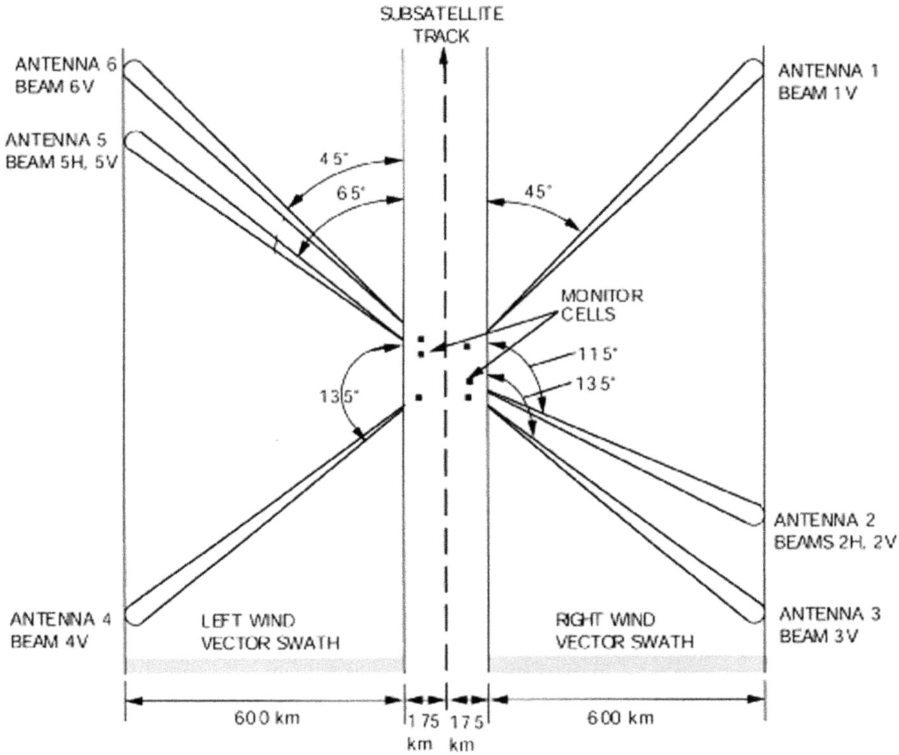

FIGURE 3.9
NSCAT viewing geometry (Courtesy: Long and Drinkwater, 2000).

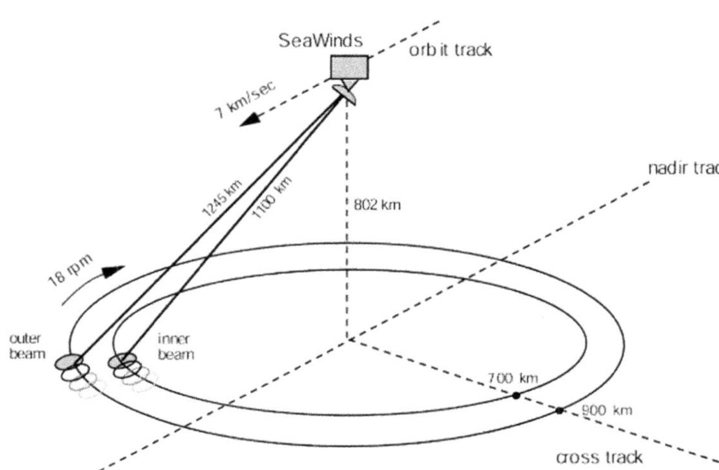

FIGURE 3.10
SeaWinds viewing geometry (Courtesy: Spence et al. 2000).

3.2.2.1.3 NSCAT: NASA Scatterometer

The NSCAT instrument flew on the Japanese Aerospace Exploration (JAXA) Advanced Earth Observing Satellite-I (ADEOS-I) that was launched on August 16, 1996. On June 30, 1997, a power failure terminated the mission. NSCAT's primary science objective was to measure surface wind speed and direction over the global oceans, with a requirement to provide coverage every 2 days under all weather and cloud conditions.

3.2.2.1.4 SeaWinds

The SeaWinds scatterometer flies on NASA's Quick Scatterometer (QuikSCAT) satellite and was launched on June 19, 1999. A similar scatterometer, SeaWinds-II, flew on JAXA's ADEOS-II satellite from December 14, 2002, to October 24, 2003, when a power failure terminated the mission. SeaWind's primary science objective is to acquire high-resolution, continuous, all-weather measurements of near-surface vector winds over the ice-free global oceans (Kramer, 1994) (Table 3.2).

3.2.2.2 Radar Altimeter

The radar altimeter is a Ku-band (13.8 GHz) nadir-pointing active microwave sensor designed to measure the return trip time for echoes from ocean and ice surfaces. The instrument transmits an electronic pulse in the microwave frequency to the Earth's

TABLE 3.2

Operating Characteristics of Spaceborne Scatterometers

	SASS (Seasat-A)	ESCAT (ERS-1/2)	NSCAT (ADEOS-I)	SeaWinds (QuikSCAT/ ADEOS-II)
Time period	July–September 1978	January 1992–present Scatterometer Climate Record Pathfinder (SCP) data are only available through January 2001	September 1996–June 1997	July 1999–present
Frequency	14.6 GHz (Ku band)	5.3 GHz (C band)	14.0 GHz (Ku band)	13.4 GHz (Ku band)
Antenna azimuth orientations	Four fixed	Three fixed	Six fixed	1 m diameter rotating dish that produces two spot beams, sweeping in a circular pattern
Polarizations	V-H, V-H	V Only	V, V-H, V	V-Outer/H-inner
Beam resolution	Fixed Doppler	Range Gate	Variable Doppler	Pencil-Beam
Resolution	50/100 km	25/50 km	25/50 km	25 km × 6 km
Swath width	750 km	500 km	600 km	1,400 km/1,800 km
Incidence angle	0–70°	18–59°	17–60°	46–54°
Orbit	Sun-synchronous 810 km altitude 106° inclination	Sun-synchronous 780 km altitude 98.52° inclination	Sun-synchronous 805 km altitude 98.7° inclination	Sun-synchronous 803 km altitude 98.6° inclination
Coverage during a 24-hour period	Variable	<41%	78%	92%

surface. The microwave pulse reflects off the surface and returns to the sensor. Altitude is determined from the pulse travel time (from transmit to receive) and from the waveform of the returned pulse. Functioning in one of the two operational modes (ocean or ice), the altimeter provides information on significant wave height, surface wind speed, sea surface elevation relating to ocean currents, the surface geoid and tides, and various parameters over sea ice and ice sheets (Table 3.3).

The altimeter operates by timing the two-way delay for a short-duration radio-frequency pulse, transmitted vertically downward. The level of accuracy required for range measurement (better than 10 cm) calls for a pulse compression (full deramp) technique. The radar emits a modulated chirp (s(t)) of duration T in the a frequency-band B (Figure 3.11) toward the Earth's surface. With a delay corresponding to the estimated return time of the emitted chirp, another which is slightly shifted in frequency. By mixing the returning and deramping chirps, the frequency shift can be estimated, which, using Fourier transforms, gives the time delay.

In ocean mode, a chirped pulse of 20 microsecond's duration is generated with a bandwidth of 330 MHz. For tracking in ice mode, an increased dynamic range is used, obtained

TABLE 3.3
Radar Altimeter Characteristics for Various Satellites

Satellite	Frequency (GHz)	Bandwidth	Wavelength (m)	Range Resolution (m)	Pulse Compression	Wave Height
ERS-1	13.50	400.00	0.02	0.10	8,000.00	
ERS-2	13.50		0.02	0.1	58,000.00	0.13 m
Geos-3	13.90	80.00	0.02	0.50	30.00	±25% (4–10 m)
GEOSAT	13.50	320.00	0.02	0.10	30,000.00	±10% (1–20 m)
GEOSAT Follow-On	13.50		0.02	0.018	58,000.00	0.035 m
SEASAT	13.50	320.00	0.02	0.10	1,000.00	±10% (1–20 m)
Skylab	13.90	100.00	0.02	1.00	13.00	1–2 m
TOPEX/Poseidon	5.3 and 13.6	320.00	0.0566 and 0.02205	0.03	58,000.00	0.13 m

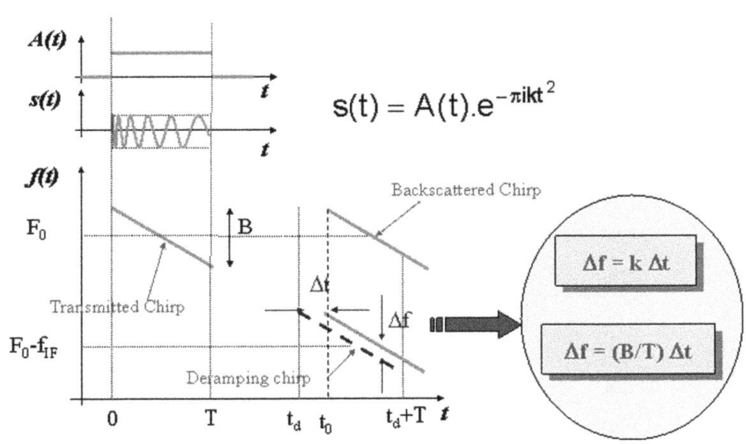

FIGURE 3.11
The concept of deramp technique (Source: Adapted from Rosmorduc et al. (2016)).

by reducing the chirp bandwidth by a factor of four to 82.5 MHz, although this also results in a coarser resolution.

3.3 Spaceborne Radars Systems

In order to provide familiarity with spaceborne active microwave data available from currently operating missions, some of the spaceborne SAR missions are very briefly discussed in the following sections.

3.3.1 RISAT Mission

The mission consists of two microwave earth observation satellites: RISAT-1 and RISAT-2.

3.3.1.1 RISAT-1

Launched on April 26, 2012, by the Indian Space Research Organization, Radar Imaging Satellite-1 (RISAT-1) is a state-of-the-art microwave remote sensing satellite carrying a SAR operating in C-band (5.35 GHz), which images the Earth in both horizontal and vertical polarization. RISAT-1 carries a multimode C-band (5.35 GHz) SAR as the payload. The choice of C-band frequency of operation and capability of imaging in HH, VV, HV, VH, and circular polarizations ensures wide applicability in the thrust areas like flood mapping, agriculture and crop monitoring, generic vegetation, forestry, soil moisture, geology, and sea ice and coastal processes. In the absence of the emergency/user request, the default mode of collection will be Medium Resolution ScanSAR (MRS) descending, left looking, with dual polarization with a repeat cycle of 25 days. The data can be acquired in all weather conditions, in both ascending (evening) and descending mode (morning) and in both right look and left look.

3.3.1.2 RISAT-2

RISAT-2 satellite was launched by the Indian Space Research Organization on April 20, 2009. Its SAR uses a large dish-like antenna to transmit and receive radar signals that can penetrate darkness and thick clouds providing images with 1 m resolution. The orbit parameters are given in Table 3.4.

TABLE 3.4

Salient Features of RISAT-2 Satellite

Frequency	X band 9.59 GHz
Velocity	7.3 km/s
Orbit	548 km
Data link rate	620 Mbps
Revisit	3 or 4 days
Repeat cycle	14 days

3.3.2 Sentinel Mission

The mission is part of the European Earth Observation Programme *"Copernicus"* which is run by the European Union together with the ESA in cooperation with the European Environment Agency (EEA). The Copernicus Space Component comprises a series of spaceborne dedicated missions called "Sentinels" and is developed and procured by the ESA. The Sentinel comprises six satellites: Sentinel-1, Sentinel-2, Sentinel-3, Sentinel-4, Sentinel-5, and Sentinel-5P.

3.3.2.1 Sentinel-1

With the objectives of land and ocean monitoring, Sentinel-1 is composed of two polar-orbiting satellites operating day and night, and performs radar imaging, enabling them to acquire imagery regardless of the weather. The first Sentinel-1 satellite was launched on a Soyuz rocket from Europe's Spaceport in French Guiana on April 3, 2014. The second Sentinel-1B satellite was launched on April 25, 2016, from same spaceport.

3.3.2.2 Sentinel-2

The objective of Sentinel-2 is land monitoring, and the mission consists of two polar-orbiting satellites providing high-resolution optical imagery. Vegetation, soil and water cover, inland waterways, and coastal areas are among the monitoring objectives. Sentinel-2 also provides disaster monitoring for emergency services. The first Sentinel-2 satellite was launched on June 23, 2015.

3.3.2.3 Sentinel-3

The primary objective of Sentinel-3 is marine observation, and it will study sea-surface topography, sea and land surface temperature, and ocean and land color. Composed of three satellites, the mission's primary instrument is a radar altimeter, but the polar-orbiting satellites will carry multiple instruments, including optical imagers. Sentinel-3 will provide ocean and global land monitoring services. The first Sentinel-3A satellite was launched on January 16, 2016, by a Eurockot Rokot vehicle from the Plesetsk cosmodrome in Russia.

3.3.2.4 Sentinel-4

The ultraviolet (305–400 nm), visible (400–500 nm), and near-infrared (750–775 nm) instrument aboard Sentinel-4 is a spectrometer carried aboard Meteosat Third Generation Sounder (MTG-S) satellite. The main objective of the Sentinel-4 mission is to monitor key air quality trace gases and aerosols over Europe in support of the Copernicus Atmosphere Monitoring Service (CAMS) at high spatial resolution and with a fast revisit time. The target species of the Sentinel-4 mission include the key air quality parameters NO_2 (nitrogen dioxide), O_3 (ozone), SO_2 (sulfur dioxide), HCHO (formaldehyde), CHOCHO (glyoxal), and aerosols. Complementarily, the low-Earth-orbiting (LEO) missions S5 and S5p will deliver the S4 target species and additionally CO (carbon monoxide), CH_4 (methane), and stratospheric O_3 (ozone) with global daily coverage for climate, air quality, and ozone/surface UV applications. It will be launched in 2023.

3.3.2.5 Sentinel-5

The Sentinel-5 UVNS instrument is a spectrometer carried aboard the MetOp Second Generation satellites. The Sentinel-5 mission consists of high-resolution spectrometer system operating in the ultraviolet to shortwave infrared range with seven different spectral bands: UV-1 (270–300 nm), UV-2 (300–370 nm), VIS (370–500 nm), NIR-1 (685–710 nm), NIR-2 (745–773 nm), SWIR-1 (1,590–1,675 nm), and SWIR-3 (2,305–2,385 nm). The instrument will be carried on the MetOp-SG A satellite. The mission aims to provide continuous monitoring of the composition of the Earth's atmosphere. It provides wide-swath, global coverage data to monitor air quality around the world. Sentinel-5 is focused on air quality and composition–climate interaction, with the main data products being O_3, NO_2, SO_2, HCHO, CHOCHO, and aerosols. Additionally, Sentinel-5 will also deliver quality parameters for CO, CH_4, and stratospheric O_3 with daily global coverage for climate, air quality, and ozone/surface UV applications.

3.3.2.6 Sentinel-5P

A precursor satellite mission, Sentinel-5P, aims to fill in the data gap and provide data continuity between the retirement of the Envisat satellite and NASA's Aura mission and the launch of Sentinel-5. The mission will perform atmospheric monitoring. The Sentinel-5 Precursor mission is a joint initiative between ESA and the Netherlands Space Office (NSO). The mission consists of one satellite carrying the TROPOspheric Monitoring Instrument (TROPOMI) UV-VIS-NIR-SWIR instrument. The main objective of the Sentinel-5P mission is to perform atmospheric measurements, with high spatio-temporal resolution, relating to air quality, climate forcing, ozone, and UV radiation. The satellite was successfully launched on October 13, 2017, from the Plesetsk cosmodrome in Russia.

3.3.3 CryoSat

ESA's Earth Explorer CryoSat is Europe's first ice mission. The aim of the mission is to determine variations in the thickness of the Earth's continental ice sheets and marine ice cover. Its primary objective is to test the prediction that Arctic sea ice is thinning due to global warming. In addition, it was important to discover the extent to which the Antarctic and Greenland ice sheets are contributing global sea level rise.

3.3.4 Soil Moisture and Ocean Salinity Mission

Real-/near-real-time soil moisture information is required for hydrological studies, and data on ocean salinity are vital for improving our understanding of ocean circulation patterns. Soil Moisture and Ocean Salinity (SMOS) mission has been designed to observe soil moisture over the Earth's landmasses and salinity over the oceans.

The SMOS, known as "Water Mission," was launched on November 2, 2009. The mission with sun-synchronous polar orbit and orbital period of 100 ± 15 minutes and local equator crossing time at 6:00 a.m. on ascending node has a repeat cycle of 23 days and a 3-day subcycle. The satellite carries an L-band (1.4 GHz) radiometer, known as Microwave Imaging Radiometer with Aperture Synthesis (MIRAS). The spatial resolution required for observing SMOS in L-band demands a huge antenna—too big for a satellite to carry. To overcome

FIGURE 3.12
Artist's rendition of SMOS mission (Credit: European Space Agency).

this challenge, the antenna for MIAS has been simulated through 69 small antennas, distributed over the three arms and central hub of the instrument (Figure 3.12).

3.3.4.1 Measurement Principle

The SMOS measures microwave radiation emitted from Earth's surface within the L-band (1.4 GHz) using an interferometric radiometer, which by way of 69 small receivers measures the phase difference of incident radiation. The technique is based on cross-correlation of observations from all possible combinations of receiver pairs. A 2D "measurement image" is taken every 1.2 seconds. As the satellite moves along its orbital path, each observed area is seen under various viewing angles. From an altitude of around 758 km, the antenna views an area of almost 3,000 km in diameter. However, due to the interferometry principle and the Y-shaped antenna, the field of view is limited to a hexagon-like shape about 1,000 km across called the "alias-free zone"—the area where there is no ambiguity in the phase difference. The MIRAS has three main operational modes: (i) dual-polarization mode, in which all receivers are switched synchronously to either H or V polarization; (ii) full polarimetric mode, in which segments of the array are switched according to a predefined sequence between H and V; and (iii) calibration mode, in which measurements of the internal load, the noise diodes, or the so-called "fringe washing function" are determined. For further details of the mission the readers may refer to Kerr et al. (2001).

3.3.5 Soil Moisture Active Passive

Launched on January 13, 2015, into a 680-km near-polar, sun-synchronous orbit, with equator crossings at 6 a.m. and 6 p.m. local time Soil Moisture Active Passive (SMAP), provides global measurements of soil moisture and its freeze/thaw state (Figure 3.13). These measurements are intended to be used to enhance understanding of processes that link the water, energy, and carbon cycles, and to extend the capabilities of weather and climate prediction models. SMAP data will also be used to quantify net carbon flux in boreal landscapes and to develop improved flood prediction and drought monitoring capabilities. The

FIGURE 3.13
Artist's rendition of the Soil Moisture Active Passive (SMAP) spacecraft in orbit (Credit: NASA/JPL-Caltech).

SMAP instrument includes a radiometer and SAR operating at L-band (1.20–1.41 GHz). The instrument is designed to make coincident measurements of surface emission and backscatter, with the ability to sense the soil conditions through moderate vegetation cover. The measurement swath width is 1,000 km, providing global coverage within 3 days at the equator and 2 days at boreal latitudes (>45° N). The specification SMAP-SAR is as follows:

- Frequency: 1.26 GHz
- Polarizations: VV, HH, HV (not fully polarimetric)
- Relative accuracy (3 km grid): 1 dB (HH and VV), 1.5 dB (HV)
- Data acquisition:
  - High-resolution (SAR) data acquired over land
  - Low-resolution data acquired globally.

On July 7, 2015, SMAP-SAR stopped transmitting data, and on September 2, 2015, NASA announced the amplifier failure. The sensor is now no more functional.

The following is the specification of SMAP radiometer:

- Frequency: 1.41 GHz
- Polarizations: H, V, 3rd and 4th Stokes
- Relative accuracy (40 km grid): 1.3 K
- Data collection:
  - High-rate (16 sub-band) data acquired over land
  - Integration time for high-rate, full-band data with averaging is typically ~40 ms
  - Low-rate data acquired globally

3.3.6 RADARSAT Mission

RADARSAT-1, the first satellite in the series with C-band SAR, was launched on November 4, 1995, and operated successfully till May 9, 2013. A schematic of the mission and operating mode are portrayed in Figure 3.14. RADARSAT-2, the follow-up of radarsat-1, was launched on December 14, 2007. It has a SAR with multiple polarization modes.

Its finest spatial resolution is 3 m with 100 m positional accuracy. The repeat cycle of RADARSAT-2 is 24 days. It has the same orbit (798 km altitude sun-synchronous with 6 p.m. ascending node and 6 a.m. descending node). It is intended to fill a wide variety of roles, including sea ice mapping and ship routing, iceberg detection, agricultural crop monitoring, marine surveillance for ship and pollution detection, terrestrial defense surveillance and target identification, geological mapping, land use mapping, wetlands mapping, soil moisture estimation, and topographic mapping (Table 3.5).

3.3.6.1 RADARSAT Constellation

Figure 3.15 provides a visual representation of the coverage area of each RADARSAT Constellation Mission (RCM) imaging mode along the subsatellite ground track (nadir), as well as the satellite velocity vector. The beam modes are shown in order of decreasing coverage area: Low Resolution 100 m Mode; Medium Resolution 50 m Mode and Low Noise/Ship Detection Mode; Medium Resolution 30 m Mode; Medium Resolution 16 m Mode and Quad-Pol Mode; High Resolution 5 m Mode; Very High Resolution 3 m Mode; and Spotlight Mode (Figure 3.16).

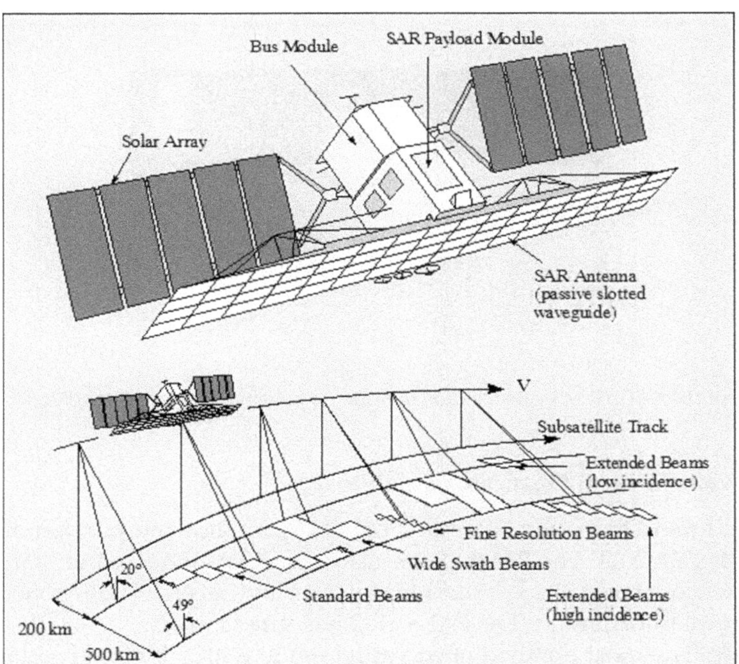

FIGURE 3.14
The RADARSAT-1 spacecraft and illustration of observation geometries (Credit: DLR).

TABLE 3.5
Various Beam Modes of RADARSAT-2 Mission

Beam Modes	Nominal Swath Width (km)	Approximate Resolution (m)
Selective polarization Transmit H or V, receive H and/or V		
Fine	50	10 × 9
Standard	100	25 × 28
Low incidence	170	40 × 28
High incidence	75	20 × 28
Wide	150	25 × 28
ScanSAR narrow	300	50 × 50
ScanSAR wide	500	100 × 100
Polarimetric Transmit H and V on alternate pulses/receive H and V on any pulse		
Fine quad-pol	25	11 × 9
Standard quad-pol	25	25 × 28
Selective single polarization Transmit H or V, receive H or V		
Ultra-fine	20	3 × 3
Spotlight	18	3 × 1
Multi-look fine	50	11 × 9

FIGURE 3.15
Artist's rendition of the RADARSAT constellation mission imaging concept (Credit: MDA, CSA).

3.3.7 The Advanced Land Observing Satellite-2

The Advanced Land Observing Satellite-2 (ALOS-2), a follow-on mission of ALOS-1, was launched on May 24, 2014. The PALSAR-2 sensor aboard ALOS-2 is an L-band SAR sensor, a microwave sensor that emits L-band radio waves and receives their reflection from the ground to acquire information. The PALSAR-2 has three modes:

Spotlight mode: The most detailed observation mode with 1 by 3 m resolution (observation width of 25 km) (Table 3.6)

Strip Map mode: A high-resolution mode with the choice of 3 m, 6 m, or 10 m resolution (observation width of 50 km or 70 km)

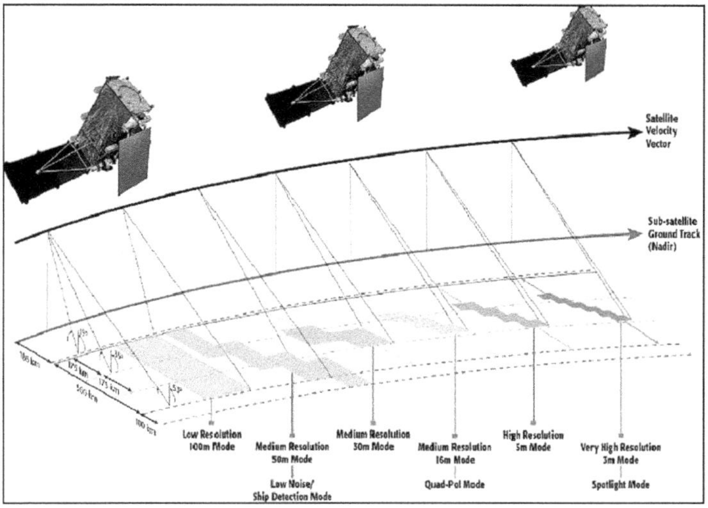

FIGURE 3.16
RADARSAT constellation imaging modes (Credit: Canadian Space Agency).

TABLE 3.6
Various Imaging Modes of Radarsat Constellation—SAR Data Acquisition

Beam Modes	Nominal Swath Width (km)	Approximate Resolution (m)
Low resolution 100 m	500	100 × 100
Medium resolution 50 m	350	50 × 50
Medium resolution 30 m	125	30 × 30
Medium resolution 16 m	30	16 × 16
High resolution 5 m	30	5 × 5
Very high resolution 3 m	20	3 × 3
Low noise	350	100 × 100
Ship detection	350	Variable
Spotlight	14 (5 km in azimuth)	1 × 3
Quad-polarization	20	9 × 9

ScanSAR mode: A broad area observation mode with observation width of 350 km or 490 km, and resolution of 100 m or 60 m, respectively.

The imaging modes of ALOS-2 are shown in Figure 3.17. It has only a microwave sensor (PALSAR-2) operating in 1.2 GHz (L-band) with several operational mode. In spotlight mode the spatial resolution is 1–3 m (Table 3.7).

3.3.8 TerraSAR-X and TanDEM-X

Launched by the German Aerospace Centre (GDR), TerraSAR-X and TanDEM-X are identical and operate in six alternate operating modes for a wide range of applications in Earth observation: Ranging from Staring SpotLight mode with a resolution of up to 25 cm at a scene size of, e.g., ~4 km × 3.7 km mainly for image intelligence applications

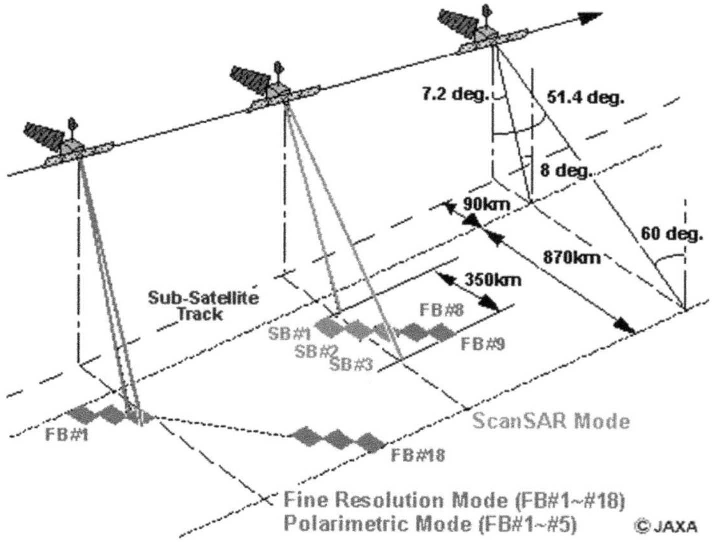

FIGURE 3.17
The imaging modes of ALOS-2 mission (Source: Adapted from Japan Aerospace Exploration Agency Earth Observation Research Center).

TABLE 3.7
Salient Features of ALOS-2 Mission

		Stripmap[a]		Full (Quad.) Polarimetry	
Observation mode	**[3 m]**	**[6 m]**	**[10 m]**	**[6 m]**	**[10 m]**
Observation mode ID (code)	UBS/UBD	HBS/HSD	FBS/FBD	HBQ	FBQ
Width (East-West) (Length of range direction)	55 km (max)	55 km (max)	70 km (max)	40–50 km	30 km
Length (North-South) (Length of azimuth direction)	70 km	70 km	70 km	70 km	70 km
Time duration of azimuth direction	10 sec	10 sec	10 sec	10 sec	10 sec
Range resolution*1	3.0 m	6.0 m	9.1 m	5.1 m	8.7 m
Azimuth resolution*1	3.0 m	4.3 m	5.3 m	4.3 m	5.3 m
Pixel spacing levels 1.5/3.1	2.5 m	3.125 m	6.25 m (2 look)	3.125 m	6.25 m (2 look)
Pixel spacing level 2.1	2.5 m/5.0 m/10.0 m	3.125 m/6.25 m/12.5 m	6.25 m/12.5 m	3.125 m/6.25 m/12.5 m	6.25 m/12.5 m
Polarization	Single (HH, HV, VH, or VV) Dual (HH+HV or VH+VV)			Full (Quad.) Polarimetry (HH+HV+VH+VV)	

[a] Strip map and full (Quad.) polarimetry modes define category names by resolution: Ultra-Fine (3 m), High-sensitive (6 m), Fine (10 m).

to Wide ScanSAR providing a resolution of 40 m at a scene size of 270 km × 200 km (extendable to a scene length of 1,500 km) particularly suitable for maritime applications (Table 3.8).

TABLE 3.8

Salient Features of TerraSAR-X and TanDEM-X Mission

Date of Launch	TerraSAR-X: June 15, 2007 TanDEM-X: June 21, 2010
Orbit height	514 km at equator
Orbit per day	15 2/11
Orbit repeat cycle	11 days
Inclination	97.44°
Ascending node/equatorial crossing time	15:00 ± 0.25 h (local)
Radar carrier frequency	9.65 GHz
Polarizations	HH, VH, HV, VV
Antenna length	4.8 m
Nominal look direction	Right
Antenna width	0.7 m
Range bandwidth	150 MHz/300 MHz

3.4 Light Detection And Ranging

LiDAR (Light Detection And Ranging, also LADAR) uses radiations in ultraviolet, visible, or near-infrared regions of the electromagnetic spectrum to image objects (Figure 3.18). The fundamental concept of a LiDAR sensor is to send a laser pulse toward a target and to measure the timing and amount (amplitude) of radiation that is scattered back from the target. The return signal timing provides measurement of the distance between the instrument and the scattering object (d):

$$t = 2d/c \tag{3.5}$$

where c is the speed of light (299.79×10^6 m/s).

The major components of a LiDAR system include (i) laser source, (ii) scanner and optics, (iii) photo-detectors and receiver electronics, and (iv) position and navigation systems: global positioning system (GPS) and an inertial navigation system (INS). Depending upon the mode of capturing backscattered energy, there are broadly two types of LiDAR, as shown in Figure 3.19. Based on the way the measurement of return signals is done, LiDAR could be grouped into two broad categories, namely discrete return LiDAR and waveform LiDAR.

3.4.1 Discrete Return LiDAR

In a discrete return scanning LiDAR system, a laser pulse is sent out from the sensor and the leading edge of the returned signal trips a response for a time measurement. For many systems currently in use, the trailing edge of the response is also used to trip a second return time. These are referred to as the "first" and "last" returns. If the first return happens to be associated with a tree canopy top and the last return the underlying ground, then this single signal can be used to provide a measurement of tree height (Figure 3.19).

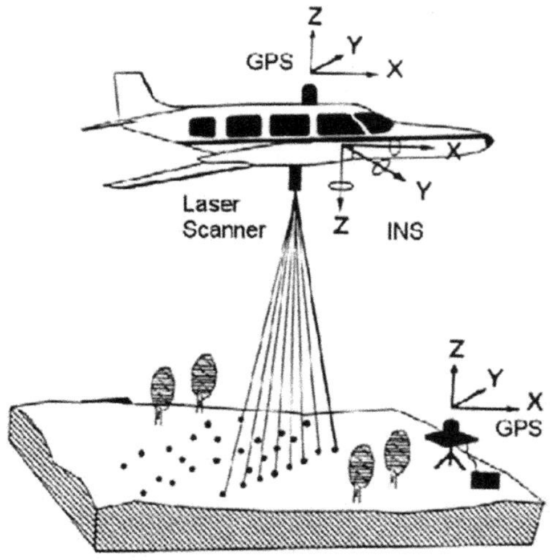

FIGURE 3.18
An airborne LiDAR system.

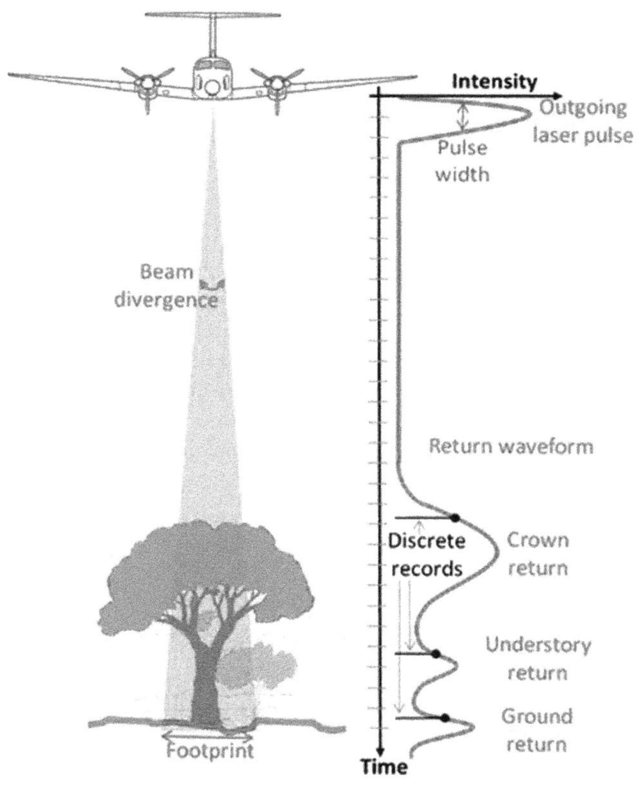

FIGURE 3.19
Observational differences between discrete-return and full-waveform LiDAR (Source: Adapted from Fernandez-Diaz (2011)).

3.4.2 Waveform LiDAR

In a waveform LiDAR, the system samples and records the energy returned for equal time intervals ("bins"). There are currently only a few such airborne systems and fewer still spaceborne instruments. Waveform LiDAR systems typically have a much larger footprint than discrete return systems, being of the order of tens of meters. This is fundamentally for signal-to-noise reasons: the quantity of backscattered energy in a small field of view is low. The energy received per unit time bin is even smaller, and so the sensor technologies need to be capable of measuring very low signal levels very quickly. For instance, in the case of LiteMapper-5600 system, the sampling interval is of the order of 1 ns, thus providing a multi-target resolution (related to bin size) of better than 0.6 m (Hug et al., 2004).

LiDAR returns are discrete observations recorded when a laser pulse is intercepted and reflected by targets. Multiple returns derive from one laser pulse intercepting multiple targets (e.g., a top of a tree, its branches, and the ground). Discrete return LiDAR returns multiple measurements from a pulse, but are not tied to the specific pulse, whereas waveform LiDAR represents multiple binned measurements associated with each pulse. The waveform data allow one to derive a continuous distribution of the laser energy for each pulse where discrete return does not.

LiDARs can detect the return signals either "incoherently" or direct energy detection, which is principally an amplitude measurement and "coherently," which is best for Doppler, or phase sensitive measurements. Coherent systems generally use optical heterodyne detection which being more sensitive than direct detection allows them to operate a much lower power but at the expense of more complex transceiver requirements.

In both coherent and incoherent LiDAR, there are two types of pulse models: micropulse LiDAR systems and high-energy systems. Micropulse systems use considerably less energy in the laser, typically on the order of 1 μJ, and are often "eye-safe," meaning thereby that they can be used without safety precautions. High-power systems are common in atmospheric research, where they are widely used for measuring many atmospheric parameters, namely the height, layering and densities of clouds, cloud particle properties, etc.

3.4.3 Scanning Mechanism

To generate coverage of the target area, some combination of moving platform and system pointing is used. There are several ways of pointing laser energy to cover an area of interest. As it is clearly seen in Figure 3.20, different kinds of mechanisms are applied by the different types of sensors; each has its own advantages and shortcomings, e.g., number of moving parts, type of rotation, etc., which lead to different kinds of error sources.

3.4.3.1 Oscillating Mirror Scanning Mechanism

Oscillating mirror scanning pattern is most popular, in which mirror rotates back and forth. This creates "Z" or zigzag lines of points on the ground (Figure 3.21). Data collection will be continuous since rotating mirror always looks terrain. However, continuous movement of mirror with variable velocity causes uneven spaced laser points on the terrain. Furthermore, the laser point density increases toward the edges of the scan field where mirror slows down.

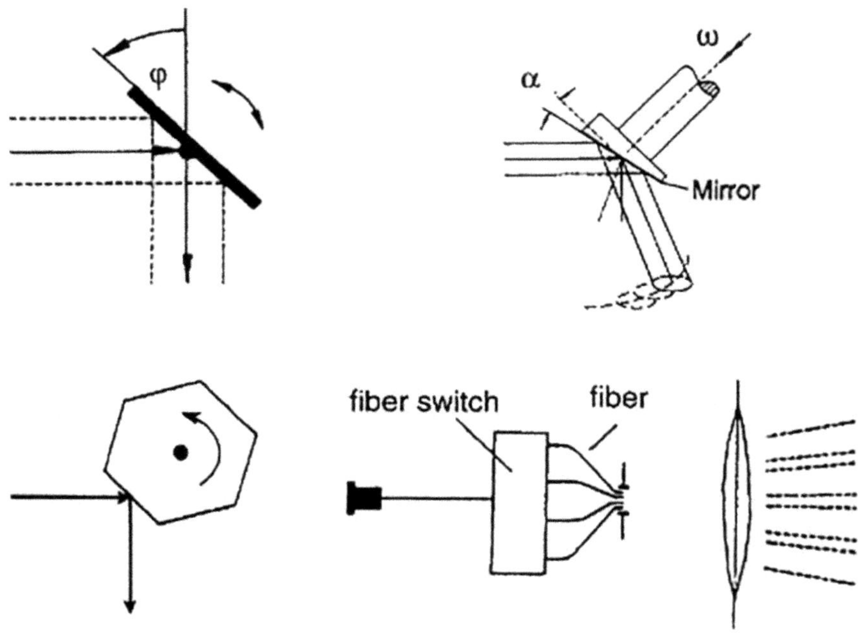

FIGURE 3.20
Various kinds of commonly used laser scanners (clock-wise) oscillating mirror scanner, Palmer scanner, fibre scanner and rotating polygon (Source: Adapted from Wehr and Lohr (1999)).

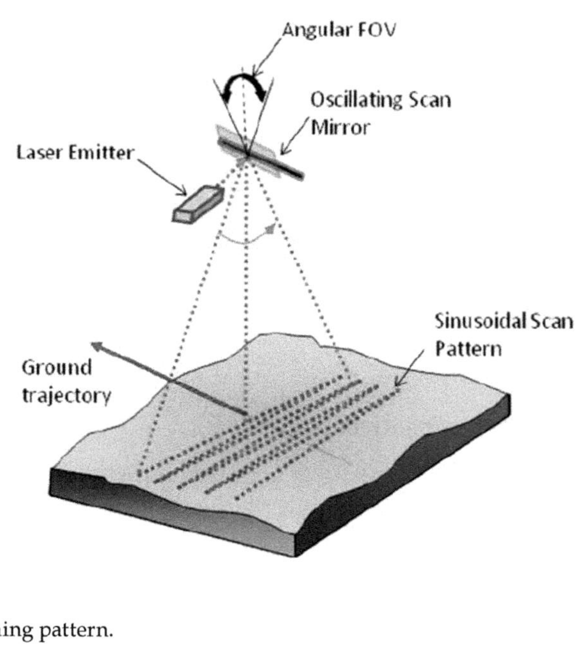

FIGURE 3.21
Oscillating mirror scanning pattern.

3.4.3.2 Rotating Polygon Scanning Mechanism

In this system, a multifaceted polygon prism or scan mirror continuously rotates around an axis of rotation. The facets of the polygon combined with its rotation direct the laser energy toward the ground. Like the oscillating system, this is generally used to sweep perpendicular to the platform trajectory generating a swath width on the ground and relying on platform motion in the along-track direction to generate coverage. However, rather than relying on an oscillating motion requiring accelerations and decelerations, the facet of the polygon controls the pointing of the continuously rotating system. As the laser energy transfers from one polygon facet to the next, there is a discontinuous and sudden jump to the opposite side of the scan, resulting in a scan pattern consisting of a series of nearly parallel scan lines (Figure 3.22).

3.4.3.3 Nutating Mirror Scanning System

In this kind of scanning, the scanning mechanism uses a nutating mirror which is inclined with reference to the light from the laser emitter, as shown in Figure 3.23. The rotation of this mirror creates an elliptical scan pattern on the ground, and the forward motion of the sensor creates coverage in the along-track direction.

3.4.3.4 Fiber Pointing System

Some of the LiDAR systems use fiber channels to direct the laser energy to the ground. The basic objective is to achieve more stable scan geometry due to the fixed relationship between the fiber channels and the other LiDAR components. In this system, the laser light is directed to the ground by a glass fiber bundle. The scan direction for a given pulse is dependent on the fiber channel it is emitted from. A similar system of fiber bundles is then used in the receiving optics (Figure 3.23).

The capabilities, i.e., repetition rate, scan frequency, scan angle, point density, of the above scanners are very similar; the main difference lies in the scanning pattern.

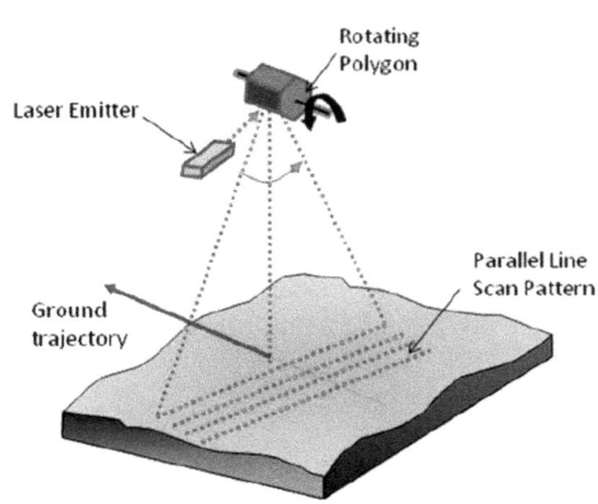

FIGURE 3.22
Rotating polygon scanning pattern.

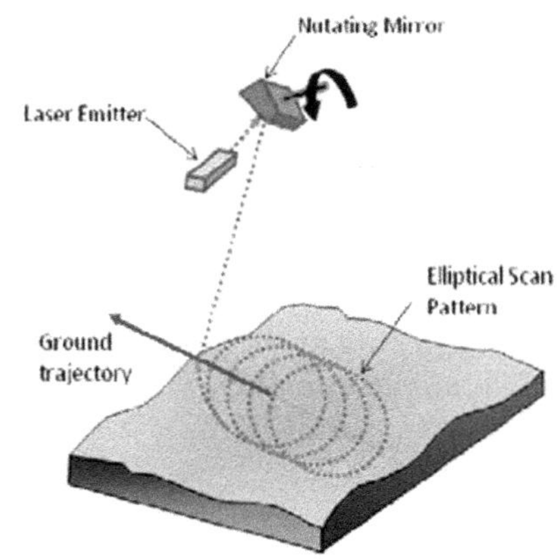

FIGURE 3.23
Nutating mirror scanning pattern.

3.4.3.5 Spaceborne LiDAR Mission

The CloudSat and the Cloud-Aerosol Lidar and Infrared Pathfinder Satellite Observation (CALIPSO) spacecraft have been launched by NASA to study the role that clouds and aerosols play in regulating Earth's weather, climate, and air quality. The satellites were launched on April 28, 2006, into a 705-km circular sun-synchronous polar orbit, and fly in orbital formation as part of the "A-Train" constellation of three other Earth Observing satellites, including Aqua, Aura, and Centre National d'Etudes Spatiales' (CNES) PARASOL. Together, the A-Train satellites are increasing our understanding of the climate system and the potential for climate change. The CloudSat mission designed lifetime is 22 months to enable more than one seasonal cycle to be observed, although the radar lifetime is expected to approach 3 years. CALIPSO is planned for 3 years of on-orbit operation. CloudSat and CALIPSO are collecting information about the vertical structure of clouds and aerosols unavailable from other Earth observing satellites. Their data are improving our models and providing a better understanding of the human impact on the atmosphere (Figure 3.24).

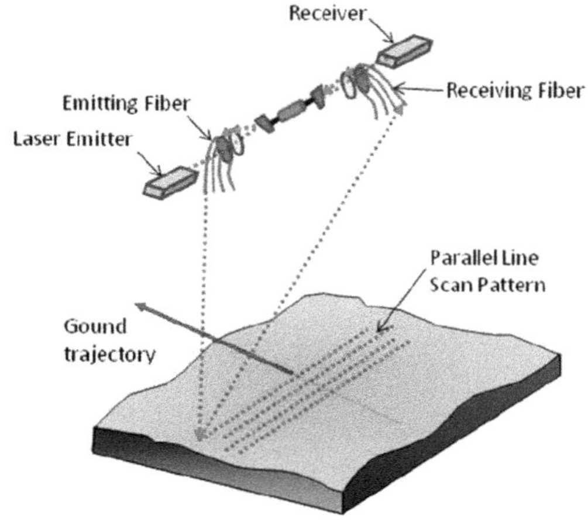

FIGURE 3.24
Fiber scanning pattern.

3.4.3.6 Cloud Profiling Radar (CPR)

CPR is a 94 GHz nadir-looking millimeter-wave radar that measures the power backscattered by clouds as a function of distance from the radar (Figure 3.25). The CPR aims at providing information on the vertical structure of all cloud systems. Clouds are weak scatterers of microwave radiation especially in contrast to the reflection of the underlying Earth's surface. Hence, maximum sensitivity of the CPR is required. Furthermore, the overriding requirement on CPR was to achieve a minimum detectable cloud reflectivity factor (Z_e) of –28 dBZ.

The design of the CPR consists of RFES Radio Frequency Electronics Subsystem (RFES), High Power Amplifier (HPA), Antenna Subsystem (Quasi-Optical Transmission Line), and DSS Digital Subsystem (DSS). The RFES consists of an up-converter which generates a pulsed signal and up-converts it to 94 GHz. The signal is amplified to about 200 mW by a state-of-the-art MMIC power amplifier. The receiver portion of the RFES down-converts the signal to an intermediate frequency (IF). The IF signal is detected using a logarithmic amplifier with high dynamic range (Figure 3.25).

The HPA amplifies the transmitted pulse to a nominal power level of 1.7 kW and consists of an extended interaction klystron (EIK) and a high-voltage power supply (HVPS). The HVPS provides 20 kV needed to operate the EIK and provides telemetry data necessary to system needs. The design uses a boost supply to minimize input current transients during the pulsing period and control EMC problems.

3.5 Conclusion

Timely and reliable information on nature, extent, magnitude, spatial distribution, and its dynamics is a prerequisite for rehabilitation and management of degraded lands. Active remote sensing technology plays a crucial role in providing such information on degraded

FIGURE 3.25
Illustration of the CloudSat spacecraft (Credit: NASA/JPL).

lands especially in regions with perpetual overcasting conditions. The SAR interferometry, as well as LiDAR technology, offers immense potential in generating precise information on terrain's slope—a key variable in assessment and modeling soil erosion and for taking up preventive and corrective measure for rehabilitation of eroded land and mining areas. In addition, the digital terrain models developed from interferometric SAR and LiDAR data would enable generating 3D models of ravines—a network of gullies (very severely eroded land) which in turn would facilitate implementation of rehabilitation program of ravines.

Land degradation is very common in dryland regions lacking adequate soil moisture due to low and scanty rainfall, for supporting vegetation growth. Dedicated missions like SMOS and SMAP would serve as valuable tools in assessment and monitoring of drought events which incidentally occur during monsoon season with perpetual cloud cover where optical sensor data very often fail to provide soil moisture status.

The rehabilitation of degraded land program would certainly benefit from advancements. Further research and developments in active sensor technologies which are underway will augment the applications of active remote sensing technology in assessment, monitoring, and modelling land degradation.

References

Fernandez-Diaz, J.C., 2011. Lifting the canopy veil—Airborne LiDAR for archeology of forested areas. *Imaging Notes*, 26(2):31–34 pp.

Franceschetti, G., and Lanari, R., 1999. *Synthetic Aperture Radar Processing.* CRC Press, BocaRaton, FL.

Henderson, F.M., and Lewis, A.J., (eds.) 1998. Principles & applications of imaging radar. *Manual of Remote Sensing*, 3rd Edition, Volume 2. John Wiley & Sons, Inc., New York.

Hug, C., Krzystek, P., and Fuchs, W., 2004. Advanced LIDAR Processing with Las Tools. IAPRS, 35, Part B, pp. 832.

Kerr, Y.H., Waldteufel, P., Wigneron, J.P., Martinuzzi, J., Font, J., and Berger, M., 2001. Soil moisture retrieval from space: The Soil Moisture and Ocean Salinity (SMOS) mission. *IEEE Transactions on Geoscience and Remote Sensing*, 39(8):1729–1735.

Kramer, H.J., 1994. *Observation of the Earth and its Environment.* Springer-Verlag, New York.

Long, D.G., and Drinkwater, M.R., 2000. Azimuth variation in microwave scatterometer and radiometer data over Antarctica. *IEEE Transactions on Geosciences and Remote Sensing*, 38(4):1857–1870.

Rosmorduc, V., Benveniste, J., Bronner, E., Dinardo, S., Lauret, O., Maheu, C., Milagro, M., Picot, N., Ambrozio, A., Escolà, R., Garcia-Mondejar, A., Restano, M., Schrama, E., Terra-Homem, M., Radar Altimetry Tutorial; Benveniste, J., and Picot, N., (eds.) 2016, European Space Agency, CNES. www.altimetry.info.

Spencer, M.W., Wu, C., and Long, D.G., 2000. Improved resolution backscatter measurements with the Sea Winds pencil-beam scatterometer. *IEEE Transactions on Geosciences and Remote Sensing*, 38(1):89–104.

Wehr, A. and Lohr, U., 1999. Airborne laser scanning—an introduction and overview. ISPRS Journal of Photogrammetry and Remote Sensing, 54:68–82.

4

Digital Image Processing

4.1 Introduction

Image processing is a method to perform some operations on an image, to either restore or enhance the image or to extract some useful information from it. It is a type of signal processing with an image as input and output in the form of an image or characteristics/features associated with that image. There are two types of methods used for image processing, namely analog and digital image processing. Analog image processing is applicable for hard copies like printouts and photographs. A digital image is a representation of a 2D image as a finite set of building blocks called picture elements or pixels. An image may be defined as a 2D function, f(x, y), where x and y are spatial (plane) coordinates and the amplitude of 'f' at any pair of coordinates (x, y) is called the intensity or gray level of the image at that point. When x, y, and the amplitude values of 'f' are all finite, discrete quantities, we call the image a digital image. In physical form, a *digital image* is a 2D array of small areas called *pixels*, or *pels*. The horizontal rows of pixels are called lines, and the vertical columns of pixels are termed *samples*. Hence, the array consists of j lines running from top to bottom and s samples running from left to right (Figure 4.1).

The digital images recorded by the sensor/(s) onboard sensor platforms represent the reflected/emitted/backscattered radiation from the earth's surface. In the case of sensors mounted aboard airborne platforms, digital image can be retrieved after completion of the aerial mission, while those recorded by spaceborne sensors are digitized onboard and transmitted to the ground station (Figure 4.2).

Owing to instability of the air- and spaceborne platforms, and nonuniform performance of sensors, the digital images acquired by the sensors aboard aerial or satellites are subject to geometric and radiometric distortions. Geometric distortions are related basically to earth's rotation, earth's curvature, and platform's attitude. Included in radiometric distortions are atmospheric effects, sensor calibration errors, and nonlinearity of sensors response. In the case of airborne platforms, geometric distortions due to earth's rotation and earth's curvature are not the major issue owing to lower altitude. Similarity, radiometric distortion due to atmospheric effects in airborne sensors data is not significant since the atmospheric column is very small for airborne platforms. Hence before utilizing the spaceborne digital images for generating information on natural resources and environment or any other terrain features, they must be processed to correct these errors so as to make them represent the faithful record of reflected/emitted/backscattered radiation from the terrain features. The process of removal of geometric and radiometric errors is referred to as digital image pre- processing.

Digital image processing is the use of computer algorithms to perform image processing on digital images. The objective of image processing is to generate a faithful

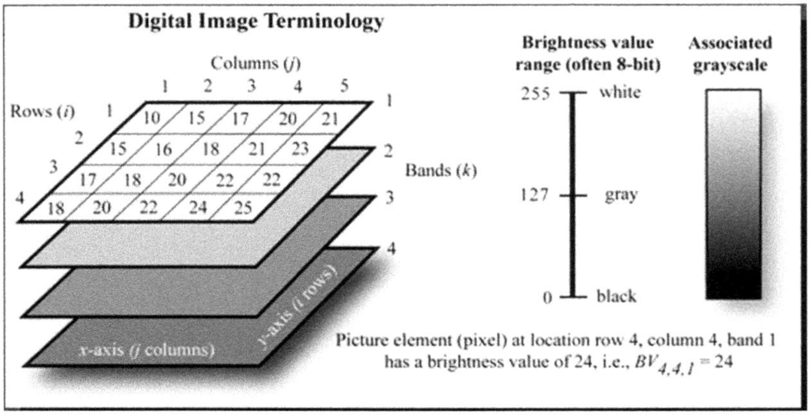

FIGURE 4.1
Digital image (Source: Adapted from Jensen (2005)).

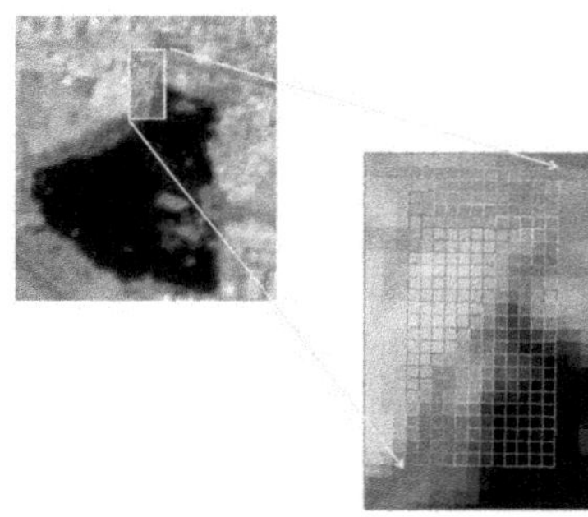

	0	1	2	3	4	5	6	7	8	9	10
0	203	206	220	210	202	200	199	215	222	222	229
1	201	219	236	227	225	221	222	235	235	249	238
2	186	185	187	189	200	201	211	230	230	246	247
3	177	168	163	167	158	155	163	163	178	194	214
4	194	190	183	172	169	168	162	162	162	160	156
5	162	169	184	193	197	198	198	195	188	179	170
6	147	151	174	178	193	208	208	211	204	194	189
7	157	181	219	212	210	210	215	214	214	210	196
8	166	194	217	224	232	232	230	230	245	239	210
9	167	201	215	226	226	235	227	232	259	253	219
10	172	213	218	224	224	222	216	211	212	202	180
11	204	221	218	218	223	221	210	183	148	120	134
12	222	221	217	217	217	212	187	165	135	113	117
13	213	216	216	212	206	194	169	166	145	112	97
14	196	214	214	212	209	198	185	189	163	117	91
15	204	204	215	201	207	208	203	196	165	141	110
16	199	199	200	183	194	197	194	171	151	153	119
17	191	203	198	182	195	191	178	142	125	121	92
18	189	207	192	190	202	189	164	117	101	93	78
19	187	189	173	182	180	170	152	102	90	81	76
20	189	175	177	187	178	170	153	110	89	81	78
21	199	181	186	194	188	184	162	122	88	82	79
22	188	171	185	197	183	179	148	111	87	79	76

Spectral band- 4(0.77-0.86 μm)

	0	1	2	3	4	5	6	7	8	9	10
0	120	118	115	114	114	117	121	121	111	111	102
1	133	128	126	120	113	112	110	101	101	97	97
2	147	140	135	128	124	124	126	116	116	95	93
3	152	140	136	133	131	126	124	124	118	104	102
4	156	150	145	145	144	140	140	140	131	122	114
5	126	139	149	161	168	163	163	161	146	141	132
6	126	143	158	163	172	170	170	160	143	131	120
7	150	180	198	191	189	189	171	149	130	106	97
8	158	191	204	197	190	190	172	145	114	89	86
9	160	181	182	177	177	173	160	145	110	88	89
10	167	178	176	173	173	172	157	137	108	89	96
11	189	191	194	194	188	177	155	126	99	85	94
12	193	195	195	195	186	168	135	106	90	82	86
13	179	184	184	186	170	141	103	90	86	79	80
14	170	175	175	176	157	122	100	95	85	79	78
15	168	168	162	143	131	120	107	98	84	81	79
16	152	152	139	110	107	121	121	103	83	84	78
17	160	151	134	102	107	131	127	99	81	81	76
18	161	149	127	101	112	127	118	90	75	75	74
19	156	126	103	96	103	106	100	80	74	74	73
20	147	110	91	89	92	91	87	78	74	74	74
21	144	107	87	88	89	91	84	75	74	75	75
22	127	94	85	88	87	86	78	72	74	74	75

Spectral band- 3(0.62-0.68 μm)

	0	1	2	3	4	5	6	7	8	9	10
0	130	128	129	130	125	126	124	121	116	116	112
1	139	138	134	131	127	125	125	119	119	116	113
2	148	142	140	142	140	136	132	122	122	119	113
3	154	143	136	134	139	136	133	133	130	127	122
4	155	147	148	151	145	141	139	139	139	133	124
5	139	138	145	153	157	158	158	156	147	138	129
6	130	139	159	163	166	164	164	155	141	135	129
7	146	164	189	182	176	176	166	147	136	123	114
8	153	178	198	194	185	185	166	148	134	115	107
9	158	176	183	178	178	170	159	148	128	112	108
10	166	176	178	173	173	164	156	143	121	108	108
11	176	182	185	185	179	168	156	139	115	106	105
12	183	187	190	190	181	165	143	124	111	103	104
13	174	176	176	181	171	150	124	114	112	104	103
14	161	165	165	167	157	141	120	115	111	104	101
15	156	156	160	151	139	133	126	118	109	104	101
16	153	153	151	130	122	128	130	122	106	103	100
17	162	151	144	125	119	128	134	124	101	98	97
18	163	156	143	121	122	131	132	113	96	96	96
19	161	143	126	115	121	118	120	106	96	97	96
20	158	130	116	112	112	111	111	103	97	96	95
21	152	123	114	112	112	111	108	101	97	97	96
22	139	113	108	111	108	108	105	98	94	97	95

Spectral band- 2(0.52-0.59 μm)

FIGURE 4.2
An illustration of the contents of a Resourcesat-2 LISS-IV digital data (digital numbers—DN values) for vegetation, soils, and water body. Note the low DN values for water body in columns 8, 9, and 10, and rows 18–20 in all the spectral bands.

record of the digital image acquired by air/spaceborne sensors. Digital image processing encompasses three major operations: (1) *image restoration or preprocessing*—to correct a degraded digital image to its intended form, (2) *image enhancement*—to improve the pictorial quality of the image or detectability of objects or patterns in a digital image for interpretation/analysis, and (3) *image classification*—to use quantitative decision rules to classify or identify objects or patterns on the basis of their spectral response patterns. For detailed discussions of digital image processing and its mathematical concepts, the readers may refer to Jensen (2007), Gao (2009), Campbell et al. (2011), Richards (2013), and Lillesand et al. (2015).

4.2 Data Storage Media

In the mid-1980s, ½- and ¼-inch magnetic tapes with a data storage capacity of 1,600 bytes per inch were commonly used for data distribution and even as media of permanent storage. These tapes were gradually replaced by more compact magnetic tapes such as 8 mm Exabyte and DAT tapes in the early to mid-1990s. These media had a much larger storage capacity (e.g., up to 1 GB) than ½- or ¼-inch tapes. Now these media have been replaced by CD, DVD, and memory sticks with a larger capacity than magnetic tapes and improved reliability and flexibility. The storage media is dealt with in detail by Gao (2009).

4.2.1 Compact Disc

Currently three major types of CDs, namely CD read-only memory (CD-ROM), CD recordable (CD-R), and CD rewritable (CD-RW), are commonly used. With a data storage capacity upto 1 GB, CD-ROM is a write-once-read-many storage medium that is compact, indelible, and highly reliable, with a life expectancy in excess of 30 years in a normal storage environment. The most common size is 670 MB. Recently, two types of CD, namely CD-R and CD-RW, have been introduced.

4.2.2 Digital Versatile Disk

Once data are recorded on CD-R it becomes permanent and the media cannot be used again. To overcome this limitation, special rewritable compact discs (CD-RW) have been developed. Like a CD, a DVD is also an optical, read-only storage, recordable and rewritable data storage media. It has the same physical dimension as a CD. A DVD has four storage capacities, namely 4.7 GB, 8.54 GB, 9.4 GB, and 17.08 GB, depending on the disk structure.

4.2.3 Memory Sticks

Universal Serial Bus (USB) memory sticks, also known as USB flash drives or pen drives, are even more portable and less subject to physical damage than a CD, because it is encapsulated inside a plastic shield with no parts moveable. The storage capacity of memory sticks usually ranges from 512 MB to 1 GB. However, larger capacities such as 2 and 4 GB, and even up to 32 GB, are also available.

4.3 Digital Data Format

A remote sensing image may be stored in one of many graphic formats depending upon the image processing system being used, as each image processing system has its own proprietary data format. Till the mid-1990s digital remote sensing data were recorded in magnetic tapes and digital audio tapes (DATs). The digital data were recorded in three formats, namely band sequential (BSQ), band interleaved by lines (BIL) and band interleaved by pixel (BPP). In the case of (BSQ) format, the DN values of all pixels in the first band are arranged one after the other until the whole band is completed. These are followed by the DNs of the second band and subsequently by the rest of the spectral bands. In BIL the DNs are organized by lines instead of bands, alternating consecutively the line for each band before starting with the next line of the image. After DNs of line 1 in band 1 are arranged, the values for line 1 in band 2 are placed, followed by line 1 of band 3 and so on until all the bands are completed. Then line 2 of the first band continues in the same manner with the rest of the image lines. On the other hand, in BIP format, instead of DNs being alternated by lines, they are arranged by pixels. Thus, the DN corresponding to the pixel of line 1, column 1 and band 1 is followed by the DN corresponding to line 1, column 1 and band 2, and line 1, column 1 and band 3, etc.

Now with the availability of new media for digital image data recording like CDs, DVDs, and memory sticks or pen drives, several new formats for data recording are in vogue (Gao, 2009). A brief overview of available image formats is given in the following sections.

4.3.1 Generic Binary

In the generic binary format, all image pixel values are represented as binary data of 0s and 1s without header information. Each pixel is represented as a byte. Ancillary image information, such as the number of rows and columns and the number of spectral bands, is stored in the header file.

4.3.2 Graphic Interchange Format

A Graphic Interchange Format (GIF) file is made up of several parts, including a screen descriptor, global color map, image descriptor, local color map, and finally raster data. The last three parts are repeated many times. The screen descriptor contains all ancillary information about the image, such as the number of bits used, image width and height, and background color.

4.3.3 JPEG

The Joint Photographic Experts Group (JPEG) format allows for storage of frame images of continuous tone in binary, grayscale, or color. This format is particularly suited for those images that must be reduced to a very small size through image compression. There are three coding systems (Gonzalez and Woods, 1992):

- A lossy baseline coding system that is adequate for most compression requirements,
- An extended coding system for greater compression, and
- A lossless independent coding system for reversible compression.

4.3.4 TIFF and GeoTIFF

The Tagged Image File Format (TIFF) format is extendable, portable, and revisable. New image types can be added without invalidating older types. Besides, the addition of new informational fields to the format will not affect the ability of older applications to read the images. This format, independent of the platform and operating system, can be used as an internal one for image editing and swapping. All TIFF images are made up of three components: the header, the image file, and the tag (Davenport and Vellon, 1987).

The GeoTIFF interchange standard is an extension of the popular TIFF format, to support geo-referenced remote sensing data (Ritter and Ruth, 1997). This standard unifies various internally represented transformations between raster data and the reference coordinate frame and guarantees accessibility to images stored in the conventional TIFF format, as well as all additional data needed for geo-referencing or geocoding independent of the TIFF image data Gao (2009).

4.3.5 Portable Network Graphics

The Portable Network Graphics (PNG) format was developed as an open format to replace GIF. It allows for a full range of color and better compression. It is used almost exclusively for web images. For photographs, PNG is not as good as JPEG, because it creates a larger file (www.ivanexpert.com/blog/2010/05/the-5-types-of-digital-image-files-tiff-jpeg-gif-png-and-raw-image-files-and-when-to-use-each-one/, accessed on June 18, 2017).

4.4 Image Preprocessing

The image restoration, also often termed *image preprocessing*, operations aim at correcting distorted or degraded image data to create a more faithful representation of the original scene. The process typically involves the initial processing of raw image data to correct for geometric distortions, to calibrate the data radiometrically, and to eliminate noise present in the data due to sensor's inconsistent performance and atmospheric effects. The nature of such corrections, obviously, varies considerably with such factors as the digital image acquisition type—digital camera/along-track scanner/across-track scanner; platform—airborne versus spaceborne satellite; and total field of view. Important image preprocessing routines include the following:

1. *Radiometric correction* for variations in the image resulting from environmental conditions (e.g., haze) or sensor anomalies,
2. *Geometric correction* to compensate for the Earth's rotation and for variations in the position and attitude of the sensor craft, and
3. *Correction for terrain's relief distortions* with the help of digital elevation model.

4.4.1 Radiometric Correction

4.4.1.1 Atmospheric Effects

Ideally, in the absence of an atmosphere, the signal measured by the sensor will be a function simply of the level of energy from the sun, actually incident on the pixel, and the reflectance properties of the target. However, the presence of the atmosphere modifies the radiance measurements made by the sensors significantly in two contradictory ways. First, it attenuates (reduces) the incoming solar radiation illuminating a ground object. Second, it acts as a reflector itself, adding a scattered, extraneous "path radiance" to the signal detected by a sensor. Thus, the composite signal observed by a sensor at any given pixel location can be expressed by

$$L_{tot} = \frac{\rho ET}{\pi} + L_p \tag{4.1}$$

where L_{tot} is the total spectral radiance measured by sensor, ρ is the reflectance of object, E is the irradiance on object, T is the transmission of atmosphere, and L_p is the path radiance. (All of the above quantities are wavelength dependent.)

Only the first term ($\rho ET/\pi$) in the above equation represents the radiation reflected by aground feature. The second term (L_p) denotes the scattered path radiance, which introduces "haze" in the image and consequently reduces image contrast.

The radiometric correction procedures to account for errors introduced in measured radiance due to atmospheric effects could be grouped into two rather broad categories: (1) *absolute atmospheric correction* and (2) *relative atmospheric correction.*

4.4.1.2 Absolute Atmospheric Correction

The atmospheric corrections can be carried out by using model atmosphere. The assumed atmosphere is calculated based on the time of year, altitude, latitude, and longitude of the study area. This approach may be successful when atmospheric attenuation is relatively small compared with the signal from the terrain being imaged (Cracknell and Hays, 1993). Another approach involves the use of a model atmosphere in conjunction with *in situ* atmospheric measurements made at the time of remote sensing data acquisition. The atmospheric model then can be calibrated using local atmospheric conditions. This is referred to as *absolute atmospheric correction.*

The atmospheric attenuation can be minimized by using multiple looks at the same object from different vantage point (e.g., fore and aft) or by looking at same object using multiple spectral bands of the spectrum. The purpose is to have the information from multiple looks or multiple bands to cancel out the atmospheric effects. The multi-look approach suffers from the fact that the atmospheric paths for multiple looks (e.g., for and aft) may not be the same. This is called *relative atmospheric correction.*

Numerous atmospheric correction models have been developed based on the use of radiative transfer principles. Some of the most important are MODTRAN, ATREM, FLAASH, 6S, and ATCOR.

4.4.1.2.1 MODerate Resolution Atmospheric TRANsmission

MODerate resolution atmospheric TRANsmission (MODTRAN) is a model for estimating atmospheric transmission of electromagnetic radiation under specified conditions. The model estimates atmospheric emission, thermal scattering, and solar scattering including Rayleigh, Mie, single, and multiple scattering, incorporating effects of molecular absorbers and scatterers, aerosols, and clouds for wavelengths from the ultraviolet region to the far infrared. It uses various standard atmospheric models based on common geographic locations and also permits the user to define an atmospheric profile with any specified set of parameters. There are several options in the model for specifying prevailing aerosols, based on common aerosol mixtures encountered in terrestrial conditions (e.g., rural, urban, maritime). Within MODTRAN, the estimate of visibility serves as an estimate of the magnitude of atmospheric aerosols (http://modtran.org; www.kirtland.af.mil/library/factsheets/factsheet.asp?id=7915).

4.4.1.2.2 ATmospheric CORection

The *ATmospheric CORection (ATCOR)* model offers options specifically tailored for many of the more common satellite and aircraft sensors, extraction of specific band ratios, and several measures of net radiation, surface flux, albedo, and reflectance. ATCOR consists of ATCOR 2, used for flat terrain, and ATCOR 3, used for rugged terrain, e.g., the 3 denotes 3 dimension. ATCOR-4 is designed for aircraft systems including both optical and thermal sensors and accommodates more rugged terrain. The model uses MODTRAN4+ (Alder-Golden et al., 1999) for calculation of look-up tables (LUT) of the atmospheric correction functions (path radiance, atmospheric transmittance, direct and diffuse solar flux).

4.4.1.2.3 Second Simulation of the Satellite Signal in the Solar Spectrum (6S)

The 6S atmospheric correction model simulates the signal observed by a satellite sensor for a Lambertian target at mean sea level. The model is widely used in a variety of atmospheric correction algorithms, including that developed for the Moderate Resolution Imaging Spectroradiometer (MODIS) surface reflectance products and the atmospheric correction algorithm (ATREM). The 6S presents a robust and vetted solution to the radiative transfer equation. The important features of the model include its ability to take into account (1) target altitude, (2) polarization by molecules and aerosols, (3) nonuniform targets, and (4) the interaction of the atmosphere and the BRDF of the target. For further details of the model readers may refer to Vermote et al. (1997) and the user manual, available online at http://6s.ltdri.org (Campbell and Wynne, 2011).

4.4.1.2.4 ATmospheric REMoval Model

The ATmospheric REMoval (ATREM) model was developed for retrieving scaledsurface reflectance from hyperspectral data using a radiative transfer model (Gao and Goetz, 1990).

The model calculates the molecular (Rayleigh) scattering present using Second Simulation of the Satellite Signal in the Solar Spectrum (6S) and a user-specified aerosol model.

Atmospheric scattering is modeled using the Second Simulation of the Satellite Signal in the Solar Spectrum (6S) radiative transfer model (Tanre et al., 1986) and a user-specified aerosol model. ATREM calculates the atmospheric absorption term using the Malkmus narrow-band model with a user-selected standard atmospheric model or a user-supplied atmospheric model (Gao et al., 1999). The amount of water vapor is derived on a pixel-by-pixel from the hyperspectral data using the 0.94 and/or 1.13 μm water vapor bands and a three channel rationing technique. The derived water vapor values are then used for modeling water vapor absorption effects over the entire 400–2,500 nm region.

4.4.1.2.5 Fast Line-of-sight Atmospheric Analysis of Spectral Hypercubes

Fast Line-of-sight Atmospheric Analysis of Spectral Hypercubes (FLAASH) is a MODTRAN4-based atmospheric correction model (Adler-Golden et al., 1999). The model provides accurate derivation of apparent surface reflectance through derivation of atmospheric properties such as surface albedo, surface altitude, water vapor column, aerosol and cloud optical depths, surface and atmospheric temperatures from hyperspectral data. FLAASH operates in the 0.4–2.5 μm spectral range. It also produces a column water vapor image, a cloud map and a visibility value for the image (Research Systems Inc., 2003).

4.4.1.3 Relative Atmospheric Correction

Another approach to atmospheric correction of remotely sensed images is based on examination of spectra of objects of known or assumed brightness recorded by multispectral imagery (Campbell and Wynne, 2011). This is known as *image-based atmospheric correction,* because it attempts to adjust for atmospheric effect solely, or mainly, from evidence available within the image itself. The image-based atmospheric correction involves identification of a very dark object or feature within the scene. Such an object might be a large water body or possibly shadows cast by clouds or by large topographic features. In the infrared portion of the spectrum, both water bodies and shadows should have brightness at or very near zero, because clear water absorbs strongly in the near-infrared spectrum and because very little infrared energy is scattered to the sensor from shadowed pixels.

A quick perusal of the histograms/DN values of these features reveals that they have the lowest values, which are not zero but some larger value. Typically, this value will differ from one band to the other. These values, assumed to represent the value contributed by atmospheric scattering for each band, are then subtracted from all digital values for that scene and that band. Thus the lowest value in each band is set to zero. Such dark objects form one of the most common examples of pseudo *invariant objects* or *pseudo invariant features* (PIFs) which are taken as approximations of features with spectral properties that do not change significantly over time, so their spectral characteristics sometimes can form the basis for normalization of multispectral scenes to a standard reference. The approach is also known as the *histogram minimum method (HMM) or the dark object subtraction* technique. Chavez (1988) proposed further modification in the strategy in which an analyst selects a relative atmospheric scattering model that estimates scattering effects across all the spectral bands, given an estimate of the value of an initial band. He proposed use of the cosine of the solar zenith angle (cosine estimation of atmospheric transmittance [COST]) to supplement the *dark object subtraction* (DOS) estimate of atmospheric transmittance (Chavez, 1996). Song et al. (2001) proposed a method that adds the effect of Rayleigh scattering to conventional DOS.

4.4.1.4 Instrumental Errors

Radiometric errors within a band and between bands can also be caused by the design and operation of the sensor system. The radiometric errors by sensor anomaly within a band can be quite severe and often require correction to render an image product useful for further analysis. The most significant of these errors is related to the detector system. The transfer characteristic (radiation in, signal out) of an ideal detector should be linear so that there is proportional increase and decrease of signal with detected radiation level. However, in real-world scenario there is some degree of nonlinearity in the sensor that will also give a small signal out even when no signal is detected. This phenomenon is known as *"dark current"* and is related to the residual electronic noise in the system at any temperature above absolute zero. It is called an "offset." The slope of the characteristic is called its transfer gain or just simply "gain."

The third approach examines the brightness of objects within each scene but attempts to utilize knowledge of interrelationships between different spectral bands. Chavez (1975) devised a procedure that paired values from each band with values from a near-infrared spectral channel. The Y intercept of the regression line is then taken as the correction value for the specific band in question. The regression technique can be applied to local areas (of possibly only 100–500 pixels each), ensuring that the adjustment is tailored to conditions important within specific regions. An extension of the regression technique is to examine the variance–covariance matrix, the set of variances and covariances between all band pairs on the data known as covariance matrix method (CMM) (Switzer et al., 1981).

4.4.2 Corrections for Solar Illumination Variation

For generating mosaics of images taken in optical region at different times or to study the changes in the reflectance of ground features at different times or locations, it is usually necessary to apply a *sun elevation correction* and an *earth-sun distance correction.* The sun elevation correction accounts for the seasonal position of the sun relative to the earth. Image data acquired under different solar illumination angles are normalized by calculating pixel brightness values assuming the sun was at the zenith on each date of sensing. The correction is usually applied by dividing each pixel value in a scene by the sine of the solar elevation angle for the particular time and location of imaging. Alternatively, the correction is applied in terms of the sun's angle from the zenith. Zenith angle is simply 90° minus the solar elevation angle. In this case each pixel value is divided by the cosine of the sun's angle from the zenith, resulting in the identical correction.

The earth-sun distance correction is applied to normalize for the seasonal changes in the distance between the earth and the sun. The earth-sun distance is usually expressed in astronomical units (approximately 149.6×10^6 km). The irradiance from the sun decreases as the square of the earth-sun distance. Ignoring atmospheric effects, the combined influence of solar zenith angle and earth-sun distance on the irradiance incident on the earth's surface can be expressed as

$$E = \frac{E_0 \cos\theta_0}{d^2} \tag{4.2}$$

where E is the normalized solar irradiance; E_0 is the solar irradiance at mean earth-sun distance; θ_0 is the sun's angle from the zenith; and d is the earth-sun distance, in astronomical units.

4.4.3 Noise Removal

Image noise is any unwanted disturbance in image data. The noise in the image can be seen for various reasons including noncompliance of the detectors, improper calibration of the detectors, interdetector variability, and the low signal-noise ratio in the image acquired by the sensor (INPE and CAST, 2004). Noise can either degrade or totally mask the true radiometric information content of a digital image. Hence, noise removal usually precedes any subsequent enhancement or classification of the image data. The objective is to restore an image to as close an approximation of the original scene as possible (Lillesand et al., 2004).

One of the radiometric distortions related to sensor failure is striping or banding: horizontal or vertical striping. Horizontal striping show up in the case of whiskbroom or across-track scanning sensors like Landsat-MSS/TM/ETM+/Terraor Aqua MODIS, whereas vertical banding is common in sensors with push-broom scanning mechanism. Missing lines in data due to sensor failure results in loss of a line of DN values—every 16th line for TM data, as there was 16 sensors for each band, scanning 16 lines at a time or 6th line for MSS (Figure 4.3). Such problems were particularly prevalent in the collection of early Landsat-MSS data. This stems from variations in the response of the individual detectors used within each band.

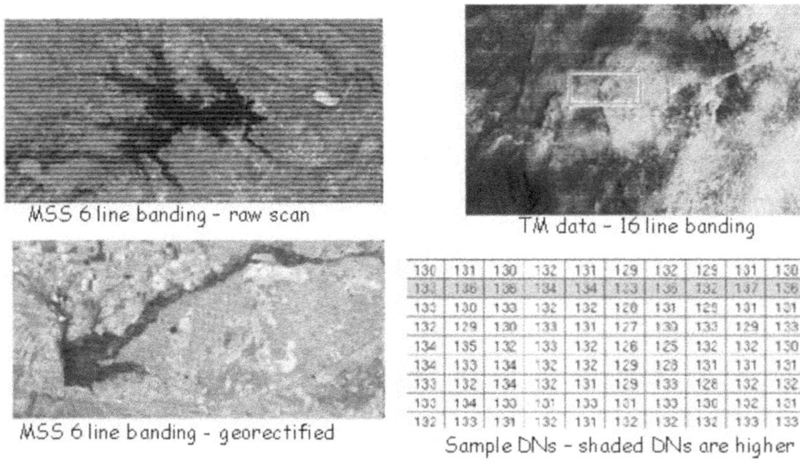

FIGURE 4.3
Stripping in satellite image and its correction.

While the six detectors used for each band were carefully calibrated and matched prior to launch, the radiometric response of one or more tended to drift over time, resulting in relatively higher or lower values along every sixth line in the image data. In this case valid data are present in the defective lines, but they must be normalized with respect to their neighboring observations. Striping noise can be removed using spatial curve functions.

As mentioned earlier in this sections, vertical striping is common in digital data captured by push-broom sensors. An illustration of vertical striping band 2 of Resourcesat-2 LISS-IV image is given in Figure 4.4.

Another line-oriented noise problem sometimes encountered in digital data is *line drop*. In this situation, a number of adjacent pixels along a line (or an entire line) may contain spurious DNs. This problem is normally addressed by replacing the defective DNs with the average of the values for the pixels occurring in the lines just above and below. Alternatively, the DNs from the preceding line can simply be inserted in the defective pixels (Figure 4.5).

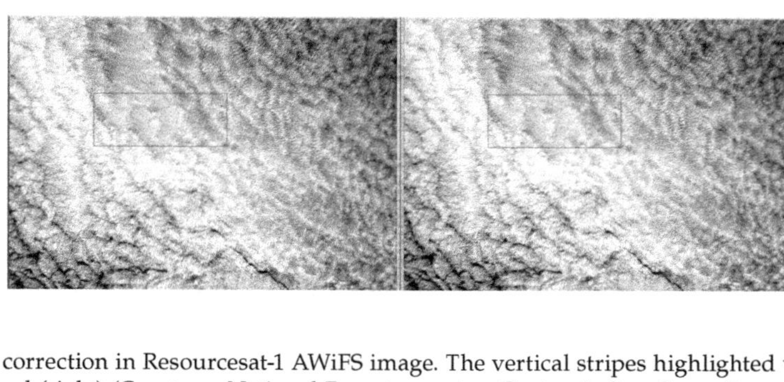

FIGURE 4.4
Vertical striping correction in Resourcesat-1 AWiFS image. The vertical stripes highlighted with red box (left) have been removed (right) (Courtesy: National Remote sensing Centre, Indian Space Research Organization, Department of Space, Government of India).

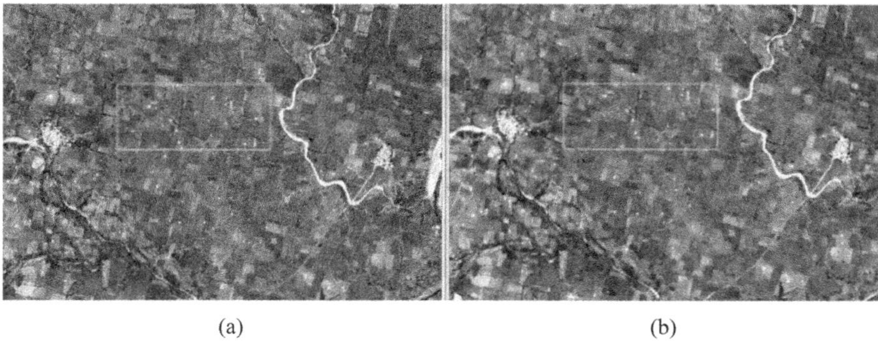

FIGURE 4.5
Noise correction in spectral band 2 of Resourcesat-2 LISS-IV image. The image in the left (a) displays vertical stripping—characteristics of push-broom sensors—and (b) shows the image after employing necessary noise corrections (Courtesy: National Remote sensing Centre, Indian Space Research Organization, Department of Space, Government of India).

Random noise problems in digital data are handled quite differently than those we have discussed to this point. This type of noise is characterized by nonsystematic variations in gray levels from pixel to pixel called *bit errors*. Such noise is often referred to as being "spikey" in character, and it causes images to have a "salt and pepper" or "snowy" appearance. Bit errors are handled by recognizing that noise values normally change much more abruptly than true image values. Thus, noise can be identified by comparing each pixel in an image with its neighbors. If the difference between a given pixel value and its surrounding values exceeds an analyst-specified threshold, the pixel is assumed to contain noise. The noisy pixel value can then be replaced by the average of its neighboring values. Moving neighborhoods or windows of 3×3 or 5×5 pixels are typically used in such procedures (Figure 4.6).

4.4.4 Geometric Image Correction

As mentioned in Section 4.1, all remote sensing images are inherently subject to geometric distortions. These distortions may be due to several factors, including: the perspective of

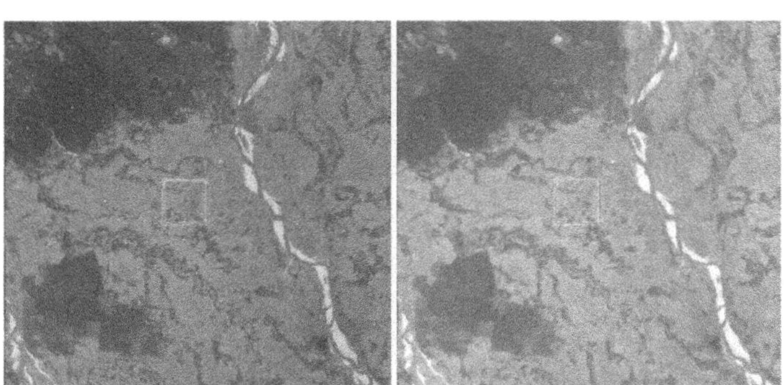

FIGURE 4.6
Pixel dropouts in Resourcesat-1 LISS-IV-band 2 image (Courtesy: National Remote sensing Centre, Indian Space Research Organization, Department of Space, Government of India).

the sensor optics; the motion of the scanning system; the motion of the platform; the platform altitude, attitude, and velocity; the terrain relief; and the curvature and rotation of the Earth. Geometric corrections are intended to compensate for these distortions so that the geometric representation of the image will be as close as possible to the real world. Many of these images are systematic or predictable in nature and can be accounted for by accurate modeling of the sensor and platform motion and the geometric relationship of the platform with the Earth. Included in the category of systemic distortions are scan skew, mirror-scan velocity variance, panoramic distortions, nonuniformity in platform velocity, earth rotation, and perspective. Besides, there are other unsystematic or random errors due to variations in platform's altitude and attitudes that cannot be modeled and corrected by modeling. Therefore, geometric registration or geo-referencing of the image to a known ground coordinate system must be performed (Levin, 1999).

4.4.4.1 Correction for Systemic Distortions

The correction procedures for various systemic distortions are described hereunder.

4.4.4.1.1 Panoramic Distortion

Panoramic distortions arising due to non-verticality of the optical axis of sensor results in squeezing at the image margins. The image is corrected such that the horizontal distance is given by

$$X = H \tan\theta \tag{4.3}$$

where H is the flying height/altitude, X is the horizontal distance, and θ is the angle of rotation of the optical axis. This correction is especially necessary in the case of aerial scanner data with a field of view (FOV) of about 50° (total angular field of view 100°). The correction is sometimes applied during preprocessing; the angle of rotation θ is related to time and therefore X can be related to a time base to produce a geometrically rectified image. In spaceborne missions the angle θ is very small (e.g., in Landsat-MSS, θ = 5.6°) due to the very high altitude, and error can often be ignored. However, in some missions like Terra/Aqua MODIS with a very large field of view (±55°), it assumes greater significance. The MODIS sensor consists of a cross-track scan mirror with a FOV of ±55°, collecting optics and individual detector elements. The swath dimensions of MODIS are 2,330 km (across track) by 10 km (along track at nadir). In contrast to other scanning sensors like AVHRR, MODIS is observing ten lines of 1 km spatial resolution (40 lines of 250 m resolution and 20 lines of 500 m resolution, respectively) within one scan. Due to this unique feature, the so-called panoramic "bow tie" effect occurs at the edges of each scene (Figure 4.7).

Three consecutive MODIS scans each consisting of ten 1 km lines are shown. Due to the panoramic "bow tie" effect, the scans are partially overlapping at off-nadir angles.

The first and third scans are represented by the light gray grids, while the second scan is shown in black (http://eoweb.dlr.de:8080/short_guide/D-MODIS.html, accessed on August 14, 2017).

The MODIS sensor with high FOV (110°) has been complicated by the so-called bow-tie effect. The source of error is panoramic distortion that is exacerbated by Earth curvature. This effect causes an overlap of the satellite scan lines producing a data repetition. MODIS

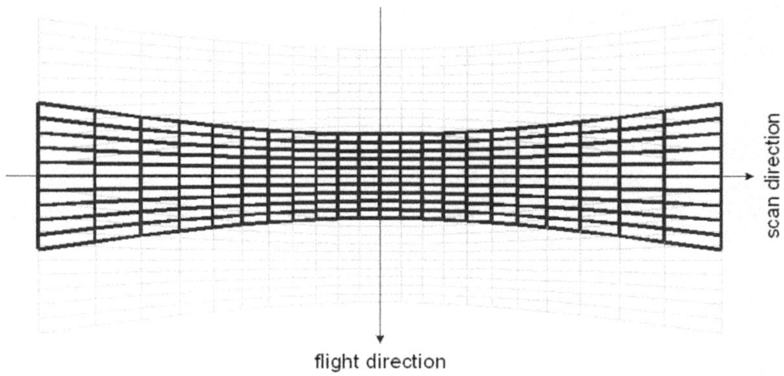

FIGURE 4.7
Bow-tie effect in Terra/Aqua MODIS image.

data have been complicated by the so-called "bow-tie effect" which causes an overlap of the satellite scan lines producing a data repetition. This effect increases with the distance from nadir and can be especially dramatic at the edge of the image. Several models have been proposed by various researchers based on geometric projection of panoramic error with Earth curvature and MODIS platform parameters (Guangmeng, 2003; National Snow and Ice Data Center, 2003; Gomez-Landesa et al., 2004); Available from http://nsidc.org/data/modis/ms2gt/.

4.4.4.1.2 Aspect Ratio Distortion

When the linear scales along the two rectangular arms of an image are not equal, aspect ratio distortion occurs. This can arise for many reasons, e.g., oversampling/undersampling, or variations in the V/H ratio of the sensor-craft. The aspect ratio distortion due to design in sampling pattern is systematic in nature. For example, the Landsat MSS sensor carried a systematic type of aspect ratio distortion. Here due to oversampling, the scale along the x-axis was longer than y-axis by a factor of 1.38. Aspect ratio distortion due to V/H variation in the spacecraft is of an unsystematic type, and is rectified by the general technique of rubber-sheet stretching.

4.4.4.2 Correction of Nonsystemic Errors

The image distortions due to sensor-craft altitude and attitude variations can be rectified by geometric corrections using ground control points (GCPs) by co-registering the distorted image with the standard topographic maps or with reference to an already geometrically rectified image. The process is termed *geo-referencing*. The geometric correction is carried out only in x- and y-axis. The distortions due to terrain's relief (z-axis) is, however, not considered. This kind of geometric error may not matter much in the case of spaceborne satellite imagery with coarse to medium spatial resolution in which topographic relief-induced shift in pixel position is negligible, or in applications in which precise geographic location is not a primary concern. In local applications (e.g., urban planning) involving very high spatial resolution satellite imagery, the geometric position of pixels need to be determined accurately as well. However, for aerial photographs, it does matter.

4.4.4.2.1 *Terrain Correction of Relief Distortions*

The geometric corrections of the image to remove terrain's relief-related geometric distortions or topographic relief-induced shift in pixel's position is termed orthorectification. For achieving required precision in geometric accuracy such image geo-referencing must take into consideration the minor shift in pixel position caused by topographic relief (Gao, 2009). *Orthorectification* is the process of transforming a central perspective image into an orthogonal image by removing positional displacement caused by topographic relief from the input image, in addition to providing the ground coordinates for all pixels. In the *central perspective* entire scene within a field of view is sensed from a single point in space, such as the geometric center of the camera lens.

4.4.4.2.2 *Methods of Image Orthorectification*

The precondition of image orthorectification is the establishment of a relationship between image coordinates (r, c) and the ground coordinates (E, N, Z). For spaceborne images, the construction of this relationship relies on the exterior and interior orientation parameters (e.g., position and orientation) of the sensor, with the assistance of 3D GCPs. Image orthorectification may be implemented *nonparametrically* or *parametrically* (Hemmleb and Wiedemann, 1997).

Nonparametric approaches such as *polynomial transformation* and *projective transformation* are very similar to the 2D polynomial-based image rectification except that the height of GCPs is also considered. No information on the sensor is utilized in the rectification, in drastic contrast to parametric approaches in which the image coordinates of all pixels are transformed to ground coordinates based on the information on the interior and exterior orientation of the sensor.

A critical preliminary step in image orthorectification is the construction of a DEM. It should cover the same geographic area as the imagery being rectified. Preferably, it should have the same spatial resolution as the imagery. For instance, the DEM should have a cell size of 30 m if the imagery being ortho-rectified is from Landsat TM.

4.4.5 Image Processing Levels

The digital image processing procedures mentioned in foregoing sections essentially help the image analyst comprehending the source/(s) of geometric/radiometric distortions and the procedures for removal of these distortions. In fact, most of the satellite data providers across the globe supply geometrically and radiometrically corrected digital data to the users. Since most remotely sensed data require similar amounts of basic processing before they are usable, image distribution agencies world over have adopted a common set of processing "levels" to describe the types of processing that have carried out to a given satellite data product. An example from satellite data available from Earth Observing System Data and Information System (EODIS) is given hereunder: The EOSDIS is a key core capability in NASA's Earth Science Data Systems (ESDS) Program. It provides end-to-end capabilities for managing NASA's Earth science data from various sources—satellites, aircraft, field measurements, and various other programs.

EOSDIS data products are processed at various levels ranging from Level 0 to Level 4. Level 0 products are raw data at full instrument resolution. At higher levels, the data are converted into more useful parameters and formats. All EOS instruments must have Level 1 products. Most have products at Levels 2 and 3, and many have products at Level 4.

Data Level	Description
Level 0	Reconstructed, unprocessed instrument and payload data at full resolution, with any and all communications artifacts (e.g., synchronization frames, communications headers, duplicate data) removed. (In most cases, the EOS Data and Operations System (EDOS) provides these data to the data centers as production datasets for processing by the Science Data Processing Segment (SDPS) or by a SIPS to produce higher-level products.)
Level 1A	Reconstructed, unprocessed instrument data at full resolution, time-referenced, and annotated with ancillary information, including radiometric and geometric calibration coefficients and geo-referencing parameters (e.g., platform ephemeris) computed and appended but not applied to Level 0 data.
Level 1B	Level 1A data that have been processed to sensor units (not all instruments have Level 1B source data).
Level 2	Derived geophysical variables at the same resolution and location as Level 1 source data.
Level 3	Variables mapped on uniform space-time grid scales, usually with some completeness and consistency.
Level 4	<RHT>Model output or results from analyses of lower-level data (e.g., variables derived from multiple measurements).

Source: http://science.nasa.gov/earth-science/earth-science-data/data-processing-levels-for-eosdis-data-products, accessed on August 4, 2016.

4.5 Image Enhancement

The range of possible image enhancement and display options available to the image analyst is virtually limitless. The enhanced images can be displayed interactively on a monitor or they can be recorded in a hardcopy format, either in black and white or in color. Interestingly, there are no simple rules for producing the single "best" image for a particular application. Often several enhancements made from the same "raw" image are necessary (Sabins, 2000). The most commonly applied digital enhancement techniques can be categorized (Lillesand and Kiefer, 1994) as follows:

- *Contrast manipulation*, which includes gray-level thresholding, color density slicing, and contrast stretching;
- *Spatial feature manipulation*, which consists of spatial filtering, edge enhancement, and Fourier analysis;
- *Multi-image manipulation*, which includes multispectral band ratioing and differencing, principal components (PCs), canonical components, vegetation components, and decorrelation stretching.

4.5.1 Contrast Modification

Contrast manipulation is a process of modifying or enlarging the range of pixel DN values in an input image in an attempt to improve its visual effectiveness or quality. In this process the digital number (DN) value of every pixel in the image is modified according to a predetermined function. The process includes *density slicing* and *contrast enhancement*, both carried out for single bands. It is a histogram-based operation wherein a pixel's DN value is modified regardless of its neighboring pixels' values. Mathematically, contrast stretching can be expressed as

$$DN_{out} = f(DN_{in}) \tag{4.4}$$

Where DN_{out} is the output DN in the contrast-stretched image, DN_{in} is the DN of the same pixel in the raw image, and f is the transformation function through which contrast is manipulated; it can be either linear or nonlinear.

4.5.1.1 Density Slicing

Also known as pixel-value thresholding, density slicing is virtually a process of discretizing the continuously varying pixel values in the input spectral band (Figure 4.7a and b). Pixel values within a certain gray-level range are amalgamated into a single value in the output image. The range of entire pixel values in the input image is reduced to a few categories of values, each corresponding to a unique range of pixel values in the input image. Thus, the potential number of pixel values is considerably reduced in the sliced image. A unique color may then be assigned to each newly created pixel value, converting a gray-level image into a pseudocolor one. However, in order to generate a meaningful pattern for the feature/phenomenon under study (e.g., levels of soil salinity and/or alkalinity), the thresholds for each discrete category must be carefully selected.

4.5.1.2 Contrast Enhancement

There are several options available for enhancing the contrast of a remote sensing image. The techniques for contrast enhancement could be grouped into *linear and nonlinear digital image enhancement techniques*. In linear enhancement the output value of a pixel is modified from its input value via a linear function. The slope of the linear function indicates the degree of stretching. Its tangent of the angle with the horizontal axis (q) represents the coefficient through which contrast is altered. If tan q > 1, then the enhanced image has a larger contrast than the original image. If tan q < 1, then the contrast of the output image is suppressed (Gao, 2009).

4.5.1.2.1 Linear Contrast Enhancement

The image enhancement resorted to improve the contrast of images with low contrast. The enhancement facilitates detection and delineation of certain features or phenomenon of interest. Figure 4.8a shows an image with low image contrast, and Figure 4.8b portrays the

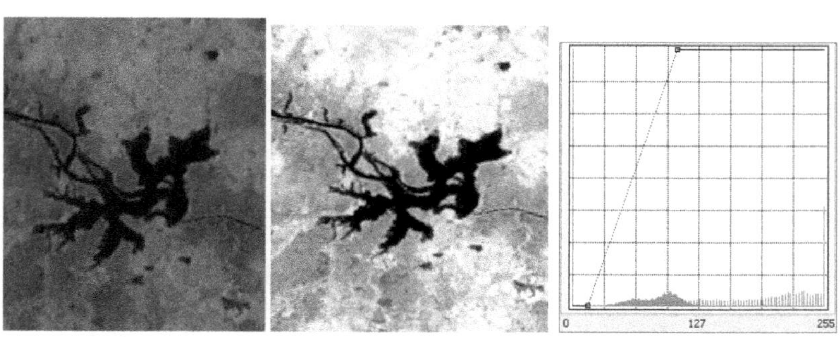

(a) (b) (c)

FIGURE 4.8
Landsat-MSS band 4 (0.7–1.1 μm) of March 2, 1985 raw digital data (a), linearly stretched data (b), and corresponding histograms of raw digital data and linear-stretched data area given in (c).

enhanced image after linear stretch. When the values in the original image are expanded uniformly to fill the total range of the output device, the transformation is called linear contrast stretching. The histogram of raw and enhanced image is appended as Figure 4.8c. Note the improvement in contrast between water body and adjoining land features.

Instead of stretching the whole range of DN values in a raw image linearly, it is also possible to linearly stretch different DN values in a piecemeal manner. That is, instead of a single stretching function for all DNs, a few linear functions are used for the stretching. Each function segment has its own slope and is applicable to a specific range of DNs. This is known as *piece-meal linear enhancement*. With the use of multiple enhancement functions, it is possible to stretch the contrast of an image at different pixel values.

4.5.1.2.2 Nonlinear Contrast Enhancement

The nonlinear contrast stretching allows some part of the input image to have a stretched contrast while contrast in some other DN ranges is suppressed in the output image. Depending on the input DN value and the nonlinear function in nonlinear stretching both contrast stretching and contrast compression can be achieved in one output image. There are a number of nonlinear functions for contrast enhancement. The logarithmic and exponential functions are most commonly used. The logarithmic function takes the following form:

$$DN_{out} = \log_{10} DN_{in} \tag{4.5}$$

In exponential contrast stretching, pixel values in the output image are adjusted according to the following form:

$$DN_{(out)} = e^{DN_{in}} \tag{4.6}$$

This exponential function has a base of e.

An example of nonlinear stretch is given in Figure 4.9. Figure 4.9a shows Landsat-MSS band 4 (0.7–1.1 μm) data acquired on March 2, 1985; Figure 4.9b nonlinear-stretched image and its histogram before and after stretching shown in Figure 4.9c).

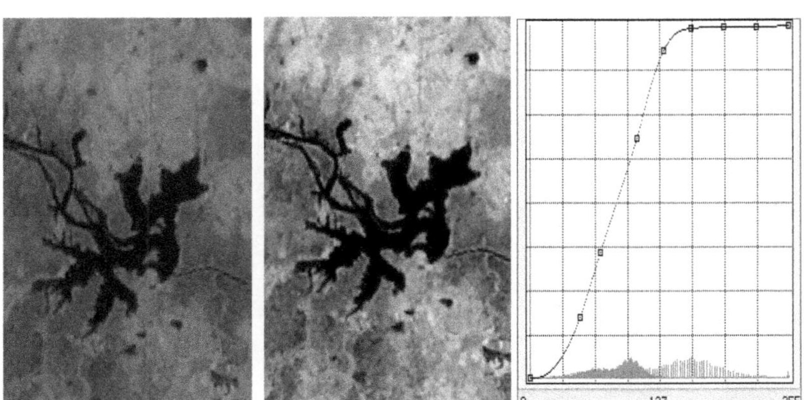

(a) (b) (c)

FIGURE 4.9
Landsat-MSS band 4 (0.7–1.1 μm) of March 2, 1985 raw digital data (a) and nonlinear-stretched data (b). The histograms of the raw digital data and nonlinear-stretched data area given in (c).

4.5.1.2.3 Histogram Equalization

The histogram of most images is more likely to be bell shaped. This kind of pixel value distribution indicates that the large majority of pixels are confined to a small range that is indicative of a low contrast. On the other hand, few very extremely bright or dark objects are likely to occupy a wide range of DNs. Consequently, there is an imbalance in the number of pixels at a given DN (Figure 4.10). For highlighting the subtle spectral variations among terrain features the predominant pixels should be intuitively represented in a wider range of DNs so that can be readily differentiated. This imbalance is ideally rectified through *histogram stretching.*

4.5.1.2.4 Histogram Matching

While developing a mosaic from satellite images/aerial photographs we come across a situation in which the tone/color of an image does not match with the adjoining image. Such a tonal/color inconsistency problem can be eliminated or reduced to a some extent through *histogram matching.* The histogram matching involves rearrangement of the pixel values in a slave image in such a way so as to achieve a distribution approximately identical to that of the master image. If both the master and slave images cover the same ground, the histogram can be created from the entire scene. Otherwise, it has to be established from a subset common to both of them. The process involves two steps in achieving this adjustment:

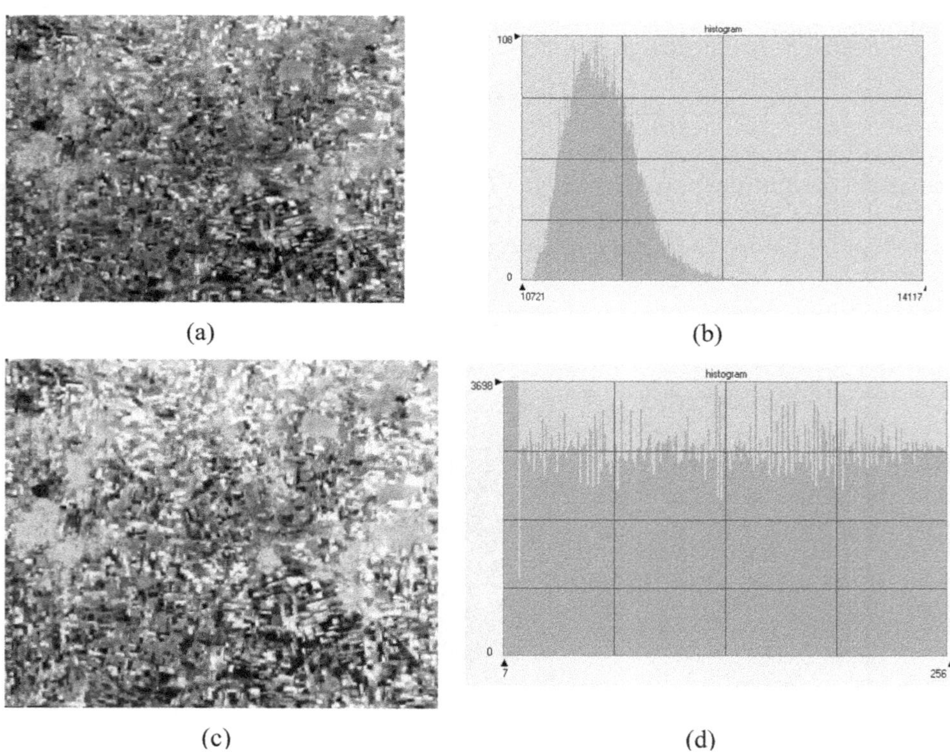

FIGURE 4.10
An illustration of histogram equalization: raw digital LISS-IV image (a), corresponding histogram (b), histogram-equalized image (c), and the histogram of the stretched image (d).

- First, a cumulative histogram c(k) is constructed for both the master image and the image to be adjusted using equation 4.7, just like in histogram equalization.

$$c(k) = \frac{1}{N}\sum_{j=0}^{k} f(DN_j) = \frac{1}{N}\sum_{j=0}^{k} n_j \tag{4.7}$$

where n_j is the number of pixels at gray levels j and k is the number of discrete gray levels.

- Second, a look-up table is constructed to determine the DN of pixels that should be reassigned to other DN levels in order to achieve the desired distribution.

For digital image fusion of panchromatic (PAN) data with finer resolution and multispectral data of coarser resolution, the multispectral data are transformed using either principal component analysis (PCA), Brovey transform, HIS transform or wavelets, one of the images of these transformed images is replaced by PAN image. While doing so due care is taken to ensure the compatibilities of DN values in these images. The process is referred to as *histogram matching*. An example of histogram matching is shown in Figure 4.11.

4.5.1.2.5 Spatial Filtering

Spatial filtering is a window-based image processing technique for altering the input pixel value based on its own value and the value of the pixels surrounding it. It requires the use of a spatial mask known as a spatial filter. Filtering is carried out to achieve several functions, such as image smoothing and feature enhancement within a neighborhood.

Low-pass filtering or low spatial frequency or *image smoothing* is a process of suppressing noise in the input image that may arise during image acquisition and transmission. Low spatial frequency features are defined as infrequent grayscale changes that occur gradually over a relatively large number of pixel distances. Raw digital image is portrayed as Figure 4.12a and the low-pass-filtered image as Figure 4.12b. Radiometric noises in an image are manifested as abnormally larger or smaller pixel values than those in

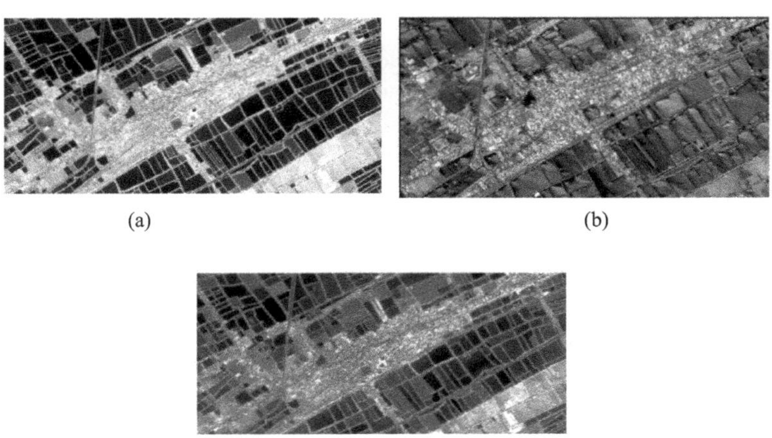

(a) (b) (c)

FIGURE 4.11
An illustration of histogram matching. The first principal component (pc1) of LISS-IV multispectral image (a), Cartosat-1 2.5 m-PAN image (b), and histogram-matched image of pc1 (c).

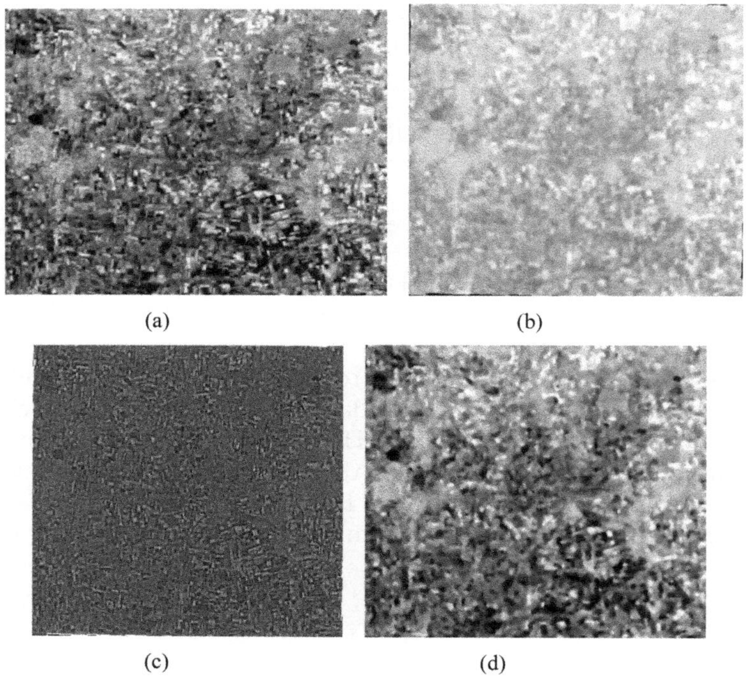

FIGURE 4.12
An example of spatial filtering of resourcesat-2 LISS-IV data (a), 3 × 3 high-pass-filtered data (b), 3 × 3 median-filtered data (c), and 3 × 3 low-pass-filtered data (d).

the neighborhood. Since the genuine pixel value is unknown, noise cannot be completely eliminated through image smoothing. Instead, this noise is suppressed to a certain degree by dividing it among all pixels within the kernel. There are several methods for suppressing the noise. A common method is to replace the noise-infected pixel with the mean of all pixel values inside the kernel.

High-pass filtering refers to a filtering operation using an operand with differential weights (Figure 4.12c). During filtering operation the difference between adjacent pixels is artificially enlarged. Unlike low-pass filtering, high-pass filtering attenuates low-frequency features (Gonzalez and Woods, 1992). As a result, high-frequency features, such as edges between homogeneous groups of pixels and other sharp details, stand out. In high-pass filtered images, large pixel values become larger and spatial frequency is increased.

Median filtering is very similar to image smoothing in that it is carried out within a window. Unlike image smoothing, it does not require an operand, or a convolution kernel. Instead, only the pixels within a neighborhood are examined, with the central pixel being the focal point. The median of these pixels is determined by sorting the nine DNs inside the window in either ascending or descending order (e.g., the fifth number in the list) (Figure 4.12d). This median is used to replace the pixel value in question in the output image.

4.5.1.2.6 Image Filtering in Frequency Domain

Apart from spatial domain, image filtering can also be implemented in the frequency domain using the common method of Fourier transformation that operates on a single band (e.g., grayscale image). The fundamental premise underlying this transformation is

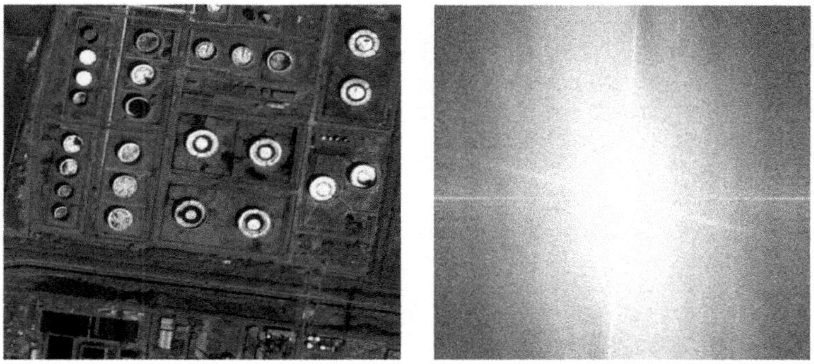

(a) (b)

FIGURE 4.13
Cartosat-2 PAN raw image (a) and Fourier-transformed image (b).

that each row of image f(x) can be approximated by a series of sinusoidal waves, each having its own amplitude, frequency, and coefficient. The transformed image can be described by the frequency of each waveform fitted to the image and the proportion of information associated with each frequency component (Mather, 2004). For satellite imagery, this generalization needs to be extended in two ways. First, the image is discrete instead of continuous; thus the transformation is termed *discrete Fourier transformation* (DFT). A highly efficient version of the DFT called fast Fourier transformation (FFT) has been developed (Gao, 2009) (Figure 4.13).

4.5.1.3 Edge Enhancement and Detection

An edge or linear feature is manifested as an abrupt change in DN along a certain direction in an image. This direction is the orientation of that feature. The manifestation becomes an extreme of the first-order derivative or a zero crossing in the second derivative. Edge detection can be based on such a discontinuity property by tracing the maximum along the bound of an area. Out of a few methods available for implementing edge detection and enhancement, two of them, self-subtraction and edge-detection templates, are introduced in the following section.

4.5.1.3.1 Edge Enhancement through Subtraction

Edge enhancement through image self-subtraction is based on the fact that non-edge features have a spatially uniform value, in sharp contrast to edges that experience a drastic and usually abrupt change in pixel value along a certain direction (Figure 4.14). A new image is created by duplicating the existing one. If this newly created image is subtracted from the source one, then nothing remains in the resultant image. No edges are detectable through this subtraction. The subtraction can be improved by slightly shifting one of the images to the left, or to the right, or above or below, or even diagonally by one or two pixels.

4.5.1.3.2 Edge-Detection Template

In this method kernels are used for edge detection. The unique feature of these kernels is the sum of all elements in a kernel is zero. These zero-sum kernels smooth out areas of low spatial frequency (e.g., absence of any edge), and cause a low output in areas of low spatial

(a) (b)

FIGURE 4.14
Edge enhancement and detection. Cartosat-1 2.5 m-PAN image (a), and the image background with edges (b).

frequency. In areas of high spatial frequency a sharp contrast results. It is possible for the edge-enhanced images to contain only edges and zeros. In areas of high spatial frequency, the disparity among pixel values is magnified as large values become larger while low values become even lower owing to the unequal element values along the first/last row/column Edges are enhanced in the filtered image as it frequently is made up of only edges and zeros.

4.5.2 Multiple Image Manipulation

The aforesaid enhancement techniques involve only a single band image in the input and in the output. In practice it is possible to generate a new image from multiple images of the same area. Usually, these images are multispectral bands from the same sensor obtained at the same time, covering an identical geographic area. If any of these conditions is not met the images need to be co-registered to the same coordinate system and resampled to the same spatial resolution using appropriate sampling methods before they can be processed.

4.5.2.1 Band Ratioing

Band ratioing refers to division of DN values of one spectral band by another from the same sensor, preferably obtained at the same time. After division, all pixel values that are expressed as a ratio between 0 and 1 may have to be rescaled to 0–255. Band ratioing is able to achieve several purposes, depending on the nature of the input bands. If the two bands are obtained at different times, band ratioing is effective at detecting changes that have taken place during the period. If the two bands are from the same sensor, then this process is effective at eliminating radiometric variations caused by topography). Figure 4.15 portrays the band-ratioed images. Note the regular-shaped very light gray to white agricultural fields in center of the image as well as upper-right and lower-left corner of image (b). Although these features are also conspicuous in band 2 by 3 image (a), they stand out much better in band 3 by 2 image.

4.5.2.2 Vegetation Indices

Vegetation index (VI) is an arithmetic disparity between pixel values in two or more spectral bands of the same multispectral image. Having acquired at the same time effectively ensures that their spatial resolution is identical and they cover the same area on the ground. VI is able to enhance the conspicuousness of vegetation through subtraction

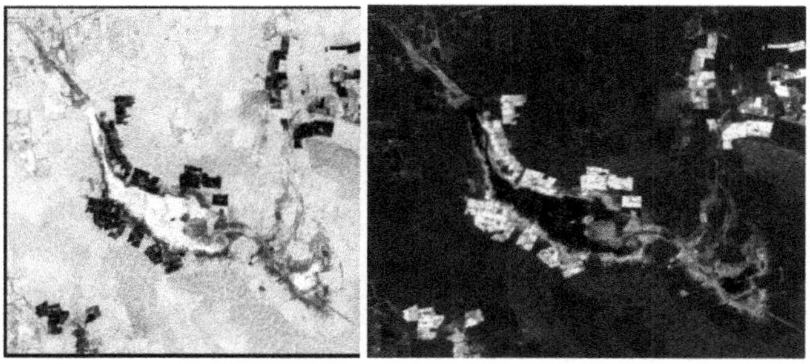

(a) (b)

FIGURE 4.15
The ratio image of Resourcesat-2 LISS-IV dada of October 1, 2015. The spectral band-2 by band-3 image (a), and spectral band-3 by band-2 image (b). Note the regular-shaped very light gray to white agricultural fields in center of the image as well as upper-right and lower-left corner of image (b). Although these features are also conspicuous in image (a), they stand out much better in band 3 by 2 image.

of one spectral band from another because of the differential reflectance of ground features over different wavelength ranges. In fact, VIs are related closely to the amount of vegetative cover present on the ground, and its greenness or biomass. These indices can serve as good surrogate measures of vegetative cover if calculated properly from the right combination of spectral bands. Furthermore, highly correlated to green-leaf density, they can be viewed as a proxy for aboveground biomass (Bannari et al., 1995; Rasmussen, 1998).

The *normalized difference vegetation index* (NDVI) is one of the most commonly used VIs. In this way, all indices obtained from images of different seasons across the globe are directly comparable to one another.

$$\text{NDVI} = \frac{\rho_{\text{NIR}} - \rho_{\text{Red}}}{\rho_{\text{NIR}} + \rho_{\text{Red}}} \tag{4.8}$$

where ρ_{NIR} and ρ_{Red} represent spectral reflectance at the near-infrared and red wavelengths, respectively (Holben, 1986).

4.5.2.3 Image Transformation

In an attempt to reduce the dimensionality of multispectral data thereby reducing the processing and/or analysis time, and to enhance some terrain features of interest improve overall image contrast, several image transforms, namely principal component transform, canonical transform, and Kauth–Thomas transforms, and digital image fusion have been developed.

4.5.2.3.1 Principal Component Analysis

The multispectral images are generally highly correlated and have redundant information in individual spectral bands. The PCA aims at generating decorrelated images thereby reducing the dimensionality of digital multispectral spectral images, and enhancing some terrain features of interest. Data redundancy among spectral bands is best visualized by plotting a scatter diagram of pixel values in these bands. If the pixels are distributed along a line, then the information content of the two bands is highly correlated. If the pixels are widely scattered without following any trend, then the information content of the two bands is uncorrelated and there is little data redundancy between them. PCA addresses the issue of data redundancy by transforming the raw data into another domain. Thus, in the context of image enhancement PCA is useful in the following:

- Ascertaining the information content of each multispectral band and in identifying the spectral band containing maximum information,
- Reducing the number of spectral bands needed to represent most of the information contained in the original spectral bands,
- Transforming the information content onto orthogonal axes thereby increasing spectral separability of certain spectrally adjacent classes that partially overlap each other in the original spectral domain.

To demonstrate the procedure, Landsat-8 OLI data (all the band except for PAN and two thermal bands) covering part of Etah district, Uttar Pradesh, northern India were taken. To begin with the spectral signature in terms of mean DN values in each spectral band, standard deviation and correlation coefficients and variance and co-variance matrices were generated, which was subsequently used for PCA. The PC transforms are shown in Figure 4.16. Eigenvalues and eigenvectors were computed for 8×8 covariance matrix. The PC loadings (eigenvectors) for 8-band data are given in Table 4.1 and eigenvalues in Table 4.2. As evident from Table 4.2 the first variance measured along the first PC (eigenvector) is 11,578,945.35 which accounts for 75.56% of the image variance, and the rest of the variance is distributed among other 7 PCs. Interestingly, the first three PCs account for 99.69% variance, while rest of the PCs represent only noise.

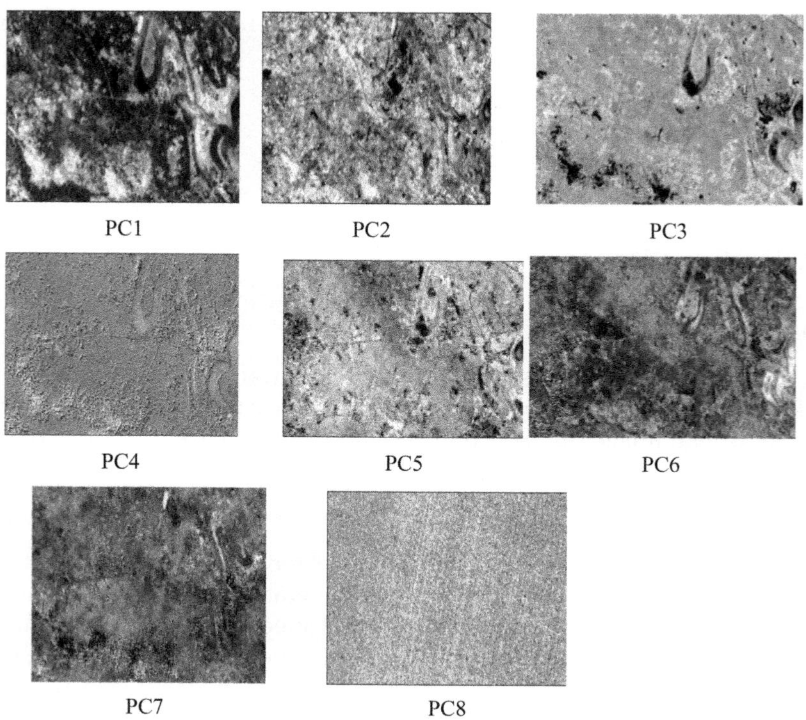

FIGURE 4.16
Principal component analysis of Landsat-8 OLI data.

TABLE 4.1

Eigen Vectors for 8-Band Landat-8 OLI Data

	Eigenvectors (PCs)							
Spectral Band	**1**	**2**	**3**	**4**	**5**	**6**	**7**	**8**
1	0.209528841	0.055025777	–0.280603983	0.240751781	0.093109527	–0.112975063	–0.891591622	0.001201801
2	0.269669565	0.091744276	–0.339741909	0.296750371	0.093467333	–0.758683242	0.361981074	–0.00304362
3	0.406534767	0.055111934	–0.426259567	0.423081848	–0.01384103	0.632605645	0.265732808	0.001134777
4	–0.176382729	0.978826955	0.025326363	0.010214658	–0.093307054	0.036531389	–0.000627094	–0.000136859
5	0.532888552	0.145861509	0.494130799	–0.087179339	0.663930544	0.042599394	0.01911293	–0.002224436
6	0.548435276	0.022291474	0.394168153	0.012687243	–0.729408765	–0.090080913	–0.055123161	0.001434708
7	0.319754447	0.075118528	–0.473248686	–0.816770467	–0.029129002	0.013250025	0.003452946	–2.30E-05
9	0.000489385	0.000578738	0.000313124	–9.58E-05	0.002698232	–0.002661957	0.001993315	0.999990489

TABLE 4.2
Eigen Values for 8-Band Landat-8 OLI Data

Eigen Values	Percent Variance (%)	Cumulative Percentage
11,578,945.35	75.56	75.56
3,066,537.27	20.01	95.57
547,176.62	3.57	99.14
85,428.73	0.55	99.69
32,324.55	0.21	99.9
8,550.77	0.0005	99.9005
4,710.39	0.0003	99.9008
86.64	0.00005	99.90085

4.5.2.3.2 Tasseled Cap Transformation

Also called the Kauth and Thomas (1976) transformation, the Tasseled Cap transformation was developed specifically for transforming the four Landsat-MSS spectral bands. Pixels in a triangle formed by the four-band feature space represent vegetation at various stages of growth (Figure 4.17). The transformation optimizes viewing the original satellite data in the feature space for certain particular purposes (e.g., maximization of vegetation difference). Through rearranging the content of the output bands, it is possible to highlight the subtle variations in crop types. Essentially, the Kauth-Thomas transformation is a rotation of axes so that the differences among the pixels are more distinguishable along the new axes (Figure 4.17). The four bands are rotated to a new space defined by four axes "brightness," "greenness," "yellowness," and "None-such." Each axis represents a unique aspect of the object of study. The brightness axis reflects mainly variations in soil reflectance. The greenness axis indicates the variation in vegetation vigor. The yellowness axis is indicative of vegetation that has reached senescence or maturity (Figure 4.17). The last axis, still orthogonal to the previous three axes that are mutually orthogonal to one another, accounts for noise in the data not related to any of the terrain features. The output of tasseled cap transformation is appended as Figure 4.18.

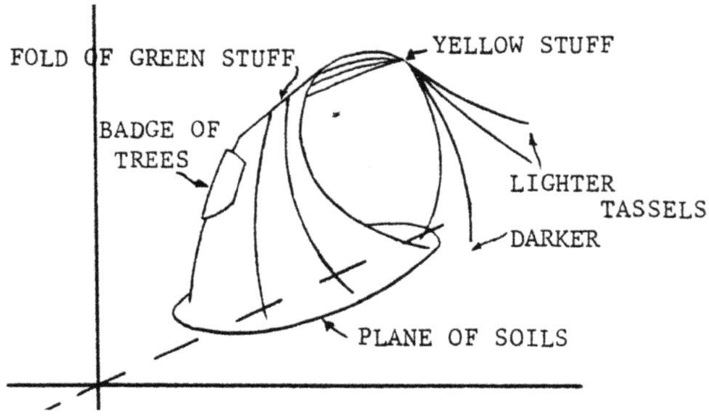

FIGURE 4.17
Kauth-Thomas transformation (Source: Adapted from Kauth and Thomas (1976)).

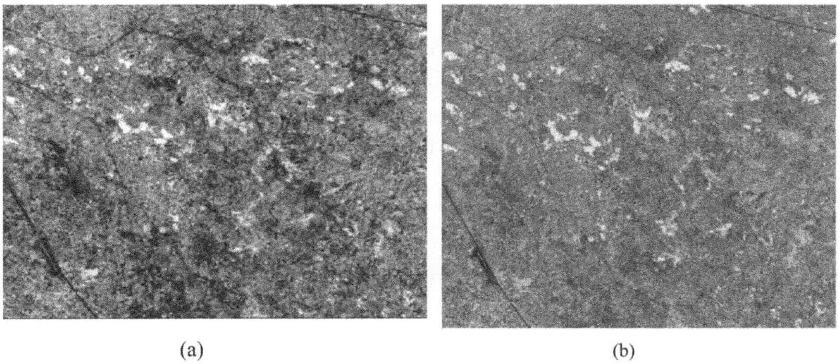

FIGURE 4.18
The standard FCC of Landsat-8 OLI data (a), and the tasseled cap-transformed image (b).

4.5.2.3.3 Digital Image Fusion

Several image fusion techniques have been developed for merging multispectral images with high spatial resolution images for various applications. For overview of image fusion techniques readers may refer to Pohl and Genderen (1998) and Jinghui et al. (2010).

4.5.2.3.4 Hue Intensity Saturation Transformation

A false-color image of three multispectral bands is visualized on a computer screen by projecting three spectral bands using three guns of blue, green, and red lights. In this process, different proportions of these primary-color lights are mixed to produce a whole range of colors, including yellow, magenta, and cyan (Figure 4.19a). This kind of image is commonly referred to as red-green-blue (RGB). If the intensities of blue, green, and red are set equal to each other, then their mixed color will become achromatic (e.g., gray).

The transformation aims at projecting pixel values in the RGB space into values in the HIS space that requires the establishment of a new reference system. In this system hue is defined as proportional to the degree of rotation about the achromatic point. Saturation is defined as the length of a vector from the achromatic point to the point (R, G, B). Intensity is the vector length from the origin. After the establishment of this system, RGB can be translated to HIS using the following algorithms (Carper et al., 1990): Quantification of color through the RGB to HIS transformation provides direct control over accurate portray

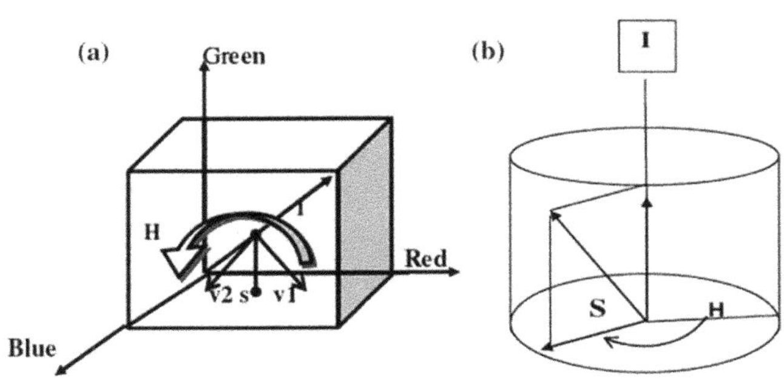

FIGURE 4.19
Models of IHS color spaces: (a) The color cube model (b)The color cylinder model (Source: Adapted from Al-Wassai (2012)).

and representation of colors. The transformation is a useful means of fusing data from multiple sensors. For instance, images of different resolutions (e.g., 1 m PAN band and 4 m multispectral bands) can be fused through RGB to HIS transformation in a procedure known as "PAN sharpening" to take advantage of the fine spatial resolution imagery. It is also possible to differentially contrast enhance the saturation and intensity components before they are transformed back to RGB.

4.5.2.3.5 *Brovey Transformation*

The Brovey transform was developed to visually improve contrast in the low and high ends of an images histogram. That is, to provide contrast in shadows, water, and high-reflectance areas such as urban features, desert terrain snow-covered areas. It is a combination of arithmetic operations and normalizes the spectral bands before they are multiplied with the PAN image (Hallada and Cox, 1983). It is very useful for generating color composite images with a higher degree of contrast in the low and high ends of the image histogram and for producing "visually appealing" images. In contrast to the IHS method, the Brovey transformation method is a ratio fusion technique that preserves the relative spectral contributions of each pixel, but replaces its overall brightness by the high-resolution PAN image. The following formula is used in the Brovey transformation:

$$
\begin{aligned}
DN_{B1\,new} &= \left[DN_{B1}/DN_{B1} + DN_{B2} + DN_{B3}\right]^{*}\left[DN_{high\,res.\,image}\right] \\
DN_{B2\,new} &= \left[DN_{B2}/DN_{B1} + DN_{B2} + DN_{B3}\right]^{*}\left[DN_{high\,res.\,image}\right] \\
DN_{B3\,new} &= \left[DN_{B3}/DN_{B1} + DN_{B2} + DN_{B3}\right]^{*}\left[DN_{high\,res.\,image}\right]
\end{aligned}
\tag{4.9}
$$

where B represents band (http://imagefusiontechniques.blogspot.com/#!, accessed on July 18, 2016). An example of the fusion of Resourcesat-2 LISS-IV 6 m multispectral image with Cartosat-1 2.5 m image (Figure 4.20).

4.5.2.3.6 *PCA Transformation*

Most of the multispectral sensors make spectral measurements in adjacent bands of the electromagnetic spectrum. This generally implies detecting redundant information, since many land covers tend to behave in a similar fashion in neighboring regions of the spectrum. The PCA transformation converts intercorrelated bands of spectral bands of multispectral data into a new set of uncorrelated components. In general, the first PC collects the information that is common to all the bands used as input data in the PCA, i.e., the spatial information, while the spectral information that is specific to each band is picked up in the other PCs (Chavez and Kwarteng, 1989). This makes PCA an ideal technique for merging multispectral and PAN images. The first component also resembles a PAN image. It is, therefore, replaced by a high-resolution PAN image for the fusion. The PAN image is fused into the low-resolution MS bands by performing a reverse PCA transform. As a result of this transformation, we obtain non-correlated new bands, the PCs. The first principal component (PC1) is substituted by the PAN image after matching the histogram of the former with that of, the latter (PAN image). Finally, the inverse transformation is applied to the whole dataset formed by the modified PAN image. The output image contains the spatial detail of PAN image incorporated into them (González-Audícana et al.,

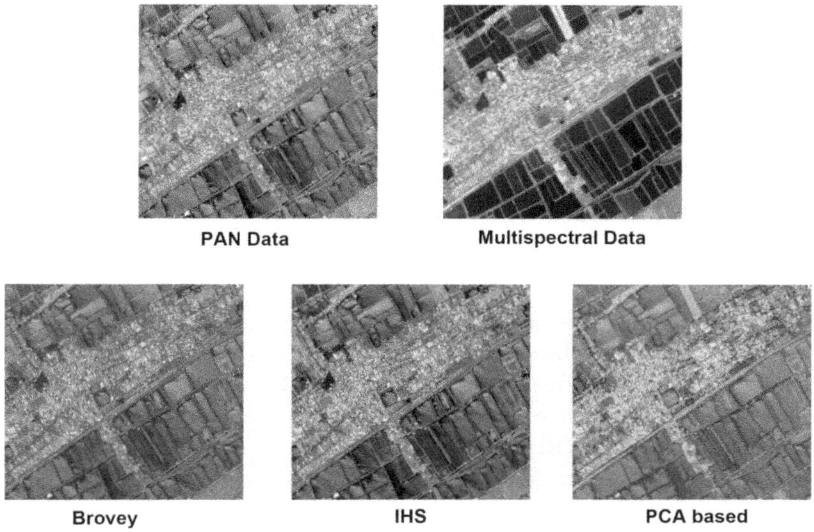

FIGURE 4.20
An illustration of digital image fusion for Cartosat-1 PAN data with 2.5 m spatial resolution and Resourcesat-2 LISS-IV image generated through the Brovey, IHS, and principal component transformations.

2004). An example of the fusion of Resourcesat-2 LISS-IV 6 m multispectral image with Cartosat-1 2.5 m image using this approach is appended as Figure 4.20.

4.5.2.3.7 Wavelet Transformation

Wavelet transform is a tool that segregates data or functions or operators into different frequency components, and then studies each component with a resolution matched to its scale Daubechies (1992). In the context of images, wavelet transformation refers to a so-called scale-space representation (Ranchin et al., 2001). The raw image is decomposed into a series of wavelets (Mallat, 1989). Initially, a "mother wavelet" is selected, and a series of wavelet functions are then derived from it by "dilation and translation, "so as to fit the spatial patterns in each row of the image." The process is then repeated for each column of the image.

In the context of wavelet transform, "dilation" refers to the stretching or compression of the wavelet, so as to represent higher or lower spatial frequencies, while translation refers to the repositioning of the leading edge of the wavelet. Hence, the wavelet transform enables the input signal to be decomposed on the basis of a series of elementary functions—known as wavelets. The output of the transform is subsequently used to recompose the original image, beginning with the coarsest levels of detail and progressing to finer levels of detail. Among various wavelet functions developed so far, those advocated by Daubechies (1992) are among the most widely used for image compression.

For merging two images with different spatial resolution their wavelet coefficients are initially determined and a transformation model can be derived to determine the missing wavelet coefficients of the lower resolution image. Using these it is possible to create a synthetic image from the lower resolution image at the higher spatial resolution. The image, thus generated, contains the preserved spectral information with the higher resolution. This method is called *"Amelioration de la Resolution spatiale par Injection de structures"* (ARSIS) (Ranchin et al., 1996).

4.6 Image Classification

Classification involves labeling the pixels as belonging to particular information classes/ features using the spectral data available. Depending on the usage of reference data or ground truth there are two broad classes of classification procedures: unsupervised classification and supervised classification.

4.6.1 Unsupervised Classification

Unsupervised classification is a mean by which pixels in an image are assigned to spectrally homogenous classes without any reference to information classes on the ground. The spectral classes, thus delineated, are subsequently associated with the information class based on available reference data, which could include maps and information from the ground truth or field visits (Richards and Jia, 2006). Several clustering algorithms have been developed (Schowengerdt, 1997; Duda et al., 2001).

4.6.1.1 Unsupervised Classification using the Chain Method

The chain method clustering algorithm operates in a two-pass mode (i.e., it passes through the multispectral dataset two times). Step 1: The program reads through the dataset and sequentially builds clusters (groups of points in spectral space). A mean vector is then associated with each cluster. Step 2: A minimum distance to means classification algorithm is applied to the whole dataset on a pixel-by-pixel basis whereby each pixel is assigned to one of the mean vectors created in step 1. The first pass, therefore, automatically creates the cluster signatures (class mean vectors) to be used by the minimum distance to means classifier.

4.6.1.2 Unsupervised Classification using the ISODATA Method

The ISODATA classifier is a variation on the minimum distance method; however, it produces results that are often considered to be superior to those derived from the basic minimum distance approach. It is a good example of a technique that shares characteristics of both supervised and unsupervised methods (i.e., "hybrid" classification) and provides evidence that the distinction between the two approaches is not as clear as idealized descriptions imply. The classification process begins with the selection of training data selected, as previously described; these data can be envisioned as clusters in multidimensional data space. All unassigned pixels (pixels to be classified) are then assigned to the nearest centroid. Thus far, the approach is the same as described previously for the minimum distance classifier. New centroids are then found for each group, and the process of allocating pixels to the closest centroids is repeated if the centroid changes position. The process is repeated until there is no change, or only a small change, in class centroids from one iteration to the next. An example of ISODATA classifier used for mapping salt-affected soils is given in Figure 4.21.

Thus, a serious practical problem with unsupervised classification is that clear matches between spectral and informational classes are not always possible. Some informational categories may not have direct spectral counterparts, and vice versa. Furthermore, the analyst does not have control over the nature of the categories generated by unsupervised classification.

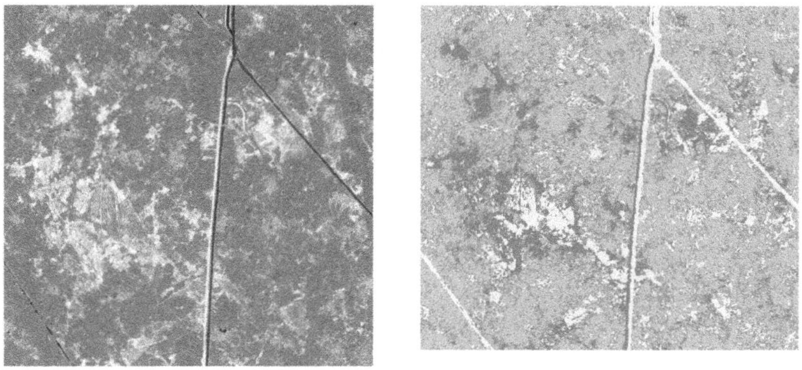

(a) (b)

FIGURE 4.21
Resourcesat-2 LISS-IV digital raw image acquired on March 7, 2017 (a) and Salt-affected soil map derived using ISODATA classifier (b). Biege color denotes severely salt-affected soils, magenta moderately salt-affected soils, yellow crop with very good vigor, and green crop with moderate vigor.

4.6.1.3 K-Means Clustering Algorithm

The k-means clustering algorithm attempts to split a given dataset into a fixed number (k) of clusters. Initially, the k number of so-called *centroids* are chosen. A *centroid* is a data point (imaginary or real) at the center of a cluster. Each centroid is an existing data point in the given input dataset, picked at random, such that all *centroids* are unique. That is, for all *centroids* c_i and c_j, $c_i \neq c_j$. These *centroids* are used to train a kNN classifier. The resulting classifier is used to classify (using k = 1) the data and thereby produce an initial randomized set of clusters. Each *centroid* is thereafter set to the arithmetic mean of the cluster it defines. The process of classification and *centroid* adjustment is repeated until the values of the *centroids* stabilize. The final *centroids* will be used to produce the final classification/clustering of the input data, effectively turning the set of initial data points into a set of data points, each with a class identity (Figure 4.21).

4.6.2 Supervised Classification

Supervised classification is the process of using samples of known identity or ground features to classify pixels of unknown identity. Samples of known identity are those pixels located within *training areas*, or *training fields*—a group of image pixels with known ground cover or information class. The analyst defines training areas by identifying regions on the image that can be clearly matched to areas of known ground features on the image. *Training areas*, or *training fields or training set* should typify spectral properties of the categories they represent and, of course, must be homogeneous in respect to the informational category to be classified.

4.6.2.1 Parallelepiped Classification

Parallelepiped classification, sometimes also known as box decision rule, or level-slice procedures, is based on the ranges of values within the training data to define regions within a multidimensional data space. The spectral values of unclassified pixels are projected into data space. Those pixels that fall within the regions defined by the training data are assigned to the appropriate categories. The procedure can be extended to as many bands, or as many categories, as necessary. In addition, the decision boundaries can be defined by the standard deviations of the values within the training sets rather than by their ranges. Advantages of this procedure for classification are accuracy, directness, and simplicity. Some of its limitations include mixing of information classes and training sets may underestimate actual ranges of classification thereby leaving large areas unclassified.

4.6.2.2 Minimum Distance Classification

This approach to classification utilizes the mean values of the spectral data that form the training sets as a means of assigning pixels to informational classes or categories. The spectral data from training fields can be plotted in multidimensional data space in the same manner illustrated previously for unsupervised classification. Values in several bands determine the positions of each pixel within the clusters that are formed by training sets for each category.

A sun classified pixels are considered for assignment to one of the several classes, the multi dimensional distance to each cluster centroid is calculated, and the pixel is then assigned to the closest cluster. Thus, the classification proceeds by always using the "minimum distance" from a given pixel to a cluster centroid defined by the training set data as the spectral manifestation of an informational class.

4.6.2.3 Maximum Likelihood Classification

In nature the classes that we classify exhibit variation in their spectral response patterns. Furthermore, spectral variability is further compounded by the effects of haze, topographic shadowing, system noise, and the effects of mixed pixels. Consequently, remote sensing images seldom record spectrally pure classes. More typically, the spectral values display a range of brightness in each band. The classification strategies considered thus far do not consider variation that may be present within spectral categories and do not address problems that arise when frequency distributions of spectral values from separate categories overlap. The *maximum likelihood* classification uses the training data as a means of estimating means and variances of the information classes, which are then used to estimate the probabilities. Maximum likelihood classification considers not only the mean, or average spectral values in assigning information class to a given pixel, but also the variability of brightness values in each class (Figure 4.22).

4.6.2.4 k-Nearest Neighbors

k-nearest-neighbors classifier (KNN) is a simple but efficient application of the distance based classification approach. It requires a set of training data representing classes present within the image. The KNN procedure examines each pixel to be classified then identifies the k-nearest training samples, as measured in multispectral data space. Typically k is set to be a relatively small integer divisor. The candidate pixel is then assigned to the class that is represented by the most samples among the k neighbors.

(a) (b)

FIGURE 4.22
Resourcesat-2 LISS-III digital raw image (a) and thematic map derived using K-mean classifier (b). Pink color indicates moderately deep to deep ravines, cyan shallow ravines, yellow cropland with very good vigor, sienne crop with moderate vigor.

4.6.2.5 Mahalanobis Spectral Distance

Also called Manhattan distance in human geography, the Mahalanobis spectral distance is the algebraic sum of the absolute differences between two pixels in the same band and in every band used in the classification. This distance, D_m, is calculated using equation 4.10. The use of the absolute value eliminates the need to worry about which pixel value should be subtracted from which pixel in deriving the difference. Since the difference is guaranteed to be positive, no differences are going to negate themselves in the summation.

$$D_m = \sum_{i=1}^{n} |DN_{Ai} - DN_{Bi}| \tag{4.10}$$

The Mahalanobis spectral distance is easier to compute and less complex than the Euclidean distance. It remains unknown which distance is able to produce more accurate results, D_e or D_m, even though the Euclidean spectral distance is used much more commonly in practice (Figure 4.23).

FIGURE 4.23
Salt-affected soil map of the area shown in Figure 4.21a developed using maximum likelihood classifier. The pink color denotes salt-affected soils, yellow crop, and blue water body.

4.6.2.6 Artificial Neural Networks

Artificial neural networks (ANNs) are computer programs that are designed to simulate human learning processes through establishment and reinforcement of linkages between input data and output data. These linkages can then be employed to link input and output in the absence of training data. ANNs are often composed of three elements: (1) an *input layer* consisting of the source data, which in the context of remote sensing are the multispectral observations; (2) the *output layer* comprising the classes required by the analyst; and (3) the *hidden*. In the context of remote sensing, associations between information classes and digital values, as expressed in the training data, strengthen weights within hidden layers that permit the ANN to assign correct labels when given spectral values in the absence of training sets.

ANNs can also be trained by *back propagation* (BP). That is, a retrospective examination of the links between input and output data in which differences between expected and actual results can be used to adjust weights. This process establishes *transfer functions*, quantitative relationships between input and output layers that assign weights to emphasize effective links between input and output layers. For example, such weights might acknowledge that some band combinations of multispectral data may be very effective in defining certain classes and others effective for other classes. In BP, hidden layers note errors in matching data to information classes and adjust the weights to minimize errors. ANNs are designed using less severe statistical assumptions than many of the commonly used classifiers like maximum likelihood, although in practice successful application requires careful application. Furthermore, ANNs have been found to be accurate in the classification of remote sensing data, although the incremental improvements in classification accuracies have generally been small or modest (Figure 4.24).

4.6.2.7 Object-Oriented Classification

Object-oriented classification uses a two-step process, namely segmentation and classification, that is intended to mimic the higher order classification processes. Each step is composed of many intermediate processes. Object-oriented classification applies a logic intended to mimic some of the higher order logic employed by human interpreters, who

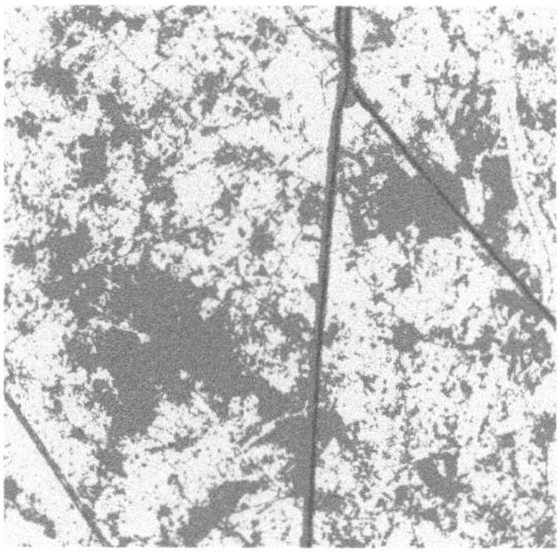

FIGURE 4.24
Salt-affected soil map of the area shown in Figure 4.21a developed using Mahalanobis spectral distance classifier. Pink color denotes salt-affected soils, yellow crop, and blue indicates water body.

can use the sizes, shapes, and textures of regions, as well as the spectral characteristics used for conventional pixel-based classification. The first process is segmentation of the image: identification of the edges of homogeneous patches that subdivide the image into interlocking regions. These regions are based on the spectral values of their pixels, as well as analyst-determined constraints. Segmentation is implemented hierarchically—regions are defined at several scales that nest within each other. These regions are the "objects" of object-oriented classification. These regions form the "objects" indicated by the term *object oriented,* to be classified. Because the regions are composed of many pixels, they have a multitude of properties, such as standard deviations, maxima, and minima, that are possible for regions but not for individual pixels. Further, object-oriented classification can consider distinctive topological properties of classes.

The second process is classification, using conventional classification procedures, usually nearest-neighbor or fuzzy classification. Each object or region is characterized by the numerous properties developed as a result of the segmentation process that can then be used in the classification process. The analyst can examine and select those properties that can be useful in classification. Though much of the conceptual and theoretical basis for object-oriented classification was developed in previous decades, it was not widely used because of the difficulties of developing useful user interfaces that enable routine use for practical applications. One of the most successful of the application programs is Definiens Professional, developed by DEFINIENS Imaging AG, a Munich-based company that has specifically devoted its resources to the development of a practical object-oriented image classification system. eCognition system provides a systematic approach and a user interface that permits implementation of concepts developed in previous decades. Object-oriented classification can be used successfully with any multispectral imagery but has proven to be especially valuable for fine-resolution imagery, which is especially difficult to classify accurately using conventional classification techniques. eCognition has been used in defense, natural resources exploration, and facilities mapping.

4.6.2.8 Spectral Angular Mapper Algorithm

The Spectral Angle Mapper (SAM) algorithms based on an ideal assumption that a single pixel of remotely sensed images represents a particular terrain feature, and can be uniquely assigned to only one terrain feature class. The SAM algorithm is based on the measurement of the spectral similarity between two spectra. The spectral similarity can be obtained by considering each spectrum as a vector in n-dimensional space, where n is the number of bands. The SAM algorithm determines the spectral similarity between two spectra by calculating the angle between the two spectra, treating them as vectors in a space with dimensionality equal to the number of spectral bands.

The SAM is a physically based spectral classifier that uses an n-dimensional angle to match pixels to reference spectra (Kruse et al., 1993). The algorithm determines the spectral similarity between two spectra by calculating the angle between the spectra, treating them as vectors in a space with dimensionality equal to the number of bands. This technique, when used on calibrated reflectance data, is relatively insensitive to illumination and albedo effects. The algorithm compares the angle between the end member spectrum vector and each pixel vector in n-dimensional space. Smaller angles represent closer matches to the reference spectrum. Pixels further away than the specified maximum angle threshold in radians are not classified (Shafri et al., 2007) (Figure 4.25).

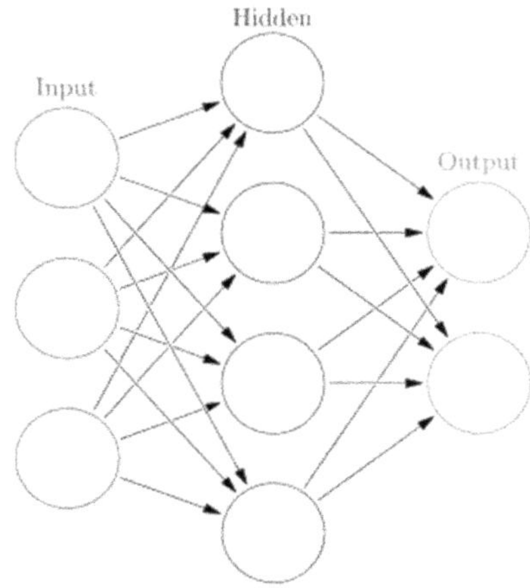

FIGURE 4.25
The concept of an artificial neural network. Each circular node represents an artificial neuron and an arrow represents a connection from the output of one neuron to the input of another (Source: Adapted from Dwivedi (2017)).

The SAM evaluates the spectra similarity in order to repress the influence of the shading to accentuate the target reflectance characteristics (Kruse et al., 1993). The mathematical formulation of SAM attempts to obtain the angles formed between the reference spectrum and the image spectrum treating them as vectors in a space with dimensionality equal to the number of bands (Boardman, 1992). The SAM may be represented as the following formulation:

$$\alpha = \cos^{-1} \frac{\sum XY}{\sqrt{\sum (X)^2 \sum (Y)^2}} \tag{4.11}$$

where α is the angle formed between reference spectrum and image spectrum, X is the image spectrum, Y is the reference spectrum. The SAM value is expressed in radians where minor angle α represents the major similarity among the curves. The angle α, determined by $\cos^{-1}$, presents a variation anywhere between 0° and 90°. The equation above can also be expressed as cos α (equation 4.11). In these conditions, the best estimate acquires values close to 1.

$$\cos\alpha = \frac{\sum XY}{\sqrt{\sum (X)^2 \sum (Y)^2}} \tag{4.12}$$

4.6.2.9 Spectral Correlation Classifier

The major limitation of SAM is the impossibility of distinguishing between negative and positive correlations because only the absolute value is considered. Table 4.1 exemplifies

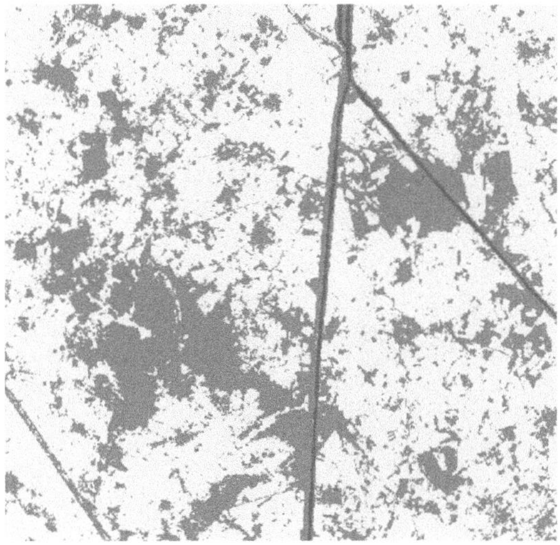

FIGURE 4.26
Salt-affected soil map of the area shown in Figure 4.21a developed using spectral angle mapper classifier. Pink color denotes salt-affected soils, yellow color cropland.

this SAM limitation when it compares the targets of the hypothetical curves in relation to a reference spectrum. On the one hand, the reference spectrum shows the opposite behaviour of the targets analyzed. The cos (SAM) values demonstrate high correlation, indicating—erroneously—the presence of the material. On the other hand, Pearson's correlation presents an excellent estimate for an analysis of high correlation being different from what is desired. The Spectral Correlation Mapper (SCM) method is a derivative of Pearson correlation coefficient that eliminates negative correlation and maintains the SAM characteristic of minimizing the shading effect resulting in better results. The SCM varies from –1 to 1 and cos (SAM) varies from 0 to 1 (Figure 4.26).

4.6.2.10 Support Vector Machines Classifier

Support vector machines (SVMs) are a relatively new supervised and have their roots in Statistical Learning Theory. They are essentially binary classifiers. However, they can be adopted to handle the multiple classification tasks common in remote sensing applications. SVMs function by nonlinearly projecting the training data in the input space to a feature space of higher (infinite) dimension by use of a kernel function. This results in a linearly separable dataset that can be separated by a linear classifier. This process enables the classification of remotely sensed datasets which are usually nonlinearly separable in the input space. In many instances, classification in high dimension feature spaces results in over-fitting in the input space; however, in SVMs over-fitting is controlled through the principle of structural risk minimization (Vapnik, 1995). The empirical risk of misclassification is minimized by maximizing the margin between the data points and the decision boundary (Mashao, 2004).

By their nature SVMs are intrinsically binary classifiers. However, there exist strategies by which they can be adopted to multiclass tasks associated with remote sensing studies. The two approaches commonly used are the One-Against-One (1A1) and One-Against-All (1AA) techniques. Whereas the 1AA technique is more predisposed to yielding

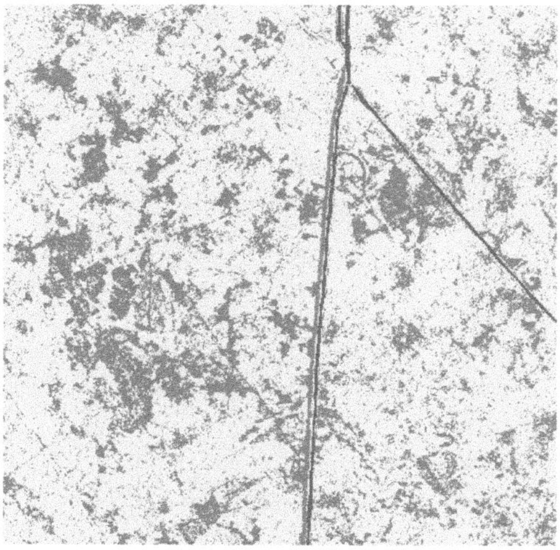

FIGURE 4.27
Salt-affected soil map of the area shown in Figure 4.21a developed using spectral correlation classifier. Pink color denotes salt-affected soils, yellow color cropland.

unclassified and mixed pixels, the resulting classification accuracy is not significantly different from 1A1 approach. The 1AA approach represents the earliest and most common SVM multiclass approach (Melgani and Bruzzone, 2004) and involves the division of an N class dataset into N two-class cases. If say the classes of interest in a satellite image include water logged areas, salt-affected soils and cropland classification would be effected by classifying waterlogged are as against non-water logged areas (i.e., salt-affected soils and cropland) or cropland against non-cropland (i.e., waterlogged areas and salt-affected soils). The 1A1 approach, on the other hand, involves constructing a machine for each pair of classes resulting in N(N–1)/2 machines. When applied to a test area, each classification gives one vote to the winning class and the point is labeled with the class having most votes.

This approach can be further modified to give weightage to the voting process. The 1AA approach has an edge over 1A1 in that its performance can be compromised due to unbalanced training datasets (Gualtieri and Cromp, 1998), Computer however, the 1A1 approach is more computationally intensive since the results of more SVM pair sought to be computed. A map showing mining areas along with other land use/land cover categories derived from Resourcesat-2 LISS-IV is appended as Figure 4.27.

4.7 Digital Change Detection

Change analysis is in essence a spatial comparison of two or more land covers of the same geographic area produced from remotely sensed data that are recorded at different times. The basic premise in using remotely sensed data for change detection is that changes in the objects of interest will result in changes in reflectance values or local textures that are separable from changes caused by other factors such as differences in atmospheric conditions, illumination and viewing angles, and soil moistures (Deer, 1995). Conventional change detection approaches are subjective, time-consuming, and cost-prohibitive, more so when a larger area needs to be addressed besides being subjective in nature. The change detection methods can be divided into three categories (Mas, 1999): (i) image enhancement-methods, (ii) multi-temporal analysis, and (iii) post-classification comparison, and other approaches combining several methods or consisting of novel methods (Lu et al., 2003). The multi-temporal nature of earth observation satellite data enables assessment of changes in the type or condition of surface features. Over the past years, several digital change detection techniques using remote sensing image have been developed (Singh, 1989; Coppin and Bauer, 1996; Blaschke et al., 2000; Richard et al., 2005).

4.7.1 Image Enhancement Techniques

4.7.1.1 Univariate Image Differencing

The technique involves subtracting one date of original or transformed (e.g., VIs, albedo) image from a second date that has been precisely registered to the first. The resultant image shows the changes that have taken period during the period in question. Image differencing is the most widely used technique for change detection and has been used in a variety of geographical environments. A critical element of the image differencing method is deciding where to place the threshold boundaries between change and no-change pixels displayed in the histogram.

4.7.1.2 Image Regression

A mathematical model that describes the fit between two multi-date images of the same area can be developed through stepwise regression. The model assumes that a pixel at time 2 is linearly related to the same pixel at time 1 in all bands of the electromagnetic spectrum acquired by the sensor. This implies that the spectral properties of a large majority of the pixels have not changed significantly during the time interval. The dimension of the residuals is an indicator of where change occurred. The regression technique accounts for differences in mean and variance between pixel values for different periods. Simultaneously, the process also reduces the adverse effects from divergences in atmospheric conditions and/or Sun angles. The regression method of change detection, pixels from time t_1 are assumed to be a linear function of the time t_2, pixels. The difference image can be defined as follows:

$$DXk_{ij} = Xk_{ij}(t_2) - XK_{ij}(t_1) \tag{4.13}$$

4.7.1.3 Image Ratioing

In ratioing two registered images from different dates with one or more bands in an image are ratioed, band by band. The data are compared on a pixel by pixel basis. A pixel that has not changed will yield a ratio value of one. Areas of change will have values either higher or lower than one. For every pair of gray values at the same location at dates t_1 and t_2 the per-pixel ratio of the two values is calculated. Like image differencing and image regression in method also, the choice of a suitable threshold level is a critical factor.

4.7.1.4 Principal Component Analysis

With this approach, it is possible to strengthen wavelength-dependent, material-specific differences. PCA can be used in different ways for change detection. Two bi-temporal spectral bands of the same location are analyzed in a 2D feature space. Consequently, all gray values are located in relation to the two PCs. Generally, the pixels with no change show a high correlation with the first PC in contrast to the changed pixel. As a consequence, the first PC contains the "unchanged" information and the second component the "change" information (Macleod and Congalton, 1998).

4.7.1.5 Multivariate Alteration Detection

The multivariate alteration detection (MAD) uses canonical correlation analysis to find relationships between two datasets. The canonical analysis provides two sets of linear combinations of the original data. The first two linear combinations (canonical variates) have the largest correlation (first canonical correlation). The second canonical variates have the second largest correlation and are orthogonal to the first canonical variates. Basis for the change detection approach is the difference between these pairs of variates. The MAD transformation is defined as follows:

$$\begin{bmatrix} X \\ Y \end{bmatrix} \rightarrow \begin{bmatrix} a_p^T X - b_p^T Y \\ \vdots \\ a_l^T X - b_l^T Y \end{bmatrix} \tag{4.14}$$

4.7.1.6 Post-Classification Comparison

Post-classification comparison sometimes referred to as "delta classification" involves independent classification of images for two time periods followed by a pixel-by-pixel or segment-by-segment comparison to detect changes in cover type. Post-classification analysis is based on a comparison of two independent classification results for at least two dates t_1 and t_2. This method allows the determination of the kind of change from one class to another. There are two basic approaches for LCLU change detection (Singh, 1989): (i) comparative analysis of independently produced classifications from different dates (post-classification comparison: map-to-map comparison) and (ii) simultaneous analysis of multi-temporal data (multidate classification and others: image-to-image comparison). However, accuracy of post-classification change detection method is lowdue to the fact that it incorporates any errors present in the original classifications (Coppin et al., 2004). The main sources of uncertainty of this method are as follows:

- Misregistration of the polygon boundaries (locational inaccuracy) in the different classifications and, therefore, the presence of border pixels with false-positive or negative changes.
- Problems derived from classification errors: a false-positive change may be recorded when no change has taken place because a polygon in one or both of the two maps is misclassified, or false-negative changes, when no change is identified but a change has taken place.

- This approach requires very good accuracy in both classifications because the accuracy of the change map is the product of the accuracies of the individual classifications (Singh, 1989; Lambin and Strahler, 1994).
- Moreover, if the images used for each classification are from different seasons of the year, the comparison can be more difficult, especially for some land use/cove categories due to vegetation phenology.

Shao (2006) summarized the main errors in change detection including errors in data (e.g., image resolution, accuracy of location and image quality), errors caused by preprocessing (the accuracy of geometric correction and radiometric correction), errors caused by change detection methods and processes (e.g., classification and data extraction error), errors in field survey (e.g., accuracy of ground reference), and errors caused by post-processing.

4.7.1.7 Artificial Neural Network-Based Change Detection

Usually, change detection involves two or more geometrically co-registered remotely sensed images acquired for the same ground area at different times. In the *change mask development (CMD) technique* only changes and non-changes are detected and no categorical change information can be directly provided, whereas in *categorical change extraction (CCE)* method complete categorical changes are extracted. In the first category, changed and non-changed areas are separated by a preset threshold when comparing the spectral reflectance values of multi-temporal satellite images. The amount of change is a function of the preset threshold. The threshold needs to be determined by experiments. The nature of the changes is unknown directly from these techniques and needs to be identified by other pattern-recognition techniques. Therefore, these techniques are only suitable for development of a change mask. Most change-detection methods fall into the first category. For example, image differencing, image ratioing, and image regression only lead to the development of a change mask.

Although plenty of successful application cases have been reported on the monitoring and detecting environmental change, there are enormous challenges on applying multi-temporal imagery to derive timely information on the earth's environment and human activities. It has been generally agreed that change detection is a complicated and integrated process. No existing approach is optimal and applicable to all cases. Furthermore, at present the degree of automation is low to prevent real-time applications (Jianya et al., 2008).

4.8 Accuracy Assessment

In the context of information extraction by image analysis, accuracy measures the agreement between a reference image or standard assumed to be correct and a classified image of unknown quality (Campbell, 2007). Classification error occurs when a pixel (or terrain feature) belonging to one category is assigned to another category. The classification error could be grouped into two categories: *errors of omission* and *error of commission*. Errors of omission occur when a feature is left out of the category being evaluated; *errors of commission* occur when a feature is incorrectly included in the category or class being evaluated. An error of omission in one category will be counted as an error in commission in another category. Thematic maps on natural resources and environment are, generally, generated at a particular point in time and indicate the status for that period. These maps are referred to as one-point-in-time (OPIT). However, for monitoring certain dynamic features like degraded lands, crop progression, forest cover, etc., thematic maps at more than one point in time are generally generated.

4.8.1 Uni-Temporal Thematic Maps

Accuracy assessment for thematic maps fora single date or OPIT is performed by comparing the map generated using remotely sensed data to a reference map based on a different information source. In order to be compared, both the map to be evaluated and the reference map must be accurately registered geometrically to each other. These maps must also have a common classification scheme, and they should have been classified at the same level of detail. One simple method of comparison is to calculate the total area assigned to a particular category in both maps and to compare the overall figures. This approach is termed *non-site-specific-accuracy*. On the other hand, *site-specific accuracy* is based on a comparison of the two maps at specific locations (i.e., individual pixels in two digital images). In this type of comparison, it is obvious that the degree to which the pixels in one image spatially align with the pixels in the second image contributes to the accuracy assessment result. Errors in classification should be distinguished from errors in registration or positioning of boundaries (www.e-education.psu.edu/geog883/node/524, accessed on June 19, 2017).

4.8.1.1 Sampling Scheme

Deciding upon the number of samples to be used for accuracy assessment and their location in the thematic map to be evaluated for accuracy is the first step. The number of samples required for accuracy assessment could be estimated based on binomial probability theory as follows:

$$N = \frac{z^2(p)(q)}{E^2} \tag{4.15}$$

where N is the sample size, p is the percent expected accuracy of the entire map, q = 110-p, E is allowable error, z = 2 (σ) covering 95.4% of map.

Another approach to estimate the number samples required for accuracy assessment is based on multinomial probability theory:

$$N = \frac{B\Pi_i(1-\Pi_i)}{b_i^2} \tag{4.16}$$

Π_i is the proportion of a population in ith class out of k classes that have proportion closest to 50%; b_i is the desired precision for this class, B is the upper (α/k) percentile of the chi square (χ^2) distribution with 1 degree of freedom, and k is the number of classes.

4.8.1.2 Accuracy Assessment

Accuracy of thematic maps is most often reported as a percentage correct. The errors in thematic maps could be grouped into two categories—the *error of commission* and *error of omission*. If the total number of correct pixels in a category is divided by the total number of pixels that were actually classified in that category, the result is a measure of commission error, or error of inclusion. This measure, also called the *user's accuracy* or reliability, is the probability that a pixel classified on the map actually represents that category on the ground.

The error of omission, on the other hand, indicates the number of pixels correctly classified in a particular category as a percentage of the total number of pixels actually belonging to that category in the image. It is also known as *producer's accuracy.* Accuracy assessment is generally made by constructing an error matrix or contingency table (Table 4.3).

	Producer's Accuracy (%)	User's Accuracy (%)
Salt-affected soils	60/73 = 90	70/88 = 80
Waterlogged areas	55/60 = 92	55/58 = 95
Cropland	99/103 = 96	99/99 = 100
Forests	37/50 = 74	37/41 = 90
Water body	121/121 = 100	121/121 = 100

The rows in the table represent categories of terrain features delineated using remotely sensed data, whereas columns represent the categories from a reference map or ground truth data. An *overall accuracy* of a classification map is determined by dividing the total classified pixels (sum of the major diagonal) by the total number of reference pixels in the error matrix (N).

4.8.1.3 Kappa Coefficient (K)

Kappa coefficient (K) can be computed as follows (Bishop et al., 1975):

$$K = \frac{N\sum_{i=1}^{r} xii - \sum_{i=1}^{r} (xi+)(x+i)}{N^2 - \sum_{i=1}^{r} (xi+)(x+i)} \tag{4.17}$$

where "r" is the number of rows in the matrix, xii is the number of observations in row i and column i (the ith diagonal elements), xi+ and x + i are the marginal totals of row "r" and column i, respectively, and N is the number of observations.

Where

N = 407

TABLE 4.3

Error Matrix for Thematic Map Accuracy Assessment

Ground Truth or Reference Information						Row Total
Remote sensing classification	Class	1	2	3	k	
	1	$X_{1,1}$	$X_{1,1}$	$X_{1,1}$	$X_{1,k}$	X_{1+}
	2	$X_{2,1}$	$X_{2,2}$	$X_{2,3}$	$X_{2,k}$	X_{2+}
	3	$X_{3,1}$	$X_{3,2}$	$X_{3,3}$	$X_{3,k}$	X_{3+}
	k	$X_{4,1}$	$X_{4,2}$	$X_{4,3}$	$X_{4,k}$	X_{k+}
	Column total	X_{+1}	X_{+2}	X_{+3}	X_{+k}	N

Source: http://web.uri.edu/ltrs/files/516AccuracyAssessment.pdf, accessed on June 19, 2017.

$$\sum_{i=1}^{r} \text{xii} = (70 + 55 + 99 + 37 + 121) = 382$$

$$\sum_{i=1}^{r} (\text{xi+})(\text{x + i}) = (88 \times 73) + (59 + 60) + (99 \times 103) + (41 \times 50) + (121 \times 121) = 36{,}792$$

$$\text{Therefore, K} = \frac{407(382) - 36{,}792}{407^2 - 36{,}792} = \frac{118{,}682}{128{,}857} = 0.921$$

Depending on whether a thematic map has been developed through visual interpretation approach—either analog (hard copy image) or digital data, or through computer-aided digital analysis—there are two types of thematic maps-polygon-based and pixel-based. Accuracy assessment is generally carried out on polygon basis. Although there are some statistical and conceptual advantages with pixel-based accuracy assessment, with the current state-of-the-art, locating consistently a pixel in an image with sufficient precision for accuracy assessment does not seem feasible. It is due to the fact that even with geometric correction algorithms, a 30 m pixel can currently be geo-rectified with a RMS error of ±0.5 pixel (Khorram et al., 1999).

In polygon-based maps too there is an element of inaccuracy. At a given map scale the smallest polygon represents the minimum mapping unit (MMU). All other units smaller than MMU are not delineated and portrayed in the map. Thus, a polygon need not be pure to be classified as a given type. For example, a salt-affected soil polygon may include a small patch of cropland and waterlogged area. Although some classification systems allow mapping even mixed categories, the basic paradigm that is assumed is that each location in the map belongs to only one category only. Other ancillary issues related to accuracy assessment are as follows: validation of historical data, the effect of the spatial scale of the remote sensing system used to assess a landscape, merging data of different spatial resolution, and boundary location of error of a polygon (Biging et al., 1999). For overview and discussions on these issues refer to Kharram et al. (1999).

4.8.2 Multi-Temporal Thematic Maps

Sampling scheme for determining the accuracy of a change map is inherently different from sampling of an *OPIT thematic map* because the change categories usually represent a very small portion of the change image or thematic map. As such they can be considered rare classes. Conventional sampling techniques often prove to be expensive and inefficient or change detection accuracy assessment. Figure 4.2 presents a comparison between an OPIT error matrix and the corresponding error matrix for change detection (Macleod et al., 1995; Macleod and Congalton, 1998). The table shows an OPIT error matrix for three vegetation or land use/land cover categories (A, B, and C). The matrix is of dimension 3 × 3. One axis represents the three categories as derived from the remotely sensed classification and the other axis shows the three categories identified from the reference data. The major diagonal of this matrix indicates correct classification. In other words, when the classification indicates the category was A and reference data agree that it is A, then the [A, A] cell in the matrix is tallied. The same logic follows for other categories B and C. Off-diagonal elements in the matrix indicate the different possibilities of error.

A change detection error matrix generated for the same three vegetation or land use/land cover categories (A, B, and C) is also shown in Figure 4.28. Now, the matrix is of 9 × 9 dimension instead of 3 × 3. The dimensions of a change detection error matrix are the square of the number of categories. The question of interest in change detection matrix is what category was this polygon (pixel) at time 1, and what it is at time 2? The answer has 9 possible outcomes for each dimension of the matrix (A at time 1, and A at time 2; A at time 1, and B at time 2; A at time 1, and C at time 2...C at time 1, and C at time 2), all of which are indicated in the matrix. It is then important to note what remotely sensed data said about the change and compare it to what reference data indicate. This comparison exactly uses the same logic as for OPIT matrix; it is just complicated by the two time periods (i.e., the change). The major diagonal indicates correct classification, while the off-diagonal elements indicate misclassification error or errors of commission or omission. Commission error occurs when the detected change is false; i.e., "false positive." Omission error occurs when change is undetected, i.e., "false negative."

A case study addressing the accuracy assessment of multi-temporal thematic map is appended as Table 4.4. There could be two scenarios when we are dealing with multi-temporal thematic maps. One, a category may change to another category, or two, there may not be any change in thematic map categories over a given period of time. Like the accuracy assessment of single-date thematic map, here too the categories delineated from remote sensing data are presented in one axis whereas corresponding categories as

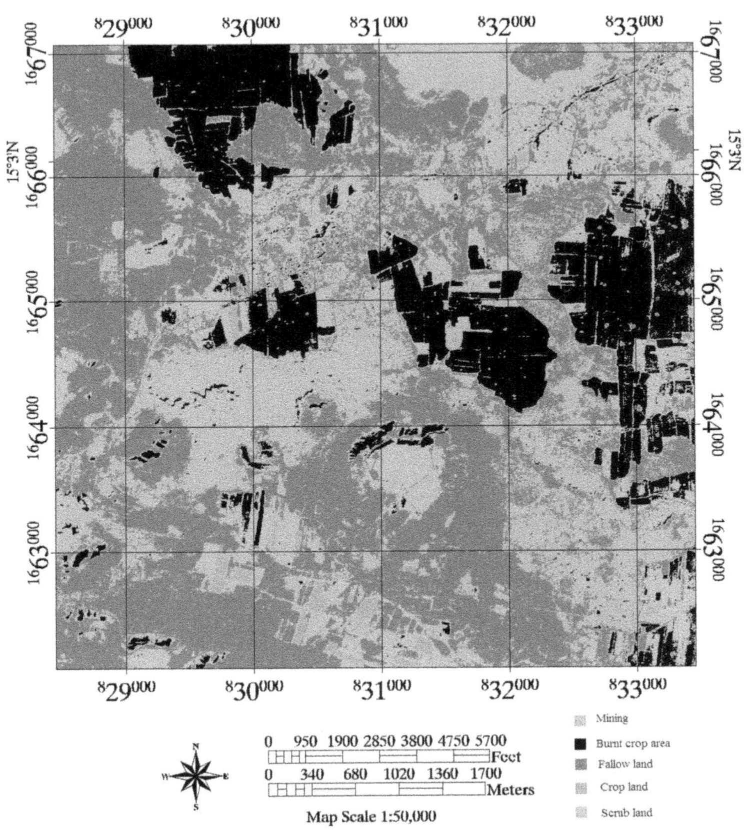

FIGURE 4.28
Map showing mining areas in part of Andhra Pradesh, southern India. The map has been developed using support vector machine algorithm.

TABLE 4.4

Error Matrix (5 × 5) for Thematic Map Accuracy Assessment

	Reference Data					
Classification	**Salt-Affected Soils**	**Waterlogged Area**	**Cropland**	**Forests**	**Water Body**	**Row Total**
Salt-affected soils	70	5	0	13	0	88
Waterlogged areas	3	55	0	0	0	58
Cropland	0	0	99	0	0	99
Forests	0	0	4	37	0	41
Waterbody	0	0	0	0	121	121
Column total	73	60	103	50	121	407

Overall accuracy = 382/407 = 93.86%.

observed in reference map on the other axis. As evident from Table 4.5, the left-half columns show no change in categories in three categories of land cover (e.g., S = Salt-affected soils. C = Cropland and W = waterlogged areas). Similarly, the first three rows indicate no change while the rest rows indicate the change in delineated categories (Figure 4.29 and Table 4.6).

TABLE 4.5

Change-Detection and No-Change/Change Error Matrices for the Post-Classification Change-Detection Technique

	Reference Data									
	No Change			Change						
	SS	**CC**	**WW**	**SC**	**SW**	**CS**	**CW**	**WS**	**WC**	**Total**
SS	201	55	106	49	59	1	7	128	11	617
CC	113	844	168	127	5	15	61	17	167	1,517
WW	106	923	8,383	547	366	86	234	237	717	11,599
SC	212	707	208	215	30	51	68	85	127	1,703
SW	28	147	108	31	4	14	5	21	37	395
CS	21	4	1	60	0	1	0	1	10	98
CW	12	129	27	14	0	11	1	2	32	235
WS	82	98	398	71	55	11	37	330	127	1,209
WC	215	873	940	211	45	93	84	284	459	3,204
Total	990	3,790	10,339	1325	664	283	497	1,105	1,687	20,580

Overall accuracy 10,438/20,580 = 50.72%.

No Change/Change Error Matrix

	No Change	**Change**	**Total**
No Change	10,899	2,834	13,733
Change	4,220	2,627	6,847
Total	15,119	5,461	20,580

Overall Accuracy 13,526/20,580 = 65.7%.

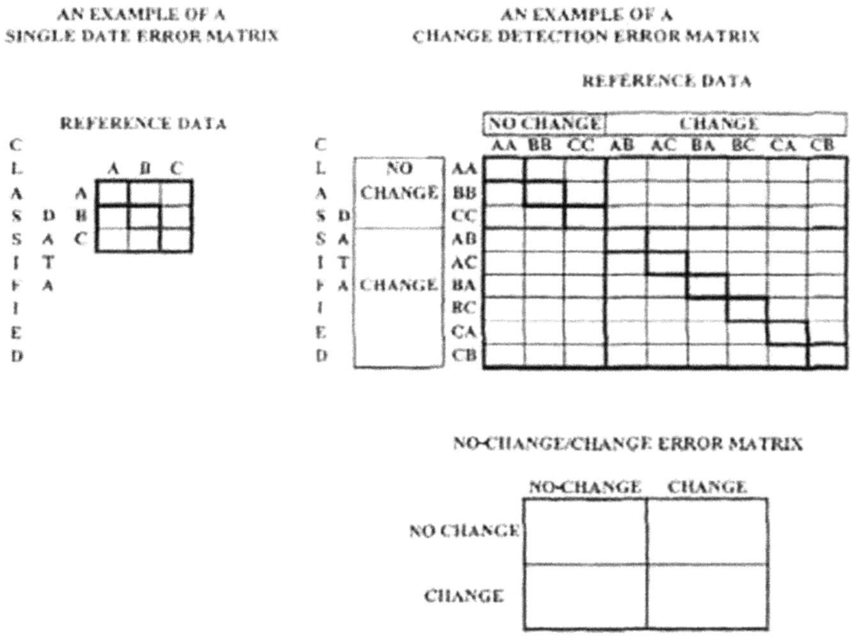

FIGURE 4.29
An illustration of the framework for accuracy assessment of single-date and multi-temporal change detection approaches (Macleod and Congalton, 1998; http://info.asprs.org/publications/pers/98journal/march/1998_mar_207-216.pdf, accessed on June 10, 2017).

TABLE 4.6
Nine Change-Detection Categories

1975 to 1995
[SS] Salt-affected soils to salt-affected soils
[CC] Cropland to Cropland
[WW] Waterlogged areas to waterlogged areas
[SC] Salt-affected soils to cropland
[SW] Salt-affected soils to water body
[CS] Cropland to salt-affected soils
[CW] Cropland to waterlogged areas
[WS] Waterlogged areas to salt-affected soils
[WC] Waterlogged areas to cropland

4.9 Conclusions

Digital image processing is a very fast developing signal processing domain. With regard to digital image analysis, several methods are available now. However, Gaussian maximum likelihood classier continues to be the most commonly used classifier. Merits and limitations of these classifiers have been brought out. Which classifier is most suitable for a particular application is very difficult to decide due to variations in terrain conditions and the complexity of the features as they appear in remote sensing images. Continued developments in the field of digital image processing may introduce newer and more efficient techniques that may help remote sensing practitioners, researchers, and academicians in deriving more meaningful information on natural resources and environment including degraded lands. The kind of processing required for a given region needs to be decided by the image analyst. For implementation of any land degradation program, detailed information on degraded land with regard to nature extent, spatial distribution, and temporal is needed. The availability of high spatial resolution multispectral data from a suite of several earth observation missions like Worldview, Quickbird, Carosat, etc. enable calls for developing newer techniques for generating such information. Experience gained while analyzing hyperspectral data from AVIRIS, Hyperion, etc. data may be utilized to develop newer and more efficient techniques for processing and analysis of such data that are likely to be available from Indo-NASA joint mission.

References

Adler-Golden, S.M., Matthew, M.W., Bernstein, L.S., Levine, R.Y., Berk, A., Richtsmeier, S.C., Acharya, P.K., Anderson, G.P., Felde, G., Gardner, J., Hoke, M., Jeong, L.S., Pukall, B., Ratkowski, A., and Burke, H.K., 1999. "Atmospheric Correction for Short-wave Spectral Imagery Based on MODTRAN4," in Summaries of the Eighth Annual JPL Earth Science Workshop, Vol. I, available at http://makalu.jpl.nasa.gov.

Al-Wassai, F.A., Kalyankar, N.V., and Al-Zaky, A.A., 2012. Multisensor images fusion based on feature-level. *International Journal of Latest Technology in Engineering, Management and Applied Science (IJLTEMAS)*, I(V):124–138.

Bannari, A., Morin, D., Bonn, F., and Huete, A.R., 1995. A review of vegetation indices. *Remote Sensing Reviews*, 13(1–2):95–120.

Biging, G.S., Colby, D.R., and Congalton, R.G., 1999. Sampling system for change detection accuracy assessment. In: Lunetta, R.S., and Elvidge, C.D., (eds.) *Remote Sensing Change Detection: Environmental Monitoring Methods and Applications*. Taylor and Francis, Milton Park.

Bishop, Y. M. M.; Fienberg, S. E.; Holland, P. W. 1975. *Discrete multivariate analysis: Theory and practice*. MIT Press, Cambridge, MA, 557 p.

Blaschke, T., Lang, S., Lorup, E., Strobl, J., and Zeil, P., 2000. Object-oriented image processing in an integrated GIS/remote sensing environment and perspectives for environmental applications. In: Cremers, A., and Greve, K., (eds.) *Environmental Information for Planning, Politics and the Public, vol. II*. Metropolis-Verlag, Marburg, pp. 555–570.

Boardman, 1992. *SIPS User's Guide Spectral Image Processing System, Version 1.2*. Center for the Study of Earth from Space, Boulder, 88 pp.

Campbell, J.B., 2007. *Introduction to Remote Sensing*, 4th edn. The Guilford Press, New York.

Campbell, J.B., and Wynne, R.H., 2011. *Introduction to Remote Sensing*, 4th edn. The Guillford Press, New York, London.

Carper, W.J., Lilesand, T.W., and Kieffer, R.W., 1990. The use of intensity–hue–saturation transformation for merging SPOT panchromatic and multispectral image data. *Photogrammetric Engineering and Remote Sensing*, 56(4):459–467.

Chavez, P.S., 1975. Atmospheric, solar, and M.T.F. corrections for ERTS digital imagery.

Chavez, Jr. P.S., 1988. An improved dark object subtraction technique for atmospheric scattering correction of multispectral data. *Remote Sensing of Environment*, 24:459–479.

Chavez, P.S., and Kwarteng, A.Y., 1989. Extracting spectral contrast in Landsat thematic mapper image data using selective principal component analysis. *Photogrammetric Engineering and Remote Sensing*, 55:339–348.

Chavez, P., 1996. Image-based atmospheric corrections—revisited and improved. *Photogrammetric Engineering and Remote Sensing*, 62:1025–1036.

Coppin, P.R., and Bauer, M.E., 1996. Digital change detection in forest ecosystems with remote sensing imagery. *Remote Sensing Reviews*, 13:207–234.

Coppin, P., Jonckheere, I., Nackaerts, K., Muys, B., and Lambin, E., 2004. Digital change detection methods in ecosystem monitoring a review. *International Journal of Remote Sensing*, 25(9), pp. 1565–1596.

Cracknell, A.P., and Hays, L.W., 1993. Atmospheric corrections to passive satellite remote sensing data. In: *Introduction to Remote Sensing*. Taylor and Francis, London, pp. 116–158.

Daubechies, I., 1992. Ten lectures on wavelets volume 61 of CMBS-NSF regional conference series in applied mathematics (Philadelphia: Society for Industrial and Applied Mathematics (S.I.A.M.)): pp 1–16.

Davenport, T., and Vellon, M., 1987. Tag image file format (Rev 4.0), www.martinreddy.net/gfx/2d/TIFF-4.txt.

Deer, P.J., 1995. Digital change detection techniques: Civilian and military applications. 2396 D. Lu et al. International Symposium on Spectral Sensing Research 1995 Report (Greenbelt, MD: Goddard Space Flight Center), www.gsfc.nasa.gov/ISSSR-95/digitalc.htm.

Duda, R.O., Hart, P.E., and Stork, D.G., 2001. *Pattern Classification and Scene Analysis*. Wiley, New York.

Dwivedi, R.S., 2017. *Remote Sensing of SAOILS*. Springer-Verlag, Germany, 532 p.

Gao, B., and Goetz, A.F.H. 1990. Column atmospheric water vapor and vegetation liquid water retrievals from airborne imaging spectrometer data. *Journal of Geophysical Research*, 95(D4):3549–3564.

Gao, B.C., Heidebrecht, K.B., and Goetz, A.F.H., 1999. *ATmospheric REMoval (ATREM) Model User's Guide, Version 3.1*. Centre for the Study of Earth from Space, Boulder, CO, 31 p.

Gao, J., 2009. *Digital Analysis of Remotely Sensed Imagery*. Mc-Graw Hill, New York.

Gonzalez, R.C., and Woods, R.E., 1992. *Digital Image Processing*. Addison-Wesley, Reading, MA.

González-Audícana, M., Saleta, J.L., Catalán, R.G., and García, R., 2004. Fusion of multispectral and panchromatic images using improved IHS and PCA mergers based on wavelet decomposition. *IEEE Transactions on Geosciences and Remote Sensing*, 42(6):1291–1299.

Gomez-Landesa, E., Rango, A., and Bleiweiss, M., 2004. An algorithm to address the MODIS bowtie effect, *Canadian Journal of Remote Sensing*, 30(4):644–650.

Guangmeng, G., 2003. Eliminating bowtie effect of MODIS Data without ephemeris, *Remote Sensing Technology and Application*, 18(3):172–175.

Gualtieri, J.A., and Cromp, R.F., 1998. Support vector machines for hyperspectral remote sensing classification. In: *Proceedings of the 27th AIPR Workshop: Advances in Assisted Recognition*, Washington, DC, October 14–16th, 1998 (Washington, DC: SPIE), pp. 221–232.

Hallada, W.A., and Cox, S. 1983. "Image sharpening for mixed spatial and spectral resolution satellite systems," *Proc. of the 17th International Symposium on Remote Sensing of Environment*, 9–13 May, pages 1023–1032.

Hemmleb, M., and Wiedemann, A., 1997. Digital rectification and generation of orthoimages in architectural photogrammetry. CIPA International Symposium, IAPRS, XXXII, Part 5C1B: 261–267, International Archives of Photogrammetry and Remote Sensing, Göteborg, Sweden.

Holben, B.N., 1986. Characteristics of maximum-value composite images from temporal AVHRR data, *International Journal of Remote Sensing*, 7:1417–1434.

Jensen, J.R., 2005. *Introductory Digital Image Processing: A Remote Sensing Perspective*, 3rd edn, Prentice Hall, Upper Saddle River, NJ, 526 p.

Jensen, J.R., 2007, *Remote Sensing of the Environment-An Earth Resources Perspective*. Prentice Hall, Upper Saddle River, NJ.

Jianya, G., Haigang, S., Guorui, M., and Qiming, Z., 2008. A review of multi-temporal remote sensing data change detection algorithms. The International Archives of the Photogrammetry, Remote Sensing and Spatial Information Sciences. Vol. XXXVII. Part B7. Beijing 2008.

Jinghui, Y., Jixian, Z., Haitao, L., Yushan, S., and Pengxian, P., 2010. Pixel level fusion methods for remote sensing images: a current review. In: Wagner, W. and Székely, B. (eds.) *ISPRS TC VII Symposium—100 Years ISPRS*, Vienna, Austria, July 5–7, 2010, IAPRS, Vol. XXXVIII, Part 7B.

Kauth, R.J., and Thomas, G., 1976. The tasselled cap—a graphic description of the spectral-temporal development of agricultural crops as seen by LANDSAT, *Proceedings of the Symposium on Machine Processing of Remotely Sensed Data*; Purdue University, West Lafayette, Indiana. 159.

Khorram, S., Biging, G., Colby, D., Congalton, R., Dobson, J., Ferguson, R., Goodchils, M., Jensen, J., and Mace, T., 1999. *Accuracy Assessment of Remote Sensing-Derived Change Detection, Photogrammetric Engineering and Remote Sensing*. ASPRS, Bethesda, MD, 64 p.

Kruse, F.A., Boardman, J.W., Lefkoff, A.B., Heidebrecht, K.B., Shapiro, A.T., Barloon, P.J., and Goetz, A.F.H., 1993. The Spectral Image Processing System (SIPS) – Interactive Visualization and Analysis of Imaging Spectrometer Data. *Remote Sensing of Environment*, 44:145–163.

Lambin, E.F., and Strahler, A.H., 1994. Change vector analysis in multitemporal space: a tool to detect and categorize land-cover change processes using high temporal resolution satellite data. *Remote Sensing of Environment*, 48:231–244.

Levin, N., (ed.) 1999. Fundamentals of remote sensing. E-book.

Lillesand, T.M., and Kiefer, R.W., 1994. *Remote Sensing and Image Interpretation*, 3rd edn. Wiley, New York.

Lillesand, T. M., Kiefer, R. W., and Chipman, J.W., 2004. *Remote Sensing and Image Interpretation*, John Wiley & Sons, New York, 763.

Lillesand, T., Kiefer, R.W., and Chipman, J., 2015. *Remote Sensing and Image Interpretation*, 7th edn. Wiley, New York.

Lu, D., Mausel, P., Brondízio, E., and Moran, E., 2003. Change detection techniques. *International Journal of Remote Sensing*, 25:2365–2407.

Macleod, R., Congalton, R., and Short, F., 1995. Using quantitative accuracy assessment techniques to compare various change detection algorithms for monitoring eelgrass distribution in Great Bay, NH generated from Landsat-TM data. *Proceedings of the Sixty First Annual Meeting of the American Society of Photogrammetry and Remote Sensing*, Charlone, N.C, Vol. 3, pp. 876–885, 1995.

Macleod, R.D., and Congalton, R.G., 1998. A quantitative comparison of change-detection algorithms for monitoring eelgrass from remotely sensed data. *Photogrammetric Engineering and Remote Sensing*, 64:207–216.

Mallat, S.G., 1989. A theory for multiresolution signal decomposition: The wavelet representation. *IEEE Transactions on Pattern Analysis and Machine Intelligence*, 11:674–693.

Mas, F.J., 1999. Monitoring land-cover changes: a comparison of change detection techniques, *International Journal of Remote Sensing*, 20:139–152.

Mashao, D., 2004. Comparing SVM and GMM on parametric feature-sets. In: *Proceedings of the 15th Annual Symposium of the Pattern Recognition Association of South Africa*, Cape Town, South Africa. November 27–29th, 2004.

Mather, P.M., 2004. Computer processing of remotely-sensed images: An Introduction models by applying environmental and climatic information: Part I. Assessment of net primary production. *International Journal of Remote Sensing*, 19(1):97–117.

Melgani, F., and Bruzzone, L., 2004. Classification of hyperspectral remote sensing images with Support Vector Machines. *IEEE Transactions on Geoscience and Remote Sensing*, 42:1778–1790.

INPE (Instituto Nacional de Pesquisas Espaciais) and CAST, Radiometric Quality Assessment of CBERS-2, 2004.

Pohl, C., and van Genderen, J.L., 1998. Multi-sensor image fusion in remote sensing: Concepts, methods and applications. *International Journal of Remote Sensing*, 19(5):823–854.

Radke, R.J., Srinivas, A., Omar, A.K., and Badrinath, R., 2005. Image change detection algorithms: A systematic survey. *IEEE Transactions on Image Processing*, 14(3):294–307.

Ranchin, T., Wald, L., and Mangolini, M., 1996. The ARSIS method: a general solution for improving spatial resolution of images by the means of sensor fusion. Fusion of Earth Data, *Proceedings EARSeL Conference, held in Cannes*, France, February 6–8, 1996, pp. 53–58.

Ranchin, T., Wald, L., and Mangolini, M., 2001. Improving the spatial resolution of images by means of sensor fusion. A general solution: the ARSIS method. In: *Remote Sensing and Urban Analysis.* Donnay, J.P., and Barnsley, M., (eds.) Taylor and Francis, Great Britain, pp. 19–34.

Rasmussen, M.S., 1998. Developing simple, operational, consistent NDVI-vegetation models by applying environmental and climatic information: Part I. Assessment of net primary production, *International Journal of Remote Sensing*, 19(1):97–117.

Research Systems Inc (RSI). 2003. *ENVI User's Guide, ENVI Version 4.0.* Research Systems Inc, Pennsylvania State University. 1084 p.

Richards, J.A., and Jia, X., 2006. *Remote Sensing Digital Image Analysis: An Introduction*, Springer-Verlag, Berlin, Heidelberg.

Richards, J.A., 2013. *Remote Sensing Digital Image Analysis.* Springer.

Ritter, N., and Ruth, M., 1997. The GeoTIFF data interchange standard for raster geographic images. *International Journal of Remote Sensing*, 18(7):1637–1647.

Sabins, F.F., 2000. *Remote Sensing: Principles and Interpretation.* W.H. Freeman & Co. New York.

Schowengerdt, R., 1997. *Remote Sensing: Models and Methods for Image Processing*, 2nd edn. Academic Press, London.

Shafri, H.Z.M., Suhaili, A., and Mansor, S., 2007. The performance of maximum likelihood, spectral angle mapper, neural network and decision tree classifiers in hyperspectral image analysis. *J. Comput. Sci.*, 3:419–423

Shao, F., 2006. Research on change information extraction of remote sensing images based on ANN. Master Dissertation Shandong University of Science and Technology, China (in Chinese).

Singh, A., 1989. Digital change detection techniques using remotely sensed data. *International Journal of Remote Sensing*, 10:989–1003.

Song, C., Woodcock, C.E., Seto, K.C., Lenny, M.P., and Macomber, S.A., 2001. Classification and change Detection using Landsat TM data: When and How to Correct Atmospheric Effects? *Remote Sensing of Environment*, 75:230–244.

Switzer, P., Kowalick, W.S., and Lyon, R.J.P., 1981. Estimation of atmospheric path-radiance by the covariance matrix method. *Photogrammetric Engineering and Remote Sensing*, 47:1469–1476.

Tanre, D., Deroo, C., Duhaut, P., Herman, M., Morchrette, J.J., Perbos, J., and Deschamps, P.Y., 1986. Simulation of the satellite signal in the solar spectrum (5S), Users's Guide (U. S. T. de Lille, 59655 Villeneuve d'ascq, France: Laboratoire d'Optique Atmospherique).

Vapnik, V.N., 1995. *The Nature of Statistical Learning Theory.* Springer-Verlag, New York.

Vermote, E.F., Tanré, D., Deuzé, J.L., Herman, M., and Morcrette, J.J., 1997. Second Simulation of the Satellite Signal in the Solar Spectrum, 6S. *IEEE Transactions on Geoscience and Remote Sensing*, 35:675–686.

5

An Introduction to Land Degradation

5.1 Introduction

Land degradation is first referred to by Aubreville (1949), who observed *"there are real deserts being born today, under our very eyes, in areas where the annual rainfall is between 700 and 1,500 mm"* (Mainguet, 1994). The term *degradation* literally means "the act or process of degrading." In the context of land, the term *degradation* refers to irreversible decline in the "biological potential" of the land. The term *land* refers to an ecosystem comprising land, landscape, terrain, vegetation, water, and climate. Land thus covers climate, landforms, water resources, soils, and vegetation (including both grassland and forests). In the soil conservation arena, the terms *soil degradation, land degradation,* and *desertification* are used interchangeably. While there is a clear distinction between "soil" and "land" there is no clear distinction between the terms *land degradation* and *desertification*. Desertification refers to land degradation in arid, semiarid, and subhumid areas due to anthropogenic activities (UNEP, 1993). The *desertification* has been defined as "degradation of land resources in arid, semiarid and dry sub-humid areas, caused by different factors including climatic variations and human activities resulting from adverse human impact" (Hurni, 1996). The definition of desertification is too narrow because severe land degradation resulting from anthropogenic activities can also occur in the temperate humid regions and the humid tropics. In fact, desertification is not a separate form of land degradation, so much as the end product of a variety of degradation processes that have adversely affected the land. Desertification is an emotive term that conjures an image of desert sand dunes advancing over adjacent areas, something that rarely happens in reality.

The word *degradation*, from its Latin derivation, implies "reduction to a lower rank" (Blaikie and Brookfield, 1987). Land degradation is the reduction in the capability of the land to produce benefits from a particular land use under a specified form of land management (Blaikie and Brookfield, 1987). Land degradation, generally, signifies a loss or reduction of land productivity as a result of human activity (UNEP, 1993). Land degradation is defined as a human-induced or natural process that negatively affects the land to function effectively. It is the temporary or permanent lowering of the productive capacity of land. Hence when land is degraded, its productivity is reduced and may continue to decline unless steps are taken to restore the lost productivity and prevent further losses. Interestingly, the concept of land degradation has evolved over time as illustrated by the changing definitions that have been used:

"Land degradation is a process which lowers the current or potential capability of soils to produce (quantitatively and/or qualitatively) goods or services". (FAO, 1979).

"Land degradation implies reduction of resource potential by a combination of processes acting on land" (UNEP, 1993).

"The reduction in the capacity of the land to perform ecosystem goods, functions and services that support society and development" (MEA, 2005).

"The reduction in the capacity of the land to provide ecosystem goods and services and to assure its functions over a period of time for its beneficiaries" (Nachtergaele et al., 2011).

In the context of latest concept of land degradation which refers to "the reduction in the capacity of the land to provide ecosystem goods and services over a period of time for its beneficiaries" the scope of land degradation further broadens. In fact, ecosystem goods refer to absolute quantities of land produce having an economic or social value for human beings (Table 5.1). They include animal and vegetal production, land availability, and water quality and quantity. Ecosystem services concern more qualitative characteristics and their impact on the beneficiaries and the environment (Ehrlich and Ehrlich, 1991; Lubchenco et al., 1993; Richardson, 1994).

5.1.1 Components of Land Degradation

There are a number of interrelated components of land degradation that lead to a decline in agricultural production. The most important ones are as follows (Douglas, 1994).

TABLE 5.1

Elements of Ecosystem Goods and Services

Elements of Ecosystem Goods and Services
Ecosystem processes
Maintenance of energy flux, dissipation, climate modulation (vegetation/water/soil)
Maintenance of hydrologic flux, hydrologic cycle, water quality (water)
Biological productivity, plant pollination (vegetation)
Maintenance of biogeochemical cycling, storage, mineral-gaseous cycles, water-air quality (soil/water)
Decomposition, weathering, soil development-stability, soil quality (soil)
Maintenance of biological diversity (biodiversity)
Absorbing, buffering, diluting, detoxifying pollutants-xenobiotics (soil/water)
Ecosystem "goods" include:
Food (economic/vegetation/social)
Construction materials (economic)
Medicinal plants (economic/biodiversity)
Wild genes for domestic plants and animals (biodiversity)
Water (water/economic)
Ecosystem "services" include:
Maintaining hydrological cycles (water)
Regulating climate (water/soil/vegetation)
Cleansing water and air (water)
Maintaining the gaseous composition of the atmosphere (vegetation)
Pollinating crops and other important plants (biodiversity)
Generating and maintaining soils (soil)
Storing and cycling essential nutrients (soil)
Absorbing and detoxifying pollutants (soil)
Providing beauty, inspiration, and recreation (social)

Source: Modified from Ehrlich and Ehrlich (1991), Lubchenco et al. (1993), and Richardson (1994).

5.1.1.1 Soil Degradation

Soil degradation refers to decline in the productive capacity of the soil as a result of soil erosion and changes in the hydrological, biological, chemical, and physical properties of the soil. It has been defined as a process which lowers the current and/or the potential capability of the soil to produce (quantitatively and/or qualitatively) goods or services (FAO, 1979). The FAO, UNEP, and UNESCO study (FAO, 1979) recognized six categories of soil degradation processes, namely water erosion, wind erosion, waterlogging and excess of salts, chemical degradation, physical degradation, and biological degradation.

The FAO, UNEP, and UNESCO study (FAO, 1979) recognized six categories of soil degradation processes: (1) water erosion, (2) wind erosion, (3) waterlogging and excess of salts, (4) chemical degradation, (5) physical degradation, and (6) biological degradation. A more recent study to assess the status of human-induced erosion in South and Southeast Asia (van Lynden, 1995) has sought to categorize the types of soil degradation in more detail, recognizing some 21 types.

5.1.1.1.1 Water Erosion

Soil erosion refers to detachment, movement, and deposition of soil particles by wind, water, gravity, and/or other agents like sea waves, human beings, and animals from one place to another place. Based on active force responsible for erosion, two types of erosions, namely water and wind erosion, have been recognized. Soil erosion by water occurs on partially vegetated/or bare soils, sloping land whenever rainfall intensities exceed the soil infiltration capacity and there is runoff. This category includes processes such as splash erosion, sheet erosion, rill and gully erosion, and mass movement (Douglas, 1994).

Splash Erosion: Splash erosion commonly initiates water erosion and occurs when raindrops fall onto the bare soil surface. The impact of raindrops breaks up the surface soil aggregates and splashes particles into the air (Figure 5.1a). On sloping land relatively more of these will fall downslope resulting in a net downhill soil movement. Some of the soil particles may fall into the voids between the surface aggregates, thereby reducing the amount of rainwater than can infiltrate into the soil and increasing runoff.

Sheet Erosion: Sheet erosion is the movement of soil from raindrop splash and runoff water. Sheet erosion refers to more or less even removal of thin layers of soils over extended area of sloping land (Figure 5.1b). It typically occurs evenly over a uniform slope and goes unnoticed until most of the productive topsoil has been lost. Deposition of the eroded soil occurs at the bottom of the slope or in low-lying areas.

Rill Erosion: Water running over sloping land, generally, tends to collect and concentrate in small streamlets, which ultimately develop into finger-like structure called "rills" (Figure 5.1c). They are small enough to not interfere with field implement and/or machinery operations. In many cases, rills are filled in each year as part of tillage operations.

Gully Erosion: Gullies, usually, follow sheet or rill erosion where volume and velocity of runoff are sufficient or concentrated. As the process of erosion progresses, rills are converted into gullies (Figure 5.1d), and with further erosion a network of gullies may be formed which is called ravines. Gullies have been defined as step-sided, eroding water courses that are subject to ephemeral flashfloods during storms.

5.1.1.1.2 Wind Erosion

Wind erosion refers to movement of soil by wind and is most common in arid and semiarid regions but occurs to some extent in subhumid regions too. Wind erosion includes both the removal and deposition of soil particles by wind action and the abrasive effects of moving

FIGURE 5.1a
Soil and water being splashed by the impact of a single raindrop (Source: https://en.wikipedia.org/wiki/Soil_erosion, accessed on 13-07-2018).

FIGURE 5.1b
In spite of across slope tillage operation sheet erosion is taking place due to the absence of adequate protective vegetation cover during rainy season.

FIGURE 5.1c
Deep gully formstion due to vertical erosion owing poor soil structure.

FIGURE 5.1d
Rill erosion as observed in the field.

particles as they are transported. It occurs when soil is left bare of vegetation as a result of cultivation, and/or overgrazing following overstocking. Not only can the wind remove topsoil from good farmland; it can result in additional damage by burying land, buildings, machinery, and fences with unwanted soil. The soil dust that winds carry can affect the Earth's atmospheric composition and contribute to climate change. The dust carried by wind also carries herbicides, sediments, soil texture residues, damaging nutrient contents, vegetation growth, and decreasing farm productivity, as well as affecting human health. Wind erosion occurs over more than one third of the Earth's surface. Aeolian transport of cultivated and grazed soils is a global problem that has generated many studies in Europe, Africa, Asia, Australia, and South America. There are three processes, i.e., surface creep, saltation, and suspension, involved in wind erosion. Characteristics of each process are outlined hereunder.

Surface Creep: In the event of wind erosion large particles ranging from 0.5 to 2 mm in diameter are rolled across the soil surface (Figure 5.2). This causes them to collide with

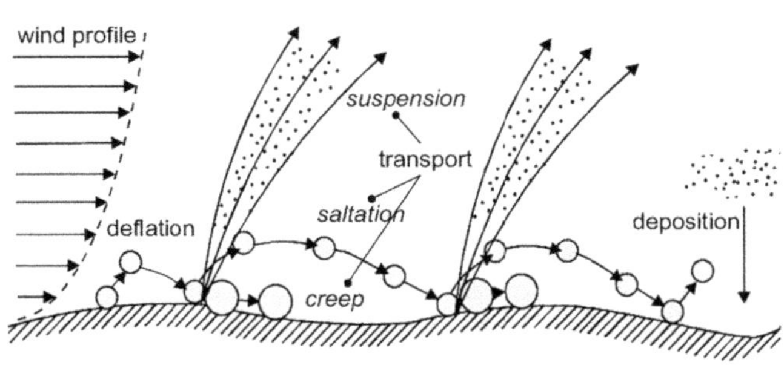

FIGURE 5.2
A schematic of wind erosion process.

and dislodge other particles. Surface creep wind erosion results in moving of these larger particles only a few meters.

Saltation: Saltation occurs among middle-sized soil particles that range from 0.05 to 0.5 mm in diameter. Such particles are light enough to be lifted off the surface but are too large to become suspended. These particles move through a series of low bounces over the surface causing abrasion on the soil surface and attrition—the breaking of particles into smaller particles. On agricultural lands, the most damaging action of wind erosion is saltation.

Suspension: Tiny particles less than 0.1 mm in diameter can be moved into the air by saltation, forming dust storms when taken further upward by turbulence. These particles include very fine grains of sand, clay particles, and organic matter. However, not all dust ejected from the surface is indefinitely carried in the air. Larger dust particles (0.05–0.1 mm) may be dropped within a couple of kilometers of the erosion site. Particles of 0.01 mm may travel hundreds of kilometers, and 0.001 mm-sized particles may travel thousands of kilometers. Fine dust may remain suspended in the air until it is washed away by rainfall.

5.1.1.1.3 Waterlogging

Waterlogging is the lowering in land productivity through the rise in groundwater close to the soil surface (FAO, 1994). In its most severe form (termed ponding), the water table rises above the surface. Waterlogging refers to a state of soil wherein there is free water in pore space. It has been defined as "An area is said to be waterlogged when the water table rises to an extent that soil pores in the root zone of a crop become saturated, resulting in the restriction of normal circulation of the air, decline in the level of oxygen and increase in the level of carbon dioxide" (Anonymous, 1991; www.civilengineeringterms.com/water-resources-irrigation-engineering/definition-of-waterloggin/). Waterlogging is the natural flooding and over-irrigation that brings water at underground levels to the surface. As a consequence, displacement of the air occurs in the soil with corresponding changes in soil processes and an accumulation of toxic substances that impede plant growth (United Nations, 1997).

5.1.1.1.4 Chemical Degradation

Salinization and/or Alkalization: Soils are termed saline or salt-affected when the concentration of salt in the root zone exceeds 4 dS/m (Richards, 1954). Soil salinity is one of the major factors affecting biomass production (Csillag et al., 1993). Soil salinization/alkalinization is one of the major factors affecting biomass production (Csillag et al., 1993). Salt-affected soils (saline/alkali) occur mostly in arid and semiarid regions, but may also be found in subhumid and coastal zones. Salt-affected soils are highly erosive and have poor structure, low fertility, and low microbial activity, which are not conducive for plant growth. Salt concentration in the soil environment has a strong impact on crop yield and agricultural production due to poor land and water management and expansion of agricultural frontiers into marginal dry lands. Crop productivity and production losses have considerable impact on farm and agricultural economies. For instance, the economic damage caused by secondary salinization was estimated at US$750 million per year for the Colorado River Basin in the United States of America, US$300 million per year for Punjab and Northwest Frontier Provinces in Pakistan, and US$208 million per year for the Murray–Darling Basin in Australia (Ghassemi et al., 1995).

Salt-affected soils contain excessive concentrations of either soluble salts or exchangeable sodium, or both. The process by which soluble salts are accumulated at the surface or near-surface of soil horizon is called salinization (Szabolcs, 1974). The build-up soluble salts in soils produce harmful effects to plants by increasing the salt content of the soil solution and by increasing the degree of saturation of the exchange materials in the soil with exchangeable sodium (US Salinity Laboratory Staff, 1954). The dominant salts in salt-affected soils consist of chlorides and sulfates of Na, Ca, and Mg. In terms of agricultural consequences, excessive salts in soil accelerate land degradation processes and increase the impact on crop yields and agricultural production. Additionally, the increase of salts in soil also affects other major soil degradation phenomena such as soil dispersion, sealing and crust formation, and structural changes, which results in unstable and compacted soil (De Jong, 1994; FAO, 1988; Metternichtand and Zinck, 2003).

Alkalinization refers to accumulation of exchangeable sodium in soils. It is a process leading to the formation of soils with high percentage of exchangeable sodium. The presence of carbonates and bicarbonates leads to formation of soils with high pH values, between 8.5 and 10 (Sparks, 1995). The main cause of alkaline reaction of soils, as described by FAO (1988), is the hydrolysis of either exchangeable cations or of salts such as $CaCO_3$, $MgCO_3$, and Na_2CO_3. Owing to the electrical charges at the surface, soil particles adsorb and retain cations on their surfaces. The adsorbed cations may be replaced by other cations that occur in the soil solution. Sodium, calcium, and magnesium cations are always readily exchangeable. In arid and semiarid regions when excess soluble salts accumulate, sodium frequently becomes the dominant cation in the soil solution. As the soil solution becomes concentrated through evaporation or water absorption by plants, the solubility limits of calcium sulfate, calcium carbonate, and magnesium carbonate are often exceeded and they are precipitated with a corresponding increase in the relative proportion of sodium. Under such conditions, a part of the original exchangeable calcium and magnesium is replaced by sodium (United States Soil Salinity Laboratory Staff, 1954).

5.1.1.1.5 Acidification

Soil pH is a measure of the number of hydrogen ions in the soil solution; the higher the concentration of hydrogen ions, the more acidic the solution. A thorough understanding of soil pH is essential for the proper soil management and optimum crop productivity. In aqueous (liquid) solutions, an acid is a substance that donates hydrogen ions (H+) to some other substance (Tisdale et al., 1993). Soil acidity is a severe soil degradation problem that can greatly reduce the production potential of farming systems. Most occurs in productive agricultural zones. It causes production losses within paddocks, and also long-term and offsite effects, including (i) poor water use by plants (leading to higher recharge and erosion), (ii) increased leaching of nutrients and aluminum, and (iii) the irreversible breakdown of layer silicate minerals in soils.

Development of acidity in soils is a natural process, especially in the high rainfall regions of southern Australia. Some soils are inherently acidic because of the high rates of leaching in these regions. Even in lower-rainfall cropping areas, some soil types have become acidic because they have no free lime in the profile. Soil acidity occurs when there is a build-up of acid-forming elements in the soil. The production of acid in the soils is a natural process; caused by rainfall and leaching, acidic parent materials and organic matter decay hence many soils in high rainfall areas are inherently acidic (McCauley et al., 2009).

Acidification is a slow process, but it is accelerated by agriculture through use of some fertilizers, like nitrogenous fertilizers, soil structure disturbance, and harvest of high yielding crops (Fageria and Baligar, 2008). The acidification of soil is caused by (i) acidic precipitation in its true sense, that is H+ ions in precipitation; (ii) the deposition from the atmosphere of acidifying gases or particles such as sulfur dioxide (SO_2), ammonia (NH_3) and nitric and hydrochloric acids (HNO_3; HCl); (iii) the application of acidifying fertilizers such as elemental sulfur (S), urea or ammonium (NH_4^+) salts and the growth of legumes such as clover; (iv) nutrient uptake by crops and root exudates; and (v) the mineralization of organic matter. As soils become more acidic, plants and crops that cannot tolerate acidic conditions do not flourish, so productivity and yields decline. When conditions become severely acidic (e.g., pH < 4, such as in inland acid sulfate soils), biogeochemical processes start to break down the layer silicates in the soil, releasing aluminum, iron, and manganese. This may lead to mineral toxicities and nutrient imbalances. For instance, the acidity affects the fertility of soils through deficiencies of P, Ca, and Mg, and the presence of phytotoxic nutrients such as soluble Al and Mn (Black, 1992).

The natural rate of acidification is accelerated by the use of acidifying fertilizers, nitrogen fertilizers, the removal of agricultural products and nitrate leaching. The management of soil acidity involves the following requirements at the farm level: (i) recognizing paddock indicators of soil acidity, (ii) monitoring soil pH, (iii) knowing crop and pasture tolerances to acidity, and (iv) treating paddocks that have acidity problems.

5.1.1.1.6 Other Chemical Degradation Processes

In addition to salinization and/or alkalization, other processes may adversely affect the chemical properties of soil, notably (after FAO, 1994):

- Loss of nutrients and organic matter;
- Fixing of soluble nutrients, so that they are "unavailable" to plants, following acidification;
- Increased leaching (when the vegetative cover is removed and the soil exposed particularly when this occurs in high rainfall tropical areas);
- Increased temperatures and oxidation of the soil organic matter (due to exposure and cultivation);
- Pollution (most commonly from improper management of industrial and mining wastes).

Plant growth may be adversely affected by the build-up of particular metals or salts to toxic levels within the soil. Aluminum toxicities may be a problem in strongly acid soils. The higher aluminum levels found in the subsoils of some tropical soils may be an additional factor in yield decline following loss of topsoil from erosion (Stocking personal communication). Calcium carbonate and gypsum (when present as calcrete or gypsum horizons) may cause nutrient imbalance. Manganese toxicity is most likely in acid soils with impeded drainage.

Acid sulfate soils are commonly developed on estuarine and marine alluvium (mangrove swamps) and contain considerable amounts of sulfides. Chemical degradation can be rapid once such areas are "reclaimed." Once drained, the sulfides oxidize to form sulfuric acid; and the pH, which is around neutral prior to drainage, drops below 3.5. The acid attacks the clay minerals causing the liberation of aluminum ions in amounts toxic for plant roots and micro-organisms.

Toxicities can also result from the presence of municipal, industrial, radioactive, or oily wastes such as occur mainly around towns, industrial areas, and mines (FAO, 1979). Such toxicities are usually of only local extent but in some countries they may lead to significant problems, for instance the extensive areas of tin tailings from former tin mining operations in Malaysia. In Vietnam (and possibly Laos and Cambodia) there are areas where the soils contain residues of the chemical defoliants used between the 1960s and 1975, during the American phase of the Indochina war.

5.1.1.1.7 Physical Degradation

Both crop and livestock production can lead to a deterioration in the physical condition of the soil. This degradation can take many forms and has a variety of consequences. Deterioration of soil structure is the most common form of physical degradation, and involves (FAO, 1994):

- Loss of stability of aggregates in surface soils, leading to crusting and compaction, and consequently poorer infiltration rates and greater run-off and erosion;
- Translocation of clay particles to subsurface layers and loss of porosity in the subsurface and deeper soil layers, with consequent loss of water transmission and storage capacity.

Physical degradation is of concern because soil structure and its stability governs soil–water relationships, aeration, crusting, infiltration, permeability, runoff, interflow, root penetration, leaching losses of plant nutrients and therefore ultimately the productive potential of a soil (Lal, 1979).

Topsoil degradation occurs when an open structure of soil aggregates is broken down by excessive tillage. Also, the impact of raindrops and/or livestock hooves may produce a continuous compacted layer or crust at the surface. Reduction in topsoil porosity, and particularly surface crusting, will result in decreased water infiltration, increased runoff, poorer seedling emergence, and often increased erosion (Lal, 1979).

Physical degradation of the subsoil may be in the form of a distinct pan, or a more general compaction, i.e., loss of the original subsoil structure and increasing bulk density with reduction in size and quantity of pore spaces. The physical effects may be decreased water storage capacity, loss of aeration, and reduced soil permeability. Waterlogging may occur in the soil above the compacted horizon and the absorptive capacity of the subsoil will be reduced, thereby increasing the amount of rainfall going as surface runoff. Plant root development will be hindered in the subsoil because compacted horizons are physically difficult to penetrate. Plant growth will be restricted because of the lower availability of air and water, and therefore nutrients, in the root zone (Lal, 1979).

Some of the blame for increased soil degradation in the region has been ascribed to a specific change in farming practice, namely the replacement of the traditional digging sticks and hoes with the plough. In part, this has facilitated the expansion of the area under cultivation thereby exposing more land to the risk of accelerated erosion. The main reason is that ploughing, using techniques imported from temperate regions in Europe and USA, increases the disturbance of the topsoil, breaking up its structure and making it less resistant to erosion. Traditional shifting cultivation systems rely on the presence of tree stumps and roots, not only for rapid restoration of the vegetation during the fallow period, but also for reducing runoff velocity and holding the soil in situ during cultivation. The conservation benefits of the traditional land-clearing techniques are lost once the plough has been adopted, as ploughing is difficult, if not impossible, unless the stumps and roots are first removed from the field (Douglas, 1994).

Subsoil compaction, particularly the formation of "plough pans," is normally associated with commercial agriculture and the use of heavy machinery such as tractors. However, evidence suggests that a cultivation pan, immediately below the cultivated layer, can occur even under traditional hand cultivation systems from the pressure of the tiller's feet and the practice of cultivating to the same depth each year. Although the low yields under low-input systems may largely be because soil nutrient and organic matter levels have reached low-level equilibrium; physical degradation in the form of a cultivation pan may also be a contributory factor.

Whereas conservation measures can be adopted to control soil erosion, and chemical fertilizers can be used to replace soil nutrients, physical degradation, particularly in the subsoil, is less easily overcome. Subsoil compaction is regarded as a form of soil degradation as it markedly reduces the yields of dryland crops and may be expensive to overcome, requiring deep ripping with a tractor. However, in areas where paddy rice is grown, farmers adopt soil compaction techniques, such as puddling, to reduce subsoil permeability thereby raising the capacity of the soil to produce rice. In such a situation, soil compaction is regarded as land improvement rather than degradation.

5.1.1.1.8 Biological Degradation

Soils used for agricultural purposes are often deficient in the biological processes which maintain both their physical structure and their ability to supply essential chemical elements to plants (Swift and Sanchez, 1984). Of particular concern is the decline in soil organic matter or humus content following cultivation. In part, this is because large amounts of biomass are harvested and removed from the site. The actual humus mineralization rates may increase due to soil temperature changes following the removal of a protective vegetative cover.

The agricultural significance of soil organic matter is greater than that of any other property with the exception of soil moisture. It improves soil structure, and thereby root penetration and erosion resistance; augments cation exchange capacity; and acts as a store of nutrients, slowly converted to forms available to plants (Young, 1976). It is possible to obtain an overall balance of soil organic matter with shifting cultivation under conditions of low population density. However, for most of the Asia Pacific region shortage of suitable land means that shifting cultivation is no longer a viable option and most of the land is cultivated each year. Under such permanent agricultural systems, decline in organic matter can be severe and rapid. Typical values of the organic matter status of tropical soils that have been under cultivation for two or more years are 30%–60% of the corresponding values under natural vegetation. Values below 50% are considered to represent an undesirable situation calling for remedial measures (Young, 1976).

Biological degradation is usually synonymous with decline in organic matter. Yet it also applies when there is a decline in the beneficial soil fauna (organisms). Certain soil organisms have the ability to influence the physical structure of soils. Some species of termites annually transport large amounts of soil through the soil profile. In so doing, they achieve a mixing of the organic and mineral components of soil and alter the porosity of the soil at the micro- and macro-pore levels. This can result in increased surface infiltration of rainwater and a consequent reduction in runoff and erosion, which (together with the stimulation to soil aeration) contribute to maintaining soil fertility (Swift and Sanchez, 1984). Earthworms, while important for the soils of the temperate parts of the Asia Pacific region, may play a similar role as termites in some soils in the tropics, although not in comparable numbers or biomass (Young, 1976).

Most mechanical forms of soil tillage bring about marked changes in the qualitative and quantitative characters of the soil fauna community. The consequences of such biological degradation are poorly understood and merit further study. However, the improved soil structure and enhanced efficiency of nutrient return from crop residues to plant, associated with conservation (minimum) tillage techniques, are believed to be associated with the maintenance of a healthy soil fauna, particularly of termites and earthworms (Swift and Sanchez, 1984).

5.1.1.2 Vegetation Degradation

Vegetation degradation is usually regarded as a reduction in the available biomass, and decline in the vegetative ground cover, as a result of deforestation and overgrazing. Such degradation is a major contributory factor to soil degradation particularly with regard to soil erosion and loss of soil organic matter. The term also applies in situations where the reduction is not in the quantity of biomass but in quality—for instance bush encroachment into rangelands and the loss of palatable pasture grasses and their replacement with nonpalatable species. In such a situation, the value of the land will have declined from an agricultural point of view with a decline in its livestock carrying capacity. However, the degraded vegetation may still be contributing to the soil in terms of ground cover and organic matter. In practice, soil degradation usually accompanies rangeland degradation but it may not be such a clear cut relationship as that associated with deforestation (shifting cultivation to be incorporated).

5.1.1.3 Water Degradation

Water degradation leads to decline in the quantity and/or quality of both surface and ground water resources. Land degradation, particularly soil and vegetation degradation, has resulted in a deterioration in the quantity and quality of both surface and groundwater resources. With less vegetative cover to protect against the impact of raindrops causing surface sealing, a decline in pore spaces resulting from loss of organic matter and loss of structural stability following cultivation, less rain infiltrates the soil. Runoff increases, stream flows fluctuate more than before (in particular stream flow storm hydrographs are likely to have sharper and greater peaks), flooding becomes more frequent and extensive. Groundwater recharge decreases, streams and springs may cease, and the water table is likely to drop so that wells and boreholes may dry up.

5.1.1.4 Climate Deterioration

Climate deterioration refers to changes in the micro- and macro-climatic conditions that increase the risk of crop failure. Loss of vegetative cover and soil degradation may also be disrupting long-term rainfall patterns and increasing the likelihood of drought (Cook, 1992). Computer models suggest three ways in which deforestation and soil degradation may reduce rainfall (Timberlake, 1985, Harrison, 1987, Milner and Douglas, 1989):

- First, intensive cultivation, overgrazing, and deforestation can all strip soil of vegetation. Bare soil and rock reflect more solar radiation back into the atmosphere than do grass, shrubs, and trees. Increased reflectivity (albedo) keeps the atmosphere warmer, disperses cloud, and reduces rain.
- Second, a general lowering of soil moisture could itself suppress rainfall. Much of the rain in tropical moist forests comes from water evaporated off the vegetation. Large-scale clearing of rain forest breaks this hydrological cycle and may well produce a drier local climate.
- Third, deforestation and loss of topsoil structure allow the wind to throw more dust into the air. This dust reduces the amount of sunshine reaching the earth surface, which would have the same rain-reducing effect as bouncing more solar radiation back off the earth's surface.

Although none of the models conclusively proves the link, it is necessary to consider the possibility that (under the stresses imposed by growing population) land degradation and climatic deterioration reinforce each other (Cook, 1992).

Apart from modulating the current climate pattern soil degradation may pose threat of global warming from increasing atmospheric concentrations of CO_2 (FAO, 1994). Terrestrial ecosystems play an important role in the global carbon budget. Furthermore, world soils, in addition to the effects of deforestation, also have a significant impact on the global carbon budget. One estimate suggests that reduction of about 1% in organic carbon content of the top 15 cm layer of soils of the tropics could lead to an annual emission of about 128 billion tons of carbon into the atmosphere (Lal, 1990). Others question the evidence for global warming, and its possible effects on food production, believing that much of it is speculative and based on massive extrapolation and doubtful assumptions (Hudson, 1992).

5.1.1.5 Losses to Urban/Industrial Development

Decline in the total area of land used, or with potential, for agricultural production occurred as a result of arable land being converted to urban, industrial and infrastructure uses. Economic development within the developing countries has led to expansion of urban and industrial land. Much of the present urban and industrial development has taken place on what was formerly good agricultural land. The uncontrolled urban and industrial development in the coastal zone (following the new policy of economic liberalization) has seen the loss of much prime agricultural land to new roads and buildings.

The expansion of urban settlements is a particularly acute problem in the small Pacific countries with little arable land. Population pressure is most acute in the capitals of the atoll nations (e.g., Tarawa in Kiribati) but is of concern even in those countries with larger land areas (e.g., Port Moresby in Papua New Guinea). Throughout the Pacific, people are gravitating from the mountains to the coastal cities, from the outer islands to the provincial or national seats of government, and from scattered rural hamlets to larger villages and towns. Under customary ownership laws, land is not readily available for housing estates. Hence squatter towns develop with health problems from overcrowding, unsanitary conditions, and water pollution; valuable agricultural resources are lost; and forests, lagoons, and reefs are degraded.

Farm households affected by urban expansion may be forced to use their remaining plots more intensively or to seek land elsewhere, which in a land scarcity situation usually means moving into marginal upland areas. Hence, urban and industrial expansion may be a contributory factor to soil degradation elsewhere.

5.2 Extent and Spatial Distribution

The land surface of the earth totals 13.0 billion hectares, of which 1.5 billion hectares are unused wastelands and 2.8 billion hectares are unused but largely inaccessible (Oldeman, 1994). Of the 8.7 billion hectares under use, most is suitable only for forest, woodland, grassland, or permanent vegetation. Only 3.2 billion hectares are potentially arable. About half of this potentially arable land is currently cropped and 41% of it is considered moderately to highly productive. Globally, 1,964.4 Mha land is affected by human-induced degradation (UNEP, 1993). Of this, 1,093.7 Mha are subject to soil erosion by water, 548.3 Mha to wind erosion, 239.1 Mha to chemical deterioration, 68.2 Mha to compaction, and 10.5 Mha to waterlogging. In addition, an estimated one billion hectares of arable land are subject to soil salinization and/or alkalinization (Ghassemi et al., 1995) (Figure 5.3).

Owing to differences in definitions and terminology, a large variation in the available statistics on the extent and rate of land degradation also exists. Two principal sources of data include the global estimates of desertification by Dregne and Chou (1994) and of land degradation by the International Soil Reference and Information Centre (ISRIC) (Oldeman et al., 1992; Oldeman, 1994). Whereas Dregne and Chou address degraded lands in dry areas, the extent of land degradation provided by Oldeman covers all climatic regions. Hence, the Oldeman statistics has been widely adopted. Globally, an estimated 1.9 billion hectares of land are subject to land degradation by all processes and all eco-regions (Oldeman, 1994) (Table 5.2). It is evident from the table that out of four processes of land degradation water and wind erosion account for 16.43 million square kilometers or 83.61% of the global affected area, followed by physical and chemical degradation. For further details readers may refer to Eshwaran et al. (2001).

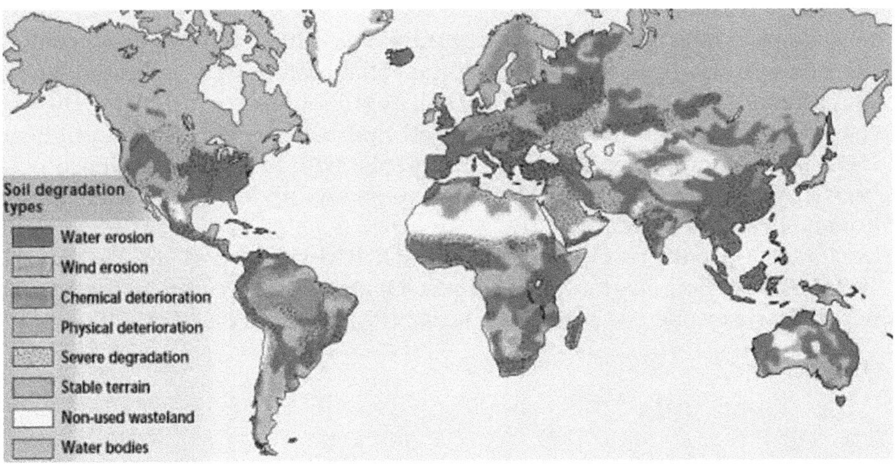

FIGURE 5.3
Human-induced soil degradation around the world (Source: Eswaran et al. (2001a)).

TABLE 5.2
Estimates of the Global Extent (in million km^2) of Land Degradation (Oldeman, 1994)

Type	Light	Moderate	Strong + Extreme	Total
Water erosion	3.43	5.27	2.24	10.94
Wind erosion	2.69	2.54	0.26	5.49
Chemical degradation	0.93	1.03	0.43	2.39
Physical degradation	0.44	0.27	0.12	0.83
Total	7.49	9.11	3.05	19.65

5.3 Land Degradation Assessment

In order to rehabilitate degraded land information on their nature, extent, spatial distribution is a prerequisite. After implementation of any preventive or curative measure/(s) toward rehabilitation of degraded land monitoring the progress and success of such programs is essential element for taking up any mid-course action, if necessary. The assessment of nature, extent, and magnitude of degraded land is the logical first step in this endeavor. The major approaches used for assessment of degraded lands can be grouped into four broad categories: (1) expert opinion, (2) satellite-derived net primary productivity, (3) biophysical models, and (4) mapping abandoned cropland. Each offers a glimpse into the conditions on the ground, but none captures the complete picture (Gibbs and Salmon, 2015).

5.3.1 Expert opinion/GLASOD Approach

Solicitation of expert opinion or Global Assessment of Soil Degradation (GLASOD) was the first approach used to map and quantify degraded lands. For generating soil degradation maps at global level Oldeman et al. (1990) developed a set of relatively uniform mapping units and then asked experts to estimate the status of soil degradation in terms of the type, extent, degree, rate, and causes of degradation within each mapping unit (roughly 1945–1990). The opinion from over 290 national collaborators were collected, and moderated by 23 regional leaders. The estimates of soil degradation are uniform in the sense that they are based upon defined mapping units and carefully structured definitions but they rely on local knowledge rather than measurements. Due to this dependence upon local experts, the GLASOD assessment is considered subjective and has been open to disparagement (Thomas, 1993; Rey et al., 1998).

The Food and Agriculture Organization (FAO) had developed the methodology for Global Assessment of Soil Degradation (GLASOD) at 1:1,000,000 scale using Landsat-MSS images which was compiled at 1:5,000,000 scale) (Figure 5.4). The GLASOD was based on

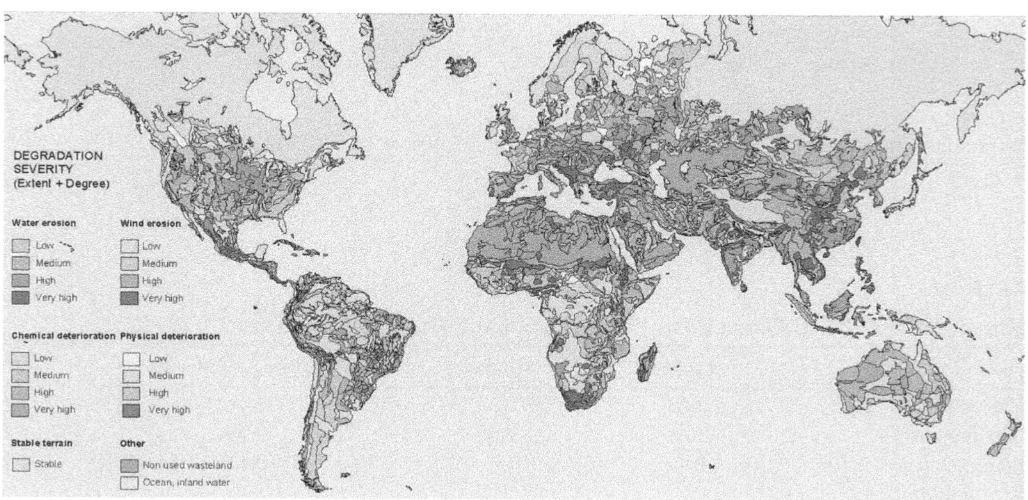

FIGURE 5.4
Global Assessment of Human-induced Soil Degradation (GLASOD) (Oldeman et al., 1991).

a structured recording of the type, severity, and extent of degradation, together with its major causes. In the late 1980s, the ISRIC, in collaboration with UNEP, undertook a global inventory of the status of human-induced soil degradation. The assessment of soil degradation was mainly based on the World Soil database at 1:10 M scale but in some parts on the Global SOTER database (at 1:5 M scale—UNEP, FAO, ISRIC, ISSS 1995–2000). The information on soil degradation, thus generated, had certain limitations like small-scale, qualitative, subjective, and complex legend.

5.3.2 Remote Sensing-Based Approach

Realizing the need for an up-to-date, quantitative, and reproducible assessment of soil degradation a project *"Land Degradation Assessment in Drylands (LADA)"* was taken up by FAO with two major in six pilot countries in 2007 objectives, i.e., to (i) develop and implement strategies, methods and tools to assess, quantify and analyze the nature, extent, severity, and impacts of land degradation (land degradation) on ecosystems, watersheds and river basins, and carbon storage in dry lands at a range of spatial and temporal scales and (ii) build national, regional, and global assessment capacities to enable the design, planning, and implementation of interventions to mitigate land degradation and establish sustainable land use and management practices (UNEP, 1993). LADA is a scientifically based approach to assessing and mapping land degradation at different spatial scales—small to large—and at various levels—local to global. It was initiated in drylands, but the methods and tools have been developed so as to be widely applicable in other ecosystems and diverse contexts with minimal required adaptation.

To avoid a negative bias due to a focus only on land degradation, LADA also assesses and maps land improvement or sustainable land management (SLM) using World Overview of Conservation Approaches and Technologies (WOCAT) tools (Food and Agriculture Organization, 2013). The LADA-WOCAT set of tools and approaches provides balanced information and mapping capabilities on land resources status and trends in any given area, as well as on their causes, impacts, and the actual and potential future responses.

The Global Assessment of Human-Induced Soil Degradation (GLASOD) information was updated by a Global Land Degradation Assessment (GLADA) based on an analysis of Normalized Difference Vegetation Indices (NDVI) considering climatic variations. The GLADA project uses Global Inventory Modeling and Mapping Studies (GIMMS) radiometer data collected by National Oceanic and Atmospheric Administration (NOAA) satellites (Bai et al., 2008). The NOAA-AVHRR data are corrected for calibration, variations in solar and view zenith angle, El Chichon and Mt Pinatubo stratospheric aerosols, and other effects not related to vegetation change, and generalized to 8 km grids for 15-day periods. Global data are currently available for the period July 1981–December 2003. The GIMMS NDVI time series has been translated to NPP using MODIS data (Running et al., 2004). NPP was estimated by correlation with MODIS 8-day NPP values for the overlapping years of the GIMMS and MODIS datasets (2000–2003), resampling the annual mean MODISNPP at 1 km resolution to 8 km resolution using nearest-neighbor algorithm. As mentioned earlier, at global level, methods were developed that evaluated long-term trends in Net Primary Production (NPP) corrected for rainfall. Also at global level, a geo-referenced inventory was made of individual indicators associated with land degradation such as slope, soil vulnerability to erosion, long-term trends in droughts, biodiversity, and socioeconomic factors (Nachtergaele et al., 2009).

The empirical relationship is as follows:

$$NPP_{MOD17}\left[kgC\,ha^{-1}\,year^{-1}\right] = 1106.37 * sum\,NDVI - 564.55 \quad (5.1)$$

(r = 0.83, n = 3 128 207)

where NPPMOD17 is annual mean NPP derived from MODIS MOD17 Collection 4 data, and sum NDVI is the 4-year (2000–2003) mean annual sum NDVI derived from GIMMS. Uncertainty is for slope ± 3.818 and for intercept ± 16.364.

Represented by the ratio of annual sum NDVI and annual rainfall, the rain-use efficiency (RUE) was calculated from the VASClimO rainfall data. VASClimO 1.1 comprises the monthly precipitation data for 1951–2000, compiled from long, quality-controlled station records, gridded at resolution of 0.5°, from 9 343 stations (Beck et al., 2004). For the period 2001–2003, these were supplemented by the GPCC full data reanalysis product (Schneider et al., 2008) to produce monthly rainfall values matching the GIMMS NDVI data. MOD17A3 is a dataset of terrestrial gross and net primary productivity computed at 1 km resolution and an 8-day interval. Though far from perfect (Plummer, 2006), MODIS gross and net primary productivity values are related to observed atmospheric CO_2 and the interannual variability associated with the ENSO phenomenon, indicating that the NPP data are reliable at the regional scale (Zhao et al., 2006). The dataset has been validated in various landscapes (Fensholt et al., 2006; Gebremichael and Barros, 2006; Turner et al., 2006).

The approach for identifying the areas of land degradation and improvement using remote sensing data involves the following steps.

5.3.2.1 Computation of NDVI Indicators

Simple NDVI indicators, namely NDVI minimum, maximum, maximum–minimum, mean, sum, standard deviation, and coefficient of variation are computed for the calendar year for the northern hemisphere, and for October to the following September for the southern hemisphere, encompassing a complete growing season.

5.3.2.2 NDVI-to-NPP Conversion

The annual sum NDVI, the aggregate of greenness over the growing season, is used as the standard surrogate for annual biomass productivity. NDVI is translated to net primary productivity by correlation with MODIS data; trends are calculated by linear regression.

5.3.2.3 Identification of the Areas Experiencing Land Degradation

In order to distinguish between declining productivity caused by land degradation and declining productivity due to other factors, rainfall variability and irrigation have been accounted for by the following:

- Identifying each pixel where there is a positive relationship between productivity and rainfall, i.e., where rainfall determines NPP;
- For those areas where rainfall determines NPP, RUE has been considered: where productivity declined but RUE increased, we may attribute declining productivity to declining rainfall; those areas are masked. In addition, urban areas are also masked;

- For the remaining areas with a positive relationship but declining RUE, and also for all areas where there is a negative relationship between NDVI and rainfall, i.e., where rainfall does not determine NPP, NDVI trend has been calculated; this is called RUE-adjusted NDVI;
- Land degradation is indicated by a negative trend of RUE-adjusted NDVI and may be quantified as *RUE-adjusted NPP*.

As an additional indicator, the residual trend of sum NDVI (RESTREND) is calculated. To take account of the significant lengthening and warming of the growing season at high latitudes and high altitudes, energy-use efficiency (EUE) is calculated for all pixels. In practice, it scarcely affects the estimation of land degradation but does affect the identification of land improvement which is indicated by a positive trend in both RUE-adjusted NPP and EUE, and is quantified as *climate-adjusted NPP*. The indices of land degradation and improvement are compared with land cover, soil and terrain, rural population density, and indices of aridity and poverty.

In order to evaluate the goods and services of the "ecosystem" a simplified scheme is proposed by LADA that reflects in accessible terms the various dimensions of land degradation. These correspond with the provisioning, regulating, supporting, and socio-cultural services developed under them but tie them down to more tangible objects of study. LADA proposes a particular focus on seven different aspects of goods and services: biomass production, yearly biomass increments, soil health, water quality and quantity, biodiversity, economic value of the land use, and social services of the land and its use.

At (sub)national level, Liniger et al. (2008) approaches were developed for both land degradation and SLM by analyzing their status, causes, and impacts in each major land use type per province within a country. At local level, a systematic inventory method on the status of the land in all its aspects (soil health, water quantity and quality, vegetation status, and biodiversity) was developed (McCauley and Jones, 2009). This information is complemented with local knowledge from land users and key informants on land use and management practices during the field survey and with results from semi-structured interviews of different categories of land users on the causes and impacts of land degradation at the community level (Nachtergaele et al., 2010).

5.3.3 Biophysical Models

Biophysical modeling has been generally used to assess the potential vegetative productivity of global land areas (e.g., Fischer et al., 2002). Common approaches include global datasets describing climate patterns and soil types to define classes of potential productivity. Such studies may incorporate empirical knowledge and expert opinion, using models to extend these into a globally consistent estimate. Recent work has indicated that these biophysical models may be combined with observations of land use to map degradation. The spatial and temporal extent, as well as the type of degradation considered, is dependent upon the input datasets for the modeling process.

5.3.4 Abandonment of Agricultural Lands

Another way to identify degraded lands is to look for areas that were once croplands but have since been abandoned because of decreased productivity, or due to political and economic reasons. The benefit of focusing on these lands is that it captures the longer time frame of the expert opinion approach but quantifies the actual conditions rather than estimating potential risk. Moreover, this approach is empirically driven, as it relies on the details of agricultural census data combined with the global consistency provided by satellite mapping. Recent satellite advances have enabled new global scale datasets of agricultural land cover, which have been developed by merging satellite derived land cover maps with ground-level agricultural inventory datasets (Ramankutty and Foley, 1999; Goldewijk, 2001; Cardille et al., 2002).

5.3.5 The Land Degradation Impact Index

Impact of land degradation is a growing concern as well in developing as in developed countries. There is a widespread belief that land degradation causes (and is caused by) poverty. This nexus results in land being degraded more and more, while people become poorer and poorer over time. This phenomenon continues until their resilience is broken and they have no other option than to migrate within countries to areas which offer better opportunities or even migrate to developed countries. This specter of mass migration caused by land degradation was discussed in the Algiers conference on desertification (Renaud and Bogardi, 2006). The effect of land degradation is also much more felt in areas where the population density is high and poverty is high, as any remedial action costs more or is more difficult to implement, while the threat to food and income security is so much greater in areas with a poor population. Therefore, in this chapter poverty is taken as one indicator against which impact of the degradation is measured (the other being the persons affected by a given level of degradation).

To measure the effect of this in each point of the globe this index is multiplied by the poverty and population levels in that point. Also, areas with no population are excluded from the calculation, as zero population means no impact of land degradation.

LDII (for IMR > 10) = (LDI * (adjusted (Population* Poverty level)))

5.4 Conclusions

The increasing demand for the food, fuel, and fiber production of ever-growing population calls for bringing more area under plough and for increasing the productivity of available cultivable lands. Rehabilitation of degreased lands and preventing fertile land from degradation seem to be possible solutions in this endeavor. Several regional- and global-level programs like LADA and GLADA have been launched for assessment and monitoring of degraded land which have certainly provided required input planning their rehabilitation program. Furthermore, Global Land Degradation Information System (GLADIS), a global-level information system for degraded land, is another very important step toward rehabilitation of degraded land. In fact, a lot more needs to be done with regard to inventory and characterization of degraded land at local level that would serve as basic input for taking up any preventive or curative measures.

Geospatial technology (e.g., remote sensing, geographic information system, global navigation satellite system, personal digital assistant (PDA), and Internet) has brought about substantial improvement in level, quality, timeliness, and cost-effectiveness of information on degraded land. Conjunctive use of geospatial technology, proximal sensing, *in situ* observations, and modeling would further enhance the level and quality on information degraded land. For instance, a ground penetrating radar (GPR) with the subsurface imaging capability on the order of less than 0.5 m in fine-textured soils to about 25 m in coarse-textured soils under unsaturated conditions (Doolitte, 1987; Tarussov et al., 1994) could be used for generating information on certain subsurface features like ground water table, the presence or otherwise of hydrological barriers or like clay pan, calcic/petrocalcic gypsic/petrogypsic horizons, etc. For precise location of *in situ* observations the Global Positioning System (GPS) with adequate accuracy may be used. By incorporating the Digital Elevation Model (DEM), namely SRTM-DEM/ASTER-DEM/SPOT-DEM/CartoDEM in the database, another dimension would be added which would be very useful for rehabilitation of degraded land.

References

Anonymous. 1991. *Waterlogging, soil salinity and alkalinity. Report of the Working Group on Problem identification in irrigated areas with Suggested Remedial Measures.* Ministry of Water Resources, Government of India, New Delhi.

Aubreville, A., 1949. Climats, Forêts et Désertification de l'Afrique Tropicale. Paris, Société d'Editions Géographiques, Maritimes et Coloniales.

Bai, Z.G., Dent, D.L., Olsson, L., and Schaepman, M.E., 2008. Global assessment of land degradation and improvement. Identification by remote sensing. Report 2008/01, ISRIC—World Soil Information, Wageningen.

Beck, C., Grieser, J., and Bruno, R., (2004). A new monthly precipitation climatology for the global land areas for the period 1951 to 2000 (ftp://ftp-anon.dwd.de/pub/data/gpcc/PDF/pdf_28_precipitation.pdf).

Black, A.B., 1992. *Soil Fertility Evaluation and Control.* Lewis Publisher, London.

Blaikie, P., and Brookfield, H., 1987. *Land Degradation and Society.* Methuen, London & New York.

Cardille, J.A., and Foley, M.H., 2002. Costa characterizing patterns of agricultural land use in Amazonia by merging satellite classifications and census data. *Global Biogeochemical Cycles,* 16(3):18–1–18–12.

Cook, M.G., 1992. Sustainable agriculture in a survival setting. In Tato, K., and Hurni, H., (eds.) *Soil Conservation for Survival.* Soil and Water Conservation Society, Ankeny, IA.

Csillag, F., Pasztor, L., and Biehl, L.L., 1993. Spectral band selection for the Characterization of salinity status of soils. *Remote Sensing of Environment,* 43:231–242.

De Jong, S.M., 1994. Applications of reflective remote sensing for land degradation studies in a Mediterranean environment. *Netherlands Geographical Studies,* 177:240 pp.

Doolitte, J.A., 1987. Using ground—penetrating radar to increase the quality and efficiency of soil surveys. In: *Soil Survey Techniques.* Soil Science Society of American Spectral Publication No. 20. Madison, Wisconsin, USA 11–32.

Douglas, M.G., 1994. Sustainable use of agricultural soils. A review of the prerequisites for success or failure. Development and Environment Reports No. 11, Group for Development and Environment, Institute for Geography, University of Berne, Switzerland.

Dregne, H.E., and Chou, N.T., 1994. Global desertification dimensions and costs. In: Dregne, H.E (ed.) *Degradation and Restoration of Arid Lands.* Texas Technical University, Lubbock.

Ehrlich, P.R., and Ehrlich, A.H., 1991. *Healing the Planet: Strategies for Resolving the Environmental Crisis.* Addison-Wesley Publishing Company, Reading, Massachusetts.

Eswaran, H., Lal, R., and Reich, P.F., 2001. Land degradation: An overview. In: Bridges, E.M., Hannam, I.D., Oldeman, L.R., Pening de Vries, F.W.T., Scherr, S.J., and Sompatpanit, S., (eds.) *Responses to Land Degradation. Proceedings of 2nd International Conference on Land Degradation and Desertification,* Khon Kaen, Thailand. Oxford Press, New Delhi, India.

Eswaran, H., Lal, R., Reich, P.F., 2001a. Land degradation: an overview. In: Bridges, E.M., Hannam, I.D., Penning de Vries, F.W.T., Scherr, S.J., and Sombatpanit, S., (eds.) *Response to Land Degradation*. Sci. Publishers, Enfield, pp. 13–19.

Fageria, N.K., and Baligar, V.C., 2008. Ameliorating soil acidity of tropical oxisols by liming for sustainable crop production. In: Sparks, D.L., (ed.) *Advances in Agronomy*. Academic Press, Brazil, pp. 345–389.

FAO, 1979. *A Provisional Methodology for Soil Degradation Assessment*. FAO, Rome.

FAO, 1994. *Cherish the Earth—Soil Management for Sustainable Agriculture and Environmental Protection in the Tropics*. Land and Water Development Division, FAO, Rome.

FAO, 1988. Salt-affected soils and their management, by Abrol, I.P, Yadav, J.S.P. and Massoud, F.I. FAO Soils Bulletin No. 39. Rome.

Fensholt, R., Sandholt, I., Rasmussen, M.S., Stison, S., and Diouf, A., 2006. Evaluation of satellite-based primary production modelling in the semi-arid Sahel. *Remote Sensing of the Environment*, 105:173–188.

Fischer, G., van Velthuizen, H.T., Shah, M.M., and Nachtergaele, F.O. (2002). *Global agro-ecological assessment for agriculture in the 21st Century: Methodology and results*. Laxenburg, Austria: International Institute for Applied Systems Analysis.

Food and Agriculture Organization, 2013. Land Degradation Assessment in Drylands: Methodology and results, Technical report.

Gebremichael, M., and Barros, A.P., 2006. Evaluation of MODIS gross primary productivity (GPP) in tropical monsoon regions. *Remote Sensing of Environment*, 100:150–166.

Ghassemi, F., Jakeman, A.J., and Nix, H.A., 1995. *Salinization of land and water resources: Human causes, extent, management and case studies, Canberra, Australia*. The Australian National University, Wallingford, Oxon.

Gibbs, H.K., and Salmon, J.M., 2015. Mapping the world's degraded lands. *Applied Geography*, 57:12–21.

Goldewijk, K.K., 2001. Estimating global land use change over the past 300 years: the HYDE database. *Global Biogeochemical Cycles*, 15(2):417–433.

Harrison, P., 1987. *The Greening of Africa. IIED-Earthscan study*. Paladin, London.

Hudson, N.W., 1992. *Land Husbandry*. Batsford, London.

Hurni, H., With the assistance of an international group of contributors, 1996. *Precious Earth From Soil and Water Conservation to Sustainable Land Management*. International Soil and Water Conservation Organization (ISCO), and Centre for Development and Environment, Berne.

Hurni, H., (eds.) 1996. *Precious earth: From Soil and Water Conservation to Sustainable Land Management*. International Soil Conservation Organization (ISCO) and Centre for Development and Environment (CDE), Berne, 89 pp.

Lal, R., 1979. Physical Characteristics of Soils in the Tropics: Determination and Management. In: Lal, R., and Greenland, D.J., (eds.) *Soil Physical Properties and Crop Production in the Tropics*. John Wiley: Chichester.

Lal, R., 1990. Managing soil carbon in tropical agro-ECOSYSTEMS. EPA Workshop on Sequestering Carbon in Soils. Corvallis Oregon USA, Feb 26–28, 1990.

Liniger, H., van Lynden, G., Nachtergaele, F., Schwilch, G., 2008. *A questionnaire for mapping land degradation and sustainable land management*. CDE/WOCAT, FAO/LADA, ISRIC.

Lubchenco, J., Risser, P.G., Janetos, A.C., Gosz, J.R., Gold, B.D., and Holland, M.M., 1993. Priorities for an environmental science agenda for the Clinton-Gore administration: Recommendations for transition planning. *Bulletin of the Ecological Society of America*, 74:4–8.

Mainguet, M., 1994. *Desertification: Natural Background and Human Mismanagement*, 2nd edn. Springer-Verlag Press, Germany.

McCauley, A., Jones, C., andJacobsen, J., 2009. Soil pH and organic matter. In: *Nutrients Managements Module*. Montana State University Extension, Bozeman, Montana, USA. 8:1–12.

Metternicht, G.I., andZinck, J.A., 2003. Remote sensing of soil salinity: potentials and constraints. *Remote Sensing of Environment*, 85:1–20.

MEA,2005. Watson R. et al. (49 authors). Living beyond our means: Natural assets and human well-being.Statement from the Millennium Ecosystem Assessment Board.

Milner, C., and Douglas, M.G., 1989. *Problems of Land Degradation in Commonwealth Africa—A Study on the Scope for Commonwealth Action*. Commonwealth Secretariat, London.

Nachtergaele, F., Biancalani, R., Bunning, S., and Hubert George, H., 2010. Land degradation assessment: The LADA approach. 19th World Congress of Soil Science, Soil Solutions for a Changing World, August 1–6, 2010, Brisbane, Australia. Published on DVD.

Nachtergaele, F., Petri, M., and Biancalani, R., 2011. Land degradation. In: Lal, R., and Stewart, B.A., (eds.) *World Soil Resources and Food Security Advances in Soil Sciences*, Taylor and Francis, CRC Press, 574 p.

Nachtergaele, F., Petri, M., Van Velthuizen, H., Biancalani, R., Van Lynden, G., and George, H., 2009. Forthcoming. Global Land Degradation Information System (GLADIS in preparation). LADA internal document. FAO.

Oldeman, L.R., Hakkeling, R.T.A., and Sombrock, W.G., et al., 1990. *World Map of the Status of Human Induced Soil Degradation. An Explanatory Note*. ISRIC/UNEP. Wageningen.

Oldeman, L.R., Hakkeling, R.T.A., Sombroek, W.G., 1991. World Map of the Status of Human-Induced Soil Degradation: An explanatory Note (rev. ed.), UNEP and ISRIC, Wageningen, 35 p.

Oldeman, L.R., Hakkeling, R.T.A., and Sombrock, W.G., 1992. *World Map of the Status of Human-induced Soil Degradation: An Explanatory Note. Global Assessment of Soil Degradation (GLASOD)*. ISRIC, Wageningen.

Oldeman, L.R., 1994. The global extent of land degradation. In: Greenland, D.J., and Szabolcs, I., (eds.) *Land Resilience and Sustainable Land Use*. CABI, Wallingford, pp. 99–118.

Oldeman, L.R., Hakkeling, R.T.A., Sombrock, W.G., and Batjes, N.H., 2014. *Global Assessment of Human-induced Soil Degradation (GLASOD)*. ISRIC—World Soil Information Dataset.

Plummer, S., 2006. On validation of the MODIIS gross primary production product. *IEEE Transactions on Geosciences and Remote Sensing*, 44(7):1936–1938.

Ramankutty, N., and Foley, J.A., 1999. Estimating historical changes in global land cover: croplands from 1700 to 1992. *Global Biogeochemical Cycles*, 13(4):997–1027.

Renaud, F.G., and Bogardi, J.J., 2006. Forced migrations due to degradation of arid lands: Concepts, debate and policy requirements. Paper presented at the Joint International Conference Desertification and the International Policy Initiative, Algiers, December 17–19, 2006.

Rey, C., Scoones, I., and Toulmin, C., 1998. Sustaining the soil: indigenous soil and water conservation in Africa. In: *Sustaining the Soil* (pp. 1–27).

Richards, L.A., (ed.) 1954. Diagnosis and improvement of saline and alkali soils. U. S. Department of Agriculture Hand Book No.60, U.S. Government Printing Office, Washington, DC.

Richardson, C.J., 1994. Ecological functions and human values in wetlands: a framework for assessing forestry impacts. *Wetlands*, 14:1–9.

Running, S.W., Heinsch, F.A., Zhao, M., Reeves, M., and Hashimoto, H., 2004. A continuous satellite-derived measure of global terrestrial production. *Bioscience*, 54:547–560.

Schneider, U., Fuchs, T., Meyer-Christoffer, A., and Rudolf, B., 2008. *Global Precipitation Analysis Products of the GPCC*. Global Climatology Centre (GPCC), DWD, Internet Publication, 1–12 pp. Available online at http://gpcc.dwd.de/.

Sparks, D.L., 1995. *Environmental Soil Chemistry*. Academic Press, Inc., San Diego, pp. 219–232.

Swift, M.J., and Sanchez, P.A., 1984. Biological management of tropical soil fertility for sustained productivity. *Nature and Resources*, 20:2-10.

Szabolcs, I., 1974. *Salt Affected Soils in Europe*. Martinus Nijhoff, The Hague, p. 63.

Tarussov, A., Bonn, F., and Judge, A., 1994. Soil survey from light research aircraft using a ground penetrating radar. *Proceedings of the First International Airborne Remote Sensing Conference and Exhibition*, Strasbourg, France, September 11–15, 1994.

Thomas, D.S., 1993. Sandstorm in a teacup? Understanding desertification. *Geographical Journal*, 159(3):318–331.

Timberlake, L., 1985. *Africa in Crisis*. Earthscan, London.

Tisdale, S.L., Nelson, W.L., Beaton, J.D., and Havlin, J.L., 1993. Soil acidity and basicity, pp. 364–404. In: *Soil Fertility and Fertilizers*, 5th edn. Macmillan Publication, New York.

United Nations, 1997. Glossary of environment statistics, studies in methods, Series F, No. 67, United Nations, New York.

Turner, D.P., Ritts, W.D., Cohen, W.B. et al., 2006. Evaluation of MODIS NPP and GPP products across multiplebiomes. *Remote Sensing of Environment* 102:282–292.

United States Salinity Laboratory Staff, 1954. *Diagnosis and Improvement of Saline and Alkali Soils.* Agriculture Hand book No. 60. United States Department of Agriculture.

van Lynden, G.W.J., 1995. Guidelines for the Assessment of the Status of Human-induced Soil Degradation in South and Southeast Asia. ISRIC Working Paper and Preprint 95/2.

Young, R.A., 1976. Managing aquifer over-exploitation: economics and policies. Selected papers on Aquifer Overexploitation, International Association of Hydro-geologists, Heise, Hannover 3: 199–222.

WRI, 1997. *The Diversity and Dynamics of Shifting Cultivation: Myths, Realities, and Policy Implications.* World Resources Institute, Washington, D.C.

Zhao, H.L., Yi, X.Y., Zhou, R.L., Zhao, X.Y., Zhang, T.H., Drake, S., 2006. Wind erosion and sand accumulation effects on soil properties in Horqin Sandy Farmland, Inner Mongolia. *Catena*, 65:71–79.

6

Water Erosion

6.1 Introduction

Soil erosion by water occurs on partially vegetated/or bare soils, sloping land whenever rainfall intensities exceed the soil infiltration capacity and there is runoff. Soil is naturally removed by the action of water or wind and it is called background soil erosion or natural soil erosion. Natural soil erosion has been occurring since the early period of earth. But accelerated soil erosion is the result of mankind's unwise actions which leave the land vulnerable during times of erosive rainfall or windstorms is of recent origin. Soil erosion refers to detachment, movement, and deposition of soil particles by wind, water, gravity, and/or other agents like sea waves, human beings, and animals from one place to another place (Judson, 1965; Merritt et al., 2003).

Eroded soils have been reported to transport pesticides, nutrients, and other harmful farm chemicals into streams and rivers; pollute surface and groundwater resources (Gallaher and Hawf, 1997); reduce productivity and crop yields (Renard et al., 1997); and cause air pollution through emissions of gases such as carbon dioxide (CO_2), methane (CH_4), and nitrous oxide (N_2O) (Cox and Madramootoo, 1998). Based on active force responsible for erosion, broadly two types of soil erosion, namely *water and wind erosion,* have been recognized. Within water erosion three stages, i.e., *sheet, rill,* and *gully erosion,* have been identified. *Sheet erosion* is the movement of soil from raindrop splash and runoff water. Sheet erosion refers to more or less even removal of thin layers of soils over extended area of sloping land. Further progression of sheet erosion in the event of continued rain leads to *rill erosion.* Water running over sloping land, generally, tends to collect and concentrate in small streamlets, which ultimately develop into finger-like structure called "rills." *Gully erosion* usually follows sheet or rill erosion where volume and velocity of runoff are sufficient or concentrated. As the process of erosion progresses, rills are converted into gullies. Gullies have been defined as step-sided, eroding water courses that are subject to ephemeral flash-floods during storms (Hudson, 1985:224).

6.2 Factors of Water Erosion

The soil erosion by water is influenced by various factors in varying degrees. These factors have been briefly discussed hereunder.

6.2.1 Climatic Factors

The rainfall is the most important factor that influences soil erosion through its two important characteristics, namely quantity and intensity. The rainfall influences the detachment of the soil particles by means of splash detachment as well as by overland flow. Therefore, the knowledge about evapotranspiration (ET) and the active role of plants on the process is required to understand runoff process and soil erosion as well as ecosystem physiology. ET is the summation of evaporation and transpiration, where the water in the soil and on the land surface gets converted into the water vapor form. The evaporation is the phenomenon that is related to the process operating on the land surface and the transpiration is related to the plants. As the time lag between precipitation and runoff will be high in large basins, there is enough scope for large amounts of water getting evaporated. Moreover, the ET rates will influence the soil moisture condition (antecedent soil moisture), which is an important parameter that will influence the infiltration rate into the soil and subsequent runoff.

6.2.2 Land Factors

Soil erosion by water involves the interaction of a complex set of physical and chemical processes governed by many factors. Soil erodibility describes the degree to which the soil surface is susceptible to soil erosion. The water erosion is influenced by the inherent soil erodibility and modulated by the slope and cover factors. The influence of these parameters is discussed hereunder.

6.2.2.1 Soil Texture and Clay Mineralogy

The clay fraction in the soil increases the soil stability by acting as a bonding agent among coarser and medium-sized soil particles, thereby improving the resistance of soils to erosion. The medium-textured soils (silty and loamy soils) are more susceptible to erosion (Ben-Hur et al., 1985). Mineralogical and chemical characteristics may modify the effect of clay particles on aggregation and stability. Though clay mineralogy influences the erosion process, it is difficult to assess its independent effect because in nature the soils contain a mixture of clay minerals. El-Swaify and Dangler (1977) showed that soils with kaolinite were less eroded than soils with montmorillonite. It is probably due to iron–kaolinite interactions producing very strong cohesive forces. Trott and Singer (1983) found that sesquioxides are the controlling parameter of disaggregation of soils. Another mechanism of aggregate breakdown is the chemical dispersion. A good relationship between soil loss and dispersion on kaolinite was reported by Miller and Baharuddin (1986). Stern et al. (1991) found that the soils with more than 5% smectite were dispersive and susceptible to erosion.

6.2.2.2 Organic Matter

The organic matter effects aggregate stability through clay–organic matter interaction and binding action of polymers, and physical protection by mulching effect. However, there are contrasting findings on the influence of organic matter content on aggregate stability and in turn on soil erosion. Panayiotopoulos and Kostopoulou (1989) found that in some cases aggregate stability is not affected by organic content variations. On the contrary, Roberson et al. (1991) and Ekwue (1990) reported an increase in the aggregate stability as the heavy fraction of carbohydrates increases with consequent reduction in splash detachment.

6.2.2.3 Sodium and Other Cations

The nature and amount of exchangeable cations on the exchange complex have profound influence on aggregate stability and soil erodibility through their effect on the clay dispersion and flocculation processes. For instance, the exchangeable sodium percentage (ESP) has been found to modulate the infiltration rate and subsequent erosion. As the ESP increases, the clay dispersion, crust formation, and erosion increases, but the effect differs from soil to soil.

6.2.2.4 Iron and Aluminum Oxides

The positive role of the sesquioxides in improving the aggregate stability is well established. A positive correlation between sesquioxides and soil structural stability has been observed by several researchers (Panayiotopoulos and Kostopoulou, 1989). The iron and aluminum on the soil exchange complex act as flocculants by lowering the zeta potential of clays and preventing the dispersion of clay particles. Similar is the case with calcium.

6.2.2.5 Antecedent Soil Moisture

Antecedent soil water plays an important role in controlling splash detachment through its effects on soil resistance to raindrop impact and the magnitude and mechanisms of aggregate breakdown on the soil surface (Le Bissonnais, 1990). When the soils are near saturation the total amount of rainfall could lead to generation of runoff, thus increases the erosion. In a dry soil, on the contrary, a portion of the rainfall will be infiltrated in to the soil depending on the time available for infiltration. Highly aggregated soils resist aggregate breakdown upon wetting and splash detachment is lower.

6.2.2.6 Soil Crusting

Soil crusting is a major problem in the medium-textured soils. When the raindrop strikes the soil surface it causes the splash detachment. The detached soil could clog the macropores, thereby leading to a situation called "surface sealing." Moore and Singer (1990) demonstrated a decrease in infiltration capacity following rainfall events because of surface sealing and crusting. Silty and loamy soils are the most prone to sealing. However, this effect is complicated because crusting may increase soil shear strength and resistance to rill formation (Bradford et al., 1987).

6.2.2.7 Topography

The steepness, length, and aspect of the slope have an important role to play in the rainfall-erosion process. Aspect is the orientation of slope, measured clockwise in degrees from 0 to 360, where 0 is north-facing, 90 is east-facing, 180 is south-facing, and 270 is west-facing. The slope steepness forms the driving force for the overland flow (runoff). As the slope steepness increases the time available for the water to percolate into the soil decreases and thereby increases the runoff, subsequently the soil detachment. The slope length influences the total overland flow of the land element in question. On longer slopes, the overland flow has enough time to get absorbed into the soil, thereby decreasing the runoff. It is vice versa with steeper slopes.

6.2.2.8 Vegetation

Vegetation has a positive role in controlling the erosion. The major effects of vegetation are interception of rainfall by absorbing part of the raindrop's kinetic energy, and thus the splash detachment. It also reduces the surface runoff by storing a part of the rainfall. In fact, apart from degree and length of slope, vegetation cover is the most important factor to control soil erosion (Lehmann, 1992).

6.3 Water Erosion Models

A model can be simply defined as "a simplified representation of a complex system." Thus, a hydrological model could be "simplified representation of hydrological system." A model could be described as logical formulation of various processes as understood by the modeler. The formulations may be representations of simple processes or combinations of processes as the modeler deems necessary to solve a specific problem. In case some of the variables included in the model are not physically measurable, these variables must be derived by a procedure of calibration for a specific case. Based on the processes that are associated with them and the end results being obtained water erosion could be classified as follows: (i) *conceptual*: if the physical processes relating the input variable(s) to produce the output variable(s) are considered in terms of physical laws, or (ii) *empirical*: models are based on observation and experiment but not on any theory. Regression models are empirical, being based on observation and not on an understanding of process. The term *physically based* model is generally being used to replace conceptual distributed model. Soil erosion models can be used for (i) farm planning, (ii) site-specific assessments, (iii) project evaluation and planning, (iv) policy decisions, and (v) research tools to study processes and the behavior of hydrologic and erosion systems (DeCoursey, 1985; Foster, 1990).

As a detailed treatment of water erosion model is beyond the scope of the chapter, only a brief introduction is given hereafter. For a detailed review of existing hydrologic and soil erosion models readers may refer to Curwick and Jennings (1982), DeCoursey (1985), Perez-Trejo (1986), Smith and Ferreira (1986), De Roo (1993), and Meritt et al. (2003).

6.3.1 Empirical Models

Included in this category of water erosion models are Universal Soil Loss Equation (USLE) and modifications thereof. The USLE has been developed to estimate inter-rill soil losses over extended time periods from areas of individual plot scale (Wischmeier and Smith, 1978). In USLE, six major factors are used to predict the soil loss (A): rainfall energy factor (R), soil erodibility factor (K), slope length factor (L), slope steepness factor (S), cropping and management factor (C), and practice factor (P). The soil loss can be computed by simple multiplication of these factors. The limitations of this model include (i) not useful for single rainfall event, (ii) does not explain soil erosion process in a watershed, and (iii) difficult to incorporate the latest developments in erosion research.

The USLE was modified by Williams and Berndt (1977) so as to predict the soil loss for single rainfall events and was termed modified USLE (MUSLE). The rainfall energy factor was replaced with a runoff factor. The runoff factor includes both the total storm runoff volume and the peak runoff rate. This model is applicable to individual storms, and eliminates the need for sediment delivery ratios, because the runoff factor represents energy used in detaching and transporting sediment. However, the MUSLE does not provide information on time distribution of sediment yield during a runoff event.

Revised Universal Soil Loss Equation (RUSLE) is a revision made to USLE by Renard et al. (1997). It updates the content and incorporates new material that has been available informally or from scattered research reports and professional journals. RUSLE has been developed to replace the USLE. However, RUSLE has the same limitations as the USLE. MUSLE-87 is a revision of MUSLE and is more catchment-oriented version of the USLE. This model has been developed by Hensel and Bork (1987) and uses the parameter "upstream catchment area" for each raster cell. Included are the modifications of the USLE to Mid-European conditions which were introduced by Schwertmann (1982).

The Differentiated Universal Soil Loss Equation (DUSLE) is a version of the modified USLE for Mid-European conditions, combined with a digital elevation model (DEM), which has the structure of a triangulated irregular network (TIN) (Flacke et al., 1990). The LS factor is differentiated for an application on complex slope geometries.

6.3.2 Physically Based Models

Several physically based soil erosion models have been developed. However, a detailed treatment of each such model is beyond the scope of this chapter. For comprehension of the concept of physically based model only a representative model is described here.

6.3.2.1 Water Erosion Prediction Project Model

It is a process-based, distributed continuous simulation model and runs on a single-storm basis. Water Erosion Prediction Project (WEPP) has several fundamental routines of infiltration, surface runoff, plant growth, residue decomposition, hydraulics, tillage, management, soil consolidation, and erosion mechanics. The model development incorporates the essence of USLE model, i.e., prediction of soil loss from small uniform slopes and extend to basin scale by dividing the basin into several such uniform units. WEPP uses equations for sediment continuity, detachment, deposition, shear stress in rills, and transport capacity. Also incorporated are the temporal variations of erodibility variables as a function of above and below ground residue, plant canopy, and soil consolidation.

6.3.3 Mixed Models

The mixed models, as the name indicates, imbibe the features of both conceptual and physically based models. Two examples of mixed models, namely Chemical runoff and Erosion from Agricultural Management Systems (CREAMS) and Areal Nonpoint Source Watershed Environment Response Simulation (ANSWERS), are described hereunder.

6.3.3.1 CREAMS

The Chemicals, Runoff, and Erosion from Agricultural Management Systems (CREAMS) is a field size model for evaluation of water quality under various farming practices (Knisel, 1980). The hydrology component estimates storm runoff using the Soil Conservation Service (SCS) curve number (CN) model when only daily rainfall data are available. When the hourly rainfall data are available, the Green/Ampt infiltration equation is used to predict runoff. Detachment is described by a modification of the USLE for a single storm event. The transport capacity of the overland and channel flow is derived from Yalin's sediment transport equation (Yalin, 1963). CREAMS cannot provide information during storms, and it can only represent variability in down slope direction. The model has limiting routing capabilities because it was specifically designed as an agricultural field scale model in contrast to a "basin" model or catchment scale model (Curwick and Jennings, 1982).

6.3.3.2 ANSWERS

The Areal Nonpoint Source Watershed Environment Response Simulation (ANSWERS) is a distributed parameter model which is designed to simulate the hydrology of agricultural watersheds during and immediately following a rainfall event. Values of variables such as slope, aspect, soil variables (porosity, moisture content, field capacity, infiltration capacity, and erodibility factor), crop variables (coverage, interception capacity, and USLE C/P factor), surface variables (roughness and surface retention), and channel variables (width and roughness) are defined for each element. Equations representing water flow and sediment transport are solved for each element using an explicit backward difference scheme with an integration time step of 1 minute. The output of upslope elements becomes the input of downslope elements. When water and sediment reach an element within a channel, they are transported to the watershed outlet. The sediment is deposited within a channel when its transport capacity diminishes. ANSWERS assume that deposited sediment is reattached to other surface particles and must, therefore, be re-detached in order to be available for surface flow. The model also assumes that channels are not erodible.

6.4 Role of Remote Sensing

Conventionally, the assessment of soil degradation by erosion has been carried out using different methods, e.g., direct field monitoring, observations and measurements obtained from distributed point datasets, identification of erosion features and soil profile truncation, assessment of Caesium-137, risk elements, and other relevant soil properties (Poreba, 2006). At a regional or global scale, information on the level of soil degradation by erosion is often based on expert knowledge involving extrapolation of acquired local data using ancillary data. Remote sensing offers immense potential in the assessment of soil degradation by erosion at a regional and global scale. The main advantage of remote sensing methods is repeated spectral measurements over larger areas at regular interval and at varying spatial resolutions.

6.4.1 Spectral Response Pattern of Eroded Soils

Soil erosion and deposition affect chemical and physical properties of upper layer of soils, which affect their spectral reflectance pattern. Eroded soils have different spectral response pattern from non-eroded "healthy" soils (Conforti et al., 2013; Schmid et al., 2016).

Latz et al. (1984) found that eroded Alfisols were to have higher spectral reflectance as compared to normal soils due to exposure of B horizon with higher iron content and low organic matter. In another study Price (1993) found high correlation between reflectance values of single Landsat TM bands, especially band 4 (0.77–0.86 μm), and erosion rates for Pinyon-Juniper woodlands. Furthermore, in arid rangelands of Australia, Pickup and Nelson (1984) successfully distinguished eroding, stable, and depositional areas using the data space defined by the 4/6 (green/NIR) and 5/6 (red/NIR) band ratios of Landsat MSS data. It was concluded that the method is dependent on the relation between erosion status and vegetation cover, hence is not suitable for humid climates. While working with eroded soils of Ethiopia Baghdadi and Gaonach (2002) developed a complex method involving Fourier scaling and multi-fractal analysis of Landsat TM and ERS-1 synthetic aperture radar (SAR) imagery, which permitted the separation of eroding lands. A review on the use of spectral reflectance data for soil erosion investigations has been provided by Wiesmiller et al. (1985).

Hyperspectral remote sensing or imaging spectrometry data have been found to be a promising tool for assessment of soil erosion. The soil properties having direct bearing on spectral response pattern of eroded soils include soil organic carbon (SOC) concentration, particle size distribution (particularly clay content), $CaCO_3$ concentration, Fe and Al oxides ratio, and soil moisture content. Other soil properties used for determination of erosion-affected areas are soil color, pH, structure, or coarse fragments content. Imaging spectroscopy or generally analysis of soil spectral characteristics has a great potential for soil erosion assessment (Ben-Dor et al., 2009; 2015). Several studies such as development of soil crust (Ben-Dor et al., 2004) or changes in soil moisture (Haubrock et al., 2005) have been reported using hyperspectral measurements. Hyperspectral data have scarcely been used for distinguishing soils influenced by soil erosion. Hill et al. (1994, 1995) used linear spectral mixture analysis and k-means clustering for classification of erosion stages based on content of skeletal material (fragments coarser than 2 mm size). Chabrillat et al. (2014) distinguished erosion classes using support vector machine classification method and development of image-derived end members for each erosion class.

6.4.2 Airborne Sensor Data

Aerial photographs have been used for both detection and monitoring of soil erosion features. Aerial photographs have been used for detection of the extent of soil erosion features and watershed management since early 1930s (Gorrie, 1935). Coldwell (1957) used aerial photographs for detection of sheet wash and rill erosion, whereas Leuder (1959) could map certain phases like slope, erosion, stoniness and drainage can be accomplished with a minimum of field checking. Similarly, based on stratified random sampling of a 5 × 5 km grid and detailed classification of erosion features from stereoscopic photographs Grieve et al. (1994) carried out survey of upland erosion in Scotland. Liddaman (2008) developed automated approaches for the identification of peat land erosion from infrared aerial photography and applied pattern analysis. Furthermore, the measurement of gullies and controlling variables on aerial photographs have been made in most available gully erosion models—especially the empirical ones, at the regional level, using multivariate regression techniques. Kamphrost and Iyer (1972) used aerial photos to differentiate four ravine classes based on depth and other morphometric features perceived through photointerpretation and parallax bar measurements. Ravines are basically network of gullies.

The analysis of spectroscopic and hyperspectral data has proven to be an effective way to characterize and monitor surface soil variables (Ben-Dor et al., 2009) that would allow studying soil erosion processes. Schmidt et al. (2016) integrated airborne hyperspectral data supported by morphological and physicochemical ground data over a Mediterranean rain-fed cultivated region to relate soil erosion processes with the land management and agricultural uses. Results show that the proposed method is successful in mapping different soil erosion and accumulation sites (SEAS) that indicate preservation partial loss, or complete loss of fertile soils, as well as downslope accumulation of different soil materials. The SEAS indicators used in this study include free iron oxide content (%), texture, $CaCO_3$ (%), soil organic matter (%), coarse fraction (% >2 mm fraction), pH, structure and consistence, and soil color.

Apart from inventory of water erosion, aerial photographs have also been used for studying the dynamics of water erosion (Pellikkaa et al., 2005) used multi-temporal aerial photographs during the period from 1955 to 1983 over Taita hills, Kenya to monitor the density of gullies, and observed a doubling of gully density since 1955.

6.4.3 Spaceborne Multispectral Data

Spaceborne multispectral measurements made by various Earth observation systems have been used to derive information on soil erosion features and its consequences, factors affecting water erosion, and input that can go into soil erosion models for predicting soil loss and the risk of soil erosion. Direct detection has been achieved through identification of individual large erosion features, discrimination of eroded areas, and assessment of erosion intensity based on empirical relations based on their spectral response patterns. Detectable effects include the damage occurred due to major erosion events, and the sedimentation of reservoirs.

6.4.3.1 Detection of Erosion Features and Eroded Areas

In the event of soil erosion the soil from surface layer is removed by fluvial process to varying degrees Landsat and SPOT imagery have been for identification of individual large and medium sized gullies (Millington and Townshend, 1984). For large gullies in Central Brazil, Vrieling and Rodrigues (2004) found that optical ASTER imagery provided better description of gully shape than Envisat Advanced Synthetic Aperture Radar (ASAR) data, when compared to a QuickBird image. An example of the potential of Indian Remote Sensing satellite (IRS) Linear Imaging Self-Scanning Sensor (LISS-III) image with 23.5 m spatial resolution in detecting various erosional features is shown in Figure 6.1. Removal of surface soil by fluvial process exposes the subsurface soil with different color that could be captured by spaceborne multispectral images (Figure 6.1). As evident from the figure, the dark-colored patch of black soil around the village Nagireddipalle, Kurnool district, southern India is a nearly level to very gently sloping plain and is experiencing the removal of a thin veneer of soil in the core area of the plain during rainy season resulting in slight soil erosion. The periphery of this pocket of land around the village is sloping radially. As we move away from the core area, the slope increases resulting thereby in an increase in the intensity of erosion and subsequent formation of rills and gullies that could be seen as figure-like structure in different shades of gray (Figure 6.1).

In another example, multi-temporal spectral characteristics have been utilized to detect areas experiencing sight erosion. Here, Resourcesat-1 LISS-III images are acquired during three cropping seasons, namely *kharif* (rainy season), November 2005; *rabi* (winter season), February 2006; and *zaid* crop April 2006 (Figure 6.2). As evident from images of November and February, the detection slightly eroded category is precluded due to the presence of crop cover. The slight erosion category is conspicuous in Aril image with bare fields, light bluish color, and smooth texture. The field photograph vividly demonstrates the status of soil erosion as captured by LISS-III sensor.

FIGURE 6.1
Sheet and rill erosion around Nagireddipalle village, Kurnool district, Andhra Pradesh, southern India as seen in Resourcesat-1 LISS-III image.

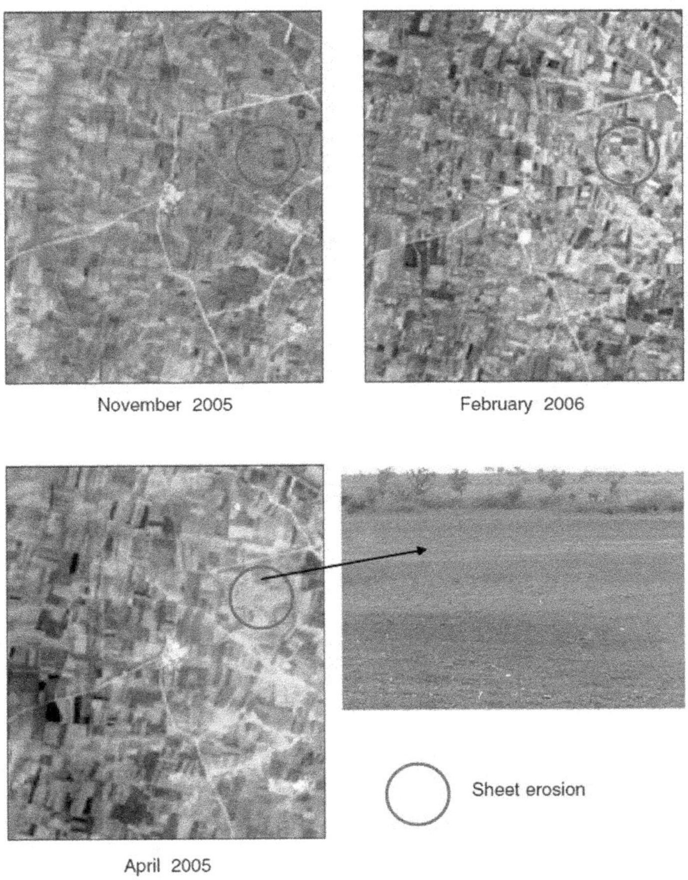

FIGURE 6.2
Sheet erosion as seen in Resourcesat-1 LISS-III images during three cropping seasons, namely *kharif* (rainy season), November 2005, *rabi* (winter season), February 2006, and *zaid* crop April 2006. The ground photograph of the area experiencing sheet erosion could be seen adjacent to April 2006 image (Courtesy: National Remote sensing Centre, Indian Space Research Organization, Department of Space, Government of India).

As evident from Figures 6.3 and 6.4 the removal of finer soil particles and organic matter and sometimes part of "A" and "B" horizon leaves a distinct mark on soil surface that impart a characteristic shape and color to the soil that helps the image analyst in detection of "rills." Rills are finger-like structure that are formed when after detachment from the surface soil particles are detached and remain suspended in water which subsequently carried away by flowing water (e.g., runoff) along the slope gradient. Terrains in both the sites have supported the development of black soils with very clay content (>35%) with montmorillonite mineralogy. These soils are rich in water dispersible clay which is highly vulnerable to water erosion. As a result of erosion, the dark-colored black soil has been removed to varying degrees leaving at many places only underlying light-colored weathered material which is seen as finger-like structures with yellowish/yellowish brown color.

Information on morphometry (length, depth, breadth, and side slope) of ravines: the network of gullies is of paramount importance for taking up any reclamative measures.

FIGURE 6.3
Rills and gullies as seen in Resourcesat-1 LISS-III images during three cropping seasons, namely *kharif* (rainy season), October 2004; *rabi* (winter season), February 2005; and *zaid* crop April 2005. The ground photograph of the area experiencing sheet erosion could be seen adjacent to April 2005 image (Credit: National Remote Sensing Centre, Indian Space Research Organization, Department of Space, Government of India).

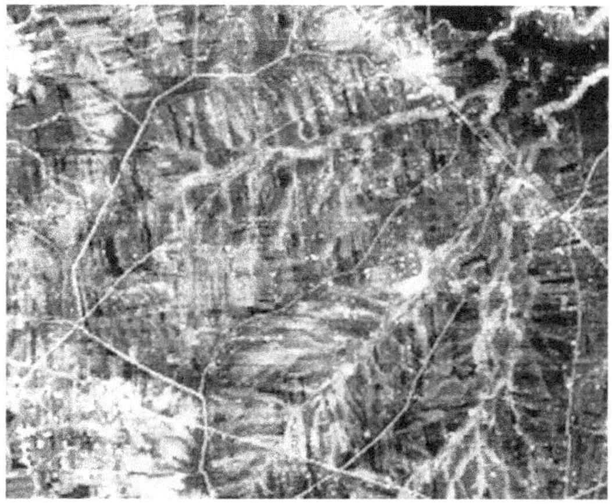

FIGURE 6.4
Rills and gullies as seen in Landsat-TM image covering part of Belgaum district, Karnataka, southern India.

The potential of satellite remote sensing in deriving information on ravines along the rivers Yamuna and Chambal in parts of Uttar Pradesh and Madhya Pradesh, northern India is demonstrated in Figures 6.5 and 6.6. It is interesting to note water erosion, in general, is lateral in nature. That is, the detachment and removal of soil material is parallel to the soil surface. However, in the case of coarser material especially along the rivers, a vertical erosion occurs instead, which leads formation of narrow gullies which

(a) (b) (c) (d)

FIGURE 6.5
(a) Shallow ravines in part of Mahoba district, Uttar Pradesh, northern India. (b) Valley land in the foreground (lower left) with fallow agricultural land amidst medium deep ravines. (c) Very deep ravines along the river Chambal bordering Uttar Pradesh and Madhya Pradesh, northern India. The elevated terrain in the background indicates the original elevation of the terrain before it had turned into ravines. Similarly, the isolated two structures- a shrine and an isolated house (d) attest the extent to which the terrain has been deformed due to very severe water erosion.

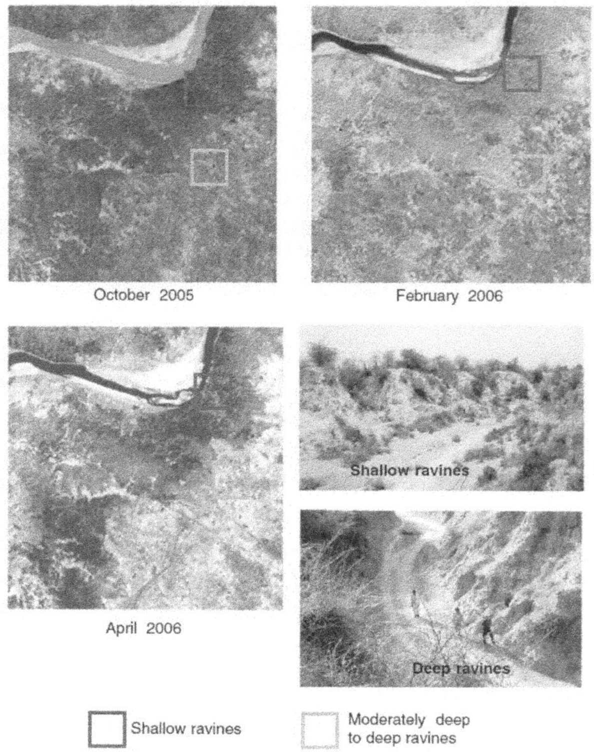

FIGURE 6.6
Ravines in parts of northern India along the river Chambal as seen in Resourcesat-1 LISS-III images during three cropping seasons, namely *kharif* (rainy season), October 2004; *rabi* (winter season), February 2005; and *zaid* crop, April 2005. The February image provides ample contrast with the agricultural crop background (seen in different hues of red color). Whereas moderately deep-to-deep ravines exhibit dark bluish green color shallow ravines confining to peripheral land show up in light bluish color. The ground photographs vividly show the magnitude of dissection (erosion) of the terrain (Credit: National Remote Sensing Centre, Indian Space Research Organization, Department of Space, Government of India).

in due course of time develop into a network of gullies called ravines. Ravines are the major threat to food security because of the fact that they develop in fluvial landscape with highly fertile and deep soils that support a variety of agricultural crops. Satellite remote sensing has been operationally utilized in detection, delineation, and monitoring ravines. While areas the adjacent ravines (e.g., farther away from river course) support very good agricultural crop ravinous land, by and large, remain either barren or support scrub vegetation. However, during rainy season grasses and shrubs are also noticed. The ground photographs of different reclamative groups of ravines are appended as Figure 6.5a-d. Remote sensing has been operationally utilized in detection, delineation and monitoring ravines. Information on morphometry (length, depth, breadth and side slope) of ravines: the network of gullies, is of paramount importance for taking up any reclamative measures. Satellite remote sensing has been operationally utilized in detection, delineation and monitoring ravines. The potential of satellite remote sensing in deriving information on ravines along the rivers Yamuna and Chambal in parts of Uttar

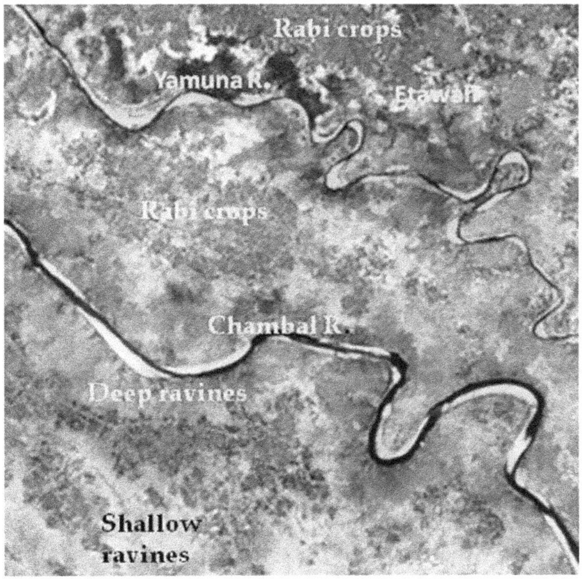

FIGURE 6.7
Ravines as seen along the river Chambal and Yamuna, in Landsat MSS image of February 28, 1975. As evident from the image ravines have devastated a fairly large areas of erstwhile fertile agricultural lands.

Pradesh and Madhya Pradesh, northern India is demonstrated in Figure 6.6. The absence of vegetation providing thereby good image contrast enables their identification. The images shown in Figures 6.6 and 6.7 vividly demonstrate this fact. As is evident from Figure 6.6, out of three cropping season Resourcesat-1 LISS-III images, *rabi* season represented by February 2006 image offers extremely good contrast which helps improve delineation of ravines.

The higher spatial resolution spaceborne multispectral data, for instance, Resourcesat-2 LISS-IV with 6 m spatial resolution permits the delineation smaller pockets of ravines and croplands along the river Yamuna (Figure 6.8).

Apart from detecting individual erosion features like rills or gullies, satellite data have been effectively used for assessing eroded areas. Extensive areas experiencing gully erosion have been mapped with visual interpretation of optical images of different sensors (e.g., Bocco et al., 1991; Dwivedi et al., 1997b). In some cases erosion classes could be separated based on vegetation cover derived from visual interpretation (Dwivedi and Ramana, 2003) or vegetation and topographic characteristics derived from additional data sources (Yuliang and Yun, 2002).

Bocco and Valenzuela (1988) applied the Gaussian maximum likelihood classifier for multispectral Landsat TM and SPOT HRV images to discern several erosion and vegetation classes, and found that the higher resolution SPOT HRV data performed better in classifying eroded areas. However, the larger number of spectral bands of Landsat TM resulted in a better classification of land cover and land use. Floras and Sgouras (1999) used the maximum likelihood classifier after principal component analysis of Landsat TM imagery to delineate eroded land. Dwivedi et al. (1997a) also found that SPOT HRV was better in classifying eroded lands than Landsat TM and MSS, but they did not use all TM bands for classification. Metternicht and Zinck (1998) performed a maximum likelihood

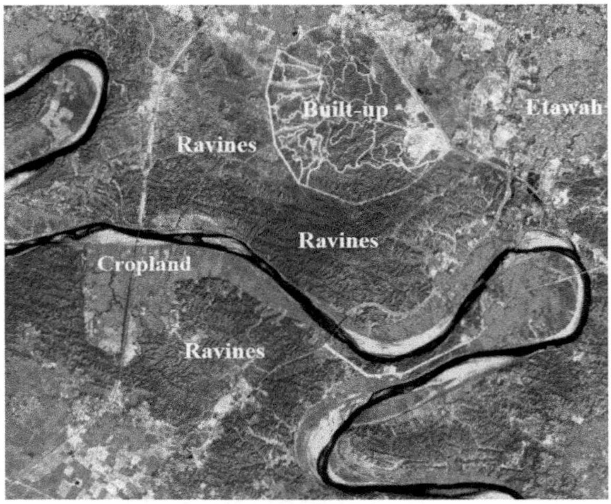

FIGURE 6.8
Resourcesat-2 LISS-IV image with 5.8 m spatial resolution and acquired on February 24, 2017 showing meandering Yamuna river in blue colour, standing winter season crops in red colour on river terraces. Deep to very deep ravines- network of gullies, with varying widths and side slopes in green and reddish brown colour indicating scrubs. Etawah town is located at the upper right corner.

classification on Landsat TM, and on the combination of Landsat TM with JERS-1 SAR data. They achieved the highest classification accuracy using the combination of both images.

6.4.3.2 Monitoring Eroded Lands

The monitoring of soil erosion using multi-temporal spaceborne multispectral data is based on the premise that changes in surface states result in variations in amplitude reflected radiation in different spectral band which in turn can provide information on incidence of water erosion. Several methods are available for change detection from spaceborne multispectral data (Coppin et al., 2004). Albedo differences between different Landsat MSS passes allowed identification of soil degradation and erosion areas in arid and semiarid environments of the USA (Frank, 1984). Karale et al. (1988) performed a bi-temporal comparison using aerial photos and Landsat TM imagery. Although a clear increase of eroded lands was found, aerial pictures allowed for a better differentiation of ravine types than satellite imagery. The delineation of eroded areas on multi-temporal images allowed an assessment of its increase (Fadul and Sgouras, 1999). Dhakal et al. (2002) showed that the spectral image differencing, principal component analysis, and spectral change vector analysis on bi-temporal Landsat TM imagery resulted in reliable detection of erosion and flooded areas, when compared to a field survey of affected and non-affected areas. The IRS-1C PAN data have been used for monitoring the spatial extent and distribution of gullied/ravinous lands in western Uttar Pradesh, northern India (Singh et al., 1998).

Repeat-pass SAR interferometry which uses the amplitude and phase information of two SAR scenes having a very similar viewing geometry and a certain time lag is another change detection technique. Interferometric coherence imagery derived from repeat-pass interferometry has most potential for erosion detection (Liu et al., 2004). Coherence between two radar signals is high when the land surface characteristics are very similar on both recording dates. Random surface changes as caused by erosion result in temporal decorrelation. Spatial decorrelation effects due to differences in satellite paths can be partly accounted for using the ratio between two coherence images (Lee and Liu, 2001). However, vegetation and soil moisture changes are also a major cause of decorrelation (e.g., Vrieling and Rodrigues, 2004), which confines erosion detection with coherence imagery to semiarid and arid environments. In spite of erosion detection possibilities, most authors stressed the need to integrate the coherence imagery with additional spatial data, like optical imagery, as an alternative to circumvent the problem of multiple causes of temporal decorrelation.

Smith et al. (2000) have demonstrated the utility of multi-temporal DEMs using SAR interferometry to assess erosion and deposition volumes during extreme events. They created DEMs from pre- and post-flood interferometric European Remote Sensing satellite (ERS) tandem pairs, and by subtraction they were able to assess volumes of erosion and deposition. Because of height accuracy DEM subtraction is limited to areas that experienced at least 4 m net erosion or deposition.

6.4.3.3 Detection of Erosion Consequences

Soil erosion is a process that detaches and transports soil particles at downstream locations, and both the transport and the deposition of soil material often cause undesired effects. Izenberg et al. (1996) determined the loss of agricultural land due to the extreme flooding of the Missouri river in 1993 with Landsat TM and SPOT HRV imagery. Thickness of deposited sand could be assessed with SIR-C L-band, as field data revealed a correlation between the sediment thickness and vegetation on deposition areas. Khan and Islam (2003) used multi-temporal Landsat data to investigate the dynamics of the Brahmaputra–Yamuna river, which is greatly influenced by the heavy sediment load originating from erosion in the Himalayas.

Most studies that erosion consequences focus on reservoirs and lakes, where sediments create important economic and ecological impacts. Using multi-temporal IRS Linear Imaging Self-Scanning Sensor (LISS-II and LISS-III) images (Jain et al., 2002) estimated the sedimentation volumes for Indian reservoirs. Images were selected of a year with maximum variation in the reservoir water levels and water-spread areas at varying depths were extracted for different dates using simple classification algorithms. Subsequently reservoir capacity was calculated with geometric equations and compared with the original capacity to determine sedimentation volumes. A comparison with a hydrographic survey revealed a slight overestimation of sedimentation rates, which was attributed to misclassification mixed pixels at the periphery of reservoir. The approach, however, permits determination of reservoir capacities only in the water fluctuation zone (Jain et al., 2002).

Erosion influences the water quality of downstream lakes and reservoirs. The suspended sediment concentration is the most important water quality parameter for erosion studies (Ritchie and Schiebe, 2000). Atmospherically corrected reflectance from surface water in the visible and infrared domain has been found to be positively related to suspended sediments (Ritchie et al., 1976). Several researchers have reported significant relationships between *in situ* determined suspended sediment concentration of inland water bodies and atmospherically corrected spectral reflectance derived from satellite remote sensing data, such as Landsat (e.g., Nellis et al., 1998), IRS LISS-I (e.g., Choubey, 1998), and multispectral SPOT HRV (e.g., Chacontorres et al., 1992).

The optimal wavelength range to determine these relations, however, depends on the sediment concentration (Curran and Novo, 1988), but often used spectral bands are between 500 and 800 nm. For detailed reviews of remote sensing for water quality assessment, including suspended sediments, the readers may refer to Dekker et al. (1995), Ritchie and Schiebe (2000), and Liu et al. (2003).

6.4.3.4 Erosion Controlling Factors

Most remote sensing studies of soil erosion focused on the assessment of erosion controlling factors, especially soil and vegetation attributes data, and to a lesser extent topography and management. Rainfall gauges are generally used for information rainfall, although Mannaerts and Saavedra (2003) proposed the use of large-scale precipitation data derived from the Tropical Rainfall Measuring Mission (TRMM).

6.4.3.4.1 Topography

Traditionally the information on slope steepness and length was derived from topographical maps. Subsequently it was replaced by aerial photographs. Nowadays, various options exist for good quality DEMs. For instance, Shuttle Radar Topography Mission (SRTM)-DEM derived from processing of SAR interferometric data is available on-line for land areas between 60-northern and 57-southern latitudes (Rabus et al., 2003). Besides, ASTER-DEM, SPOT-DEM (vertical accuracy <20 m) are also downloadable. For the Indian subcontinent, Carto-DEM derived from 2.5 m spatial resolution stereo images is available to the user community. Khawlie et al. (2002) calculated slope gradient from an ERS SAR interferometric DEM. Also ERS SAR interferograms, an interferometric product obtained before the DEM construction step, have been used directly for slope extraction (Liu et al., 2000). SPOT HRV-derived DEMs were used to aid visual interpretation of erosion features (Bishop and Shroder, 2001) and to extract slope and topographic curvatures (Haboudane et al., 2002). Stereo-data of MOMS-2 allowed DEM extraction and subsequent calculation of slope length and gradient (Reusing et al., 2000).

6.4.3.5 Soil Erosion Risk

Two qualitative mapping methods resulting in five erosion risk classes were evaluated against field-based erosion risk estimates. The first method derived slope classes from the SRTM DEM and fractional vegetation cover (FVC) classes from the Landsat image and FVC field estimates. Data integration was achieved with a decision tree. The second method did not apply field data for the map construction, but combined five equally sized NDVI classes and five slope classes using the minimum operator. Resulting maps showed a high and comparable accuracy, 80% for Method 1, and 81% for Method 2, when allowing a maximum difference of one class with field erosion risk estimates (Vrieling et al., 2006; Datta and Schack-Kirchner, 2010).

6.4.3.5.1 Soils

Soil surface characteristics, namely soil surface roughness, soil porosity, soil texture, and initial moisture content, among others, have a major influence on infiltration, thus affecting the quantum of runoff to basin outlets and streams. Most of these parameters are not directly accessible by satellite and only indirect or empirical information is obtainable. Soils differ in their resistance to erosion, which is a function of a range of soil properties such as texture, structure, soil moisture, roughness, and organic matter content. The susceptibility of soil to erosion agents is generally referred to as *soil erodibility* (Lal,

2001). Soil classifications are often used to account for spatial differences in erodibility. Important factors on the basis of which soils can be classified include soil properties, climate, vegetation, topography, and lithology. These factors can be potentially mapped with satellite remote sensing (McBratney et al., 2003). The relation between soil classes and erodibility was determined using equations of Wischmeier and Smith (1978). Soil classification by visual interpretation of optical satellite imagery has been used to assess differences in soil erodibility (Sharma and Singh, 1995). The success of this mapping must be contributed to the existence of a clear link between natural vegetation and soil types in the studied area.

Surface state conditions that are important for erosion are surface crusting and the uncovering of subsoil. Crusted soil can sometimes have distinct spectral properties as compared to uncrusted soil, due to an increase in clay particles at the soil surface, and a decrease of surface roughness (Ben-Dor et al., 2003), but this is largely dependent on the crust type and soil type. Satellite applications using this relation for erosion assessment are limited to analysis of crust dynamics of bare soil surfaces for Northern France with multi-temporal SPOT HRV imagery (Mathieu et al., 1997).

Soil surface roughness affects various hydrologic and erosion processes, namely surface depression storage, water infiltration, overland flow velocity as well as overland flow organization (Darboux et al., 2002). Soil properties that can be assessed with SAR systems are surface roughness, soil moisture, and texture (Ulaby et al., 1979). Baghdadi et al. (2002) have demonstrated the possibility of deriving surface roughness (in three classes) for bare soils from the backscattering coefficient for high-incidence radar images (Figure 6.3). Moran et al. (2002) have also shown that SAR was sensitive to differences in soil roughness (related to tillage) for dry bare soils. This parameter can be used in distributed runoff and erosion models.

6.4.3.5.2 Vegetation

Vegetation cover provides protection of the soil against erosion processes. To account for vegetation in erosion assessments, a cover and management factor (C-factor) has often been used. The C-factor is defined as the ratio of soil loss from land cropped under specified conditions to the corresponding clean-tilled continuous fallow (Wischmeier and Smith, 1978). Many satellite remote sensing studies of soil erosion focus on the assessment of vegetation cover. In many regions of the world, vegetation cover shows a high temporal dynamics. These studies need to account somehow for the temporal variation, and consequently image timing is highly important, although not always sufficiently highlighted.

Depending on the purpose of the study, sometimes a mono-temporal assessment can be sufficient. However, especially for physically based models (see Section 5.1), careful matching of satellite imagery with rainy periods and crop phenology (development) is required, which demands a time series of remote sensing images to account for seasonal variability (e.g., Cyr et al., 1995).

Vegetation indices are a specific class of spectral band ratios. A wide range of such indices exist. Often they exploit the fact that green vegetation has high reflectance in the NIR and low reflectance in the red part of the spectrum. A common index is the normalized difference vegetation index (NDVI), which is defined as the NIR reflection minus red reflection divided by the sum of the two (Tucker, 1979). NDVI has been used directly as an indicator of the protective cover of vegetation (Thiam, 2003) or was related to vegetation cover with regression analysis (Symeonakis and Drake, 2004).

Main problems with indices like the NDVI are the effect of soil reflectance (Escadafal, 1994), and the sensitivity to the vitality of the vegetation (De Jong, 1994b). To account for soil reflectance, several soil-adjusted vegetation indices have been developed, like the transformed soil-adjusted vegetation index TSAVI (Baret and Beasley, 1991). Cyr et al. (1995) showed that TSAVI performed better for the assessment of low vegetation covers than NDVI. Negative TSAVI values have been related to potentially degraded areas (Hochschild et al., 2003). However, soil-adjusted indices have difficulty in accounting for spatially variable soil types (Hill et al., 1994).

6.4.3.5.3 Conservation Practices and Tillage

Especially in agricultural areas, conservation practices such as contouring, strip cropping, or terracing, reduce soil losses. The effectiveness of such practices is often analyzed with a support practice factor (P-factor), defined as the ratio of soil loss with the practice applied and up- and downslope cultivation (Wischmeier and Smith, 1978). P-values have been assigned to land use classes that were derived from a classification of remote sensing imagery, using literature values (Lee, 2004) or expert opinion (Bhuyan et al., 2002) for practices that commonly occur for the land use class in the studied area. Analysis/interpretation of high-resolution multispectral images from WorldView-2, -3, and -4 and from Cartosat-2 series of satellites (Cartosat-2, -2A, -2B, -2C, -2D, and -2E) allows the detection of many conservation measures.

Tillage practices differ in their effect on surface roughness and amount of crop residues, which forms the basis of their detection. Image timing is very important, because tillage operations are performed during a specific time of the year. DeGloria et al. (1986) employed visual interpretation technique to monitor land under conventional and conservation tillage practices in California using 5-year Landsat MSS data. Various authors used logistic regression techniques to separate tillage practices in Landsat imagery (Bricklemyer et al., 2002; Gowda et al., 2001; Van Deventer et al., 1997). Although several band combinations were used in the different studies, all authors included TM band 5 (SWIR), which has been related to crop residue.

To assess the effect of autumn tillage on erosion in Norway, ERS-1 SAR imagery allowed a good separation of grain stubble and ploughed fields, whereas SPOT HRV imagery performed worse (Leek and Solberg, 1995). Using Radarsat imagery, McNairn et al. (1998) separated classes with different erosion potential based on the effect of residue cover and tillage operations on radar return. They stress that multi-temporal or multi-polarization SAR imagery is required for effective class separation. As radar return depends on other factors than those affected by tillage, such as soil moisture and vegetation density, assessment of tillage practices is not always straightforward. For example, effects of tillage row direction on radar backscatter have found to be as significant as the differences between tillage implements for grain stubble fields (Brisco et al., 1991). Moran et al. (2002) demonstrated that integration of optical and SAR imagery provides more information on tillage and other surface characteristics than the separate analysis of both data sources.

6.4.3.6 Assimilation of Remote Sensing Data into Runoff and Erosion Models

In several empirical erosion models four factors, topography, land use, intrinsic characteristics of the soil, and characteristics of the rainfall events, are used as basic input (De Roo, 1993). The role of remote sensing in such models is confined to facilitating the updation of the land use components that drives part of the mechanism generating runoff and erosion. Maps derived from such a qualitative analysis allow the detection and ranking of the degree of intensity and the probability of occurrence of erosion risk at the catchment scale. SPOT and Landsat data have been operationally used to derive land cover information in order to assess regional soil erosion risk in Canada (Cyr et al., 1995) and India (Dwivedi et al., 1997a; 1997b).

Satellite imagery offers immense potential in providing regional spatial data for several input parameters of erosion models (e.g., Pelletier, 1985). However, in most of the studies optical satellite data have been used to assess the vegetation component. Additional spatial data are generally extracted from rainfall gauges, readily available soil maps and DEMs, topographic maps, available aerial photographs, and field measurements. Conservation practices are sometimes considered, but usually assumed to be not present. Ideally, the spatial scale of a model is in balance with the various erosion processes that operate in a given region.

The Soil Erosion Model for Mediterranean regions (SEMMED) is the only model which was developed with the intention to use satellite data (De Jong, 1994a). The model is based on the Morgan et al. (1984) model. Modifications were subsequently made to model the erosion process in a spatially distributed approach and to enable the input from satellite imagery and DEMs. SEMMED uses optical imagery to assess the crop cover factor (same as USLE C factor) and the rainfall interception factor of vegetation at different moments. Other models that have been used in combination with satellite data include the Thornes model (Thornes, 1985), the Agricultural Nonpoint Source Pollution model (AGNPS: Young et al., 1989), and the ANSWERS model (Beasley et al., 1980).

In another study, Landsat TM data were used to derive information on landform, drainage, soil, land use, and land cover for three watersheds (320–1,020 km^2) in the arid zone of northwest India that were utilized for prediction of soil loss using the ANSWERS model (Sharma and Singh, 1995). Paringit and Nadaoka (2003) compiled their own physical model from existing equations using a Landsat image to derive vegetation parameters. While working in a catchment in South Africa using a combination of soil erosion models, Flugel et al. (2003) initially delineated terrain units with homogeneous process dynamics from aerial photographs and a Landsat-TM image. Depending on the stages of water erosion (e.g., sheet, rill erosion, and gully erosion), the RUSLE and a gully erosion model were applied to estimate sediment yields.

6.5 Conclusion

Water erosion continues to be the major land degradation process affecting a fairly large area of agricultural land and quality and quantity of water in water bodies especially multipurpose reservoirs. Extensive research has been carried out on various aspects of water erosion, namely causative factors, the mechanism; impact on agriculture, environment, and livelihood; modeling; and remedial measures and their impact assessment. Geospatial technology plays a key role in deriving the information required for rehabilitation of eroded land, their management and in projecting future scenario. With the advancements in sensor and data processing and analysis technology substantial improvements in the level and accuracy of derived information on eroded land have been achieved. For instance, precise information on soil cover factor (C) during rainfall event is of paramount importance from soil conservation view point. The images of terrain captured in microwave region provide a plausible solution for assessment of vegetation cover under perpetual cloud cover conditions. Although several models have been developed for soil loss estimation, for implementation of any soil conservation program the qualitative information on water erosion in terms of spatial extent and magnitude may serve the purpose.

For deciding upon the kind of soil conservation measures to be employed in a given region, apart from soil attributes the information on relief with respect to length, steepness of slope and aspect is very crucial. The DEMs, namely SRTM-DEM. ASTER-DEM and SOPT DEM could be effectively utilized for deriving such information. For Indian subcontinent, apart from aforesaid DEMs, the Cartosat-1 DEM with around 10 m in z-axis resolution is available from Department of Space, Government of India (www.nrsc.gov.in).

In spite of tremendous improvements in the level of information water erosion that can be derived from geospatial technology, there are a few challenging areas requiring research and development efforts. Included among them are (i) the automatic detection of individual erosion features like gullies or medium-sized rills, (ii) the spatial and temporal distribution of rainfall characteristics, (iii) evaluation of differential suspended sediment across water bodies in a landscape, (iv) the use of geostatistics for soil erodibility or C-factor mapping, and (v) precise mapping of soil properties and antecedent soil moisture in a wide range of environments (Vrieling, 2006; Shoshany et al., 2013).

References

Baghdadi, N., King, C., Bourguignon, A., Remond, A., 2002. Potential of ERS and Radarsat data for surface roughness monitoring over bare agricultural fields: application to catchments in Northern France. *International Journal of Remote Sensing* 23(17):3427–3442.

Baret, F., Guyot, G., 1991. Potentials and limits of vegetation indices for LAI an APAR assessment. *Remote Sensing of Environment* 35(2–3):161–173.

Beasley, D.B., Huggins, L.F., Monke, E.J., 1980. ANSWERS: A model for watershed planning. *Transactions of the ASAE* 23:938–944.

Beaulieu, A., Gaonac'h, H., 2002. Scaling of differentially eroded surfaces in the drainage network of the Ethiopian Plateau. *Remote Sensing of Environment* 82(1):111–122.

Ben-Dor, E., Goldshleger, N., Benyamini, Y., Agassi, M., Blumberg, D.G., 2003. The spectral reflectance properties of soil structural crusts in the 1.2- to 2.5-Am spectral region. *Soil Science Society of America Journal* 67(1):289–299.

Ben-Dor, E., Chabrillat, S., Demattê, J.A.M., Taylor, G.R., Hill, J., Whiting, M.L., Sommer, S., 2009. Using Imaging spectrometry to study soil properties. *Remote Sensing of Environment* 13(1):S38–S55.

Ben-Dor, E., Demattê, J.A.M., 2015. Remote sensing of soil in the optical domains. In: Thenkabail, P.S., (ed.) *Land Resources Monitoring, Modeling, and Mapping with Remote Sensing*. CRC Press, Boca Raton, FL, pp. 733–787.

Ben-Dor, E., Goldshleger, N., Braun, O., Kindel, B., Goetz, A.F.H., Bonfil, D., Margalit, N., Binaymini, Y., Karnieli, A., Agassi, M., 2004. Monitoring infiltration rates in semiarid soils using airborne hyperspectral technology. *International Journal of Remote Sensing* 25:2607–2624.

Ben-Hur, M., Shainberg, I., Bakker, D., Keren, R., 1985. Effect of soil texture and CaC03 content on water infiltration in crusted soil as related to water salinity. *Irrigation Science* 6:281–294.

Bhuyan, S.J., Marzen, L.J., Koelliker, J.K., Harrington Jr., J.A., Barnes, P.L., 2002. Assessment of runoff and sediment yield using remote sensing, GIS, and AGNPS. *Journal of Soil and Water Conservation* 57(6):351–364.

Bishop, M.P., Shroder Jr., J.F., 2001. Remote sensing and geomorphometric assessment of topographic complexity and erosion dynamics in the Nanga Parbat massif. In: Khan, M.A., Treloar, P.J., Searle, M.P., Jan, M.Q., (eds.) *Tectonics of the Nanga Parbat Syntaxis and the Western Himalaya, vol. 170*. Special Publication, Geological Society, London, pp. 181–200.

Bocco, G., Palacio, J.L., Valenzuela, C.R., 1991. Gully erosion modelling using GIS and geomorphological knowledge. *ITC Journal* 1991(3):253–261.

Bocco, G., Valenzuela, C.R., 1988. Integration of GIS and image processing in soil erosion studies using ILWIS. *ITC Journal* 4:309–319.

Bradford, J.M., Ferris, J.E., Remley, P.A., 1987. Interrill soil erosion processes. II: Relationship of splash detachment to soil properties. *Soil Science Society of America* 51:1571–1575.

Bricklemyer, R.S., Lawrence, R.S., Miller, P.R., Battogtokh, N., 2002. Predicting tillage practices and agricultural soil disturbance in north central Montana with Landsat imagery. *Agriculture Ecosystems and Environment* 114(2-4):210–216.

Brisco, B., Brown, R.J., Snider, B., Sofko, G.J., Koehler, J.A., Wacker, A.G., 1991. Tillage effects on the radar backscattering coefficient of grain stubble fields. *International Journal of Remote Sensing* 12(11):2283–2298.

Chabrillat, S., Milewski, R., Schmid, T., Rodriguez, M., Escribano, P., Pelayo, M., Palacios-Orueta, A. Potential of hyperspectral imagery for the spatial assessment of soil erosion stages in agricultural semi-arid Spain at different scales. In: *Proceedings of the 2014 IEEE Geoscience and Remote Sensing Symposium*, Quebec City, QC, Canada, 13–18 July 2014; pp. 2918–2921.

Chacontorres, A., Ross, L.G., Beveridge, M.C.M., Watson, A.I., 1992. The application of SPOT multispectral imagery for the assessment of water quality in Lake Patzcuaro, Mexico. *International Journal of Remote Sensing* 13(4):587–603.

Choubey, V.K., 1998. Laboratory experiment, field and remotely sensed data analysis for the assessment of suspended solids concentration and Secchi depth of the reservoir surface water. *International Journal of Remote Sensing* 19(17):3349–3360.

Coldwell, A.E., 1957. Importance of channel erosion as a source of sediment. *American Geophysical Union Transactions* 38:908–912.

Conforti, M., Buttafuoco, G., Leone, A.P., Aucelli, P.P.C., Robustelli, G., Scarciglia, F., 2013. Studying the relationship between water-induced soil erosion and soil organic matter using VIS-NIR spectroscopy and geomorphological analysis: A case study in southern Italy. *Catena* 110:44–58.

Coppin, P., Jonckheere, I., Nackaerts, K., Muys, B., Lambin, E., 2004. Digital change detection methods in ecosystem monitoring: a review. *International Journal of Remote Sensing* 25(9):1565–1596.

Cox, C., Madramootoo, C., 1998. Application of geographic information systems in watershed management planning in St. Lucia. *Computers and Electronics in Agriculture* 20(3):229–250.

Curran, P.J., Novo, E.M.M., 1988. The relationship between suspended sediment concentration and remotely sensed spectral radiance: a review. *Journal of Coastal Research* 4(3):351–368.

Curwick, P.B., Jennings, M.E., 1982. Evaluation of rainfall-runoff models for the hydrologic characterization of surface-mined lands. In: Applied modeling in catchment hydrology: *Proceedings of the International Symposium on Rainfall — Runoff modeling* held May 18-21, 1981 at Mississippi State University, Mississippi State, Mississippi, pp. 259–274.

Cyr, L., Bonn, F., Pesant, A., 1995. Vegetation indices derived from remote sensing for an estimation of soil protection against water erosion. *Ecological Modelling* 79(1–3):277–285.

Darboux, F., Davy, Ph., Gascuel-Odoux, C., 2002. Effect of depression storage capacity on overland-flow generation for rough horizontal surfaces: water transfer distance and scaling. *Earth Surface Processes and Landforms* 27(2):177–191.

Datta, P.S., Schack-Kirchner, H., 2010. Erosion relevant topographical parameters derived from different DEMs—A comparative study from the Indian lesser Himalayas. *Remote Sensing* 2:1941–1961.

De Jong, S.M., 1994a. Applications of reflective remote sensing for land degradation studies in a Mediterranean Environment. PhD thesis, Utrecht University, The Netherlands, 237 pp.

De Jong, S.M., 1994b. Derivation of vegetative variables from a Landsat TM image for modelling soil erosion. *Earth Surface Processes and Landforms* 19(2):165–178.

de Roo, A.P.J., 1993. Validation of the ANSWERS catchment model for runoff and soil erosion simulation in catchments in the Netherlands and the United Kingdom. In: *Applications of Geographic Information Systems in Hydrology and Water Resources Management. Proceedings of International Conference*, Vienna, 1993, pp: 465–474.

DeCoursey, D.G., 1985. Mathematical models for nonpoint water pollution control. *Journal of Soil and Water Conservation* 40(5):408–413.

DeGloria, S.D., Wall, S.L., Benson, A.S., Whiting, M.L., 1986. Monitoring conservation tillage practices using Landsat multispectral data. *Journal of Soil and Water Conservation* 41(3):187–190.

Dekker, A.G., Malthus, T.J.M., Hoogenboom, H.J., 1995. The remote sensing of inland water quality. In: Danson, F.M., and Plummer, S.E., (eds.) *Advances in Environmental Remote Sensing*. Wiley, Chichester.

Dhakal, A.S., Amada, T., Aniya, M., Sharma, R.R., 2002. Detection of areas associated with flood and erosion caused by a heavy rainfall using multi-temporal Landsat TM data. *Photogrammetric Engineering and Remote Sensing* 68(3):233–239.

Dwivedi, R.S., Kumar, A.B., Tewari, K.N., 1997a. The utility of multi-sensor data for mapping eroded lands. *International Journal of Remote Sensing* 18(11):2303–2318.

Dwivedi, R.S., Ramana, K.V., 2003. The delineation of reclamative groups of ravines in the Indo-Gangetic alluvial plains using IRS-1D LISS-III data. *International Journal of Remote Sensing* 24(22):4347–4355.

Dwivedi, R.S., Sankar, T.R., Venkaratnam, L., Karale, R.L., Gawande, S.P., Rao, K.V.S., Senchaudhary, S., Bhaumik, K.R., Mukharjee, K.K., 1997b. The inventory and monitoring of eroded lands using remote sensing data. *International Journal of Remote Sensing* 18(1):107–119.

Ekwue, E.L., 1990. Effect of organic matter on splash detachment and the process involved. *Earth Surface Processes and Landforms* 15:175–181.

El-Swaify, S.A., Dangler, E.W., 1977. Erodibility of selected tropical soils in relation to structural and hydrologic parameters. In: *Soil Erosion: Prediction and Control*, Special Publication. 21.

Escadafal, R., 1994. Soil spectral properties and their relationships with environmental parameters: Examples from arid regions. In: Hill, J., Me´gier, J., (eds.) *Imaging Spectrometry—a Tool for Environmental Observations*. Kluwer Academic Publishers, Dordrecht, pp. 71–87.

Flacke, W., Auerswald, K., Neufang, L., 1990. Combining a modified universal soil loss equation with a digital terrain model for computing high resolution maps of soil loss resulting from rain wash. *CATENA* 17:383–397.

Floras, S.A., Sgouras, I.D., 1999. Use of geoinformation techniques in identifying and mapping areas of erosion in a hilly landscape of central Greece. *International Journal of Applied Earth Observation and Geoinformation* 1(1):68–77.

Flugel, W.A., Ma¨rker, M., Moretti, S., Rodolfi, G., Sidorchuk, A., 2003. Application of geographic information systems and remote sensing for quantifying patterns of erosion and water quality. *Hydrological Processes* 17:929–942.

Foster, G.R., 1990. Process-based modelling of soil erosion by water on agricultural land. In: Boardman, J., Foster, I.D.L., Dearing, J.A., (eds.) *Soil Erosion on Agricultural Land*. John Wiley & Sons Ltd, Hoboken, NJ, 429–445.

Frank, T.D., 1984. Assessing change in surficial character of a semiarid environment with Landsat residual image. *Photogrammetric Engineering and Remote Sensing* 50(4):471–480.

Gallaher, R.N., Hawf, L., 1997. Role of conservation tillage in production of a wholesome food supply. Partners for A, 23.

Gorrie, R.M., 1935. The use and misuse of land. In: *Oxford Forestry Memoirs*, pp. 37–47 and 66–67.

Gowda, P.H., Dalzell, B.J., Mulla, D.J., Kollman, F., 2001. Mapping tillage practices with Landsat thematic mapper based logistic regression models. *Journal of Soil and Water Conservation* 56(2):91–96.

Grieve, I. C., Hipkin, J.A., and Davidson, D.A., 1994. Soil erosion sensitivity in upland Scotland. Scottish Natural Heritage Research Survey and Monitoring Reports 24. Scottish Natural Heritage, Edinburgh.

Haboudane, D., Bonn, F., Royer, A., Sommer, S., Mehl, W., 2002. Land degradation and erosion risk mapping by fusion of spectrally-based information and digital geomorphometric attributes. *International Journal of Remote Sensing* 23(18):3795–3820.

Haubrock, S., Chabrillat, S., Kaufmann, H., 2005. Application of hyperspectral imaging for the quantification of surface soil moisture in erosion monitoring and modeling, Proceeding of the 4th Workshop on Imaging Spectroscopy: New Quality in Environmental Studies, 197–206.

Hensel, H., Bork, H.R., 1987. EDV—gestutzte Bilanzierung von Erosion und akkumulationin kleine Einzugsgebieten unter Verwendung der modifizierten Universal Soil Loss Equation. Landschaftsoko logisches Messen und Auswerten 2.2/3, Braunschweig.

Hill, J., Mehl, W., Altherr, M., 1994. Land degradation and soil erosion mapping in a Mediterranean ecosystem. In: Hill Jr., J., Mégier, J., (eds.) *Imaging Spectrometry—A Tool for Environmental Observations*. Kluwer Academic Publishers, Dordrecht, pp. 237–260.

Hill, J., Mégier, J., Mehl, W., 1995. Land degradation, soil erosion and desertification monitoring in Mediterranean ecosystems. *Remote Sensing Reviews* 12:107–130.

Hochschild, V., Märker, M., Rodolfi, G., Staudenrausch, H., 2003. Delineation of erosion classes in semi-arid southern African grasslands using vegetation indices from optical remote sensing data. *Hydrological Processes* 17(5):917–928.

Hudson, N.W., 1985. A world view of the development of soil conservation. *Agricultural History* 59:327–39.

Izenberg, N.R., Arvidson, R.E., Brackett, R.A., Saatchi, S.S., Osburn, G.R., Dohrenwend, J., 1996. Erosional and depositional patterns associated with the 1993 Missouri River floods inferred from SIR-C and TOPSAR radar data. *Journal of Geophysical Research* 101(E10):23149–23167.

Jain, S.K., Singh, P., Seth, S.M., 2002. Assessment of sedimentation in Bhakra Reservoir in the western Himalayan region using remotely sensed data. *Hydrological Sciences Journal* 47(2):203–212.

Judson, S. 1965. *Physical Geology*, 3rd edn. Prentice Hall, Upper Saddle River, NJ, pp. 143–144.

Kamphrost, A., Iyer, H.S., 1972. Application of aerial photo-interpretation of ravine surveys in India. Twelfth Congress of International Society of Photogrammetry. Ottawa, 1972.

Karale, R.L., Saini, K.M., Narula, K.K., 1988. Mapping and monitoring ravines using remotely sensed data. *Journal of Soil and Water Conservation, India* 32(1,2):75.

Khan, N.I., Islam, A., 2003. Quantification of erosion pattern in the Brahmaputra-Jamuna River using geographical information system and remote sensing techniques. *Hydrological Processes* 17(5):959–966.

Khawlie, M., Awad, M., Shaban, A., Kheir, R.B., Abdallah, C., 2002. Remote sensing for environmental protection of the eastern Mediterranean rugged mountainous areas, Lebanon. *ISPRS Journal of Photogrammetry and Remote Sensing* 57(1–2):13–23.

Knisel, W.G., 1980. *CREAMS: A Field Scale Model for Chemicals, Runoff and Erosion from Agricultural Management Systems*. USDA, Washington, D.C.

Lal, R., 2001. Soil degradation by erosion. *Land Degradation & Development* 12(6):519–539.

Latz, K., Weismiller, R.A., Scoyoc, G.E.V., Baumgardner, M.F., 1984. Characteristic variations in spectral reflectance of selected eroded Alfisols. *Soil Science Society of America Journal* 48(5):1130–1134.

Le Bissonnais, Y., 1990. Experimental study and modelling of soil surface crusting process. *Catena Supplement* 17:13–28.

Lee, S., 2004. Soil erosion assessment and its verification using the universal soil loss equation and geographic information system: A case study at Boun, Korea. *Environmental Geology* 45:457–465.

Lee, H., Liu, J.G., 2001. Analysis of topographic decorrelation in SAR interferometry using ratio coherence imagery. *IEEE Transactions on Geoscience and Remote Sensing* 39(2):223–232.

Leek, R., Solberg, R., 1995. Using remote sensing for monitoring of autumn tillage in Norway. *International Journal of Remote Sensing* 16(3):447–466.

Lehmann, R., 1992. Terrace degradation and soil erosion on Naxos Island, Greece. In: Wicherek, S., (ed.) *Farm Land Erosion in Temperate Plains Environment and Hills. Proceedings of the Symposium on Saint-Cloud*, Paris, pp. 429–450.

Leuder, D.R., 1959. *Aerial Photographic Interpretation (Principles and Applications)*. McGraw-Hill Book Company, Inc., New York.

Liddaman, L., 2008. Geospatial patterning and the hydro-morphological functioning of eroding blanket peat, Unpublished PhD Thesis, Department of Geography, University of Manchester. 275p.

Liu, Y., Anisul Islamand, M.d., Gao, J., 2003. Quantification of shallow water quality parameters by means of remote sensing. *Progress in Physical Geography* 27(1):24–43.

Liu, J.G., Hilton, F., Mason, P., Lee, H., 2000. A RS/GIS study of rapid erosion in SE Spain using ERS SAR multi-temporal interferometric coherence imagery. In: Owe, M., Zilioli, E., D'Urso, G., (eds.) *Remote Sensing for Agriculture, Ecosystems, and Hydrology II, Proceedings of SPIE*, vol. 4171. SPIE International, Barcelona, Spain, pp. 367–375.

Liu, J.G., Mason, P. Hilton, F., Lee, H., 2004. Detection of rapid erosion in SE Spain: A GIS approach based on ERS SAR coherence imagery. *Photogrammetric Engineering and Remote Sensing* 70:1179–1185.

Mannaerts, C.M., Saavedra, C.P., 2003. Regional scale erosion modelling and monitoring using remotely sensed data: some spatial data scale issues. In: Gabriels, D., Cornelis, W., (eds.) *Proceedings of the International Symposium on 25 Years of Assessment of Erosion*. Ghent University, Ghent, Belgium, pp. 433–440.

Mathieu, R., King, C., Bissonnais, Y.L., 1997. Contribution of multi-temporal SPOT data to the mapping of a soil erosion index: The case of the loamy plateaux of northern France. *Soil Technology* 10(2):99–110.

McBratney, A.B., Mendon Santosb, M.L., Minasny, B., 2003. On digital soil mapping. *Geoderma* 117(1-2):3–52.

McNairn, H., Wood, D., Gwyn, Q.H.J., Brown, R.J., Charbonneau, F., 1998. Mapping tillage and crop residue management practices with RADARSAT. *Canadian Journal of Remote Sensing* 24(1):28–35.

Merritt, W.S., Letcher, R.A., Jakeman, A.J., 2003. A review of erosion and sediment transport models. *Environmental Modeling and Software* 18:761–799.

Metternicht, G.I., Zinck, J.A., 1998. Evaluating the information content of JERS-1 SAR and Landsat TM data for discrimination of soil erosion features. *ISPRS Journal of Photogrammetry and Remote Sensing* 53(3):143–153.

Miller, W.P., Baharuddin, M.K., 1986. Relationship of soil dispersibility to infiltration and erosion of southeastern soils. *Soil Science* 142:235–240.

Millington, A.C., Townshend, J.R.G., 1984. Remote sensing applications in African erosion and sedimentation studies. In: Walling, D.E., Foster, S.S.D., Wurzel, P., (eds.) *Challenges in African Hydrology and Water Resources: Proceedings of the Harare Symposium*. IAHS Publication, vol. 144. IAHS Press, pp. 373–384.

Moore, D.C., Singer, M.J., 1990. Crust formation effect on soil erosion processes. *Soil Science Society of America Journal* 54:1117–1123.

Moran, M.S., Hymer, D.C., Qi, J., Kerr, Y., 2002. Comparison of ERS-2 SAR and Landsat TM imagery for monitoring agricultural crop and soil conditions. *Remote Sensing of Environment* 79(2–3):243–252.

Morgan, R.P.C., Morgan, D.D.V., Finney, H.J., 1984. A Predictive Model for the Assessment of Soil Erosion Risk. *Journal of Agricultural Engineering Research* 30:245–253.

Nellis, M.D., Harrington, Jr., J.A., Wu, J., 1998. Remote sensing of temporal and spatial variations in pool size, suspended sediment, turbidity, and Secchi depth in Tuttle Creek Reservoir, Kansas: 1993. *Geomorphology* 21(3–4):281–293.

Panayiotopoulos, K.P., Kostopoulou, S., 1989. Aggregate stability dependence on size, cultivation and various soil constituents in red Mediterranean soils (alfisols). *Soil Technology* 2:79–89.

Paringit, E.C., Nadaoka, K., 2003. Sediment yield modelling for small agricultural catchments: land-cover parameterization based on remote sensing data analysis. *Hydrological Processes* 17(9):1845–1866.

Pelletier, R.E., 1985. Evaluating nonpoint pollution using remotely sensed data in soil erosion models. *Journal of Soil and Water Conservation* 40(4):332–335.

Pellikkaa, P.K.E., Clarka, B.J.F., Sirviöa, T., Masalina, K., 2005. Environmental change monitoring applying satellite and airborne remote sensing data in the Taita Hills, Kenya. Remote sensing and geo-information processing in the assessment and monitoring of land degradation and desertification, Trier, Germany, 2005.

Perez-Trejo, F., 1986. *Review of Existing Watershed Models for the Delft Hydraulics Laboratory*. Colorado State University, Fort Collins, CO, p. 20.

Pickup, G., Nelson, J., 1984. Use of Landsat radiance parameters to distinguish soil erosion, stability, and deposition in arid Central Australia. *Remote Sensing of Environment* 16(3):195–209.
Porêba, G.J., 2006. Caesium-137 as a soil erosion tracer: a review. *Geochronometria* 25:37–46.
Rabus, B., Eineder, M.A.R., Bamler, R., 2003. The shuttle radar topography mission: a new class of digital elevation models acquired by spaceborne radar. *ISPRS Journal of Photogrammetry and Remote Sensing* 57(4):241–262.
Renard, K.G., Foster, G.R., Weesies, G.A., McCool, D.K., Yoder, D.C., 1997. *Predicting Soil Erosion by Water: A Guide to Conservation Planning with the Revised Universal Soil Loss Equation (RUSLE).* Agricultural Handbook No. 703, U. S. Department of Agriculture, Washington DC, p. 384.
Reusing, M., Schneider, T., Ammer, U., 2000. Modelling soil loss rates in the Ethiopian Highlands by integration of high resolution MOMS-02/D2-stereo-data in a GIS. *International Journal of Remote Sensing* 21(9):1885–1896.
Ritchie, J.C., Schiebe, F.R., 2000. Water quality. In: Schultz, G.A., Engman, E.T., (eds.) *Remote Sensing in Hydrology and Water Management*. Springer-Verlag, Berlin, pp. 287–303, and 351–352.
Ritchie, J.C., Schiebe, F.R., McHenry, J.R., 1976. Remote sensing of suspended sediment in surface water. *Photogrammetric Engineering and Remote Sensing* 42(12):1539–1545.
Roberson, E. B., Sarig, S., Firestone, M.K., 1991. Cover crop management of polysaccharides-mediated aggregation in an orchard soil. *Soil Science Society of America Journal* 55:734–739.
Schmid, T., Manuel Rodríguez-Rastrero, M., Escribano, P., Alicia Palacios-Orueta, A., Ben-Dor, E., Plaza, A., Milewski, R.Huesca, M., Ashley Bracken, A., Víctor Cicuéndez, V., Marta Pelayo, M. and Sabine Chabrillat, S., 2016. Characterization of soil erosion indicators using hyper-spectral data from a Mediterranean rainfed cultivated region. *IEEE Journal of Selected Topics In Applied Earth Observations and Remote Sensing* 9(2):845–860.
Schwertmann, U., 1982. Die Vorausschatzung des Bodenabtrags durch Wasser in Bayern. Bayer. Staatsmin. Landwirtsch. Forsten, Munchen, pp. 126.
Sharma, K.D., Singh, S., 1995. Satellite remote sensing for soil erosion modelling using the ANSWERS model. *Hydrological Sciences Journal* 40(2):259–272, DOI:10.1080/02626669509491408.
Shoshany, M., Goldshleger, N., Chudnovsky, A., 2013. Monitoring of agricultural soil degradation by remote-sensing methods: A review. *International Journal of Remote Sensing* 34:6152–6181.
Singh, A.N., Sharma, Y.K., Singh, S., 1998. *Evaluation of IRS-1C PAN Data for Monitoring Gullied and Ravinous Lands of Western U.P. Remote Sensing and Geographic Information System for Natural Resources Management*. A Joint Publication of Indian Society of Remote Sensing and NNRMS, Department of Space, Government of India, Bangalore, pp.149–160.
Smith, R.E., Ferreira, V.A., 1986. Effect of soil hydraulic properties on root zone transport processes. *In: Proceedings of AGU Front Range Branch Hydrology Days*. pp. 123–137. Ft. Collins: Hydrology Days Publications. Colorado State University.
Smith, L.C., Alsdorf, D.E., Magilligan, F.J., Gomez, B., Mertes, L.A.K., Smith, N.D., Garvin, J.D., 2000. Estimation of erosion, deposition, and net volumetric change caused by the 1996 Skeiðara´rsandur jo¨ kulhlaup, Iceland, from synthetic aperture radar interferometry. *Water Resources Research* 36(6):1583–1594.
Sreenivas, K., Dwivedi, R.S., 2010. Personal communication.
Stern, R., Ben-Hur, M., Shainberg, I., 1991. Clay mineralogy effect on rain infiltration, seal formation and soil losses. *Soil Science* 152:455–462.
Symeonakis, E., Drake, N., 2004. Monitoring desertification and land degradation over sub-Saharan Africa. *International Journal of Remote Sensing* 25(3):573–592.
Thiam, A.K., 2003. The causes and spatial pattern of land degradation risk in southern Mauritania using multitemporal AVHRR-NDVI imagery and field data. *Land Degradation and Development* 14(1):133–142.
Thornes, J.B., 1985. The ecology of erosion. *Geography* 70:222–234.
Trott, K.E., Singer, M.J., 1983. Relative erodibility of 20 California range and forest soils. *Soil Science Society of America Journal* 47:753–749.
Tucker, C.J., 1979. Red and photographic infrared linear combinations for monitoring vegetation. *Remote Sensing of Environment* 8(2):127–150.

Ulaby, F.T., Bradley, G.A., Dobson, M.C., 1979. Microwave backscatter using GIS and geomorphological knowledge. *ITC Journal* 1991(3):253–261.
van Deventer, A.P., Ward, A.D., Gowda, P.H., Lyon, J.G., 1997. Using Thematic Mapper data to identify contrasting soil plains and tillage practices. *Photogrammetric Engineering and Remote Sensing* 63(1):87–93.
Vrieling, A., 2006. Satellite remote sensing for water erosion assessment: A review. *Catena* 65:2–18.
Vrieling, A., Rodrigues, S.C., 2004. Erosion assessment in the Brazilian Cerrados using multi-temporal SAR imagery. *Proceedings of the 2004 Envisat and ERS Symposium*. SP-572. ESA, Salzburg, Austria.
Weismiller, R.A., Van Scoyoc, P.E., Pazar, S.E., Latz, K., Baumgardner, M.F., 1985. Use of soil spectral properties for monitoring soil erosion. In: *Soil Erosion and Conservation*. Soil Conservation Society of America, Ankeny, IA.
Williams, J.R., Brendt, H.D., 1977. Sediment yield prediction based on watershed hydrology. *Transactions of the ASAE* 20:1100–1104.
Wischmeier, H., Smith, D.D., 1978. *Predicting Rainfall Erosion Losses: A Guide to Conservation Planning*. Agriculture Handbook 537. U.S. Government Printing Office, Washington, DC.
Yalin, M.S., 1963. An expression for bed load transportation. *Journal of Hydraulics Division, American Society of Civil Engineers* 98(HY3):221–250.
Young, R.A., Onstad, C.A., Bosch, D.D., Anderson, W., 1989. AGNPS: a nonpoint-source pollution model for evaluating agricultural watersheds. *Journal of Soil and Water Conservation* 44(2):168–173.
Yuliang, Q., Yun, Q., 2002. Fast soil erosion investigation and dynamic analysis in the Loess Plateau of China by using information composite technique. *Advances in Space Research* 29(1), 85–88.

7

Wind Erosion

7.1 Introduction

Erosion is the process of dislodgement and transport of soil particles by the two main agents of erosion—water and wind. It is often difficult to determine which the dominant agent on a particular site is. In general, however, wind erosion is a long-term process of gradual removal, while water erosion is often very rapid. Wind erosion refers to movement of soil by wind and is most common in arid and semiarid regions but occurs to some extent in subhumid regions too. Wind erosion has both on-site and off-site impacts. On-site impacts include soil and nutrient losses such as soil organic matter and carbon stores (Leys et al., 2008). Off-site impacts of wind erosion include reduced air quality, leading to increased health risk of respiratory disease (Rutherford et al., 2003). In terms of climate change, the dust (aerosol) generated during wind erosion process holds the potential to modulate global air temperatures through radiative forcing. The net effect of this remains, however, to be ascertained and modeled (Rotstayn et al., 2010).

7.2 Background

The term *wind* is derived from a Greek word "aeolus." Historically, the wrath of Aeolus (the Greek God of wind, hence "aeolian" phenomena) has had catastrophic effects on the human biological and cultural environment, manifested in hurricanes and typhoons in the tropics, in blowing icicles and blizzards in the colder regions and in dust storms in arid areas. Wind is air in movement, caused by a gradient in the air density generated by differences in atmospheric pressures and/or temperatures. The erosive power of wind, as in the case of water, increases exponentially with velocity, but, unlike water, wind is not affected by the force of gravity. However, the length of unobstructed terrain (fetch) over which the wind flows is important in allowing the wind to gain momentum and to increase its erosive power. Air movement must attain a certain velocity or become an "efficient wind"—with enough speed to generate visible movement of particles at the soil level, before it can generate deflation and transport of particles. Winds with velocities of less than 12–19 km/h at 1 m above the ground seldom impart sufficient energy at the soil surface to dislodge and put into motion sand-sized particles. Drifting of highly erosive soil usually begins when the wind attains a forward velocity of 25–30 km/h at a height of

30cm above the soil surface (Hopkins et al., 1937). The wind velocity at which sediment begins to move is termed the *threshold wind velocity.*

The process consists of three distinct phases, namely initiation of movement, transportation, and deposition. Finer soil fractions like silt, clay, and organic matter are removed as dust and carried away by the wind leaving coarse particles like sand and gravel. Wind erosion often results in the formation of dunes. Wind erosion occurs as the wind interacts with the soil surface to cause detachment (termed entrainment), transport, and finally deposition of soil particles. Detachment occurs as the wind exerts drag and lift forces to overcome the gravitational and cohesive forces that hold particles to the soil surface. Detachment also occurs as rolling or bouncing particles cause other particles to be released by impacts or abrasion. Winds are considered erosive when they reach a speed of about 6 m/sat 0.3m above the soil.

A feel of the terrain experiencing wind can be had from a few ground photographs portraying desert terrain ranging from fresh sand deposition to partially and for fully stabilized dunes supporting agricultural crops (Annexure 7a–f).

7.2.1 Wind Erosion Processes

The wind erosion involves three sequential processes, namely *surface creep, saltation,* and *suspension.*

Surface Creep: Initially, large particles ranging from 0.5 to 2mm in diameter are rolled across the soil surface. This causes them to collide with and dislodge other particles.

Saltation occurs among middle-sized soil particles that range from 0.05 to 0.5mm in diameter. Such particles are light enough to be lifted off the surface but are too large to become suspended. These particles move through a series of low bounces over the surface causing abrasion on the soil surface and attrition (the breaking of particles into smaller particles).

Suspension occurs when very fine dirt (0.1–0.01mm) and dust (smaller than 0.01mm) particles are lifted into the air. They can be thrown into the air through impact with other particles or by the wind itself.

With constant wind speed, the rate of erosion changes with the distance downwind from the point where the erosion starts, as well as by time, over an erosion period lasting several hours. The soil surface is continually modified as it is eroded. The amount of soil eroded increases with distance downwind until it reaches a maximum. For sandy soils, erosion can start within about 0.2m and reach the maximum erosion rate within about 5m from where the erosion started, e.g., at the edge of a paddock. This has implications for erosion control. Any method of erosion control should be intended to stop saltation before it reaches its maximum rate; thus, the non-erodible roughness (clods or vegetation) should be no more than a few meters apart. Measurements of dust in the air up to 3m above the soil surface at Jodhpur, India, showed that on a stormy day the amount of dust blowing varied between 50 and 420 kg/ha (Mann, 1985). In the Jaisalmer region of India, where wind speeds generally are higher, an average soil loss of 511 kg/ha was recorded (Nordstrom and Hotta, 2004).

7.2.2 Causative Factors

The amount of soil lost due to wind erosion depends on several factors including wind velocity (USDA, 1961). Wind Erosion Equation (WEQ) which is designed to predict long-term average annual soil erosion by wind based on a specific set of climatic and field conditions is expressed as follows:

$$E = f(IKCLV) \tag{7.1}$$

where E is the estimated annual soil loss (mass/unit area/time period), f indicates a non-linear functional relationship among the variables, I is the soil erodibility index (mass/unit area), K is the soil surface roughness factor, C is the climatic factor, L is the unsheltered distance, and V is the vegetative over factor.

7.2.2.1 Soil Erodibility

Soil erodibility, "I" in equation 7.1, is the susceptibility of soil particles to detachment and transportation by erosive agents, i.e., wind and water, is an essential parameter required for the prediction of soil erosion (Bryan, 1968; Geeves et al., 2000). Soil erodibility is defined as the amount of soil loss per unit exogenic force or erosivity, which is controlled by the intrinsic properties of soil (Raupach and Lu, 2004). It is generally considered an inherent soil property with a constant value for a given soil type (Webb et al., 2009). The erodibility of soils depends primarily on the soil texture or relative proportion of sand, silt, and clay, water holding capacity, structure, and degree of water dispersion within the soil grains. Other factors such as surface roughness, fetch distances and wind speeds, size and stability of soil aggregates, and crusting are also important for soil's susceptibility to wind erosion. Potential susceptibility of soil to wind erosion is mainly due to soil conditions, but is also affected by the climatic conditions of the area.

7.2.2.2 Soil Surface Conditions

If the soil surface is stony, forming a "pavement," the risks of wind erosion are lower—as, for example, in regs (desert pavement). A rough surface, left by cloddy tillage or ridges perpendicular to the prevailing wind direction, slows down the wind at ground level, thus reducing saltation. Crusts and immobile aggregates serve as strong modulators of wind erosion. Crust index (CI) is computed using remote sensing data using the spectral reflectance patterns of the terrain in blue and red spectral bands. CI (equation 7.2) is based on the normalized difference between the red and the blue spectral values of the imagery (Karnieli, 1997). CI, when applied to the remote sensing imagery, is capable of detecting and differentiating between different lithologic/morphologic units, such as active crusted sand areas, which are expressed in the topography. Karnieli (1997) demonstrated that CI is a powerful index to map geological features and is more sensitive to ground features than the wavebands comprising the original image. The CI value ranges between 0 and +2, but typically is between 0 and 1.

$$CI = 1 - \left(\frac{Red - Blue}{Red + Blue} \right) \tag{7.2}$$

The strength of crusts and aggregates depends on both the formation processes and the soil composition. The breakdown of crusts and aggregates by abrasion depends on their strength and the abrader discharge rate (Hagen, 2008).

7.2.2.3 Soil Texture

The loamy sand soils rich in particles between 10 and 100 microns in size are more vulnerable to wind erosion. More clayey soil is much stickier, better-structured, and thus more resistant. Coarse sand and gravelly or rocky soils are also more resistant since the particles are too heavy to be removed by wind erosion. The optimum size soil particles to be transferred by wind are about 80 microns. Biological soil crusts, consisting of cyanobacteria, green algae, lichens, and mosses, are important in stabilizing soils in semiarid and arid lands.

An extremely important parameter is the threshold velocity for dust production. This parameter is dependent on effects of vegetative residue, roughness of the soil, live standing plants, soil texture, and the effect of atmospheric precipitation. Silt and clay fractions are removed by wind first, leaving coarser minerals like sand and gravel. This sorting action over many years makes soil progressively coarser until nothing remains but infertile skeletal material that forms shifting sand dunes and gravelly pavements (Lyles, 1975).

7.2.2.4 Climate

Wind erosion demonstrates that regional climatic factors such as effective soil moisture and wind power have significant influence upon wind erosion rates measured by dust storm frequencies (Mctainsh et al., 1990). The aridity of an environment is often evaluated by the Budyko dryness ratio. The general Budyko (Budyko, 1958) equation can be written as

$$\frac{ET}{P} = F\frac{ET_0}{P} \tag{7.3}$$

where ET/P is the evapotranspiration ratio and ET_0/P is the aridity index. F is the function to be determined.

The dryness ratio at a given site indicates the number of times the net radiative energy could evaporate the mean annual precipitation. Semiarid zones where wind erosion is likely to be a serious problem have a dryness ratio of between 2 and 7. Areas with dryness ratios larger than 7 are in the desert and desert margin zones. Aridity can indicate a considerable water deficit in the top layer of the soil. Wind can easily remove and transfer soil particles when the top layer of the soil is totally dry and has lost the water content necessary for bonding. Wind erosion can also take place in high-rainfall climates during particularly dry months of the year in case if the soil is tilled with techniques that crush the soil surface to fine particles.

7.2.2.5 Vegetation

Vegetation plays an important role in entrainment and transport of sand by the wind. Vegetation protects the soil surface, trapping dust particles and, more importantly, slowing down the momentum of the wind. Also, it additionally protects the soil surface by direct coverage, trapping of particles, and most importantly by extracting momentum from the airflow (Figure 7.1). Based on its influences on wind erosion intensity, vegetation coverage

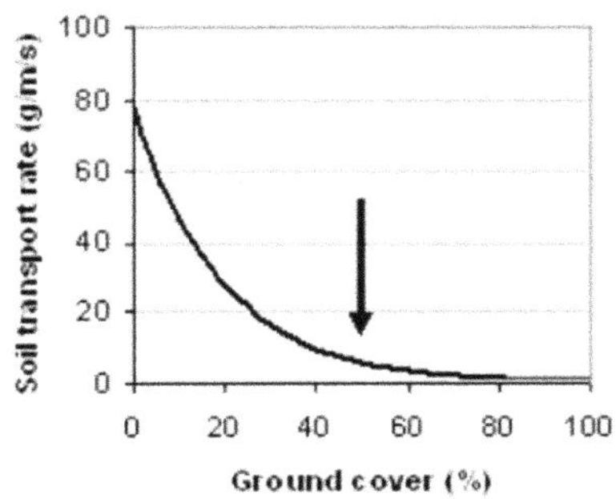

FIGURE 7.1
The effect of vegetation cover on wind transport.

can be grouped into three intensity categories: 60% is non-erodible to slightly erodible, 60–20% is moderately erodible, and 20% is severely (Dong et al., 1996).

7.2.2.6 *Soil Moisture*

Soil moisture is one of the most important factors influencing resistance to wind erosion. The increase in intrinsic factor in soil resistance due to moisture content is the cohesive force of soil water. As soil moisture increases, it decreases the wind erosion rate. When the moisture content reaches more than 4%, the rate erosion of decreases and slows down and become almost constant with successive increments of moisture. It was shown that erodibility by wind is a function of the cohesive force of adsorbed water films surrounding the soil particles (Chepil, 1955).

7.3 Global Scenario

Wind erosion, transport, and deposition occur over more than one third of the land surface of the Earth (Chen and Fryrear, 1996). Wind erosion occurs over more than one third of the Earth's surface. The soil dust that winds carry can affect the Earth's atmospheric composition and contribute to climate change. It also carries herbicides, sediments, soil texture residues, damaging nutrient contents, vegetation growth, and decreasing farm productivity, as well as affecting human health. The frequency of blowing dust may be caused by natural processes and is common in desert environments, but agricultural, grazing, and human activities that disturb the soil can greatly increase the frequency and amount of airborne dust. Aeolian transport of cultivated and grazed soils is a global problem that has generated many studies across the globe including Europe, Africa, Asia, Australia, and South America (Figure 7.2). Erosion and transport of soil has adverse effects both within and outside of source areas, and dust's capability to be transported to great distances makes the problem of both national and international scopes. For example, by the mid-1980s, off-site costs associated with wind erosion in the State of New Mexico were estimated at $466 million per year (Nordstrom and Hotta, 2004). Wind erosion is a chronic problem in areas where marginal lands are cultivated, and more mechanized faring is practiced (Nordstrom and Hotta, 2004).

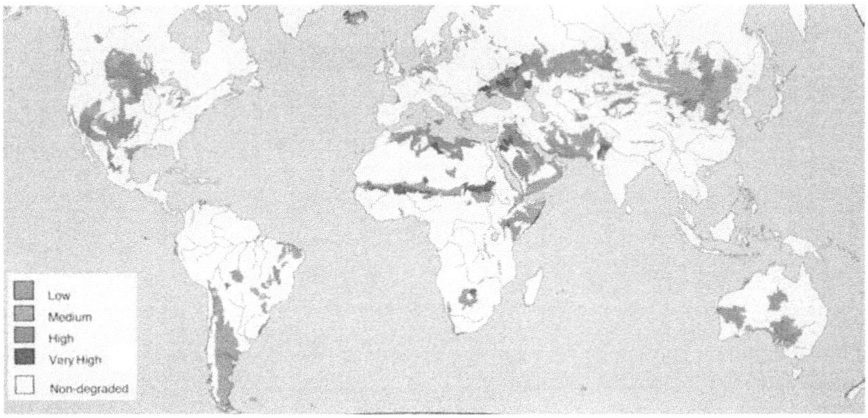

FIGURE 7.2
Degraded dry land area susceptible to wind erosion. Note that the white areas are non-degraded: the Sahara sand desert e.g. is not considered as being degraded (Source: GLASOD; UNEP (1997)).

7.4 Role of Remote Sensing

Both airborne and orbital sensor spectral measurements have been used for wind erosion assessment, monitoring dune dynamics, and wind erosion modeling. A brief overview of the potential of geospatial technology in this endeavor is presented hereunder:

7.4.1 Airborne Sensors Data

Air photos have been used for detection and delineation, and studying the temporal behavior of features associated with wind erosion. Using 1:25,000 scale aerial photographs over coastal New Zealand for 1956 and 1979, Stephens and Cock (1991) observed a significant positive change in vegetation cover over erstwhile sand-covered areas. Pandey et al. (1964) studied the movement of sand dunes in the central Luni basin, Rajasthan, India using 1:40,000 scale aerial photographs. Additionally, a comparison of aerial photographs acquired at different time intervals over Qatar has enabled Embabi and Asour (1992) estimating the annual mean rate of movement of sand.

In addition, several researchers have generated digital topographic data from hardcopy aerial photographs in order to resolve 3D changes associated with dune activity (Brown and Arbogast, 1999; Ojeda et al., 2005). Brown and Arbogast (1999) applied digital photogrammetric methods to scanned stereographic aerial photographs in order to quantify erosion and deposition at a dune complex along the eastern shore of Lake Michigan. Despite relatively poor accuracy in the digital elevation models produced by the photogrammetric models, the data provided a reasonable estimate of coarse-scale topographic changes during the timespan between photographs (1965–1987) (Brown and Arbogast, 1999). However, photogrammetrically generated digital elevation models of sand dunes from hardcopy aerial photographs with high accuracy are rare. Thus, accounting for errors in subsequent analyses with these data is of paramount importance for qualifying interpretations.

Apart from aerial photographs airborne Light Detection And Ranging (LiDAR) data were also used for wind erosion studies. In one of the studies, Wolfe and Hugenholtz (2009) used high-resolution airborne LiDAR data to detect morphological signatures of barchan dunes that evolved to parabolic dunes in south-western Saskatchewan, Canada. The regional extent of these features was only apparent in the high-resolution elevation model produced from the LiDAR data.

7.4.2 Orbital Sensor Data

As mentioned in Section 7.1, wind erosion involves detachment, transport, and deposition of soil particles from source location to farther destinations or even local redistribution of detached soil particles depending on wind velocity and vector. Synoptic perspective of terrain provided by orbital sensors in terms of spectral measurements offers immense potential in deriving information on (1) dune features, (2) their dynamics, and (3) processes. In a study, Scheidt et al. (2010) used data from the Advanced Spaceborne Thermal Emission and Reflection (ASTER) radiometer to map soil moisture dynamics and sand availability across the White Sands Dune field, New Mexico, USA. The proliferation of RS data and SA techniques in the Earth sciences has been enhanced further by the development and distribution of standardized global geospatial datasets, many of which are available to users worldwide free of charge (Table 7.1). This availability enables the compilation of digital atlases of sand seas, such as the Namib Sand Sea Digital Atlas (Livingstone et al., 2010) or the World Digital Quaternary Atlas. While the quantitative study of aeolian dunes with remote sensing and spatial analysis is still in its infancy, several key contributions have been made that demonstrate the potential of remote sensing and spatial analysis for addressing fundamental research questions about the morphometry, dynamics, and organization of dunes. A brief overview of the potential of orbital sensor data is provided hereunder: For further details readers may refer to Hugenholtz et al. (2012).

TABLE 7.1

Some of the Sources of Free or Low-Cost Remote Sensing Data Suitable for Wind Erosion Studies

Data	Description	Period	Spatial Resolution (m)	Spectral Bands
Corona[a]	Earth observation photographs	1960–1972	2–7.5	Panchromatic
Landsat[a,b]	Earth observation imagery	1972–present	15–120	up to 8 bands
MODIS[a,b]	Earth observation imagery	1999–present	250–1000	36 bands
ASTER[a,b]	Earth observation imagery	1999–present	15–90	14 bands
SRTM[b]	Earth elevation data		90	
ASTER GDEM[b]	Earth elevation data		30	
HiRISE DTM	HiRISE Digital Terrain Modelled Mars elevati-on data		1–2	

[a] http://earthexplorer.usgs.gov.

[b] https://wist.echo.nasa.gov/~wist/api/imswelcome/.

7.4.2.1 Detection of Wind Erosion Features and Eroded Areas

By virtue of providing perspective view of a larger area in discrete spectral bands of the electromagnetic spectrum at regular intervals orbital sensor data hold a great promise in providing timely and reliable information on wind erosion features and their dynamics. Figure 7.3 shows the prevalence of active barchans sand dunes in part of *Thar* desert, Rajasthan, western India. The figure shows Resourcesat-1 LISS-III images acquired in October, January, and April during 2006. The active barchans (crescent-shaped) sand dunes can be seen as bright yellowish white color (with green and blue circle). Note the establishment of vegetation cover as different hues of red color in October 2006 (post-monsoon) image. The ground photographs of sand dunes are seen adjacent to April 2006 image. In another example, Figure 7.4 illustrates the manifestation of partially stabilized sand dunes (longitudinal dunes), finger-like structures in Thar desert in temporal Resourcesat-1 LISS-III images. Of the three-period (October, January, and April images during 2005–2006) LISS-III images, the post-monsoon (October, 2005) image shows vegetation in light pinkish color.

Several researchers have utilized spaceborne multispectral data or indices generated there from to study wind erosion features and their dynamics. Using brightness and redness indices derived from the Nimbus Coastal Zone Colour Scanner (CZCS) data over northern Africa, Escadafal (1992) distinguished sand seas with different sand types (pale to reddish) apart from other features. In India, several researchers have studied the separability of terrain features related to wind erosion. Dwivedi et al. (1992) used Landsat-TM data and its various transforms, namely principal component, normalized difference

FIGURE 7.3
Active barchan (**crescent**-shaped) dunes in part of *Thar* desert, Rajasthan, western India as captured by Resourcesat-1 LISS-III sensor in October, January, and April images during 2005–2005. The establishment of vegetation cover seen adjacent to April 2006 image. Of the three-period LISS-III images, the post-monsoon (October 2005) shows vegetation in light pinkish color. Blue circle indicates sand sheet and green color unstabilized barchans dunes (Courtesy: National Remote Sensing Centre, Department of Space, Government of India).

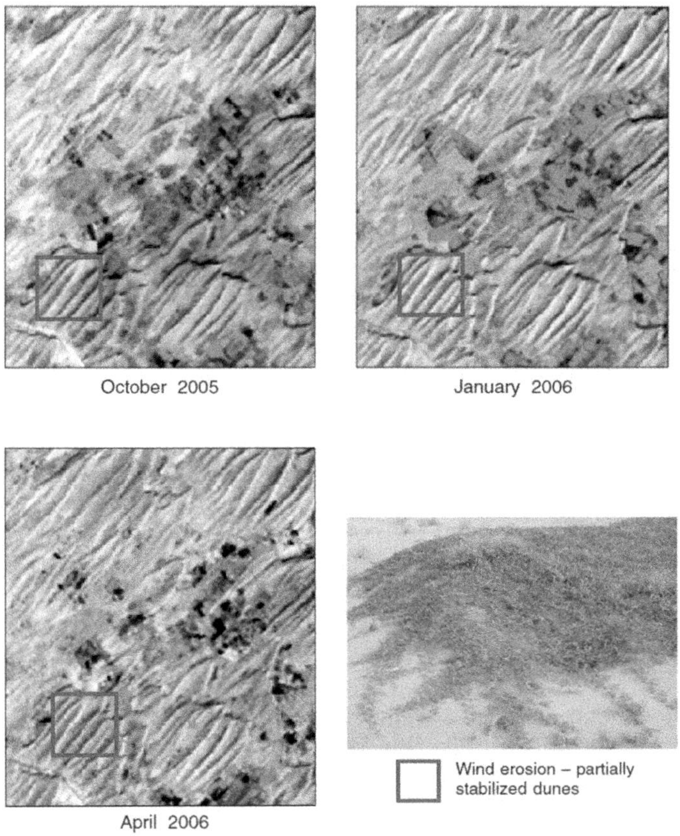

FIGURE 7.4
Partially stabilized longitudinal dunes (finger-like structures) in part of Thar desert, Rajasthan, western India as captured by Resourcesat-1 LISS-III sensor in October, January, and April images during 2005–2006. The establishment of vegetation cover (ground photo) seen adjacent to April 2006 image. Of the three-period LISS-III images, the post-monsoon (October 2005) image shows vegetation in light pinkish color (Courtesy: National Remote Sensing Centre, Indian Space Research Organization, Department of Space, Government of India).

vegetation index (NDVI), soil brightness index (SBI), and perpendicular vegetation index (PVI), derived there from over central Luni basin in Rajasthan, and have shown that the information on sand dunes and other associated features generated by the conjunctive use of the standard data product, i.e., false color composite (FCC) alone prints generated from green, red, and near-infrared spectral bands, and the transforms, far exceeds the one derived from standard FCC alone.

In another study, Chavez et al. (2002) detected dust storms with Geostationary Operational Environmental Satellite (GOES) visible band image by employing three approaches, namely (1) a change procedure to generate digital change images, (2) a change-visualization procedure, and (3) movies. In the first two cases, two GOES images collected at approximately the same time of day on two different dates are used in the analyses—one imaged during the windy day when airborne dust is potentially present, and the second imaged on a calm day within a day or two of the windy day image.

While working in the Indira Gandhi canal command area in Rajasthan, western India, Goossens et al. (1992) could discriminate parabolic dunes from interdunal depressions using Landsat MSS data of 1973 and 1986 coinciding with the before and after commissioning of the canal. Additionally, Singh et al. (1992) and Mitra and Bhoj (1992) could delineate several landforms in western Rajasthan associated with the wind erosion using space-borne multispectral data. Kharin (1986) has reviewed the potentials of space images in studying the wind erosion with special reference to Kara Kum desert.

7.4.2.2 Characterization of Dune Activity

Dune activity is typically viewed as an index of aeolian sand transport potential that spans a continuum: from bare dunes characterized by uninhibited erosion and deposition (active), to fully stabilized dunes exhibiting no surface change (stabilized). Stabilized dunes are commonly covered by extensive vegetation. Lancaster (1999) provided a robust framework for comprehending dune activity through the three components of *sand supply, sand availability,* and *wind transport capacity.* Transport capacity refers to the potential sand transport rate of the wind; sand supply is the quantity of source material; and sand availability is the susceptibility of surface grains to be entrained by the wind. Monitoring dune activity, therefore, relates to observing and quantifying surface change on a dune related to the aeolian sand transport potential. The variable states of activity are associated with different sand fluxes on the individual dunes. The complex interplay between these components forms an important control on the level of aeolian sand transport occurring in dune fields.

The presence of vegetation or crusts serves as protective cover to soils inhibits the sand availability and controls dune activity (Buckley, 1987). Field studies have demonstrated that the presence of vegetation effectively reduces aeolian sand transport, and vice versa (Levin et al., 2008). The complex interplay and feedbacks between vegetation growth and sand transport drives the dune system toward one of the two end-member states: active or stabilized. Thus, a biogeomorphic approach is central to the study of dune activity; investigators must consider both biologic and geomorphic processes (Hugenholtz and Wolfe, 2005).

7.4.2.3 Measuring Sand Availability

In semiarid environments, sand availability is commonly limited by surface vegetation and/or crusting. Techniques and data used to measure the spatial extent and distributions of both have undergone considerable recent evolution. Although the high-resolution and/or multispectral images are quite promising from a detail perspective, the dynamics of sand availability typically change on a scale where multi-temporal studies involving long baselines and older spatial datasets, such as historical maps and analog aerial photographs, continue to be important. In fact, short baselines generally provide an incomplete picture of the morphodynamic response of dunes to changes in sand availability. Archived aerial photographs and historical maps provide opportunities to develop a more complete understanding of the dynamics and effects of sand availability. While much of these spatial data are accessible only in hardcopy format, the proliferation of image analysis and Geographic Information System (GIS) software in the early to mid-1990s increased the objectivity and quantitative rigor of measurements made with these data. By scanning the hard copy photographs or negatives and digitally rectifying the internal (e.g., camera coordinates) and external (e.g., ground coordinates) image parameters, it became possible to create planimetrically true images suitable for spatial measurements.

Land Cover Dynamics: The proliferation of image analysis software in the earlytomid-1990s ushered in a new era in improving the objectivity and quantitative rigor of measurements made with aerial photographs. This emerging geospatial framework, coupled with a move away from reductionist attempts to understand dune field dynamics, enhanced the shift away from the single-dune studies popular in the 1980s and 1990s to more synoptic-scale investigations of dune fields involving both remote sensing and spatial analysis (Livingstone et al., 2007). In the context of dune activity, one of the earliest accomplishments was the adaptation of existing image analysis techniques to discriminate vegetation and crusts, so as to deduce the degree of sand availability.

Both aerial photographs and satellite images have been used to discriminate vegetation cover on dunes and quantify its changes over time. For some dune fields, archival black-and-white aerial photographs have been used to great advantage, providing long temporal baselines and reasonably high spatial resolution (Levin, 2011). When the photos are converted to digital images and rectified they can be subjected to image analysis routines that generalize or classify the scene into the dominant surface covers, namely vegetation-covered dunes, bare, open sand. Bare dunes appear as relatively bright areas (high digital number) and vegetation-covered sand appear as darker areas (low digital number).

Digital analysis of spaceborne multispectral data enables deriving information not only on the vegetation density, structure, and species but also the distribution of different vegetation functional groups (e.g., trees, shrubs, grasses) within a dune field. The multispectral data can also be used to develop various vegetation indices (VIs), which further improve the identification and spatial variation of types and densities of photosynthetically active dune vegetation. Levin et al. (2006) found that the NDVI and the soil adjusted vegetation index (SAVI) explained up to 78% of the aeolian transport rate on coastal dunes. They also developed a predictive model that incorporated SAVI and topographic variables in order to quantify the spatial pattern of sand transport and dune activity.

In some dune fields, the effectiveness of vegetation in controlling sand availability is further enhanced by the presence of biological soil crusts. Measurements of the spectral characteristics of biological soil crusts have found that, in cases, crusts have a unique spectral signature (Levin et al., 2007). However, the spectral signature of biological soil crusts can be modulated by the presence of moisture (O'Neill, 1994), the relative abundance and types of cyanobacteria, and absorption of iron oxide and clay minerals. Tsoar and Karnieli (1996) demonstrated that the spectral reflectance of crust can be used to assess the activity of dunes.

Dune Dynamics: Observable movement of a dune form provides an objective indicator of active sand transport. Although it is relatively straightforward to qualitatively identify dune movement, devising a rigorous quantitative method that can be unambiguously applied requires much more care, especially given the diversity of dune forms that exist in natural environments. Measuring the distance between two lines representing successive positions of the dune (Gay, 1999), also termed *nose to nose* (Bailey and Bristow, 2004) is the most widely used method for estimating dune movement. Typically, the edges of slip faces, brink lines, or cuspate vegetation marks are used to represent the position of the dune (Levin et al., 2009). Several arrows/lines are drawn between the lines and the average length recorded. However, challenges can exist in reproducibility of results (Goudie, 1994). In addition, the method assumes that all the different parts of the dune advance at the same rate, which may not be the case in stabilizing dunes. This method can result in an overestimation of the calculated advance rate (Levin and Ben-Dor, 2004).

Recent advances in image-based cross-correlation methods demonstrate the potential to derive more objective measurements of dune form movement (Vermeesch and Drake, 2008;

Necsoiu et al., 2009). The technique involves sub-pixel correlation of co-registered images, which yields a velocity field of dune displacements, and can also be used to derive an estimate of sand flux. Another remote sensing-based method with demonstrated potential for inferring dune involves exploration of changes in the dune surface. Typically, the only process that moves sand on dune surfaces is aeolian transport (mass wasting is minimal, although possible; see Hugenholtz et al., 2007). Consequently, surface changes can be used as a reliable proxy for aeolian sand transport. Substantial advancements have been made in evolving and adapting traditional remote sensing techniques such as photogrammetry, and integrating newer remote sensing technologies such as LiDAR.

7.4.2.4 Erosion Control Measures and Impact Assessment

As mentioned in Section 7.2.2, apart from wind velocity, vegetation cover plays a very crucial role in the incidence of wind erosion. Establishment of vegetation cover and development of wind breaks are the two very commonly used erosion control measures. In order to establish vegetation cover, prevention from human and animal encroachments, plantation of trees, and introduction or expansion of irrigated agriculture is resorted to. Spaceborne multispectral measurements hold great promise in identifying wind control structures, inventory of irrigated agriculture and establishment of other vegetation cover resulting from wind erosion control measures like prevention from gazing, and tree/shrub felling.

The establishment of vegetation cover over fully or partially stabilized dunes is illustrated in Figure 7.5. In pursuance of expanding vegetation cover which acts as a protective cover for soils which prevents it from erosion, in a few desert terrain canal irrigation has been introduced. Furthermore, efforts have also been made to prevent the affected area from grazing and felling trees/shrubs.

Erosion Control Structures: Wind breaks are generally established across the prevailing wind, consists of trees with reasonably good height and canopy or sometimes hedges of dry woods and grasses are constructed for the same purpose. The terrain experiencing wind erosion, being devoid of substantial vegetation cover, generally exhibits very reflectance look bright while wind breaks either vegetative or non-vegetative are manifested as medium gray to dark gray features on black-and-white image. In standard false color image vegetative wind breaks appear as linear features with different hues of red color. The non-vegetative windbreaks

FIGURE 7.5
Shelterbelt in part of Ganganagar district, Rajasthan, western India, for protection of crops from wind erosion as captured by Resourcesat-1 LISS-IV (Courtesy: National Remote Sensing Centre, Indian Space Research Organization, Department of Space, Government of India).

appear as medium to dark black-colored linear features. An illustration of the potential of high-resolution remote sensing data (Resourcesat-1 LISS-IV data with 5.8 m spatial resolution) in the detection of vegetative windbreak is given in Figure 7.5.

Prevention of Encroachments: In the area around Rampura there was hardly any vegetation cover in 1975. As evident from the figure, there has been substantial increase in vegetation cover through 1980, 1990, 1995, and 2005. Fencing of a large area around the village has resulted in the establishment of scrubs and grasses seen as green and cyan and green colors. The vegetation cover, thus established, provides the protection to surface soil from wind erosion (Figure 7.6).

Introduction of Irrigated Agriculture: The introduction of irrigation results in an increase in irrigated agriculture leading to the establishment of protective cover. In an example of the potential of spaceborne multispectral data in providing the information on dynamics

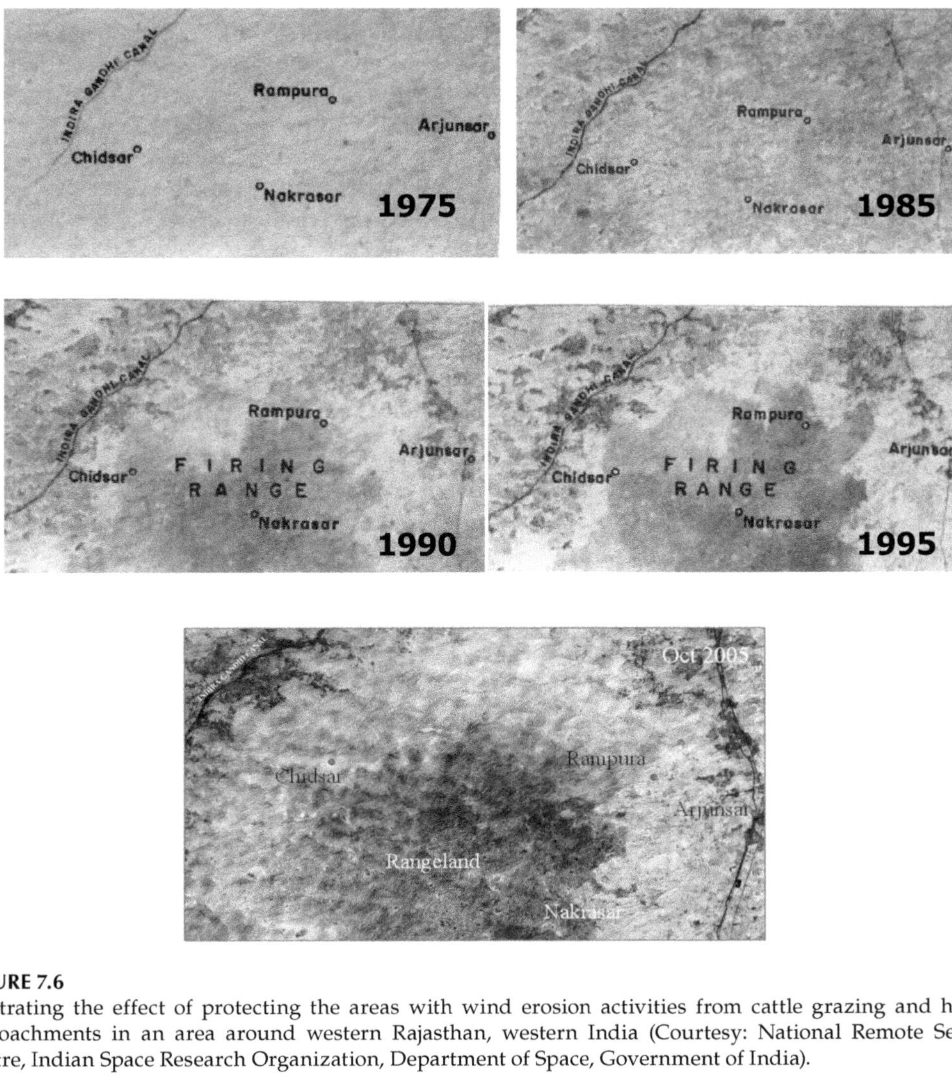

FIGURE 7.6
Illustrating the effect of protecting the areas with wind erosion activities from cattle grazing and human encroachments in an area around western Rajasthan, western India (Courtesy: National Remote Sensing Centre, Indian Space Research Organization, Department of Space, Government of India).

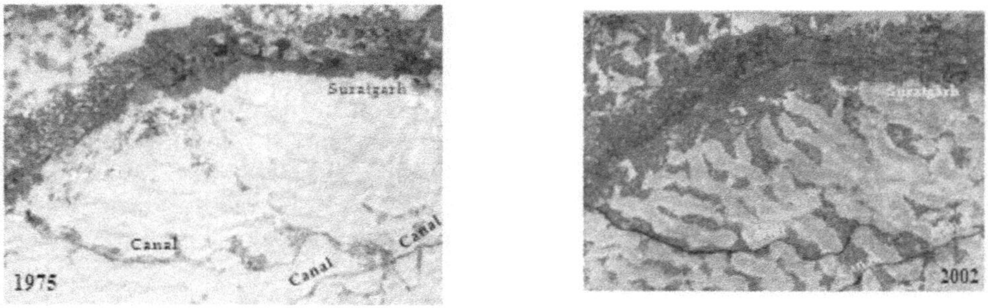

FIGURE 7.7
Development of crop land due to introduction of canal irrigation around Suratgarh, part of Ganganagar district, Rajasthan, western India.

of agriculture (Figure 7.7). Agriculture is confined in the area to interdunal depressions and in the palaeochannel (old river bed of the river Saraswati). As evident from the figure, during a span of 27 years (March 1975 to March 2002) a fairly large area has been brought under agriculture. The red color of different hues indicates *rabi* (winter crop mostly wheat).

Waterlogging Dynamics: By virtue of providing synoptic perspective of a larger in indiscrete spectral bands of the electromagnetic spectrum at regular intervals, spaceborne multispectral data offer immense potential inventory and monitoring of waterlogged areas. In a study carried out in parts of the Indira Gandhi Canal command area in Ganganagar district of Rajasthan, western India, the dynamics of waterlogging was studied during the period 1975–1995. Since the area receives very meager rain—mean annual rainfall 316.6 mm (https://en.wikipedia.org/wiki/Sri_Ganganagar)—salts which are released during mineral decomposition in the soil are not leach down to deeper subsurface and different kinds of impervious layers called pans. It may be clay pan, *kankar* pan, calcium carbonate pan, duripan, etc. These pans act as hydrological barrier to the subsurface movement of water. When irrigation is introduced in such areas the surplus irrigation water moves vertically down. However, due to presence of lenses of clay and calcium carbonate (*kankar*) at shallow depth this water does not percolate down below impervious layer and gradually ground water table starts building up.

The Landsat Multi-Spectral Scanner (MSS) and Thematic Mapper (TM), and Resourcesat-2 Linear Imaging Self-Scanning Sensor (LISS-III) data for the year 1975, 1985, 1990, 1995, and 2002 were used to study the dynamics of waterlogging. Vegetation condition was assessed by generating the NDVI from multispectral digital data. Waterlogged areas were delineated through a systematic visual interpretation approach. The area under cropland has almost doubled during 27 years (Figure 7.8a–e). The expansion in cropland has been only in interdunal depressions. As against 39,473 hectare during 1975, the area under cropland has risen to 1,15,911 hectare during the year 2002 (Table 7.2). Another land cover category that has undergone substantial changes is the waterlogged areas (cyan color in Figure 7.8a–e) which showed a steady increase in its spatial extent during 27 years period. While there was no sign of waterlogging during 1975, it had expanded over an area of 4,172 hectare in 2002. Such areas could be seen between Suratgarh and Hanumangarh and around Baropal village. Contrastingly, owing to uncertainties in the release of surplus water from canal into interdunal depression, no regular trend in the spatial extent and distribution pattern (Figure 7.8a–e) of water bodies could be noticed (Table 7.2).

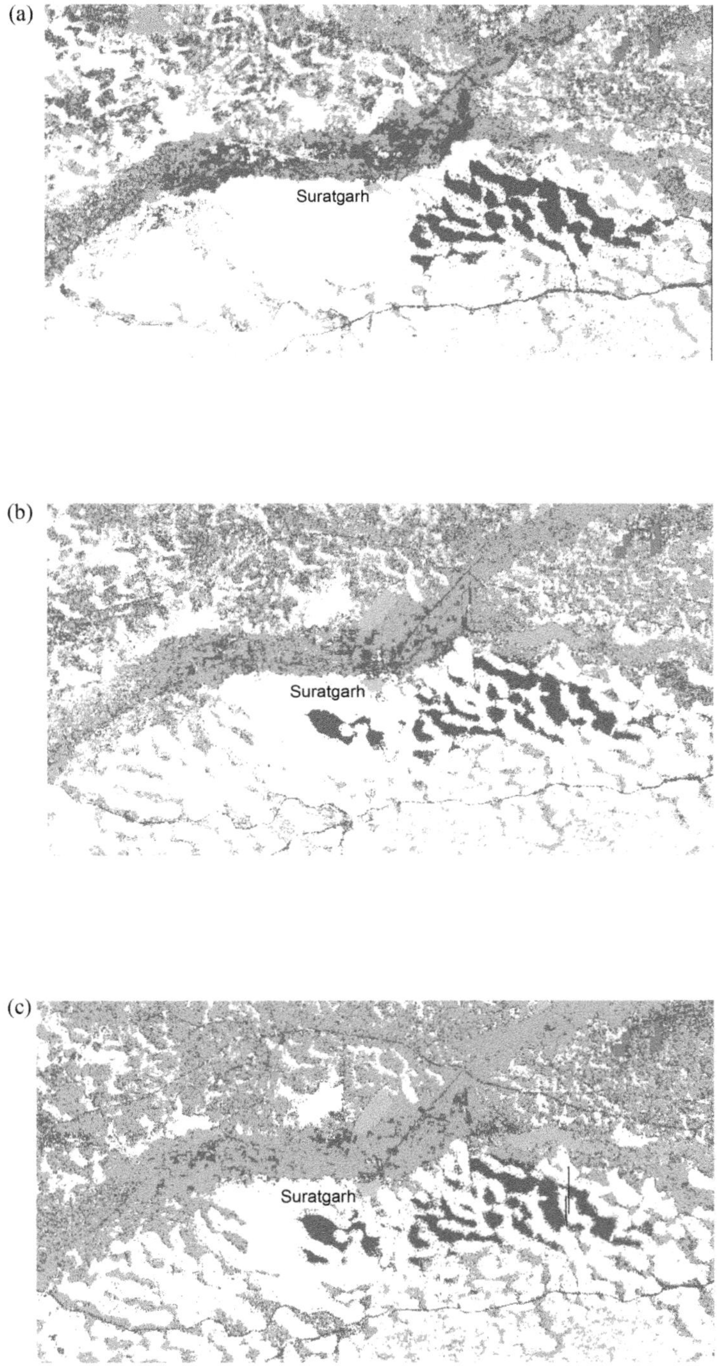

FIGURE 7.8
Waterlogged areas and other land use/land cover categories in (a) 1975, (b) 1985, (c) 1990, (d) 1995, and (e) 2002.

(Continued)

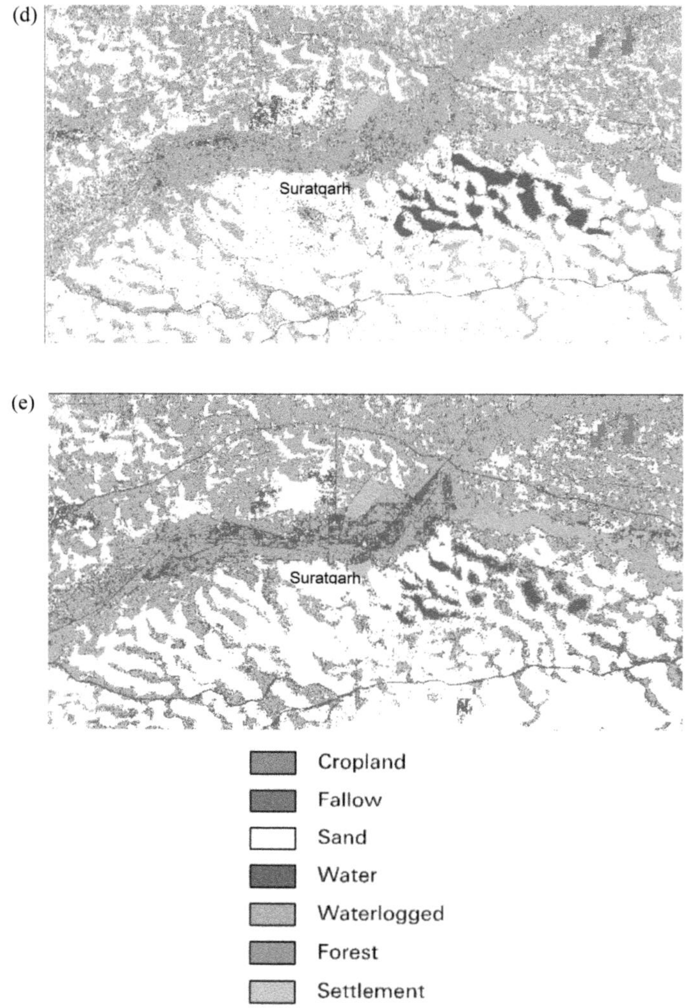

FIGURE 7.8 (CONTINUED)
Waterlogged areas and other land use/land cover categories in (a) 1975, (b) 1985, (c) 1990, (d) 1995, and (e) 2002.

TABLE 7.2
Aerial Extent of Waterlogged Areas and Other Land Use/Land Cover Categories

Sl. No.	Land Use/Cover Class	1975	2002
1	Cropland	38,563	93,370
2	Sand dune/sheet	1,86,503	1,21,043
3	Water bodies	7,127	1,503
4	Fallow land	15,651	8,218
5	Waterlogged area	0	3,405
6	Forests	545	545
7	Settlements	668	3,273
8	Others	502	17,608

Spectral Indices: Apart from analysis of digital air/spaceborne multispectral data for deriving information on wind erosion features and their dynamics, some spectral indices have also been developed. They are described hereunder.

Normalized Difference Sand Dune Index: To identify and assess the spatial extent of sand accumulations and sand spread, Fadhil (2009) suggested an index called normalized difference sand dune index (NDSDI) using Landsat ETM+ data. The index based on the normalized difference between the RED and the short-wave infrared (SWIR) spectral values. This index is aimed at differentiating between sand dune accumulations, bare soils, and other types of soil. It can be expressed as follows:

$$\text{NDSDI} = 1 - \left(\frac{\text{R} + \text{SWIR}}{\text{R} - \text{SWIR}}\right) \tag{7.4}$$

where R is the reflectance of red (band 3: 0.63–0.69 mm) and SWIR is the short wavelength infrared (band 7: 2.08–2.35 mm) of the Landsat ETM+ sensor.

NDSDI mainly distinguishes vegetation and non-vegetation, water and arid surface, sandy and bare soil, while reflectance in the red and SWIR bandscan discriminate the mineral and rock types as it is sensitive to the moisture content of soil and vegetation. The value of the NDSDI ranges between –1 and +1. Whereas the sand dune accumulation sand drifting sands often give values below zero, vegetative cover produces values greater than zero.

Grain Size Index (GSI): The GSI was specifically designed for using Landsat TM/ETM+ data. The GSI is computed as follows:

$$\text{GSI} = \frac{(\text{Red} - \text{Blue})}{(\text{Red} + \text{Green} + \text{Blue})} \tag{7.5}$$

where R, B, and G are the reflectance of the red, blue, and green bands of the Landsat TM and ETM+ sensors. GSI value is close to 0 or a smaller value in vegetated area, whereas for a water body it is a negative value (Xiao et al., 2006). Higher positive values of GSI represent the sand affected region. The difference between band R and B in the GSI equation illustrates the difference between vegetated or water surface and bare soil, while the accumulation of the reflectance in the R, G, and B bands helps in differentiating the topsoil with different grain size composition (www.nrcs.usda.gov/wps/portal/nrcs/detail/vt/soils/?cid=nrcs142p2_054008, accessed on August 20, 2017.

The Dust Storm Index (DSI): The DSI provides a composite measure of the contributions of: local dust events, moderate dust storms and severe dust storms using weightings for each event type, based upon dust concentrations inferred from visibility reduction during each of these event types. The DSI provides a measure of the frequency and intensity of wind erosion activity at continental scale from observations made at Bureau of Meteorology (BoM) stations McTainsh (1998). The DSI is calculated using the following equation:

$$\text{DSI} = \sum_{i=1}^{n} [(5 \times \text{SD}) + \text{MD} + (0.05 \times \text{LDE})]i \tag{7.6}$$

where DSI = dust storm index at n stations where i is the ith value of n stations for i = 1 to n. The number of stations (n) is the total number of stations recording a dust event observation in the time period; SD = Severe dust storm (BoM daily maximum weather codes: 33,

34, and 35); MD = Moderate dust storm (daily maximum weather codes: 09, 30, 31, 32, and 98); LDE = Local dust event (daily maximum weather codes: 07 and 08).

The DSI was developed by McTainsh (1998) using uncorrected BoM weather code data. Further refinements have been made in the model to overcome certain weaknesses. DSI_2 was developed in response which, among other things, uses independently collected visibility data to define event types. DSI_3 is a further development, in which local dust events are re-assigned, based on the visibility criteria, into severe dust storms, moderate dust storms, or local dust events. The accuracy of the model is limited by spatial density of meteorological stations and reliability of meteorological observations.

7.5 Modeling Wind Erosion

In order to address wind erosion at spatial scales ranging from farm field to regional and global scales, several models have been developed. These models are briefly described hereunder. For further details readers may refer to Webb (2008).

7.5.1 Field Scale Wind Erosion Models

These models represent a range of approaches for assessing wind erosion from farm fields at different spatial and temporal scales. With the advancement in computing facilities there have been progressive developments in quantification of land erodibility in each model from early empirical approach with limited computing capability, to more recent process-based studies.

7.5.1.1 Wind Erosion Equation

The WEQ was developed by Woodruff and Siddoway (1965) to predict annual soil erosion (kg/ha) from farm fields in the USA. The model was designed for both the analysis and management of wind erosion, with an emphasis on the application of the model to determine the effects of field conditions and erosion mitigation strategies on erosion rates. The model uses a relationship among five generalized factors that account for the major controls on wind erosion in cultivated environments. These factors are combined into five inputs for the model in the form:

$$E = f(I', C', K', L', V) \tag{7.7}$$

where I' is the soil and knoll erodibility, C' is the local wind erosion climatic factor, K' is the soil ridge roughness factor, L' is the field length, and V is the equivalent quantity of vegetation cover.

7.5.1.2 Revised Wind Erosion Equation

The Revised Wind Erosion Equation (RWEQ) was designed to predict soil loss due to wind erosion at sub-annual time scales. The model combines empirical and process-based model components. While WEQ provides estimates of annual soil loss, RWEQ can be applied to calculate the sediment transport mass at specific field lengths, as well as average and maximum soil loss within a field. A major development in RWEQ was the incorporation of management factors to describe soil surface conditions, vegetation state, and soil moisture changes due to irrigation (Fryrear et al., 2000). RWEQ can be expressed as follows:

$$Q_{max\ p} = 109.8(WF \cdot EF \cdot SCF \cdot K' \cdot COG). \tag{7.8}$$

where $Q_{max\ p}$ is the maximum amount of soil that can be transported downwind in an event. Input variables of RWEQ include ER, the soil erodible fraction computed from soil properties; SCF, a soil crust factor computed from soil clay and organic matter content; WF, a weather factor; field size; COG, the crop type and orientation; and Hills, a factor used to modify wind speeds depending on field slope and height.

7.5.1.3 Wind Erosion Prediction System

The Wind Erosion Prediction System (WEPS) was developed as a process-based, continuous, daily time-step wind erosion model that simulates weather, field conditions management, and erosion (Hagen, 1991). The model can be run to simulate single erosion events or soil loss over multiple years. It uses a database of inputs to define land surface characteristics. Inputs include data on climate, soils, management, and crop conditions. The inputs are processed by submodels to compute soil and land erodibility components that are used by the erosion scheme. The WEPS erosion model operates with a scheme that computes entrainment when the wind friction velocity over a field surface (u_*) exceeds a threshold friction velocity (u_{*t}) for particle mobilization. For further details of the model readers may refer to Webb (2008).

7.5.1.4 Texas Erosion Analysis Model

The Texas Erosion Analysis Model (TEAM) was developed to have a low dependence on field-measured inputs and empirical relationships. The model simulates the detachment, maximum transport rate, and emission of dust particles (Singh et al., 1999). Like WEPS, the model erosion scheme operates on the principle that entrainment occurs when u_* exceeds u_{*t}. The schemes used to compute land surface conditions draw upon external studies of the effects of land surface processes on u_{*t}. The model computes both a static threshold friction velocity required for the initiation of entrainment, and a dynamic threshold for sustained sediment transport. Input variables for TEAM include Wind speed, relative humidity, soil particle size distribution, clay content, residue and vegetative cover, soil aggregate cover, field length, and windbreak height and porosity are the input variables for the model. The threshold friction velocity is calculated for the average particle size d (not including the clay fraction) and surface moisture conditions of the field undergoing simulation.

7.5.1.5 Wind Erosion Stochastic Simulator

The Wind Erosion Stochastic Simulator (WESS) is the wind erosion module of the EPIC (Environmental Policy Integrated Climate) erosion model (Van Pelt and Zobeck, 2004). It is a process-based model that predicts wind erosion on an event or periodic basis. The model provides assessments of soil loss at user-specified distances within individual fields. The model uses inputs of soil surface data including texture, soil surface moisture and drying rate, erodible soil thickness, surface soil bulk density, and soil roughness parameters for large aggregates (random roughness), and ridge height and spacing (oriented roughness) in conjunction with local wind speed data and a stochastic wind speed perturbation factor to simulate dust emission (Potter et al., 1998). WESS simulates wind erosion based on daily wind speed distributions adjusted by soil, surface roughness, vegetation cover, and eroding field length factors:

$$YW = (FI)(FR)(FV)(FD)\int_{0}^{t} \frac{YWR}{WL} dt \tag{7.9}$$

where YW is the wind erosion estimate (kg/m^2), YWR is the erosion rate (kg/ms), WL is the unsheltered field length (m), FI is the soil erodibility factor, FR is the surface roughness factor, FV is the vegetation cover factor, FD is the field length factor (based on WL), and t is the duration of wind greater than a threshold velocity. YWR is calculated using the approach of Skidmore (1986).

7.5.2 Regional Scale Models

Regional scale models are essentially process-based wind erosion models.

7.5.2.1 Wind Erosion on European Light Soils

The Wind Erosion on European Light Soils (WEELS) model was developed to map wind erosion risk and to predict long-term, wind-induced soil loss under different climate and land use scenarios (Böhner et al., 2003). The model was developed from the EROKLI model (Beinhauer and Kruse, 1994) and can be run at local scales in domains up to 5 × 5 km in size, using input data at a 25 × 25 m resolution. It has a modular structure, and is run in a GIS environment. WEELS computes soil erodibility then independently accounts for soil moisture and vegetation cover factors by integrating these into the wind erosivity component of the model. Soil erodibility is quantified using a dimensionless factor K which expresses the intrinsic susceptibility of a dry, freshly cultivated sandy soil to erosion. Values for K are computed as a function of the mean weighted diameter of the texture (by particle size distribution), organic matter content, and the proportion of aggregates > 0.63 mm:

$$\log K = 1.24 - 4.21(MWD) - 0.04(\%DA) \tag{7.10}$$

where MWD is the soil particle mean weighted diameter (mm) and %DA is the weight percentage of aggregates > 0.63 mm (%) calculated as follows:

$$\%DA = -2.42 + 8.6\left[\log(\%OM)\right] + 44.5\left[\frac{(\%Slit + \%Clay)}{\%Sand}\right] \tag{7.11}$$

where %DA is the aggregate portion > 0.63 mm (%), %OM is the weight percent soil organic matter, and %Silt, %Clay, %Sand are the soil silt, clay, and sand fractions (%). Both equations

3.18 and 3.19 are regression equations derived from field experimentation in Germany (Böhner et al., 2003). Field measurements of %DA can also be used as input to the model.

7.5.2.2 Wind Erosion Assessment Model

Developed by Shao et al. (1994), the Wind Erosion Assessment Model (WEAM) is the first integrated model developed in Australia to quantify wind erosion rates. The model was designed to operate at a moderate spatial resolution and predict dust emissions from the Murray-Darling Basin in south-eastern Australia. WEAM combined current theories on the effects of climate, soil, vegetation, and land use on wind erosion. It quantifies land erodibility through the threshold friction velocity, with erosion occurring when $u_* > u_{*t}$. The model computes stream-wise sediment flux (Q) and vertical dust flux (F) and uses the saltation model developed by Owen (1964), a sand transport equation, to compute Q, which is adjusted using soil particle size distributions. The wind shear velocity (u_*) is determined from climatic conditions and the surface roughness height (Shao et al., 1996). The threshold friction velocity is calculated as a function of soil particle size, frontal area index of surface roughness elements, soil moisture, and the hardness of the surface crust:

$$u_{*t}(d_s, \lambda, w, c) = \frac{u_{*t}(d_s, 0, 0, 0)}{R(\lambda)H(w)M(s)} \tag{7.12}$$

where $u_{*t}(d_s, 0, 0, 0)$ was approximated from the model of Greeley and Iversen (1985), R(L) is the ratio the bare threshold velocity over the covered (rough) threshold velocity, H(w) is the ratio of the threshold velocity of the wet surface over the threshold velocity of the dry surface, and M(s) is a mobility coefficient describing the influence of the state of soil aggregation and crusting, and the chemical binding strength which maintains this state.

7.5.2.3 Integrated Wind Erosion Modeling System

The Integrated Wind Erosion Modeling System (IWEMS) is an extension of WEAM and was developed by coupling WEAM with climate and land surface simulators within a GIS framework. It was developed with the capacity to receive input data from a weather prediction model. The input GIS database for land surface conditions has a spatial resolution of ~25 × 25 km at surface. The atmospheric component has 10–31 vertical layers, and a horizontal resolution ranging from 5 × 5 to 75 × 75 km. The atmospheric model provides forcing for both the dust emission and transport models (Lu and Shao, 2001). In addition to the spatial expansion of model application, and input upgrades, the model estimation of u_{*t} was also revised. The revisions built upon the model presented by Shao et al. (1996) and Shao and Leslie (1997), with the inclusion of factors to account for multiple nonerodible roughness element layers, and the erodible fraction of the exposed surface.

7.5.3 Global Scale Models

These models address the prediction of land erodibility in continental and global scale dust emission and transport models, and, in general, use erosion schemes developed for smaller-scale models, with adaptations to suit their application environments.

7.5.3.1 Dust Production Model

Developed by Marticorena and Bergametti (1995), the Dust Production Model (DPM) was designed with a dust emission scheme that accounts for spatial variations in dust source erodibility. Early regional to global scale models compute dust emission as a function of wind velocity. The DPM computes emissions in a manner similar to WEAM and IWEMS, using a relationship among soil particle size distribution, surface roughness, and u_{*t}. Importantly, the approach allows for the direct and relative contributions of various source areas to global dust emissions to be quantified.

The basis of the dust emission scheme is that land erodibility is strongly dependent on soil texture and surface roughness characteristics. Values for u_{*t} are assigned to soil input maps based on the size distribution of particles in different textured soils (Marticorena and Bergametti, 1995). Semiempirical equations of Iversen and White (1982) were then modified to obtain a relationship between particle or aggregate diameter and u_{*t}. The relationship was developed as a piecewise function dependent on the Reynolds number (here denoted B):

$$u_{*t}(D_p) = \frac{0.129K}{\left(1.928(aD_p^x + b)^{0.092} - 1\right)^{0.5}} \quad \text{for } 0.03 < B < 10$$

$$u_{*t}(D_p) = 0.129K\left[1 - 0.0858\exp\left(-0.0617\left((aD_p^x + b) - 10\right)\right)\right] \quad \text{for } B > 10 \tag{7.13}$$

where D_p is the size of the soil surface aggregates (cm). The Reynolds number (B) is described by the term:

$$B = aD_p^x + b \tag{7.14}$$

where $a = 1331$, $b = 0.38$, and $x = 1.56$.

The influence of surface roughness on the loss of wind momentum is accounted for using a scheme developed by Marticorena and Bergametti (1995) and Marticorena et al. (1997). The drag partitioning scheme of Raupach et al. (1993) requires a measure of the frontal area and estimation of the m parameter, so an alternate specification of drag was included in DPM and is based on the roughness length z_0.

$$f_{eff}(Z_0, z_{0s}) = 1 - \left(\frac{\ln(Z_0/z_{0s})}{\ln\left(0.35(10/z_{0s})^{0.8}\right)}\right) \tag{7.15}$$

where z_0 is the aerodynamic roughness length of the overall surface (cm) and z_{0s} is the aerodynamic roughness length of the erodible part of the surface (cm). The following expression is then used for the computation of u_{*t}:

$$u_{*t}(D_p, Z_0, z_{0s}) = \frac{u_{*t}(D_p)}{f_{eff}(Z_0, z_{0s})} \tag{7.16}$$

7.5.3.2 Dust Entrainment and Deposition Model

The dust entrainment and mobilization scheme used in DEAD is similar to that of Marticorena and Bergametti (1995). The model emission scheme is particle size dependent, with u_{*t} being computed as a function of soil particle size distributions. Furthermore, modifications are made to u_{*t} by accounting for soil moisture effects using the scheme of Fécan et al. (1999), and surface roughness effects using the drag partitioning scheme of Raupach et al. (1993). The model considers the Owen effect, a positive feedback of saltation on the surface roughness length and friction velocity, using the model developed by Gillette et al. (1998).

Dust source area soil erodibilities in DEAD are defined by soil particle size distributions from a global soil texture dataset (Zender et al., 2003). Land surface and geographic constraints are applied to the dust source areas in addition to the roughness and moisture parameterizations in the emission scheme. The fraction of bare soil surface is described by a parameter which is a function of total vegetation, snow, lake, and wetlands cover. Vegetation cover is derived from satellite imagery of leaf area index (LAI) at a 1 × 1 km spatial resolution. These factors are considered constant during simulations. A further adjustment is made for dust source area erodibility S. This adjustment has a basis in global dust source area characterizations reported by Ginoux et al. (2001), identified using Total Ozone Mapping Spectrometer (TOMS) aerosol optical thickness image data. Results demonstrated that simulations of global dust emissions are highly sensitive to dust source area characterizations.

$$\text{TOPO} = \left(\frac{z_{max} - z}{z_{max} - z_{min}} \right) \tag{7.17}$$

where TOPO is the erodibility factor, z is the elevation of the relevant grid point, and z_{max} and z_{min} are the highest and lowest points in the surrounding area (10× 10 lat./long.).

7.5.4 Other Global Dust Models

A number of global dust emission models have been developed, and there are similarities in their erosion and dust emission schemes. In general, three approaches have been taken to model wind erosion (Zender et al., 2003). The first type parameterizes mobilization in terms of the third or fourth power of the wind speed or friction velocity, then impose size distribution factors on the emitted dust (Perlwitz et al., 2001). These models are reliant on assumptions about the general characteristics of dust source areas and do not account for microphysical entrainment processes.

The second type uses a microphysical specification of the land surface to predict size-resolved saltation mass flux and dust emission (e.g., Marticorena and Bergametti, 1995; Shao, 2001). Due to the complexity of inputs required for these models, this model type has typically been applied in regional scale modeling where spatial data requirements are often better met and inherent assumptions built into the model are less likely to violate the ranges of conditions seen in global dust source areas.

The third type represents those models that employ microphysical parameterizations with a number of simplifying assumptions to account for global dust source characteristics. In general, these models are not able to accommodate fine scale soil erodibility factors like the field to regional scale models. These models include the Community Aerosol and Radiation Model for Atmospheres (CARMA), the Global Ozone Chemistry Aerosol Radiation and Transport (GOCART), and DEAD (Ginoux et al., 2001; Luo et al., 2003).

Recent advances in the application of remote sensing to detect atmospheric aerosols and global source areas have enabled significant improvements to be made to global dust emission models. An example of this is in the DEAD model, where global characterizations of dust source areas by a range of geomorphic, topographic, and hydrological factors have enabled more advanced dust source parameterizations to be employed. These source area factors represent a shift forward from broad global dust source characterizations, and result in improved estimations of global dust loads.

7.6 Conclusion

As evident from foregoing, wind erosion is an immensely complex problem. A precise understanding of the wind erosion processes is key in developing effective wind erosion control strategies. Information on the nature, extent, and magnitude spatial distribution dynamics of areas affected by wind erosion is a prerequisite for planning and implementation of effective wind erosion control measures. Both aerial and spaceborne remote sensing data have been employed with varying success to derive such information. Detection of wind erosion features like sand deposition, morphometry dynamics of sand dunes, the presence of vegetation cover/crust, information erosion control measures employed (to some extent), and their impacts have been possible with the state-of-the-art remote sensing technology. In spite of several advancements made on wind erosion studies using *in situ*, and remote sensing technology, there are a few key challenges that need to be addressed. These include (1) resolving "mixed pixel" problem, (2) the effects of topography on reflection, (3) image commensurability, and (4) identification of areas susceptible to wind erosion. Since water, through the effect of gravity, moves according to a determined pattern, the areas susceptible to erosion may be predicted. In the case of wind, however, its horizontal and vertical movements are not predictable; hence, the areas susceptible to wind erosion cannot be forecast with reasonable accuracy. Besides, resolving the issue of farm-field level modeling of wind erosion using high spatial resolution (fraction of a meter) satellite data in another issue that merits attention.

From the aeolian geomorphological view point, it is the issue of form and process, and how to reconcile the spatiotemporal scales involved in dune field evolution with the limitations of our brief observational record and empirical perspective that needs to be addressed (Hugenholtz et al., 2012). While geospatial technologies do not provide an immediate, or the only, solution to this challenge, they offer a synoptic perspective and platform from which to help guide field-based research and validate modeling efforts. It will soon be possible to use these technologies to develop a coupled topography–climate–sediment–ecosystem model to predict real-world dune field morphodynamics. Whereas current models simulate artificial topography, we believe the next generation of models will use real topography to simulate the effects of different climate change scenarios.

ANNEXURE 7A
A ground photo showing severe wind erosion encroaching boundary wall in village in western Rajasthan, western India (Courtesy: Dr D.C. Joshi, Central Arid Zone Research Institute, Indian Council of Agriculture, Jodhpur, Rajasthan).

ANNEXURE 7B
A ground photo stabilized dunes in western Rajasthan, western India (Courtesy: Dr D.C. Joshi, Central Arid Zone Research Institute, Indian Council of Agriculture, Jodhpur, Rajasthan).

ANNEXURE 7C
A ground photograph sowing mixed pearl millet crop on a stabilized sand dune in part of Thar desert, Rajasthan, western India (Courtesy: Dr D.C. Joshi, Central Arid Zone Research Institute, Indian Council of Agricultural Research Institute, Jodhpur, Rajasthan).

ANNEXURE 7D
Mobile dune encroaching the village in part of Thar desert, Rajasthan, western India (Courtesy: Dr D.C. Joshi, Central Arid Zone Research Institute, Indian Council of Agricultural Research, Jodhpur, Rajasthan).

ANNEXURE 7E
Barchan dunes in part of Thar desert, Rajasthan, western India (Courtesy: Dr D.C. Joshi, Central Arid Zone Research Institute, Indian Council of Agricultural Research, Jodhpur, Rajasthan).

ANNEXURE 7F
Fresh sand deposition in an active wind erosion terrain in the periphery of *Thar* desert, Rajasthan, western India.

References

Bailey, S.D., Bristow, C.S., 2004. Migration of parabolic dunes at Abberffraw, Anglesey, north Wales. *Geomorphology* 59: 165–174.

Beinhauer, R., Kruse, B., 1994. Soil erosivity by wind in moderate climates. *Ecological Modelling* 75–76: 279–287.

Böhner, J., Schafer, W., Conrad, O., Gross, J., Ringeler, A., 2003. The WEELS model: methods, results and limitations. *CATENA* 52(3–4): 289–308.

Brown, D.G., Arbogast, A.F., 1999. Digital photogrammetric change analysis as applied to active coastal dunes in Michigan. *Photogrammsetric Engineering and Remote Sensing* 65: 467–474.

Bryan, R.B., 1968. The development, use and efficiency of indices of soil erodibility. *Geoderma* 2: 5–26.

Buckley, R., 1987. The effect of sparse vegetation cover on the transport of dune sand by wind. *Nature* 325: 426–428.

Budyko, M.I., 1958. *The Heat Balance of the Earth's Surface*. Department of Commerce, Weather Bureau, 144–155.

Chavez, Jr., P.S., Mackinnon, D.J., Reynolds, R.L., Velasco, M.G. 2002. Monitoring dust storms and mapping landscape vulnerability to wind erosion using satellite and ground-based digital images. *Arid Lands News Letter*, no. 51.

Chen, W., Fryrear, D.W., 1996. Grain-size distributions of wind eroded material above a flat bare soil. *Physical Geography* 17: 554–584.

Chepil, W.S., 1955. Influence of moisture on erodibility of soil by wind. *SSSAJ* 20(2): 288–292.

Dwivedi, R.S., Ravisankar, T., Venkataratnam, L. and Rao, D.P. 1992. Detection and delineation of various desert terrain features using Landsat-TM derived image transforms. *Journal of Arid Environment* 25: 151–162.

Embabi, N.S., Asour, M.M., 1992. Sand dunes in Qatar. *Proceedings of the International Symposium on Evolution of Deserts*. Ahmedabad, India, February 11–19, 1992.

Escadafal, R. 1992. Remote sensing of north African sand seas with medium resolution satellite imagery (Nimbus-CZCS). *International Symposium on Evolution of Deserts*. Ahmedabad, India. February 11–19, 1992.

Fadhil, A.M., 2009. Land degradation detection using geo-information technology for some sites in Iraq. *Journal of Al-Nahrain University* 12(3): 94–108.

Fécan, F., Marticorena, B., Bergametti, G., 1999. Parameterization of the increase of theaeolian erosion threshold wind friction velocity due to soil moisture for arid and semiarid areas. *Annales Geophysicae* 17(1): 149–157.

Fryrear, D.W., Bilbro, J.D., Saleh, A., Schomberg, H.M., Stout, J.E., Zobeck, T.M., 2000. RWEQ: improved wind erosion technology. *Journal of Soil and Water Conservation* 55(2): 183–189.

GayJr., S.P., 1999. Observations regarding the movement of barchan sand dunes in the Nazca to Tanaca area of southern Peru. *Geomorphology* 27: 279–293.

Geeves, G.W., Leys, J.F., McTainsh, G.H., 2000. Soil erodibility. In: Charman, P.E.V., Murphy, B.W. (eds.), *Soils: Their Properties and Management*. New York: Oxford University Press, 205–220.

Gillette, D.A., Marticorena, B., Bergametti, G., 1998. Change in the aerodynamic roughness height by saltating grains: experimental assessment, test of theory, and operational parameterization. *Journal of Geophysical Research* 103(D6): 6203–6210.

Ginoux, P., Chin, M., Tegen, I., Prospero, J., Holben, B., Dubivik, O., Lin, S.J., 2001. Sources and distributions of dust aerosols simulated with the GOCART model. *Journal of Geophysical Research* 106: 20,273–20,555.

Goossens, E.B., de Roover, Goossens, R., 1992. A digital approach to separation of parabolic sand dune area from interdunal using Landsat MSS data. *Proceedings of the International Symposium on Evolution of Deserts*. Ahmedabad, India, February 11–19, 1992.

Goudie, A., 1994. *Geomorphological Techniques*, 2nd edn, edited for the British Geomorphological Research Group. London and New York: Routledge, 570 pp.

Greeley, R., Iversen, J.D., 1985. *Wind as a geological process on Earth, Mars, Venus and Titan*. Cambridge: Cambridge University Press, 333 pp.

Hagen, L.J., 1991. A wind erosion prediction system to meet user needs. *Journal of Soil and Water Conservation* 46: 106–111.

Hagen, L.J., 2008. Updating soil surface conditions during wind erosion events using the wind erosion prediction system (WEPS). *American Society of Agricultural and Biological Engineers* 51(1): 129–137.

Hopkins, E.S., Palmer, A.E. and W.S. Chepil, 1937. Soil drifting control in the prarie. provinces, Canada Department of Agriculture Farmers Bulletin, 32. 51 pp.

Hugenholtz, C.H., Levin, N., Barchyn, T.E., Baddock, M.C., 2012. Remote sensing and spatial analysis of aeolian sand dunes: review and outlook. *Earth-Science Reviews* 111: 319–334.

Hugenholtz, C.H., Wolfe, S.A., 2005. Recent stabilization of active sand dunes on the Canadian prairies and relation to recent climate variations. *Geomorphology* 68: 131–147.

Hugenholtz, C.H., Wolfe, S.A., Moorman, B.J., 2007. Sand-water flows on cold-climate eolian dunes: environmental analogs for the eolian rock record and Martian sand dunes. *Journal of Sedimentary Research* 77: 1–8.

Iversen, J.D., White, B.R., 1982. Saltation threshold on Earth, Mars and Venus. *Sedimentology* 29: 111–119.

Karnieli, A., 1997. Development and implementation of spectral crust index over dune sands. *Int J Remote Sensing* 18(6): 1207–1220.

Kharin, N.G. 1986. Multipurpose desertification maps and a methodology for compiling them from space images. *Soviet Journal of Remote Sensing* 5: 69–80.

Lancaster, N., 1999. Geomorphology of desert sand seas. In: Goudie, A.S., Livingstone, I., Stokes, S. (eds.). *Aeolian Environments, Sediments and Landforms*. Chichester: John Wiley & Sons, 49–70.

Levin, N., 2011. Climate-driven changes in tropical cyclone intensity shape dune activity on Earth's largest sand island. *Geomorphology* 125: 239–252.

Levin, N., Ben-Dor, E., 2004. Monitoring sand dune stabilization along the coastal dunes of Ashdod-Nizanim, Israel, 1945–1999. *Journal of Arid Environments* 58: 335–355.

Levin, N., Kidron, G.J., Ben-Dor, E., 2006. The spatial and temporal variability of sand erosion across a stabilizing coastal dune field. *Sedimentology* 53: 697–715.

Levin, N., Kidron, G.J., Ben-Dor, E., 2007. Surface properties of stabilizing coastal dunes—combining spectral and field analyses. *Sedimentology* 54: 771–788.

Levin, N., Kidron, G.J., Ben-Dor, E., 2008. A field quantification of the adaptations of costal dune plants to erosion and deposition of sand. *Sedimentology* 55: 751–772.

Levin, N., Tsoar, H., Herrmann, H.J., Claudino-Sales, V., Maia, L.P., 2009. Modeling the formation of arcuate vegetated dune ridges behind barchan dunes in NE Brazil. *Sedimentology* 56: 1623–1641.

Leys, J.F., McTainsh, G., Strong, C., Heidenreich, S., Biesaga, K., 2008. DustWatch: using community networks to improve wind erosion monitoring in Australia. *Earth Surface Processes and Landforms* 33(12): 1912–1926.

Livingstone, I., Wiggs, G.F.S., Weaver, C.M., 2007. Quantifying controls on aeolian dune processes and dynamics: a review of current understanding. *Earth-Science Reviews* 80: 239–257.

Livingstone, I., Bristow, C., Bryant, R.G., Bullard, J., White, K., Wiggs, G.F.S., Baas, A.C.W., Bateman, M.D., and Thomas, D.S.G., 2010. The Namib sand sea digital database of aeolian dunes and key forcing variables. *Aeolian Research*, 2: 93–104.

Lu, H., Shao, Y., 2001. Toward quantitative prediction of dust storms: an integrated wind erosion modelling system and its applications. *Environmental Modelling and Software* 16(3): 233–249.

Luo, C., Mahowald, N.M., del Corral, J., 2003. Sensitivity study of meteorological parameters on mineral aerosol mobilzation, transport and deposition. *Journal of Geophysical Research* 108(D15): 4447.

Lyles, L., 1975. Possible effects of wind erosion on soil productivity. *The Journal of Soil and Water Conservation* 30(6): 279–283.

Mann, H.S., 1985. Wind erosion and its control. FAO Conservation Guide No. 10. Rome, FAO.

Marticorena, B., Bergametti, G., 1995. Modeling the atmospheric dust cycle: 1. Design of a soil-derived dust emission scheme. *Journal of Geophysical Research* 100(D8): 16,415–16,430.

Marticorena, B., Bergametti, G., Aumont, B., Callot, Y., N'Doume, C.N., Legrand, M., 1997. Modeling the atmospheric dust cycle: 2. Simulation of Saharan dust sources. *Journal of Geophysical Research* 102(D4): 4387–4404.

McTainsh, G.H., 1998. Dust storm index. In: *Sustainable Agriculture: Assessing Australia's Recent Performance, Report of the National Collaborative Programme on Indicators for Sustainable Agriculture*, 56–62.

McTainsh, G.H., Lynch, A.W., Burgess, R.C., 1990. Winderosion in Eastern Australia. *Australian Journal of Soil Research* 28(2): 323–339.

Mitra, D.S. & R. Bhoj. 1992. Dune forms in space images-A case study from Shahgarh and Myajlar areas of western Rajasthan. *Proceedings of the International Symposium on Evolution of Deserts*. Ahmedabad, India, February 11–19, 1992.

Necsoiu, M., Leprince, S., Hooper, D.M., Dinwiddie, C.L., McGinnis, R.N., Walter, G.R., 2009. Monitoring migration rates of an active subarctic dune field using optical imagery. *Remote Sensing of Environment* 113: 2441–2447.

Nordstrom, K.F., Hotta, S., 2004. Wind erosion from cropland in the USA: a review of problems, solutions and prospects. *Geoderma* 121: 157–167.

Ojeda, J., Vellejo, I., Malvarez, G.C., 2005. Morphometric evolution of the active dunes system of the Doñana National Park, Southern Spain (1977–1999). *Journal of Coastal Research SI* 49: 40–45.

O'Neill, A.L., 1994. Reflectance spectra of microphytic soil crusts in semi-arid Australia. *International Journal of Remote Sensing* 15: 675–681.

Owen, P.R., 1964. Saltation of uniform grains in air. *Journal of Fluid Mechanics* 20: 225–242.

Pandey, S., Singh, S., Ghosh, B., 1964. Orientation, distribution and origin of sand-dunes in the central Luni basin. *Proceedings of the Symposium on Problems of Indian Arid Zone*, Jodhpur (Rajasthan), November 23 to December 2, 1964.

Perlwitz, J., Tegen, I., Miller, R.L., 2001. Interactive soil dust aerosol model in the GISS GCM, 1, Sensitivity of the soil dust cycle to radiative properties of soil dust aerosols. *Journal of Geophysical Research* 106(D16): 18167–18192.

Potter, K.N., Williams, J.R., Larney, F.J., Bullock, M.S., 1998. Evaluation of EPIC's wind erosion submodel using data from southern Alberta. *Canadian Journal of Soil Science* 78: 485–492.

Raupach, M.R., Gillette, D.A., Leys, J.F., 1993. The effect of roughness elements on wind erosion threshold. *Journal of Geophysical Research* 98(D2): 3023–3029.

Raupach, M.R., Lu, H., 2004. Representation of land-surface processes in aeolian transport models. *Environmental Modelling and Software* 19: 93–112.

Rotstayn, L.D., Collier, M.A., Dix, M.R., Feng, Y., Gordon, H.B., O'Farrell, S.P., Smith, I.N., Syktus, J., 2010. Improved simulation of Australian climate and ENSO-related rainfall variability in a global climate model with an interactive aerosol treatment. *International Journal of Climatology* 30: 1067–1088.

Rutherford, S., Macintosh, K.I., Jarvinen, K.A., McTainsh, G.H., Chan, A., 2003. Characterisation and health impacts of the October 2002 dust event in Queensland: a preliminary investigation. In: *National Clean Air Conference: Linking Air Pollution Science, Policy and Management*, November 23–27, 2003, Newcastle, NSW. Newcastle: Clean Air Society of Australia and New Zealand.

Scheidt, S., Ramsey, M., Lancaster, N., 2010. Determining soil moisture and sediment availability at White Sands Dune field, New Mexico, from apparent thermal inertia data. *Journal of Geophysical Research* 115: F02019.

Shao, Y., 2001. A model for mineral dust emission. *Journal of Geophysical Research* 106(D17): 20239–20254.

Shao, Y., Leslie, L.M., 1997. Wind erosion prediction over the Australian continent. *Journal of Geophysical Research* 102(D25): 30,091–30,105.

Shao, Y., Raupach, M., Short, D., 1994. Preliminary assessment of wind erosion patterns in the Murray-Darling Basin. *Australian Journal of Soil and Water Conservation* 7(3): 46–51.

Shao, Y., Raupach, M.R., Leys, J.F., 1996. A model for predicting aeolian sand drift and dust entrainment on scales from paddock to region. *Australian Journal of Soil Research* 34: 309–342.

Singh, S., Kar, A., Joshi, D.C., Ram, B., Kumar, S., Vats, P.C., Singh, N., Kolarkar, A.S., Dhir, R.P., 1992. Desertification mapping in western Rajasthan. *Annals of Arid Zone* 31(4): 237–246.

Singh, U.B., Gregory, J.M., Wilson, G.R., 1999. Texas erosion analysis model: theory and validation. In: Skidmore, E.L., Tatarko, J. (eds.), *Wind Erosion – Proceedings of an International Symposium/ Workshop*, June 3–5, 1997, United States Department of Agriculture (USDA), Agricultural Research Service, Wind Erosion Research Unit, Kansas State University, Manhattan, Kansas, 23.

Skidmore, E.L., 1986. Wind erosion climate erosivity. *Climatic Change* 9: 195–208.

Stephens, P., Cock, J., 1991. Changes in vegetation and sand-dunes from digitized aerial photographs and SPOT data: changes over a period of 23years. In: Murai, S.(ed.), *Applications of Remote Sensing in Asia and Oceania: Environmental Change Monitoring*. Japan: Asian Association on Remote Sensing, 175–179.

Tsoar, H., Karnieli, A., 1996. What determines the spectral reflectance of the Negev–Sinai sand dunes. *International Journal of Remote Sensing* 17: 513–525.

UNEP (1997). World Atlas of Desertification. Arnold, London.

USDA-Agricultural Research Service, 1961. A universal equation for measuring wind erosion. USDA-ARS, 22–69.

Van Pelt, R.S., Zobeck, T.M., 2004. Validation of the Wind Erosion Equation (WEQ) for discrete periods. *Environmental Modelling and Software* 19(2): 199–203.

Vermeesch, P., Drake, N., 2008. Remotely sensed dune celerity and sand flux measurements of the world's fastest barchans (Bodélé, Chad). *Geophysical Research Letters* 35: L24404.

Webb, N.P., McGowan, H.A., Phinn S.R. et al., 2009. A model to predict land susceptibility to wind erosion in western Queensland, Australia. *Environmental Modelling and Software* 24: 214–227.

Webb, N.P., 2008. Modelling land susceptibility to wind erosion in western Queensland, Australia. The University of Queensland in August 2008. The University of Queensland in August 2008.

Wolfe, S.A., Hugenholtz, C.H., 2009. Barchan dunes stabilized under recent climate warming on the northern Great Plains. *Geology* 37: 1039–1042.

Woodruff, N.P., Siddoway, F.H., 1965. A Wind Erosion Equation. *Soil Science Society of America Proceedings* 29(5): 602–608.

Xiao, J., Shen, Y., Tateishi, R., Bayaer, W., 2006. Development of topsoil grain size index for monitoring desertification in arid land using remote sensing. *International Journal of Remote Sensing* 27(12): 2411–2422.

Zender, C.S., Bian, H., Newman, D., 2003. The Mineral Dust Entrainment and Deposition (DEAD) model: description and 1990s dust climatology. *Journal of Geophysical Research* 108(D14): 4416–4439.

Zhibao, D., Weinan, C., Guangrong, D., Guangting, C., Zhenshan, L., Zuotao, Y., 1996. Influences of vegetation cover on the wind erosion of sandy soil. *Acta Scientiae Circumstantiae*. 1996-0422.

8

Soil Salinization and Alkalinization

Coauthored by Dr. Jamshid Fareftih
Faculty of Bioscience Engineering, Department of Biosystems, Katholieke Universiteit Leuven, Heverlee, Belgium

8.1 Introduction

Soil salinization/alkalinization is one of the major factors affecting biomass production (Csillag et al., 1993). Salt-affected soils (saline/alkali) occur mostly in arid and semiarid regions, but may also be found in subhumid and coastal zones. Salt-affected soils are highly erosive, and have poor structure, low fertility, low microbial activity, which are not conducive for plant growth. Salt concentration in the soil environment has a strong impact on crop yield and agricultural production due to poor land and water management and expansion of agricultural frontiers into marginal dry lands. Crop productivity and production losses have considerable impact on-farm and agricultural economies. For instance, the economic damage caused by secondary salinization was estimated at US$750 million per year for the Colorado River Basin in the United States of America, US$300 million per year for Punjab, and Northwest Frontier Provinces in Pakistan, and US$208 million per year for the Murray–Darling Basin in Australia (Ghassemi et al., 1995).

Salt-affected soils contain excessive concentrations of either soluble salts or exchangeable sodium, or both. The process by which soluble salts are accumulated at the surface or near-surface of soil horizon is called salinization (Szabolcs, 1974). The build-up soluble salts in soils produce harmful effects to plants by increasing the salt content of the soil solution and by increasing the degree of saturation of the exchange materials in the soil with exchangeable sodium (US Salinity Laboratory Staff, 1954). The dominant salts in salt-affected soils consist of chlorides and sulfates of Na, Ca, and Mg. In terms of agricultural consequences, excessive salts in soil accelerate land degradation processes and increase the impact on crop yields and agricultural production. Additionally, the increase of salts in soil also affects other major soil degradation phenomena (Figure 8.1) such as soil dispersion, sealing and crust formation, and structural changes, which results in unstable and compacted soil (FAO, 1988; De Jong, 1994; Metternichtand and Zinck, 2003).

Alkalinization refers to accumulation of exchangeable sodium in soils. It is a process leading to the formation of soils with high percentage of exchangeable sodium. The presence of carbonates and bicarbonates leads to formation of soils with high pH values, between 8.5 and 10 (FAO, 1988; Sparks 1995). The main cause of alkaline reaction of soils, as described by FAO (1988), is the hydrolysis of either exchangeable cations or of salts such as $CaCO_3$, $MgCO_3$, and Na_2CO_3. Owing to the electrical charges at the surface, soil particles adsorb and retain cations on their surfaces. The adsorbed cations may be replaced by other cations that occur in the soil solution. Sodium, calcium, and magnesium cations are always readily exchangeable. In arid and semiarid regions when excess soluble salts accumulate, sodium

FIGURE 8.1
Excessive soil degradation caused by soil salinity, Southeast Iran (Farifteh, 1988).

frequently becomes the dominant cation in the soil solution. As the soil solution becomes concentrated through evaporation or water absorption by plants, the solubility limits of calcium sulfate, calcium carbonate, and magnesium carbonate are often exceeded and they are precipitated with a corresponding increase in the relative proportion of sodium. Under such conditions, a part of the original exchangeable calcium and magnesium is replaced by sodium (United States Soil Salinity Laboratory Staff, 1954).

Salt-affected soils can be found worldwide under almost all climatic conditions; however, they are most widespread in the arid and semiarid regions. In general, human-induced factors and unfavorable natural conditions are considered to cause and accelerate soil salinity. The consequent changes in land use mainly due to the common policy of agricultural intensification, improper use of land, and quality of the water used for irrigation are among the human-induced factors causing secondary type of soil salinity. Geological formations, climate factors, and geomorphology are among environmental factors considered to be responsible for accumulation of salt in soil. Geological formation consisting of salt-accumulated sediments can provide large volume of salts to be carried by surface or ground water. Climate factors such as low precipitation and high evaporation can increase the concentration of salts in soil and water, contributing to salinity problems. Geomorphology has substantial effects on accelerating salinization/alkalinization within closed basin or lowland areas, where restricted drainage largely contributes to the salinization of soils due to low permeability, lack of outlet and very low slope gradient also causing an increase in a high ground-water table (United States Salinity Laboratory Staff, 1954).

8.2 Origin of Salts

As mentioned in Section 7.1, the soluble salts in soils consist mostly of various proportions of the cations sodium, calcium, and magnesium, and the anions chloride and sulfate. Constituents that ordinarily occur only in minor amounts are the cation potassium and the anions bicarbonate, carbonate, and nitrate. The original and, to some extent, the direct sources of all the salt constituents are the primary minerals found in soils and in the exposed rocks of the earth's crust. Clarke (1924) has estimated that the average chlorine and sulfur content of the earth's crust is 0.05% and 0.06%, respectively, while sodium, calcium, and magnesium each occur to the extent of 2% or 3%. During the process of chemical weathering, which involves hydrolysis, hydration, solution, oxidation, and carbonation, these constituents are gradually released and made soluble. Bicarbonate ions form as a result of the solution of carbon dioxide in water. The carbon dioxide may be of atmospheric or biological origin (United States Salinity Laboratory Staff, 1954). The released salts are transported away from their source of origin through surface or groundwater. The salts in the groundwater are gradually concentrated as the water with dissolved salts moves from the more humid to the less humid and relatively arid areas.

The predominant ions near the site of weathering in the presence of carbon dioxide will be carbonates and hydrogen-carbonates of calcium, magnesium, potassium, and sodium; their concentrations, however, are low. As the water with dissolved solutes moves from the more humid to the arid regions, the salts are concentrated and the concentration may become high enough to result in precipitation of salts of low solubility. Apart from the precipitation, the chemical constituents of water may undergo further changes through processes of exchange, adsorption, differential mobility, etc., and the net result of these processes invariably is to increase the concentration in respect of chloride and sodium ions in the underground water and in the soils.

8.3 Nature of Salt-Affected Soils

The term *soil* is used in several senses by agriculturists. Soil, generally, is considered to be a 3D piece of landscape having shape, area, and depth (Soil Survey Staff, 1951). The concept of a soil as a profile having depth but not necessarily shape or area is also a common use of the term (Soil Survey Staff, 1951). Szabolcs (1974) grouped salt-affected soils in two main categories: (1) soils affected by neutral sodium salts (mainly sodium chloride and sodium sulfate) and (2) soils affected by sodium salts capable of alkali hydrolysis ($NaHCO_3$, Na_2CO_3, and Na_2SiO_3). Soils belonging to the first group have mainly been termed saline, and those of the second as alkali.

Based on pH, electrical conductivity (EC), and exchangeable sodium percentage (ESP), these soils have been grouped into three categories, namely saline, alkali, and non-saline-alkali soils. *Saline* is used in connection with soils for which the conductivity of the saturation extract is more than 4 mmhos/cm (Table 8.1) at 25°C, the ESP is less than 15, and the pH, ordinarily, is less than 8.5 (United States Salinity Laboratory Staff, 1954). The chief anions are chloride and sulfate, and sometimes nitrate. The term *saline-alkali refers* to soils for which the conductivity of the saturation extract is greater than 4 dSm^{-1} at 25°C and the ESP is greater than 15. Under conditions of excess salts, the pH values are seldom higher than 8.5 and the particles remain flocculated. *Non-saline-alkali soils* are those soils for which the ESP is greater than 15, the pH readings usually are between 8 and 10 and

TABLE 8.1

Soil Salinity Classes in Terms of ECe (Richards, 1954)

Salinity Class	ECe ($dS\ m^{-1}$)	Salinity Effects on Crops
Nonsaline	<2	Salinity effects are negligible
Slightly saline	2–4	Yields of very sensitive crops may be restricted
Moderately saline	4–8	Yields of many crops restricted
Very saline	8–16	Only tolerant crops yield satisfactory
Extremely saline	>16	Only a few very tolerant crops yield satisfactorily

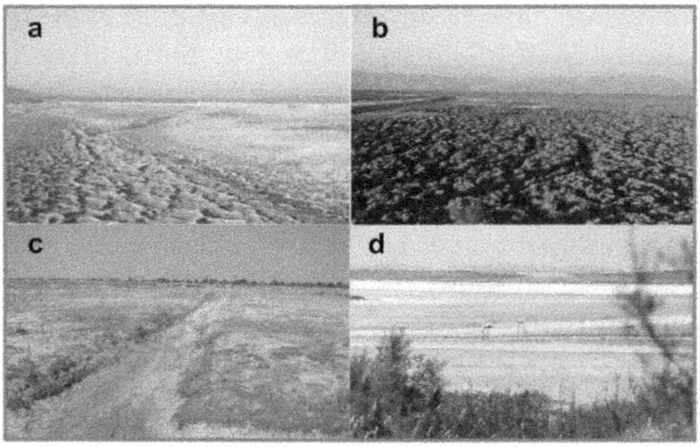

FIGURE 8.2
Severely salt-affected soils in (a and b) Dashat-e-Kavir, Iran, (c) Northeast of Thailand, (d) South of Spain; Laguna de Fuente de Piedra (Farifteh, 2007).

the conductivity of the saturation extract is less than 4 dSm^{-1} at 25°C. Examples of severely salt-affected soils are shown in Figure 8.2.

8.4 Extent and Spatial Distribution

Soil salinity is a prevalent environmental hazard in arid and semiarid regions around the world (Hillel, 2000). The salt-affected soils are found worldwide (Figure 8.3) including Africa, Asia, Australia, Europe, Latin America, Near East, and North America (Koohafkan, 2012).

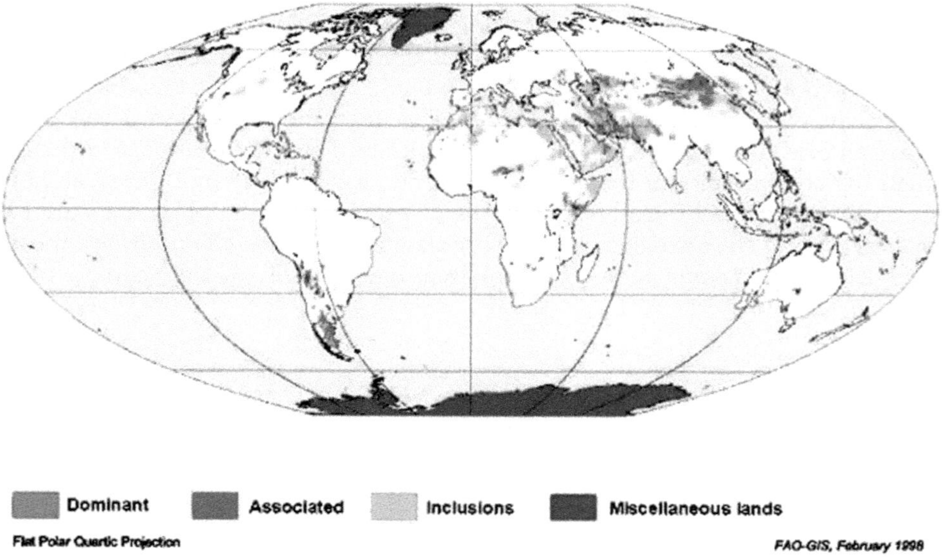

FIGURE 8.3
Global distribution of solanchalks based on WRB and FAO/UNESCO soil map of the world (FAO, 1998).

TABLE 8.2

Extent of Salt-Affected Soils (Szabolcs, 1979)

Regions	Area (million ha)
North America	15.7
Mexico and Central America	2.0
South America	129.2
Africa	80.5
South America	84.8
North and Central Asia	211.7
South-east Asia	20.0
Australia	357.3
Europe	50.8
Total	952.0

Globally, an estimated 831 million hectares including 392 million hectares of saline soils and 434 million hectares sodic soils occur in arid and semiarid regions (FAO, 2000). Besides naturally occurring salt-affected soils, 76.3 million hectares has been salinized as consequence of human activities (Oldeman et al., 1991). It is considered that about 25 Mha of lands have been salinized through human intervention since the development of irrigated agriculture (WRI-IIED-UNEP, 1988; Szabolcs, 1989). In India, 6.727 million hectares of land have been affected by soil salinization and/or alkalinization (National Remote Sensing Agency, 2008). Continent-wise spatial extent of salt-affected soils is given in Table 8.2.

8.5 Soil Salinity Symptoms

The incidence of salinization and/or alkalization manifests both in soils and plants.

8.5.1 Surface Manifestation

As mentioned earlier, soil salinity is manifested at the surface as salt efflorescence or salt encrustation (Figure 8.4). However, the extent and magnitude of manifestation is highly dynamic and varies spatially as well temporally depending on physicochemical properties of soil such as soil moisture content, organic matter, soil texture, types of clay color and surface roughness, EC, and exchangeable sodium content. During dry season the salt encrustation/salt efflorescence with higher spectral reflectance in visible and near-infrared (NIR) region will be quite extensive. Schmid et al. (2007) observed that due to smooth surface saline soils with crust reflect strongly in the visible and NIR bands compared to non-saline soils. This, in turn, would facilitate the detection of salt-affected soils from remote sensing data. On the contrary, owing to the presence of moisture during rainy season the manifestation of salt crust/salt efflorescence will be often poor. As a result, the spectral response pattern will be relatively lower compared to dry season thereby making the delineation of salt-affected soils difficult.

Apart from seasonal variations in surface manifestation, the image contrast between salt-affected soils and normal soil background helps in delineation of salt-affected soils using remotely sensed images. Salt-affected soils, in general, occur as pockets of varying

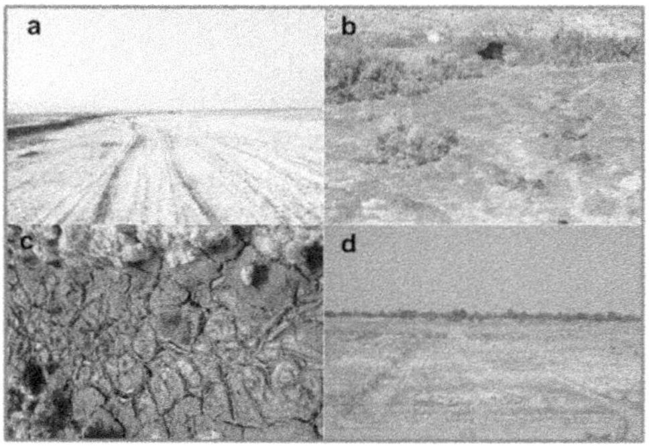

FIGURE 8.4
Surface features formed as the results of excessive salt accumulation in soil: (a) Dashat-e-Kavir, Iran; (b) South Spain; (c) Tedej, Northeast Hungary; and (d) Northeast of Thailand.

sizes within the agricultural fields (Figure 8.5). Being invariably barren (devoid of vegetation), salt-affected soils are seen as pockets of white to gray color of different sizes and provide very good image contrast in remotely sensed images that enables delineation of salt-affected soils. Because of the effects of salt features on the soil spectral reflectance, they have been considered good direct indicators of soil salinity. For example, Fernandez-Buces et al. (2006) used surface features to predict soil salinity. They found that the correlation coefficient between surface colors, EC and the sodium adsorption ratio (SAR) were statistically significant, which suggested that efflorescence color is a promising surface indicator with which to estimate soil salinity.

8.5.2 The Presence of Halophytic Plants

Salt-affected soils are usually devoid of vegetation, but at lower concentrations some plant communities can be present in natural conditions (e.g., *Salicornia* spp.), and salt-tolerant crops such as barley, cotton, and alfalfa, can be cultivated. The presence of hydrophytic (salt-tolerant) plants that thrive well in soils with excess salt is a good indirect indicator of the occurrence of salt-affected soils (Glenn et al., 1999). However, not all halophytes have been found to be good indicators of soil salinity. For instance, Metternicht (1998a) observed high absorption in the visible range and high reflectance in the NIR range of halophyte *Chenopodiaceae* in

FIGURE 8.5
Pocket of saline soils encapsulated in agricultural fields: (a) Southwest Australia, (b) Northeast Thailand, (c) Southeast Iran, and (d) South Spain.

Bolivia, to be similar to that of chlorophyll-rich vegetation. In contrast, due to lower chlorophyll content the spectral reflectance curve of *Cynodon dactylon*, also a halophyte, increased continuously in the visible and NIR bands. This study concluded that halophytic plants were permissive indicator to distinguish saline areas from non-affected ones.

8.5.3 Crop Performance

The performance of some crops that can be grown on saline soils, such as alfalfa, barley, and cotton, reflects the severity of soil salinity. For example, based on the high correlations between the normalized difference vegetation index (NDVI) values of cotton, sugar-cane crops and the EC Wiegand et al. (1994, 1996) could successfully assess the severity and extent of soil salinity in terms of the economic impact on crop production, and also distinguished saline soils from non-affected soils. However, the relationship most likely exists only where soil salinity is the major factor that causes crop yield variability. Therefore, the delineation of salt-affected soils is likely to result in large errors. Hence the possible use of tolerant crop as an indicator for determining the level of soil salinity must be verified carefully. To overcome this issue, some researchers have proposed using average crop production over a series of years to mask out the noise from non-soil factors that differ from year to year. For example, they obtained a strong correlation between yield losses and soil salinity by using a 6-year temporal series of satellite images of yield and concluded that yield loss in agricultural crops could be primarily due to several factors, including soil salinity.

8.6 Proximal Sensing

Information on the nature, extent, and spatial distribution of salt-affected soils is a prerequisite for assessment of the progress and success of any reclamation program. Conventional soil surveys which are cost-and labor-intensive and time-consuming have been traditionally employed for deriving such information. With the availability of aerial photographs and subsequent developments in proximal and remote sensing techniques, inventory and monitoring of salt-affected soils is carried out using remote sensing techniques. An overview of proximal and remote sensing is presented hereafter.

Proximal soil sensing is, in general, a collection of technologies that employ a sensor close to, or in direct contact with the soil to directly or indirectly measure a soil property. Viscarra Rossel et al. (2011) provide a description of proximal soil sensing, sensing technologies, and what soil properties these technologies can measure. The spectral response of any feature is fundamental to its detection and/or delineation.

8.6.1 Spectral Measurements in Laboratory

Spectroscope/spectro radiometers are used for generating laboratory spectra which are used for identification of substances through the spectrum emitted from or absorbed in them and also for interpretation/analysis of air/spaceborne spectral data. Spectroscopy is the study of the interaction between matter and radiated energy, and is used to refer to the measurement of radiation intensity as a function of wavelength and is often used to describe experimental spectroscopic methods. Spectral information enables the identification of matter based on spectral absorption features of the material's specific absorption of spectrally active groups, and has been found very useful in many terrestrial and marine applications (e.g., Clark and Roush, 1984). Although the soil spectrum may look somewhat monotonous, compared with the spectra of rocks and pure minerals, it contains significant information about many soil constituents. A comprehensive review of soil spectroscopy is provided by Ben-Dor et al. (1999) and Ben-Dor (2002).

The salt-affected soil spectra and the effect of salts on soil reflectance are illustrated in Figure 8.6 (Farifteh et al., 2006). Most of the features seen in spectrum of saline minerals (400–2500 nm) can be attributed to internal vibration modes of certain molecular groups, particularly carbonate, borate, hydroxyl anion groups, and neutral water molecules (Hunt, 1980; Crowley, 1991).

Using several salt contents and types, namely bischofite, halite, sylvite, arcanite, epsomite, and thenardite, Farifteh (2007) observed significant spectral changes such as water absorption bandsat around 1,400 and 1,900 nm based on the hydration of salts, and albedo changes. Spectral-based models to identify salt types and to estimate salt contents in the soil samples were developed using multivariate analysis. The study concluded that irrespective of the sensor type and platform, the soil property analyzed must carry direct spectral absorption features or, at least, be correlated to a soil's physical/chemical property

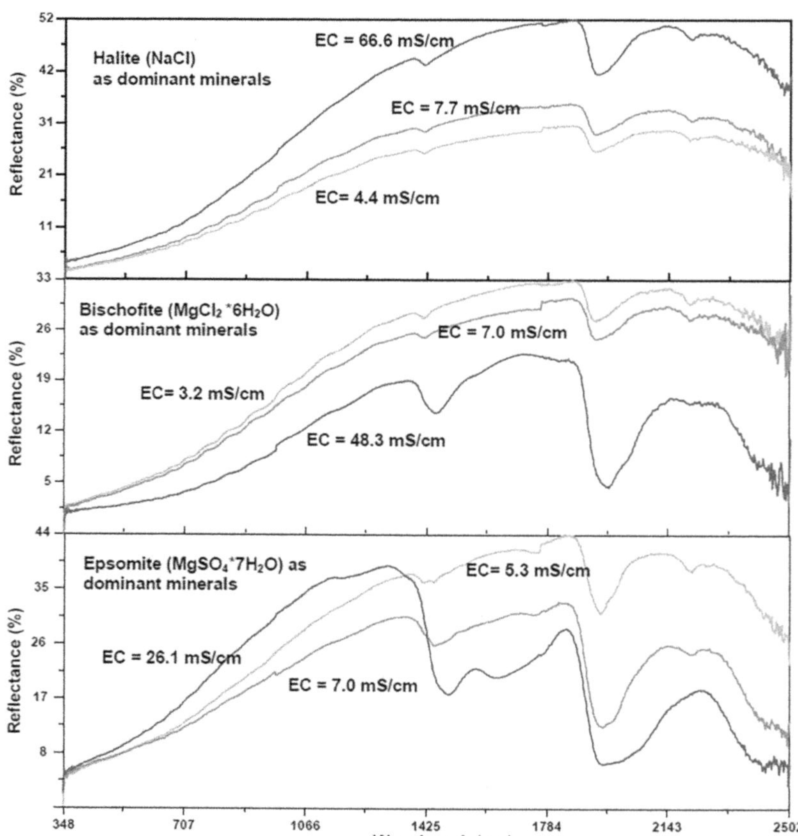

FIGURE 8.6
Laboratory spectra of salt-affected soils from soil materials impregnated by different evaporate minerals (Source: After Farifteh et al. (2006)).

(indirect relationship). Furthermore, spectroscopy is limited in assessing directly the content and, obviously, the type of salt (Farifteh et al., 2008; Farifteh, 2011).

vThe broad salt mineral classes exhibit unique thermal infrared emissivity spectra as a result of their different mineral structures (Figure 8.7). As evident from the figure, within each mineral class, the structurally associated cation alters the spectrum and produces additional variations in the emissivity spectra of a single class of minerals, as shown for carbonate.

8.6.2 *In situ* Spectral Measurements

Spectral reflectance studies are fundamental to understanding the spectral behavior of salt-affected soils, which facilitates their recognition and delineation. The spectral information collected from soils can be used to identify salt-affected soils (Dehaan and Taylor, 2002; Howari et al., 2002) or to relate spectral properties and abundance of salts quantitatively (Shuya et al., 2005; Farifteh et al., 2007a,b, 2010).

For surface sensing of salt-affected areas, spectroradiometers are used, such as the field Spec Pro FR, Field Spec JR, and Field Spec VNIR (Analytical Spectral Device, Inc.); the GER3700 and GER2600 (Geophysical and Environmental Research Corporation); and the PIMA SP (Integrated Spectronics). These instruments collect continuous, narrow, band width (e.g., 3–30 nm) in the VIS and NIR to SWIR regions of the spectrum, usually within the range of 350–2,500 nm, with an instantaneous field of view ranging from 1° to 22°. The spectra collected are used primarily to create spectral libraries of surface features (soils, vegetation, and minerals), which are further used as end members to classify air- or satellite-borne data (Shepherd and Walsh, 2002), to identify the presence and spectral location of primary diagnostic spectral features of salt minerals (Howari et al., 2002; Farifteh et al., 2007b), and to calibrate air or satellite imagery acquired within the same spectral range.

Ground observations and radiometric measurements indicate that the main factors affecting the reflectance are quantity and mineralogy of salts, moisture content, color, and surface roughness (Mougenot et al., 1993; Metternicht and Zinck, 2003). Salt-affected soils with salt encrustation at the surface are, generally, smoother than non-saline surface and cause high reflectance in the visible and NIR bands (Everitt et al., 1988; Rao et al., 1995).

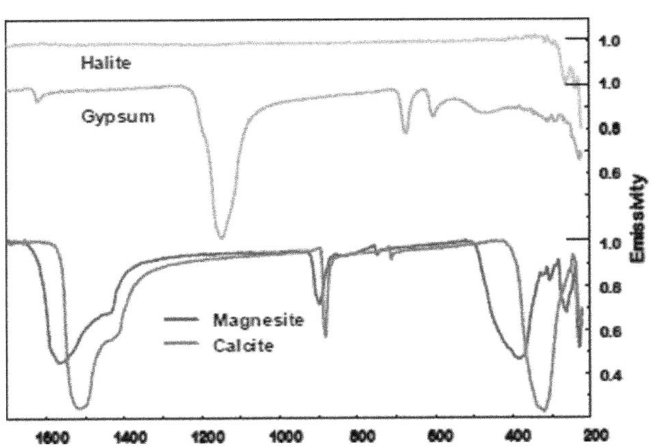

FIGURE 8.7
Thermal infrared emissivity laboratory spectra of salt minerals including chloride (halite), sulfate (gypsum), and carbonate (magnesite and calcite). Spectra are offset for clarity (Source: After Lane (2002)).

Eventhough it is well known that some of the salt minerals such as halites are featureless (Hunt et al., 1980), but the presence and amount of salt (e.g., halite) in soils have significant impact on the water absorption bands at around 1.4 and 1.9 µm (Hick and Russell, 1990; Farifteh et al., 2008; Farifteh, 2011). Carbonatesin soil exhibit absorption features in the thermal region (between 11 and 12 µm) due to internal vibration of the CO_3^{2-} group whereas anions have an absorption band near 10.2 µm caused by overtones or combination tones of internal vibration of constitutional water molecules (Siegel and Gillespie, 1980; Mulders, 1987). Csillag et al.(1993) found the visible (550–570 nm), NIR (900–1030 and 1270–1520 nm) and middle infrared bands (1940–2150, 2150–2310, and 2330–2400 nm) portion of the spectrum at 20, 40, and 80 nm spectral resolution as the key spectral bands in characterizing the salinity status of soils.

In addition to field-based hyperspectral images covering the visible to short-wave infrared (SWIR), thermal images can also be acquired in the field across the emitted spectral range (3–5 and 8–14 µm). Soil salinity causes moisture stress resulting in the reduction in transpiration with a consequent increase in the thermal infrared radiation. In addition, the thermal infrared region registers features caused by energy absorption of sulfates, phosphates, and chlorides (Siegal and Goetz, 1977; Mulders, 1987).

Apart from thermal infrared images, microwave data have also been used for identification of soil salinity. The significant difference in the imaginary part of dielectric constant between pure water and saline water at microwave frequencies less than 7 GHz has been used to derive information on soil salinity (Ulaby et al., 1986). Besides, dielectric properties of salt-affected soils have been studied using dielectric probe (Sreenivas et al., 1995). Brightness temperature decreases with an increase in soil salinity especially at low frequencies especially L-band (1.44 GHz) under varying terrain conditions. Furthermore, better results are obtained at lowest moisture content (Chaturvedi et al., 1983).

8.6.3 Frequency-Domain Electromagnetic Techniques

Being a sub-surface phenomenon currently operating optical do not and microwave sensors do not provide information on 3D view of the soil in regard to salinity. Additional tools such as penetrating optical fiber head (3S-HeD) are required to explore salinity process in the soil profile. Therefore, remote sensing technologies based on active EM radiation are being widely adopted. EM induction is based on the principle that electric currents can be applied to the soil through induction and that the magnitude of the induced current loops is directly proportional to the depth weighted EC_a of the soil (McNeill, 1980; Rhoades and Corwin, 1981). Ground-based electromagnetic (EM) methods measure apparent electrical conductivity (EC_a) in substratum horizons and can thus recognize salinity anomalies in the field before salinization approaches the surface (Farifteh et al., 2006; Farifteh, 2008). Among the geophysics-based approaches that employed in ground water and soil investigation, the EC imaging based on a Frequency-Domain Electromagnetic Techniques (FDEM) technique was ranked as the best out of ten techniques, including passive remote sensing means (Allen, 2004).

An EC imaging instrument is a FDEM sensor that works within a range of 30 cm to 5 m depth and performs best while scanning the area from about 1 m above the ground. The bulk EC_a, as measured by EM devices, is affected by the conductors found in the ground as well as the physical properties of the soil matrix. In general, when dealing with soil salinity that occurs at 0–3 m depth, the tow FDEM is the most suitable device because of its sensitivity to depths ranging from 30 cm to 5 m, depending on the waveband. Apparently, airborne EM systems are not appropriate for soil mapping because they sense at a subsurface level deeper than most soil salinity surveys require for analyzing salinity effects on crops (root zone). Sensing with a FDEM device precludes separating the surface layer or the root zone volume that is located within the first 25 cm of the soil profile.

In agriculture, the EM technique was introduced in the late 1970s with work conducted by Rhoades et al. (1976), De Jong et al. (1994), and Rhoades and Corwin (1981). Subsequently several studies were conducted to assess the spatial and temporal changes in soil salinity (Lesch et al., 1998; Cassel et al., 2005; Nogue et al., 2006). Hallet (1994) and Street and Duncan (1992) observed a good correlation between the EC_a as measured by an electromagnetic induction (EMI) meter mounted on an aircraft, and EC from field samples. However, the results were not encouraging in areas where rock substratum was covered with a thin soil layer (Metterincht, 1998a,b). The EM techniques are most successful in areas where the subsurface properties are reasonably homogenous, the effects of one property dominates over those of the other properties, and variations in the EM response can be related to changes in the dominant properties (Rhoades and Corwin, 1990; Cook and Walker, 1992).

8.6.4 Ground Penetrating Radar Measurements

The ground penetrating radar (GPR), a noninvasive device, uses short radar pulses to image the sub-surface. It aims at detection of reflected signals from sub-surface structures. The transmitting antenna radiates short pulses of high-frequency (usually polarized) radio waves into the ground. When the wave hits a buried object or a boundary with different dielectric constants, the receiving antenna records the variations in the reflected return signals. The depth range of GPR is limited by the EC of the ground and transmitting frequency. As conductivity increases, penetration depth decreases. This is because the EM energy is more quickly dissipated into heat energy, causing a loss in signal strength at depth. Higher frequencies do not penetrate as far as lower frequencies, but give better resolution. Optimal depth penetration is achieved in dry sandy soils or massive dry material such as granite, limestone, and concrete, where the depth of penetration is up to 15 m. In moist soils and/or clayey soils and in soils with high EC, penetration is sometimes only a few centimeters. Usually, standard GPR antennas are in direct contact with the ground to obtain the strongest signals, while GPR horn antennas can be used at 0.3–0.6 m above the ground. GPR images can be analyzed for deriving information on electrical properties and sub-surface characteristics as well as for spatial mapping of water content. The resolution range is a function of the sub-surface dielectric constants and the wave frequency.

8.7 Inventory and Monitoring of Salt-Affected Soils

Spectral measurements, thus made, are interpreted/analyzed using visual interpretation or computer-aided image analysis techniques to derive information on natural resources and environment including salt-affected soils. The effects of excess salts and exchangeable sodium in soils are manifested as salt efflorescence/salt encrustation (thick fluffy salt layer of varying colors, namely white or gray color of different shades) on the surface (Teggi et al., 2012; Matinfar et al., 2013). In addition, the presence of halophytes (salt tolerant vegetation), and performance level of salt-tolerant crops also serves as an indicator for the presence of excess salts in the soil (Alhammaadi and Glenn, 2008; Zhang et al., 2011).

8.7.1 Airborne Sensors Data

8.7.1.1 Aerial Photographs, Videography, and Digital Multispectral Camera Images

By virtue of the presence of salt crust/salt efflorescence on the surface, aerial photographs of different types, namely black-and-white, infrared, and color-infrared enable detection of salt-affected soils. Besides, because of the sensitivity of NIR radiation to vegetation vigor, color-infrared photographs are very useful in identification of salt-stressed crops. However, in both the cases, namely bare salt-affected soils as well as salt-stressed crops, the period of aerial flying assumes a significant role. Ideally, aerial photographs acquired during the peak vegetative growth period of crops are preferred as they provide very good image contrast between normal fertile soils/salt-stressed crops and normal fertile soils and healthy crops. The delineation of salt-affected soils from aerial photographs interpretation begins with the delineation of physiographic/landform features and grey tones/colors. Ground truth observations and chemical analysis of soil samples allow relating salt-affected soils, thus delineated, to variations in salt content.

Studies on salt-affected soils with respect to their spatial extent and temporal behavior have been carried out from aerial photographs of different types and scales (Hilwig and Karale, 1973; Manchanda and Khanna, 1979). Karale and Venugopal (1970) brought out a soil map for part of Meerut district, Uttar Pradesh, northern India, depicting various categories of salt-affected soils through a systematic stereoscopic visual interpretation of aerial photographs. Following similar approach, Nagar and Singh (1979) prepared a reconnaissance soil map for part of Hardoi district of Uttar Pradesh, northern India, wherein various types of salt-affected soils having surface manifestation in the form of salt efflorescence were delineated. With respect to the period of aerial flying for mapping salt-affected soils, Dale et al. (1986) observed that the color-infrared aerial photographs taken in autumn with high red reflectance provide maximum discrimination between the classes of salt marsh in southeast Queensland, Australia. Besides, Ghassemi et al. (1995) found color aerial photographs a good means for recording multi-temporal and crop-to-crop responses to salinity and management practices. Furthermore, multispectral video systems, recording the spectral response pattern of vegetation in the visible and NIR regions of the spectrum, have been used for mapping crop variations due to soil salinity.

Color-infrared aerial photographs, with a sensitive spectral range of 380–900 nm, were used to discriminate barren saline soils (in white) and salt-stressed crops (in reddish-brown) from other soil surface and vegetation features, and for mapping small areas requiring high spatial resolution and for validating satellite data (Manchanda and Iyer, 1983; Everitt et al., 1988; Rao and Venkataratnam, 1991; Weigand et al., 1994). Besides, color-infrared photographs have also proven useful to identify variations in salinity-induced plant stress owing to the response of crops to salinity severity which usually varies from season to season and from crop to crop (Ghassemi et al., 1995).

Apart from aerial photographs, airborne digital multispectral cameras and videography, usually with three to four bands in the VIS and NIR, were also used together with color-infrared photographs to identify and assess soil salinity in agricultural areas of the USA in the 1980s and 1990s. Everitt et al. (1988) used narrow-band videography to detect and estimated the extent of salt-affected soils in Texas. Wiegand et al. (1991, 1992) analyzed and mapped the response of cotton to soil salinity using color-infrared photographs and videography in three spectral bands (840–850, 640–650, and 540–550 nm), with a spatial resolution of 3.4 m by relating video and field data such as soil EC, plant height, and percent bare area these researchers determined the interrelations between plant, soil salinity, and spectral observations. Wiegand et al. (1994) analyzed and mapped the response of cotton to soil salinity using color-infrared photographs and videography.

Airborne sensors, such as the DMSI systems developed by SpecTerra Services of Perth, Western Australia, operating in the VIS_NIR regions of the spectrum and variable spatial resolution as a function of the flying height (0.25–2 m) have shown potential for qualitative monitoring and detection of vegetation degradation in agricultural landscapes (Metternicht and Zinck, 2003).

Bell et al. (2001) have used Small Perturbation Model (SPM), Physical Optics, and Dubois Dielectric Retrieval models for mapping soil salinity in the Alligator river region of the Northern Territory, Australia using C-(5.33 GHz), L-(1.25 GHz), and P-(440 MHz) bands airborne SAR data.

8.7.1.1.1 Hyperspectral Sensor Data

The hyperspectral measurements made in several narrow and contiguous spectral bands offer immense potential in delineation of salt-affected soils (Taylor et al., 1994; Ben-Dor et al., 2002; Metternicht and Zinck, 2003; Farifteh et al., 2007a). Taylor et al. (1994) were the first to demonstrate the potential of airborne hyper spectral (Geoscan) data covering VIS-NIR-SWIR regions to map salinity in soils at pyramid Hill, Victoria, Australia. The hyperspectral images facilitated the differentiation of salt-affected soils based on the mapping of halophytic vegetation as an indirect indicator (Honey, 1989; Derriman and Agar, 1990). Using the HyMap (Cocks et al., 1998), airborne sensor that acquires images over the spectral range of 450–2500 nm in 128 bands, Dehaan and Taylor (2002, 2003) mapped saline areas with salt scalds, halophytic vegetation, and soils with different salinity degrees and types, at the pyramid Hill test site (Australia). Likewise, Ben-Dor et al. (2002) report promising results in using the hyperspectral DAIS-7915 sensor, equipped with 70 bands across the VIS-NIR-SWIR spectral regions to derive quantitative information on soil salinity.

8.7.1.1.2 Microwave Sensor Data

Microwave-based salinity studies are essentially based on the differential behavior of the real and imaginary parts of the dielectric constant to detect salinity (Sreenivas et al., 1995; Shao et al., 2002). While the real part is insensitive to the presence of salts but strongly responds to soil moisture, the imagery part is highly sensitive to variations in soil EC, which varies according to the salt content of the soil. Microwave sensing of salt-affected soils was undertaken in an experimental phase by the AirSAR-TOPSAR sensors of the JPL-NASA. Taylor et al. (1996), Schmullius and Evans (1997), Shao et al. (2002), and De Valle et al. (2009) highlight the adequacy of the L-band, and that of the C-band and P-band to a lesser extent, for detecting salinity features such as saline seeps, surface soil features, and halophytic vegetation associated with saline landscapes. Based on the complex dielectric properties of saline soils, Shao et al. (2002) found L-band and P-band SAR effective in soil salinity detection. This finding enabled Taylor et al. (1996) to map dielectric soil properties at various depths using multifrequency, quad-polarized airborne SAR data. These dielectric maps accurately delineated the extent of surface manifestations of soil salinity at the Pyramid Hill test site (Australia) and the paths of buried saline palaeochannels that are abundant in the region.

8.7.2 Orbital Sensor Data

The multispectral measurements made in the optical, thermal, and microwave regions of the spectrum have been used to detect and map soil salinity and/or alkalinity employing either monoscopic visual interpretation of satellite imageries or computer-aided digital analysis. The broad framework for both the approaches is discussed hereunder.

8.7.2.1 Multispectral Visible, NIR, and Thermal IR Sensor Data

8.7.2.1.1 Visual Interpretation

Various steps involved in deriving information on salt-affected soils through visual interpretation of hard copy imageries are discussed below.

8.7.2.1.1.1 Data Selection Since the delineation of salt-affected soils lends heavily on surface manifestation of soil salinity and/or alkalinity in the form of salt efflorescence/encrustation, satellite images acquired during the season of the year when salt encrustation is maximum need to be used for mapping salt-affected soils. It coincides with winter/summer season. Furthermore, in general, salt-affected soils occur as parcels of barren lands/pockets of standing crop with very poor vigor and density. In order to delineate salt-affected soils satellite images acquired during the maximum vegetative growth are usually preferred. The kind of satellite data to be used depends on the intended scale of mapping (e.g., regional scale or local level or detailed level). Another factor in satellite data selection for mapping salt-affected soils is the spatial resolution of satellite image that has direct bearing on scale of mapping. For example, Landsat-8 Operational Linear Imager (OLI) data with 30 m spatial resolution or Resourcesat-2 Linear Imaging Self-Scanning Sensor (LISS-III) data with 24 m spatial resolution is suited well for reconnaissance level of mapping (1:250,000 scale). or preparing salt-affected soil maps at larger scale, i.e., 1:50,000 scale (e.g., 1:5,000 scale or larger Resourcesat-2 LISS–IV data with 5.8 m spatial resolution or IKONOS-multispectral data with 4 m spatial resolution, Cartosat-2 data with 1.6 m multispectral and 30 cm PAN and Worldview-4 data with 1.24 m in multispectral mode and 31 cm in PAN mode suit well).

8.7.2.1.1.2 Data Preparation In the context of image interpretation/analysis, data preparation refers to restoring the radiometric or geometric fidelity of the data, culling out area of interest (AOI) or digital image fusion in the event of using two datasets with varying spatial resolutions. For one-time mapping of salt-affected soils or single date satellite image geo-referencing is carried out taking standard topographic map as base. In case of geo-referencing of multi-temporal data taking geo-referenced image of t_1 other dates images are co-registered using image-image routines available in image analysis system.

8.7.2.1.1.3 Geo-referencing To begin with, the spaceborne multispectral data of the AOI is digitally co-registered to standard topographic scale compatible with the scale of mapping on a digital image analysis system, and is resampled to the same pixel size as of satellite digital image using nearest neighbor/cubic convolution sampling algorithm using map-to-image registration module. In case the objective is to monitor the changes in the extent and magnitude of soil salinity and/or alkalinity, the t_1 image is first co-registered to standard topographic map using map-to-image routine of the image analysis system. In either case the positional error is maintained at better than 0.5 pixel.

8.7.2.1.1.4 Radiometric Normalization Radiometric normalization is required when satellite images have been acquired at different point of time. The purpose of radiometric normalization is to make the multi-temporal images radiometrically compatible. This is done to ensure that whatever changes an image analyst is indicating should accrue from the changes in the spatial extent of salt-affected soils only. Furthermore, it is also done to bring down the variations in spectral reflectance due to sun elevation differences and radiometric gain settings. Initially, the digital numbers (DNs) values are converted to radiance values using the gain settings and saturation radiance values provided in the satellite data header. Later, corrections for sun elevation angle variations using cos θ correction (where θ is the sun elevation angle) are made. Subsequently all the radiance values are scaled using a common linear scaling factor.

$$L\lambda = LMIN\lambda + \left(\frac{LMAX\lambda}{QCALMAX} \right) QCAL \tag{8.1}$$

where

QCAL = calibrated and quantified scaled radiance in units of DNs
LMINλ = spectral radiance at QCAL =0
LMAXλ = spectral radiance at QCAL = QCALMAX
QCALMAX = range of rescaled radiance in DN
Lλ = spectral radiance

8.7.2.1.1.5 Preliminary Visual Interpretation Based on image elements, namely tone, color, texture, pattern, shape, size, shadow, and association, tentative land cover features, namely crop lands, forests, scrubs, settlements, water bodies, and waste lands are delineated on hard copies of the false color composite image. Alternatively, if image analysis software is available the same can be accomplished by displaying the digital image onto the screen. This is called heads-up or on-screen visual interpretation.

The manifestation of salt-affected soils on different sensors data and during different cropping season varies significantly. Figures 8.8–8.11 attest the fact.

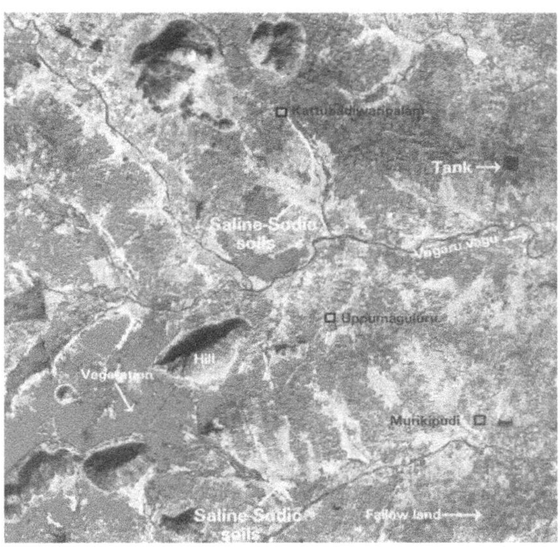

FIGURE 8.8
Salt-affected soils in black soils (Vertisols) as seen in IRS-1C LISS-III image with 24 m spatial resolution in part of Guntur district, Andhra Pradesh, southern India. Here salt-affected soils are confined to the stream beds. The source rock for sodium bearing mineral (plagioclase feldspar) that impart salinity and/or sodicity to the soils are located in the upper slope. After weathering of rock the mineral is released and is carried away by fluvial activities (Courtesy: National Remote Sensing Centre, Indian Space Research Organization, Department of Space, Government of India).

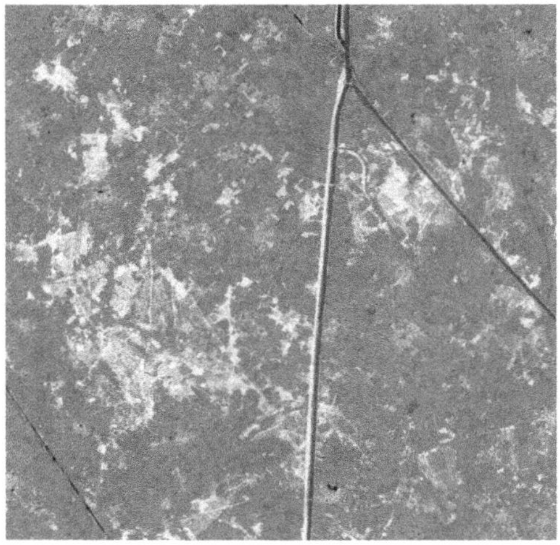

FIGURE 8.9
Resourcesat-2 LISS-IV image over an alluvial plain, part of Etah district, Uttar Pradesh, northern India. Salt-affected soils could be seen in different shades white color. The light reddish brown color represents salt-affected soils under different stages of reclamation (Courtesy: National remote Sensing Centre, Indian Space Research Organization, Department of Space, Government of India).

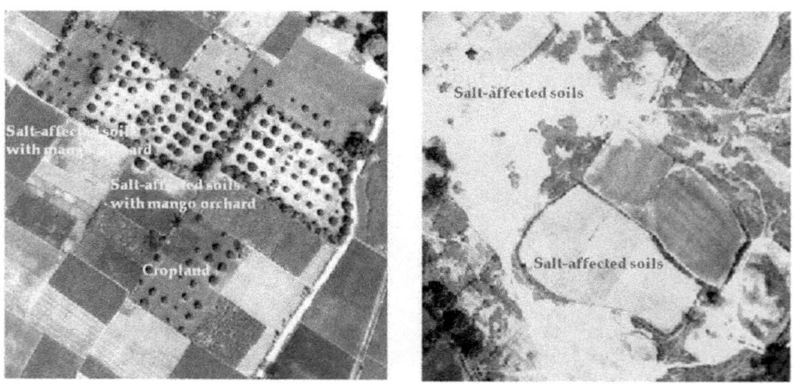

FIGURE 8.10
Salt-affected soils developed on the Indo-Gangetic alluvium as seen in IKONOS-2 image in part of Sitapur district, Uttar Pradesh, northern India (after Dwivedi, 2008). By virtue of higher spatial resolution, even individual mango tree is also seen.

8.7.2.1.1.6 Ground truth collection After broad delineation of salt-affected soils, sample areas for field verification are identified and located onto topographical maps. During the field visit a reconnaissance traverse of the terrain was made to finalize the sample points based on accessibility and variation in salt-affected soils. Subsequently, in each sample area, sample points are precisely located with the help of topographic maps. Additionally, terrain features like land use/land cover, surface drainage, irrigation sources, depth of ground water, etc. are also recorded. Soil profiles are excavated at representative sites and after recording morphological features soil samples are collected for chemical analysis in the laboratory. Besides, auger bores and surface samples are studied to account for variations within the unit, if any.

FIGURE 8.11
QuickBird image of Northeast Thailand. The letter 'S' indicates saline soils.

Soil samples collected from the field are analyzed in the laboratory following the standard procedure. Alternatively, researchers have used spectroscopy for estimation of certain parameters related to soil salinity and/or alkalinity. Goldshlegera et al. (2013) explored the possibility of using spectroradiometer operating in the 400–2,500 nm region to quantitatively assess foliar Cl^- and Na^+ concentration as indicators to assess salinity in tomato plants. The relationship between the Na and Cl contents of tomato plants growing in various saline environments and soil spectral reflectance was determined using partial least squares regression. The results indicate that Cl-content model was more accurate for determining leaf salinity ($R^2 = 0.92$, root mean square error of prediction (RMSEP) = 0.2%) than the Na-content model ($R^2 = 0.87$, RMSEP = 0.6%). They concluded that reflectance spectroscopy is potentially useful for characterizing the key properties of salinity in growing vegetation and assessing its salt quality.

8.7.2.1.1.7 Post-field Interpretation Based on chemical analysis data and morphological characteristics, salt-affected soils are classified according to norms for PH, EC, and ESP given in Table 8.3. The salt-affected soil boundaries drawn earlier are modified, wherever needed, and were transferred onto base maps prepared from standard topographical maps of matching scale.

TABLE 8.3
Keys to Degree of Soil Salinity and/or Alkalinity

	Salinity		Alkalinity
	ECe (dS m^{-1})	pH (1:2.5)	ESP
Slight	4–8	8.5–9.0	<15
Moderate	8–30	9.0–9.8	15–40
Strong	>30	>9.8	>40

Source: National Remote Sensing Agency (2008).

8.7.2.1.1.8 Error Assessment For quantitative estimates of the classification error of salt-affected soil maps generated from sample areas representing different types of salt-affected soils are selected randomly (Congalton et al., 1983). An adequate number of sample points representing different types of salt-affected soils are identified on the maps for accuracy estimation. A one-to-one comparison of the categories of salt-affected soils mapped and ground-truth data and the information available in the published report is made. An accuracy estimation in terms of overall accuracy, errors of omission, and errors of commission, and Kappa coefficient (2) is subsequently made after generating confusion matrix. The Kappa coefficient (K) is computed as follows (Bishop et al., 1975):

$$K = \frac{N\sum_{i=1}^{r} xii - \sum_{i=1}^{r}(xi+)(x+i)}{N^2 - \sum_{i=1}^{r}(xi+)(x+i)} \tag{8.2}$$

where r is the number of rows in the matrix; xii is the number of observations in row i and column i (the ith diagonal elements); xi+ and x+i are the marginal totals of row r and column i, respectively; and N is the number of observations.

8.7.2.2 Computer-Assisted Digital Analysis

The computer-aided digital analysis is based on the premise that each land cover feature exhibits a characteristic spectral response pattern. For mapping salt-affected soils an attempt is made to correlate the spectral response pattern observed in the digital image with the features on the ground. With respect to selection, preprocessing of satellite data, the procedures mentioned in visual interpretation are followed. However, with digital multispectral data as input in computer-aided digital analysis approach, the analyst is provided with various options for image enhancements for improving the image contrast or enhancing certain features of interest. This also provides an opportunity to use several band combinations for delineation of current features.

With respect to the study of dynamics of salt-affected soils in terms of spatial extent and magnitude of the problem, multi-temporal data are used. For objective assessment of changes that have taken place over a given period appropriate procedures for radiometric normalization, as mentioned in section visual interpretation are employed.

8.7.2.2.1 Preliminary Digital Analysis

In order to study the spectral variation among different cover types of the area and to delineate broadly salt-affected soils, the digital data is displayed onto color monitor of the image analysis system. Sample areas to be studied extensively in the field are identified and digital clipping of sample areas are taken to the field for establishing the relationship between the spectral response pattern and the nature of salt-affected soils. Furthermore, sample areas are also located onto topographic maps for locating them on the ground. Generally, clustering algorithms are used to delineate a spectrally homogenous areas. Subsequently based on available ancillary like topographic maps, published soil maps and an attempt is made to relate spectral classes with the broad cover types, namely salt-affected soils, crop land, forest, water bodies, etc.

8.7.2.2.2 Ground Truth Collection

For ground truth collection the procedure enumerated under *visual interpretation* section is followed.

8.7.2.2.3 Finalization of the Map

The group/cluster of homogenous pixels representing various categories of salt-affected soils and other cover types were located after carefully studying the spectral variation in the digital data. The spectral response pattern in terms of mean, standard deviation, correlation coefficient, and co-variance matrix are generated. Entire image is analyzed and each pixel in the image is assigned to either normal soils or into different categories of salt-affected soils. Subsequently error assessment is carried out. After assessing the performance of thematic map, training areas were refined, wherever necessary, and the classification process is repeated. A sample salt-affected soil map, thus prepared, is appended as Figure 8.12.

8.7.2.2.4 Error assessment

For error assessment the procedure outlined under visual interpretation section is followed.

8.7.3 State-of-the-Art

Though the spaceborne multispectral data became available with the launch of the first Landsat satellite in July 1972, mapping of salt-affected soils was already attempted earlier using Apollo-9 (Wiegand et al., 1971) and Skylab (Everitt et al., 1977) data. Some of the important studies on delineation and mapping of salt-affected soils carried out by various researchers are reviewed hereunder:

In India, the Landsat-MSS data were used to delineate salt-affected soils through computer-assisted digital analysis approach (Singh et al., 1977). Subsequently, Landsat-MSS (Singh and Dwivedi, 1989), Landsat-TM (Metterricht and Zinck, 1997), SPOT-MLA

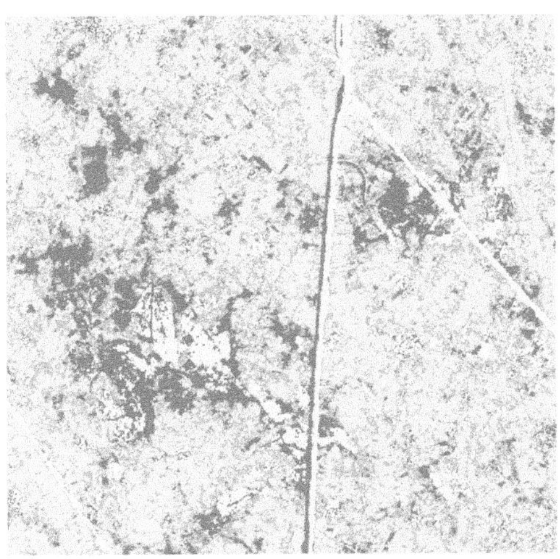

FIGURE 8.12
Salt-affected soil map derived from Resourcesat-2 LISS-IV digital data of March 7, 2017 using ISODATA classifier. Beige color denotes severely salt-affected soils, magenta moderately salt-affected soils, yellow crop with very good vigor and green crop with moderate vigor.

(Leonardo et al., 1996), and the Indian Remote Sensing Satellite (IRS-1A/1B, Linear Imaging Self-Scanning Sensor (LISS-I and II) data (Dwivedi and Venkataratnam, 1992; Sharma et al., 2000) were used for mapping salt-affected soils. Landsat TM and SPOT multispectral data were used by Ahmed and Andraianasolo (1997) to map salinity at semi-detailed level.

Mougenot et al. (1993), however, was of the view that the possibility for broadband multispectral satellite images (e.g., Landsat, SPOT) to detect salt-affected areas becomes limited when salt content falls below 10%–15%. Utilizing visual interpretation approach the salt-affected soils of India have been mapped at 1:250,000 scale from Landsat TM standard false color composite (FCC) images (National Remote Sensing Agency, 2008).

8.7.3.1 Temporal Behavior of Salt-Affected Soils

The temporal behavior refers to change detection in respect of extent and magnitude of soil salinity and/or alkalinity. Change analysis is in essence a spatial comparison of two or more land covers of the same geographic area produced from remotely sensed data that are recorded at different times. The basic premise in using remotely sensed data for change detection is that changes in the objects/features of interest will result in changes in reflectance values or local textures that are separable from changes caused by other factors such as differences in atmospheric conditions, illumination and viewing angles, and soil moistures (Deer, 1999). Several change detection techniques, namely PCA transformation, multi-temporal analysis, spectral mixture model, and post-classification comparison, have been developed and employed for change detection (e.g., Singh, 1989; Lu et al., 2003).

Traditionally, three methods (i) comparing the map content with the real world on location, (ii) direct comparison of the map database and an up-to-date orthophoto/(s)/orthoimage, and (iii) using two neighboring photos/images from the flight campaign and generating a stereo model have been used for manual change detection. Multi-temporal orbital sensor data acquired at two points of time during similar cropping season or similar phonological condition of vegetation to ensure that the changes that are documented are exclusively from changes the soil salinity and/or alkalinity. Variations in the manifestation salt-affected soils during different cropping seasons, representing intra-annual variations in spectral response pattern and long-term changes are portrayed in Figures 8.13 and 8.14, respectively. It is quite clear from Figure 8.13 that the contrast of salt-affected soils with the background (*rabi* crop seen in red color of different hues) is optimal in the month of March indicating thereby the ideal period satellite date acquisition for delineation of salt-affected soils. The long-term spatial changes in salt-affected sols are portrayed in Figure 8.14. With the passage of time due to reclamative methods employed by the farmers, the spatial extent of salt-affected soils has sown a considerable shrinkage.

Figure 8.15 shows the impact of rehabilitation program of salt-affected soils taken up by the farmers. Due to reclamative efforts the area under salt-affected soils which was 29,121 ha in 1975 has shrunken to 12,872 ha in 1992 with an attendant increase in cropland to the tune of 17,808 ha (National Remote Sensing Agency, 1995). In another study, the temporal behavior of salt-affected soils in Periyar–Vaigai irrigation command, Tamil Nadu state, southern India was studied through a systematic visual interpretation of temporal Landsat images (Figure 8.16). Variations in different categories of salt-affected soils was observed during the period 1986, 1990, and 1995.

In yet another study that was taken up in black soils (Vertisols) region, part of Andhra Pradesh, southern India using Landsat MSS data for 1973 and the Indian Remote Sensing Satellite (IRS-1A) LISS-I data with 72.5 m spatial resolution of 2001 (Figure 8.17). An increase to the tune of 4,188 hectares in salt-affected soils was observed during the intervening

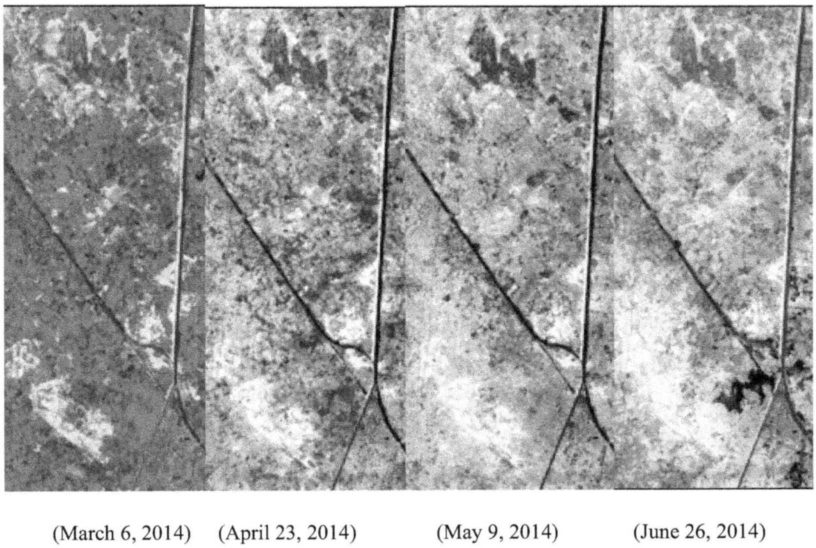

FIGURE 8.13
An illustration of intra-annual variations in spectral response patterns of salt-affected soil in part of Etah district, Uttar Pradesh, northern India. Note the manifestation of a pocket of salt-affected soils in the lowerleft of image acquired on March 6, 2014 seen as white color (adjacent to canal-linear feature in blue color) on three other dates (April 23, May 9, and June 26, 2014). During the month of March when *rabi* (winter season) crop is in its maximum vegetative growth stage it provides very good image contrast that helps in improved delineation of these soils.

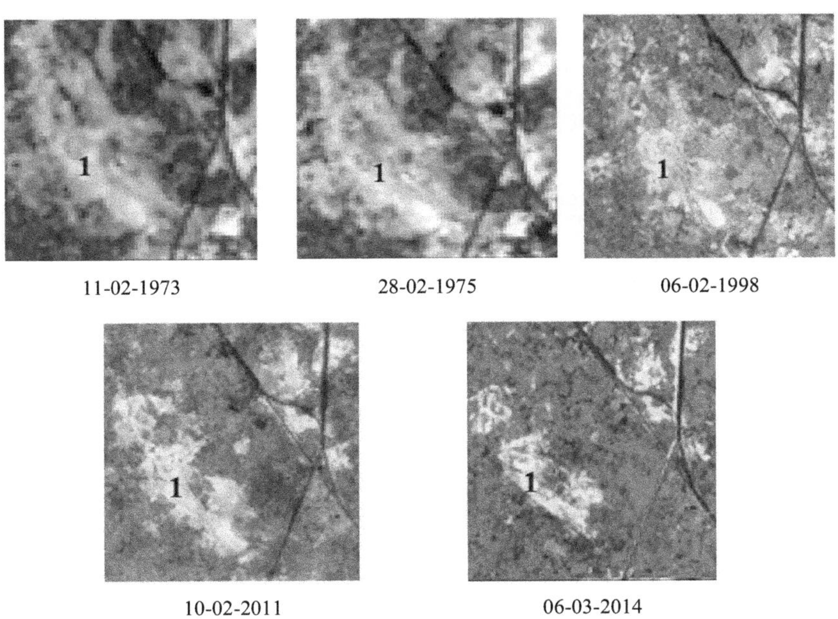

FIGURE 8.14
Temporal behavior of salt-affected soil as seen in Landsat images for 1973, 1975, 1998, 2011, and 2014. The numeral '1' indicates salt-affected soils. The red color background shows standing winter crop and the linear features are irrigation canals. As evident from the unclassified images (raw images) of different years there has been substantial shrinkage in the spatial extent of salt-affected soils during 41 years period.

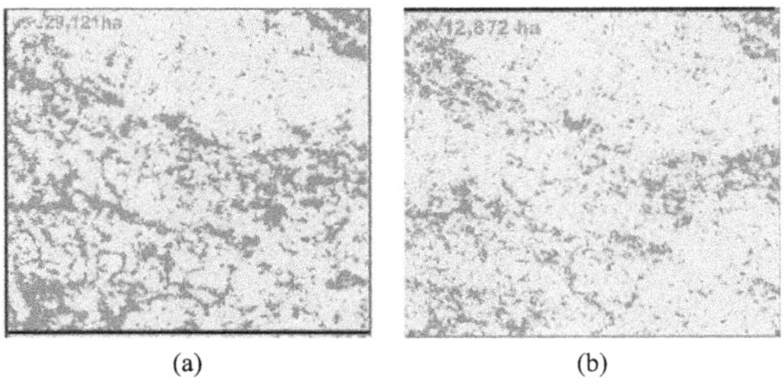

FIGURE 8.15
Temporal behaviour of salt-affected soils in part of Jaunpur and Varanasi distiricts of Uttar Pradesh, northern India as seen in thematic maps derived from Landsat MSS data of March, 1975 (a) and Landsat TM data of March, 1992 (b). Yellow colour indicates cropland and purple colour salt-affectes soils.

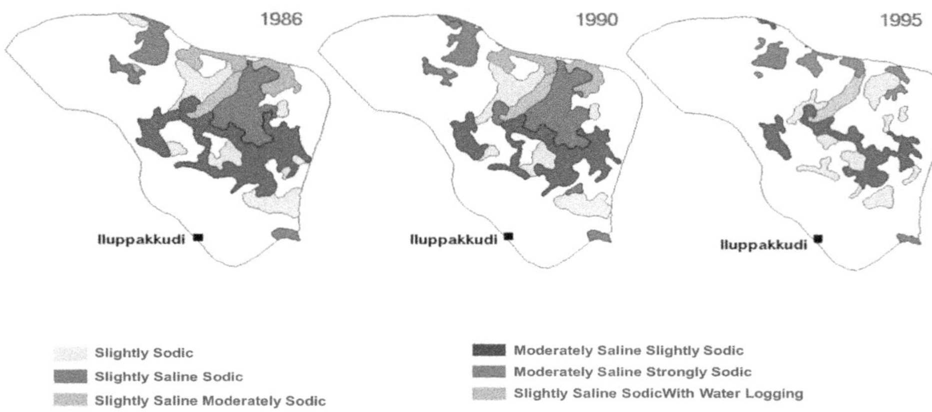

FIGURE 8.16
Spatio-temporal behavior of salt-affected soils in Periyar–Vaigai command area, part of Tamil Nadu, southern India.

28-year period. The increase in the area under salt-affected soils has been attributed to the increased in the salt regime in the surface layer of soils owing to changing hydrological regime due irrigation with canal water.

Apart from using satellite data with improved spatial, spectral, and radiometric resolutions attempts have been made to develop newer techniques for improving the mapping accuracy or to derive additional information on sat-affected soils. For mapping salt-affected soils in the Indo-Gangetic alluvial plains Dwivedi and Rao (1992) identified a three-band combination from Landsat-TM data, namely bands –1 (0.45–0.52 μm), –3 (0.63–0.69 μm), and –5 (1.55–1.75 μm). In South America case study Metternicht (2001)mapped salinity distribution using an approach that integrates multi-temporal classification of remotely sensed data, physical and chemical soil properties, and landform attributes. Three rule-based expert systems using fuzzy sets and fuzzy linguistic rules to formalize the expert knowledge about the actual possibility of changes to occur were designed and implemented within a geographical information system (GIS).

To improve mapping accuracy, Metternicht and Zinck (1997), Bell et al. (2001), and Castaneda and Herrero (2009) have tried approaches focusing on the fusion of data from different parts

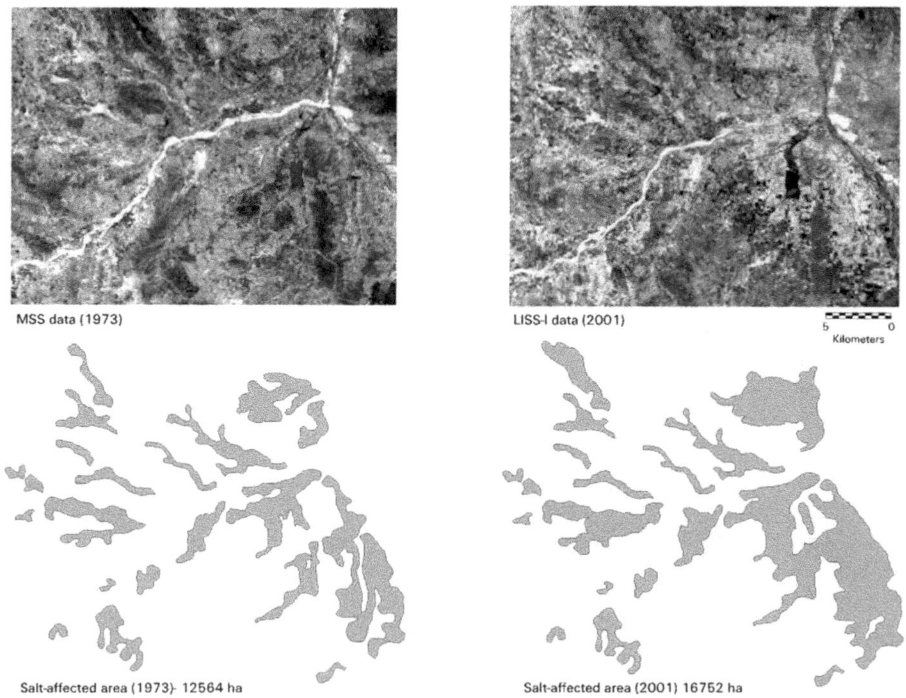

FIGURE 8.17
Monitoring salt-affected soils around Kanekallu village, part of Anantapur district, Andhra Pradesh.

of the spectrum (e.g., VIS-NIR and microwave), whereas Lobell et al. (2007) propose the integration of multi-sensor satellite and multiyear yield data for mapping salinity at subsurface level. The digital image fusion technique (intensity-hue-saturation image transforms) was used to merge LISS-III digital data with 23.5 m spatial resolution and Panchromatic data with 5.8 m resolutions to map salt-affected areas in the Indo-Gangetic plains of northern India with more than 85% accuracy (Dwivedi et al., 2001). In another study, Dwivedi et al. (2008) evaluated the relative performance of IKONOS multispectral imagery (4 m spatial resolution) and IRS-ID LISS-III image (23.5 m spatial resolution) and PAN image with 5.8 m spatial resolution) -merged data for mapping salt-affected soils. Using different image transformation and classification techniques, an overall accuracy of 92.4% was obtained when using IKONOS data against an overall accuracy of 78.4% and 84.3% achieved when using LISS-III multispectral data or a fusion of LISS-III panchromatic and multispectral data.

Brunner et al. (2004) derived uncalibrated salinity maps from atmospherically corrected multispectral ASTER images In a recent study on the Aibi Lake in Xinjiang, China, Shuya et al. (2005) identified saline areas dominated by sodium sulfates by using Terra ASTER imagery and the grey system theory.

Thermal sensors can extract the soil and salt emissivity property, which holds better spectral variations than the reflected spectral region. Verma et al. (1994) demonstrated that the incorporation of the thermal range (e.g., Landsat TM band 6) to the VIS-NIR channels helped overcome issues of spectral similarity of saline soils. Emittance can provide sub-surface information that reflected radiation cannot. This can be further interpreted for the root zone, rather than the upper soil crust. Overall mineral bands are distinguishable with exception of chloride minerals that exhibit relatively flat infrared emissivity spectra (Lane, 2002).

Landsat 7 ETM+ data covering upper Colorado River drainage were digitally analyzed for delineation of gypsic and natric soil areas (Nield et al.,2007). Gypsic soil areas were mapped using the normalized difference ratio of Bands 5 and 7 whereas natric soil areas using the normalized difference ratio of Bands 5 and 4. While gypsic soil areas could be mapped with reasonably good accuracy (82%), the accuracy of delineation of nitric soils areas was poor. While working with Landsat TM data for detection of salt-affected soils, Zhang et al. (2007) concluded that the neural network is superior to the maximum likelihood method for detailed mapping of the Taibei Salt Field where salty water bodies are spectrally uniform and spatially extensive on the image with clear-cut boundaries among them.

While working in southeastern Oregon, USA, Elnaggar and Noller (2010) had observed that the maximum-likelihood supervised classification of the Landsat images yields two levels of soil salinity: non-saline soils (EC < 4 dSm^{-1}) with an accuracy of 97%, and saline soils (EC > 4 dSm^{-1}) with 60% accuracy. Addition of decision-tree analysis (DTA) resulted in successful mapping of five classes of soil salinity and an overall accuracy of about 99%. In another study, Lobell et al. (2010) used 7-year MODIS enhanced vegetation index (EVI) and NDVI for a regional-level inventory of salt-affected soils in Red River Valley, North Dakota and Minnesota, USA, and observed that the average of the MODIS enhanced vegetation index (EVI) for a 7-year period exhibited a strong relationship with soil salinity, and outperformed the NDVI.

Various spectral indices, namely BI (Brightness Index), NDSI (Normalized Difference Salinity Index), SI (Salinity Index), ASI [(Aster Salinity Index (Agriculture)], Index of Salinity (using GIS Geographic Information System and remote sensing), and the SSSI "Soil Salinity and Sodicity Index" derived from the Landsat ETM+ (Enhanced Thematic Mapper) were utilized for mapping salt-affected soils (Dehni and Lounis, 2012).

8.7.3.2 Spaceborne Microwave Sensor Data

Apart from optical sensor data, microwave data have also been used to delineate salt-affected soils. Applying a fuzzy classification approach on JERS-1 data, Metternicht (1998b) distinguished saline from non-saline surfaces using microtopographic variations of the terrain as indirect indicator of salinity occurrences. As mentioned in Section 8.5, salt-affected soils are generally characterized by the absence of vegetation and the presence of salt encrustation which lends these soils with smooth surface. Taking advantage of the differences in surface roughness of salt-affected soils and their counterpart, as revealed in ERS C-band image, Castaneda and Herrero (2009) could delineate the saline wetlands in Spain Furthermore, Yun et al. (2003) reported a correlation of 0.69 between the backscattering coefficients of a Radarsat-1 SAR image (C-HH) and the έ component of the dielectric constant measured on saline soil samples, and concluded that the C-band HH polarization SAR images can be useful for monitoring soil salinity. An innovative study by Aly et al. (2007) proposes to detect soil salinity from series of polynomial regressions of the έ computed on saline soil samples, simulation models, and radarsat-1 SAR images acquired in the standard beam (Si) modes (i = 1, 3, 5). Their parametric model infers salinity (έ) from the measurements of the Si modes of Radarsat-1 SAR without the need to apply backscattering models.

The use of radar images for the detection of salt-affected soils is a promising domain of remote sensing (Taylor et al., 1996; Metternicht, 1998a; Aly et al., 2004). The basic principle rests on the relationship among the amount of salt in the soil, soil moisture content, and the dielectric property of this mixture. Advances in the field of L-band microwave remote sensing have stimulated the development of new techniques with predictive mapping capabilities (Sreenivas et al., 1995).

8.7.3.3 Spaceborne Hyperspectral Sensor Data

Currently there are two experimental sensors, namely Hyperion (30 m spatial resolution and 242 bands covering the 400–2,500 nm spectral range) on board the Earth Observing-1 (EO-1) (NASA) and CHRIS (19 bands at a spatial resolution of 18 m or 63 bands at a spatial resolution of 36 m, within the 410–1,050 nm range) on board the PROBA platform (ESA). Dutkiewicz (2006) evaluated the performance of the Hyperion imagery for mapping surface symptoms of dryland salinity using a partial unmixing technique called mixture-tuned matched filtering. A kappa accuracy of 0.50 for the mapping of saltpans, small depressions with extreme soil salinity, and 0.38 for the mapping of samphire vegetation (*Crithmum maritimum*) that grows in areas of high to very high salinity was achieved. The study concluded that hyperspectral satellite imagery is unable to map accurately salinity indicators related to slight or moderate soil salinity levels. Interestingly, hyperspectral airborne instrument (e.g., CASI and HyMap) allow mapping saline soils with more accuracy than hyperspectral satellite instruments do (e.g., Hyperion). The performance of the CHRIS-PROBA sensor for cartography of salt-affected areas is still poorly documented and limited so far to soil degradation mapping in semiarid wetlands (Schmid et al., 2007). A mosaic of the hyperspectral data over part of Australia is appended as Figure 8.18.

The future hyperspectral earth observation mission include a Hyperspectral Imager EnMAP, a German earth observation mission, of push broom type with two separate spectral channels: one for VNIR range from 420 to 1,000 nm, and the other for the SWIR range from 900 to 2,450 nm with a swath width of 30 km. It is slated for launch in 2020. The mission will have a sun-synchronous orbit and a +/−30° off-nadir pointing feature, 4-day revisit capability and a Ground Sampling Distance (GSD) of 30 m (www.enmap.org/?q=sensor). The Canadian space agency's Hyperspectral Environment and Resource Observer (HERO), is another hyperspectral sensor with more than 200 bands in the range of 400–2,500 nm with a temporal resolution of 7 days and a spatial resolution of 30 m (Hollinger et al., 2006).

FIGURE 8.18
Airborne hyperspectral image (HyMap) of saline soils in Toolbin lake in Western Australia.

8.8 Solute Transport Modeling

In arid and semiarid regions, many soils and often groundwater are potentially saline. Any disturbance in the delicate hydrological balance may lead to the mobilization of the inheritably available salt. Information on vertical distribution of salute (salt) is of paramount importance from soil profile salt regime management point of view. The salt transport mechanism is often highly complex, the understanding of which necessitates the use of computer modeling, preferably in combination with field studies and geostatistics (Jolly, 1998; Pishkar, 2003). Solute transport modeling technique enables predicting the salt distribution in the subsoil. It has the advantage of providing subsoil information on dynamics of the salt movement regimes. This technique provides complementary data on salt movement in the soil profile which can be used in combination with remote sensing data (Farifteh et al., 2006) (Figure 8.19).

Mathematical models of 1D, 2D, or 3D type, or of unsaturated–saturated flow, and/or of transport of a particular solute are often used to simulate water flow and solute transport processes and to determine or estimate the average concentration of the salt in soil profile (Van Dam et al., 1997). For instance, Jury (1982) used a transfer function model to predict average values of solute concentration as a function of depth and time, through highly variable field systems. Following an integrated approach, El-Kadi et al. (1994) utilized a numerical model in a GIS environment to study the solute transport. Besides, using a stochastic–convective model (Butters and Jury, 1989), and a multiple linear regression model with four soil salinization factors: soil permeability, depth to the groundwater, groundwater quality, and leaching fraction as input, Corwin et al. (1989) in conjunction with GIS successfully estimated the solute transport and its leading edge migration.

8.9 Conclusion

As evident from the foregoing, remote sensing offers tremendous potential in deriving information on salt-affected soils that is very useful in taking up curative and preventive

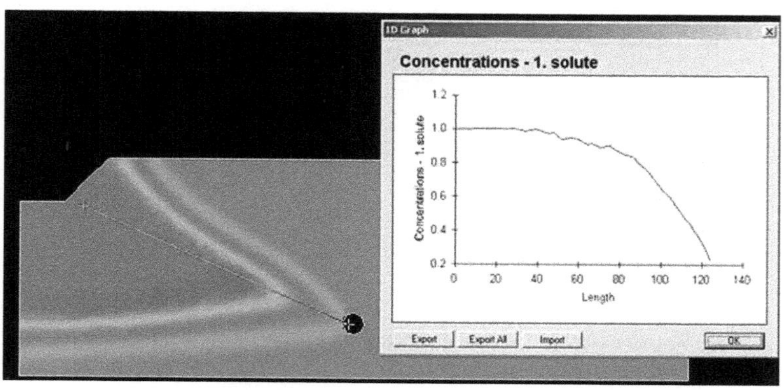

FIGURE 8.19
Illustrating the estimation of solute concentrations in sub-surface (Source: Hydrus (2017), reproduced with permission).

measures. In practice, most of these studies using remote sensing data have focused mainly on severely salt-affected soils with surface manifestation of soil salinity and/or alkalinity in the form of salt crust/salt efflorescence. However, less attention has been paid to the detection of slightly or moderately affected soils having very little or no surface manifestation. These soils with ample salt reserves in the sub-surface could not be detected due to limitations associated with the nature of available spaceborne images, which do not allow extracting information from the third dimension of the soil body, e.g., where salts concentrate in subsoil. In order to address the identification of slightly and moderately salt-affected soils and incipient salinity geophysical techniques, namely EMI, GPR, and solute transport modeling have been attempted on an experimental basis and have been found quite promising.

For deriving reliable information on sub-surface salinity and slight to moderate category of salt-affected soils the information on surface topography (since salt accretion takes place in local depressions and in lower elements of slopes), lithology (providing information on kind of minerals present), and hydrology need to be integrated with the remote sensing observations, geophysical measurements and solute transport models in a GIS environment tin different terrain and agro-climatic conditions. Further studies on the potential of hyperspectral remote sensing may be taken up to identify specific absorption spectral band/(s) to identify a particular type of salts and its abundance. Similarly, extensive laboratory and field studies are required to develop a dedicated sensor specifically for detection of soil salinity and/or alkalinity.

It is amply clear from the foregoing that any feature or phenomenon unless otherwise exhibits some surface manifestation, its detection using remote sensing seems very limited. Taking further on, Farifteh et al. (2006) postulated that remote sensing-based thematic mapping requires enables identifying three types of variables, namely the measurable, the retrievable, and the hidden ones. The quantum of thematic information derived from remote sensing depends on relative proportions of these variables. For example, in an extreme case, if the behavior of the system under study is entirely explained by hidden variables, then there is very little scope to derive useful information using remote sensing techniques.

References

Ahmed, I., and Andraianasolo, H., 1997. Comparative of multisensory data suitability in study of the soil salinity using remote sensing and GIS in the Fordwah irrigation division, Pakistan. In: *Proceedings of the Geosciences and Remote Sensing symposium*, Toronto, Canada. 4, 1627–1629.

Alhammadi, M.S., and Glenn, E.P., 2008. Detecting date palm trees health and vegetation greenness change on the Eastern Coast of the United Arab Emirates Using SAVI. *International Journal of Remote Sensing*, 29(6):1745–1765.

Allen, D., 2004. A review of geophysical instruments applied to ground water and soil investigation. A report to the ANVIS/sustainable irrigation travel fellowship. Available at http://downloads.lwa2.com/downloads/final_reports/Davis/Allen_Felloship_report_04–05.pdf (accessed Nov 6, 2007).

Aly, Z., Bonn, F.J. and Magagi, R., 2007. Analysis of the backscattering coefficient of salt-affected soils using modeling and RADARSAT-1 SAR data. *IEEE Transactions on Geoscience and Remote Sensing* 45(2):332–341.

Bell, D., Menges, C., Ahmad, W., and van Zyl, J.J., 2001. The application of dielectric retrieval algorithms for mapping soil salinity in a tropical coastal environment using polarimetric SAR. *Remote Sensing of Environment*, 75:375–384.

Ben-Dor, E., Irons, J. R., and Epema, G., 1999. Soil reflectance studies in manual of remote sensing. In: Rencz, A.N., (ed.) *Remote Sensing for the Earth Sciences, Manual of Remote Sensing, Vol. 3*, 3rd edn. Wiley, New York, pp. 111–189.

Ben-Dor, E., Patkin, K., Banin, A., and Karnieli, A., 2002. Mapping of several soil properties using DAIS7915 hyperspectral scanner data. A case study over clayey soils in Israel. *International Journal of Remote Sensing*, 23:1043–1062.

Bishop, Y., Fienberg, S., and Holland, P., 1975. *Discrete Multivariate Analysis-Theory and Practices.* MIT Press, Cambridge, MA, 575p.

Brunner, P., Kinzelbach, W., and Li, H., 2004. Generating large-scale soil salinity maps with geophysics and remote sensing. *Geophysical research Abstracts*, 6:04745.

Butters, G.L., Jury, W.A., 1989. Field scale transport of bromide in an unsaturated soil: 2. Dispersion modelling. *Water Resources Research*, 25:1583–1589.

Cassel, F., Adhikari, D., and Gorahoo, D., 2005. *Salinity Mapping of Fields Irrigated with Winery Effluents.* Irrigation Association, Falls Church, VA.

Castaneda, C., and Herrero, J., 2009. Application of Landsat and ERS imagery to the study of saline wetlands in semi-arid agricultural areas of northeast Spain. In: Graciela, M., and Zinck, A., (eds.) *Remote Sensing of Soil Salinization: Impact on Land Management*. CRC Press, Boca Raton, pp. 175–198.

Chaturvedi, L., Carver, K.R., Harlan, J.C., Hancock, G.D., Small, F.V., and Dalstead, K.J., 1983. Multispectral remote sensing of saline seeps. *IEEE Transactions on Geosciences and Remote Sensing*, 21(3):239–250.

Clark, R.N., and Roush, T.L., 1984. Reflectance spectroscopy: Quantitative analysis techniques for remote sensing applications. *Journal of Geophysical Research*, 89:632–634.

Clarke, F.W., 1924. The data of geochemistry. U.S. Geological Survey Bulletin 770, 841pp, Ed.5.

Cocks, T., Jensen, R., Stewart, A., Wilson, I., and Shields, T., 1998. The HypMap airborne hyperspectral sensor, the system, calibration and performance. In: *Proceedings of the 1st EARSEL Workshop on Imaging Spectrometry*, Zurich, Switzerland, pp. 37–42.

Cook, P.G., and Walker, G.R., 1992. Depth profiles of electrical conductivity from linear combinations of electromagnetic induction measurements. *Water Resources Research*, 28:2953–2961.

Corwin, D.L., Sorensen, M., Rhoades, J.D., 1989. Field-testing of models which identify soils susceptible to salinity development. *Geoderma*, 45(1):31–64.

Congalton, R., Oderwald, R., and Mead, R., 1983. A quantitative method to test for consistency and correctness in photo interpretation. *Photogrammetric Engineering & Remote Sensing*, 49(1): 69–74.

Crowley, J.K., 1991. Visible and near-infrared (0.4–2.5 μm) reflectance spectra of playa evaporite minerals. *Journal of Geochemical Research*, 96(B10):16231–16240.

Csillag, F., Pasztor, L., and Biehl, L.L., 1993. Spectral band selection for the characterization of salinity status of soils. *Remote Sensing of Environment*, 43:231–242.

Dale, P.E.R., Hulsman, K., and Chandica, A.L. 1986. Classification of reflectance on colour infrared aerial photographs and sub-tropical salt-marsh vegetation types. *International Journal of Remote Sensing* 7(12):1783–1788.

De Jong, S.M., 1994. Applications of reflective remote sensing for land degradation studies in a Mediterranean environment. *The Netherlands Geographical Studies*, 177.

de valle, H.F., Blanco, P.D., Sione, W., Rostango, C.M., and Elissalde, N.O., 2009. Assessment of salt-affected soils using multisensory radar data: A case study from Northeastern Patagonia, Argentina. In: Graciela, M., and Zinck, A., (eds.) *Remote Sensing of Soil Salinization: Impact on Land Management.* CRC Press, Boca Raton, pp. 155–174.

Dehaan, R., and Taylor, G.R., 2003. Image-derived spectral end members as indicators of salinization. *International Journal of Remote Sensing*, 24(4):775–794.

Dehaan, R., and Taylor, G.R., 2002. Field-derived spectra of salinized soils and vegetation as indicators of irrigation-induced soil salinization. *Remote Sensing of Environment*, 80:406–417.

Derriman, M., and Agar, R.A., 1990. Gold and base metal exploration in the Pelbara craton, Western Australia using the Geoscan airborne multispectral scanner. In: *Proceedings of the International Geoscience and Remote Sensing Symposium*, Washington, DC, pp. 1715–1718.

Dehni, A., and Lounis, M., 2012. Remote sensing techniques for salt affected soil mapping: Application to the Oran region of Algeria. *Procedia Engineering*, 33: 188–198.

Dutkiewicz, A., 2006. Evaluating hyperspectral imagery for mapping the surface symptoms of dry-land salinity. PhD dissertation, University of Adelaide, Australia.

Dwivedi, R.S., Ramana, K.V., Thammappa, S.S., Singh, and A.N., 2001. The utility of IRS-1C LISS-III and PAN-merged data for mapping salt-affected soils. *Photogrammetric Engineering and Remote Sensing*, 67(10):1167–1175.

Dwivedi, R.S., and Rao, B.R.M., 1992. The selection of the best possible Landsat—TM band combinations for delineating salt-affected soils. *International Journal of Remote Sensing*, 13(11):2051–2058.

Dwivedi, R.S., Ramana, K.V., and Singh, A.N., 2008. Generation of farm-level information on salt-affected soils using IKONOS-II multispectral data. In: Graciela, M., and Zinck, A., (eds.) *Remote Sensing of Soil Salinization: Impact on Land Management*. CRC Press, Boca Raton, pp. 73–89.

Dwivedi, R.S., and Venkataratnam, L., 1992. Land degradation studies using IRS-1A data. In: Karale, R.L., (ed.) *Natural Resources Management. A New Perspective*. NNRMS, Bangalore, pp. 153–169.

El-Kadi, A.I., Oloufa, A.A., Eltahan, A.A., and Malik, H.U., 1994. Use of a geographic information system in site-specific groundwater modelling. *Ground Water*, 32:617–625.

Elnaggar, A.A., and Noller, J.S., 2010. Application of remote-sensing data and decision-tree analysis to mapping salt-affected soils over large areas. *Remote Sensing*, 2:151–165.

Everitt, J.H., Gertermann, H., and Cuellar, J.Q. 1977. Distinguishing saline from non-saline range lands with Skylab. *Photogrammetric Engineering and Remote Sensing* 43:1041–1047.

Everitt, J., Escobar, D., Gerbermann, A., and Alaniz, M., 1988. Detecting saline soils with video imagery. *Photogrammetric Engineering and Remote Sensing*. 54:1283–1287.

FAO, 1988. Salt-affected soils and their management, by I.P. Abrol, J.S.P. Yadav & F.I. Massoud. FAO Soils Bulletin No. 39. Rome.

FAO, 2000. Extent and causes of salt-affected soils in participating countries. Global network on integrated soil management for sustainable use of salt-affected soils. FAO-AGL website.

Farifteh, J., Farshad, A., and George, R.J., 2006. Assessing salt-affected soils using remote sensing, solute modelling and geophysics. *Geoderma*, 130:191–206.

Farifteh, J., 2007. Imaging spectroscopy of salt-affected soils: Model-based integrated method, International Institute for Geo-information Science and Earth Observation (ITC) and Utrecht University, 215p (PhD thesis).

Farifteh, J., 2008. Model-based integrated methods for quantitative estimation of soil salinity from hyperspectral remote sensing data. In: Metternicht, G., and Zinck, J.A., (eds.) *Remote Sensing of Soil Salinization: Impact and Land Management*. CRC Press, Taylor & Francis group, New York, pp. 305–339.

Farifteh, J., 2011. Interference of salt and moisture on soil reflectance spectra. *International Journal of Remote Sensing*, 32(23): 8711–8724.

Farifteh, J., Van der Meer, F., Atzberger, C., and Carranza, E.J.M., 2007a. Quantitative analysis of salt-affected soil reflectance spectra: A comparison of two adaptive methods (PLSR and ANN). *Remote Sensing of Environment*, 110(1):59–78.

Farifteh, J., Van der Meer, F., and Carranza, E.J.M., 2007b. Similarity measures for spectral discrimination of salt-affected soils. *International journal of remote sensing*, 28(23): 5273–5293.

Farifteh, J., van der Meer, F., Van der Meijde, M., and Atzberger, C., 2008. Spectral characteristics of salt-affected soils: A laboratory experiment. *Geoderma*, 145(3–4):196–206.

Farifteh, J., Tolpekin, V., Van der Meer, F., and Sukchan, S., 2010. Salinity modelling by inverted Gaussian parameters of soil reflectance spectra. *International Journal of Remote Sensing*, 31(12): 3195–3210.

Fernandez-Buces, N., Siebe, C., Cram, S., and Palacio, J.L., 2006. Mapping soil salinity using a combined spectral response index for bare soil and vegetation: A case study in the Former Lake Texcoco, Mexico. *Journal of Arid Environments*, 65(4):644–667.

Ghassemi, F., Jakeman, A.J., and Nix, H.A., 1995. *Salinization of Land and Water Resources: Human causes, extent, Management and case studies.* The Australian National University, Canberra, Australia and CAB International, Wallingford.

Glenn, E.P., Jed Brown, J., and Blumwald, E., 1999. Salt tolerance and crop potential of halophytes. *Critical Reviews in Plant Sciences*, 18(2):227–255.

Goldshlegera, N., Chudnovskyb, A., and Ben-Binyamina, R., 2013. Predicting salinity in tomato using soil reflectance spectra. *International Journal of Remote Sensing*, 34(17): 6079–6093.

Hallet, M., 1994. Groundwater and salinity studies using airborne geophysics. *Australian Journal of Soil and Water Conservation*, 7:24–27.

Hick, R.T., and Russell, W.G.R., 1990. Some spectral considerations for remote sensing of soil salinity. *Australian Journal of Soil Research*, 28:417–431.

Hillel, D., 2000. *Salinity Management for Sustainable Irrigation: Integrating Science, Environment, and Economics.* World Bank Publications, Washington, DC. http://dx.doi.org/10.1596/0-8213-4773.

Hilwig, F.W., and Karale, R.L., 1973. Physiographic systems and elements of photo-interpretation as applied to soil survey in Ganges plain. *Journal of Indian Society of Soil Science*, 21(2):205–212.

Hollinger, A., Bergeron, M., Maskievicz, M., Qian, S., Othman, H., Staenz, K., Neville, R., and Goodenough, D., 2006. Recent developments in Hyperspectral Environment and Resource Observer (HERO) mission. In: *Proceedings of the Geosciences and Remote Sensing Symposium*, Denver, CO, pp. 1629–1623.

Honey, F., 1989. Technical specifications of the Geoscan MarkII AMSS, Unpublished report of Geoscan Pvt. Ltd.

Howari, F.M., GoodelL, P.C., and Miyamoto, S., 2002. Spectral properties of salt crust formed on saline soils. *Journal of Environmental Quality*, 31:1454–1461.

Hunt, G.R., 1980. Electromagnetic radiation: The communications link in remote sensing. In: Siegal, B.S., and Gillespie, A.R., (eds.) *Remote Sensing in Geology*. Wiley, New York, p. 545.

Jolly, I.D., 1998. The impact of flooding on modeling salt transport processes to streams. *Environmental Modelling and Software*, 13:87–104.

Jury, W.A., 1982. Simulation of solute transport using a transfer function model. *Water Resources Research*, 18:363–368.

Karale, R.L., and Venugopal, K., 1970. Soil survey of Ganges plain in Meerut. Project Report, Indian Institute of Photo-interpretation and Remote Sensing (Indian Institute of Remote Sensing), National Remote Sensing Agency, Hyderabad, India.

Koohafkan, P., 2012. *Water and Cereals in Drylands*. The Food and Agriculture Organization of the United Nations and Earthscan, Rome.

Lane, M.D., 2002. Upcoming THEIS investigation of salt of Mars. Lunar and Planetary Science XXXII 1749.

Leonardo, P.M., Weigand, C.L., Hector, S.S., and Carlos, R.Z., 1996. Using Landsat—TM and SPOT satellite images to identify salinity in irrigated areas of Mexico. *Proceedings of 16th International Congress on Irrigation Drainage*, Cairo Egypt, Sep 15–22, 1996.

Lesch, S.M., Herrero, J., and Rhoades, J.D., 1998. Monitoring temporal changes in soil salinity using electromagnetic induction techniques. *Soil Science Society of America Journal*, 62:232–242.

Lobell, D.B., Leschb, S.M., Corwinc, D.L., Ulmerd, M.G., Andersone, K.A., Pottsf, D.J., Doolittleg, J.A., Matosh, M.R., and Baltes, M.J., 2010. Regional-scale assessment of soil salinity in the Red River Valley Using Multi-Year MODIS EVI and NDVI. *Journal of Environmental Quality*, 39(1):35–41.

Lu, D., Mausel, P., Brondi´zio, E., and Moran, E., 2003. Change detection techniques, *International Journalof Remote Sensing*, 25(12):2365–2407.

Lobell, D.B., Ortiz-Monasterio, J.I., Gurrola, F.C., and Valenzuela, L., 2007. Identification of saline soils with multiyear remote sensing of crop yields. *Soil Science Society of America Journal*, 71(3):777–783. http://dx.doi.org/10.2136/sssaj2006.0306.

Manchanda, M.L., and Iyer, H.S., 1983. Use of Landsat imagery and aerial photographs for delineation and categorization of salt-affected soils of part of north-west India. *Journal of Indian Society of Soil Science*, 31: 263–271.

Manchanda, M.L., and Khanna, S.K., 1979. Use of aerial photographs for study of soil salinity and landscape relationship in parts of Haryana. *Journal of Indian Society of Remote Sensing*, VII(1): 27–34.

Matinfar, H.R., et al., 2013. Detection of soil salinity changes and mapping land cover types based upon remotely sensed data. *Arabian Journal of Geosciences*, 6(3):913–919.

McNeill, J.D., 1980. *Electromagnetic Terrain Conductivity Measurements At Low Induction Numbers*. Technical noteTN-6Geonics Ltd., Mississauga, ON.

Metternicht, G., 2001. Assessing temporal and spatial changes of salinity using fuzzy logic, remote sensing and GIS. *Ecological Modelling*, 144(2–3): 163–179.

Metternicht, G.I., 1998a. Fuzzy classification of JERS-1SAR data: An evaluation of its performance for soil salinity mapping. *Ecological Modelling*, 111:61–74.

Metternicht, G.I., 1998b. Analyzing the relationship between ground based reflectance and environment indicators of salinity processes in the Cochabamba Valley (Bolivia), *International Journal of Ecology and Environmental Sciences*, 24(4): 359–370.

Metternicht, G.I., and Zinck, J.A., 2003. Remote sensing of soil salinity: Potentials and constraints. *Remote Sensing of Environment*, 85:1–20.

Metternicht, G.I., and Zinck, J.A., 1997. Spatial discrimination of salt- and sodium affected soil surfaces. *International Journal of Remote Sensing*, 18:2571–2586.

Mougnet, B., Pouget, M., and Epima, G.F., 1993. Remote sensing of salt-affected soils. *Remote Sensing Reviews*, 7:241–259.

Mulders, M.A., 1987. *Remote Sensing in Soil Science, Developments in Soil Science*. Elsevier, Amsterdam, p. 379.

Nagar, S.C., and Singh, D., 1979. Semi-detailed soil survey of part of district Hardoi, Uttar Pradesh. Project Report Indian Photo-interpretation Institute (Indian Institute of Remote Sensing), National Remote Sensing Agency, Hyderabad, India.

National Remote Sensing Agency, 1995. Study of land degradation problems in Sharda Sahayak command area for sustainable agriculture development. Project report. National Remote Sensing Agency, ISRO, Department of Space, Government of India.

National Remote Sensing Agency, 2008. Mapping salt-affected soils of India. Project report, National Remote Sensing Agency, ISRO, Hyderabad, India.

Nield, S.J., Boettinger, J.L., and Ramsey, R.D., 2007. Digitally mapping gypsic and natric soil areas using Landsat ETM data. *Soil Science Society of America Journal*. 71:245–252.

Nogue, J., Robinson, D.A., and Herrero, J., 2006. Incorporating electromagnetic induction methods into regional soil salinity survey of irrigation districts. *Soil Science Society of America Journal*, 70:2075–2085.

Oldeman, L.R., Hakkeling, R.T.A., and Sombroek, W.G., 1991. *Word Map of the Status of Human-Induced Soil Degradation, Andexplanatory Notes*. Global Assessment of Soil Degradation (GLASOD), ISRIC, Wageningen.

Pishkar, A.R., 2003. Analysis of the relationship between soil salinity dynamics and geopedologic properties, a case study of the Goorband area, Iran. MSc thesis, ITC, Enschede, The Netherlands.

Rao, B.R.M., and Venkataratnam, L., 1991. Monitoring of salt-affected soils: A case study using aerial photographs and Salyute-7 space photographs, and Landsat-TM data. *Geocarto International*, 6:5–11.

Rao, B.R.M., Ravi Sankar, T., Dwivedi, R.S., Thammappa, S.S., Venkataratnam, L., Sharma, R.C., and Das, S.N., 1995. Spectral behaviour of salt-affected soils. *International Journal of Remote Sensing*, 16(12):2125–2136.

Rhoades, J.D., and Corwin, D.L., 1981. Determining soil electrical conductivity—depth relations using an inductive electromagnetic soil conductivity meter. *Soil Science Society of America Journal*, 45:255–260.

Rhoades, J.D., and Corwin, D.L., 1990. Soil electrical conductivity: Effects of soil properties application to salinity appraisal. *Communications in Soil Science and Plant Analysis*, 21:837–860.

Rhoades, J.D., Raats, P.A.C., and Prarther, R.J., 1976. Effects of liquid-phase electrical conductivity, water content and surface conductivity on bulk soil electrical conductivity. *Soil Science Society of America Journal*, 40:651–655.

Richards, L.A., (ed.) 1954. *Diagnosis and Improvement of Saline and Alkali Soils.* U. S. Department of Agriculture Hand Book No.60 U. S. Government Printing Office, Washington, DC.

Schmid, T., Gumuzzio, J., Koch, M., Mather, P., and Solana, J., 2007. Characterizing and monitoring semi-arid wetlands using multiangle hyperspectral and multispectral data. In *Proceedings Envisat Symposium 2007*, ESA SP-636, Montreux Switzerland, pp. 1–6.

Schmullius, C., and Evans, L., 1997. Table summary of SIR-C/X-SAR results: SAR frequency and polarization requirements for applications in ecology and hydrology. In: *Proceedings of the IGARSS'97*, Singapore, pp. 1734–1736.

Shao, Y., Guo, H., Hu, Q., and Han, C., 2002. Study on complex dielectric properties of saline soils. In: *Proceedings of the Geosciences and Remote Sensing symposium*, Toronto, ON, Vol.3, pp. 1541–1541b.

Sharma, R.C., Saxena, R.K., and Verma, K.S., 2000. Reconnaissance mapping and management of salt-affected soils using satellite images. *International journal of Remote Sensing*, 21: 3209–3218.

Shephered, K.D., and Walsh, M.G., 2002. Development of reflectance spectra libraries for characterization of soil properties. *Soil Science Society of America Journal*, 66:988–998.

Shuya, H., Qinhuo, L., and Xiaowen, L., 2005. Spectral model of soil salinity IN Xinjiang of China. In: *Proceedings of the Geosciences and Remote Sensing symposium*, Seoul, Korea, Vol.6, pp. 4458–4460.

Siegal, B.S., and Goetz, A.F.H., 1977. Effects of vegetation on rock and soil type discrimination. *Photogrammetric Engineering and Remote Sensing*, 43(2):191–196.

Siegel, B., and Gillespie, A. 1980. *Remote Sensing in Geology*. John Wiley, New York.

Singh. A., 1989. Digital change detection techniques using remotely-sensed data. *International Journal Remote Sensing*, 10(6):989–1003.

Singh, A.N., and Dwivedi, R.S., 1989. Delineation of salt-affected soils through digital analysis of Landsat-MSS data. *International Journal of Remote Sensing*, 10(1):83–92.

Singh, A.N., Baumgardner, M.F., and Kristof, S.T., 1977. Delineating salt-affected soils in part of Indo- Gangetic plain by digital analysis of Landsat data, Tech. Report 111477, LARS, Purdue University, West Lafayette, Indiana.

Soil Survey Staff, 1951. *Soil Survey Manual*, Agriculture Hand book No.18, USDA. Soil Survey Staff, Washington, DC.

Sparks, D.L., 1995. *Environmental Soil Chemistry*. Academic Press, San Diego, CA.

Sreenivas, K., Vennkatratnam, L., and Rao, P.V.N., 1995, Dielectric properties of salt-affected soils. *International Journal of Remote Sensing*, 16: 641–649.

Street, G.J., and Duncan, A.C., 1992. The application of airborne geophysical surveys for land management. In: *7th ISCO Conference Proceedings—Helping People Protect Their Land*, pp. 762–770.

Szabolcs, I., 1974. *Salt Affected Soils in Europe*. Martinus Nijhoff, The Hague, The Netherlands, p. 63.

Szabolcs, I., 1979. Review of research on salt affected soils. Natural Resources Research XV, UNESCO, Paris.

Szabolcs, I., 1989. Amelioration of soils in salt-affected areas. *Soil Technology*, 2(4):331–344.

Taylor, G.R., Mah, A.H., Kruse, F.A., Kierein-Young, K.S., Hewson, R.D., and Bennett, B.A., 1996. Characterization of saline soils using airborne radar imagery. *Remote Sensing of Environment*, 57:127–142.

Taylor, G.R., Bennett, B.A., Mah, A.H., and Hewson, R.D., 1994. Spectral properties of salinized land and implications for interpretation of 24-channel imaging spectrometry. In: *Proceedings of the First International Airborne Remote Sensing Conference and Exhibition*. Strasbourg, France, Sep 11–15, 1994.

Teggi, S., et al., 2012.SPOT 5 Imagery for Soil Salinity Assessment in Iraq. In: *Proceedings of SPIE—Earth Re-sources and Environmental Remote Sensing/GIS Applications III*, Vol. 8538, 2012, pp. 85380V-85380V–12.

Ulaby, F.T., Moore, R.K., and Fung, A.K., 1986. *Microwave Remote Sensing Active and Passive-From Theory to Applications. Vol. 3,* Artech House, Dedham, MA.
United States Salinity Laboratory Staff, 1954. *Diagnosis and Improvement of Saline and Alkali Soils.* Agriculture Hand book No. 60. United States Department of Agriculture.
Van Dam, J.C., Huygen, J., Wesseling, J.G., Feddes, R.A., Kabat, P., van Walsum, P.E.V., Groenendijk, P., and Van Diepen, C.A., 1997. *Theory of SWAP version 2.0. Simulation of Water Flow, Solute Transport and Plant Growth in the Soil–Water–Atmosphere–Plant Environment,* Technical Document, Vol. 45. DLO Winand Staring Centre, Wageningen, The Netherlands.
Verma, K.S., Saxena, R.K., BArthwal, A.K., and Deshmukh, S.K., 1994. Remote sensing techniques for mapping salt-affected soils. *International Journal of Remote Sensing,* 15:1901–1914.
Viscarra Rossel, R.A., Adamchuk, V.I., Sudduth, K.A., McKenzie, N.J., and Lobsey, C., 2011. Proximal soil sensing: An effective approach for soil measurements in space and time, Chapter 5. *Advances in Agronomy,* 113:237–283.
Wiegand, C., Rhoades, J.D., Escobar, D.E., and Everitt, J.H., 1996. Photographic and videographic observations for determining and mapping the response of cotton to soil salinity. *Remote Sensing of Environment,* 49(3):212–223.
Wiegand, C.L., Escobar, D.E., and Lingle, S.E., 1994. Detecting growth variation and salt stress in sugarcane using videography. In: *Proceedings of the Biennial Workshop on Colour Aerial Photography and Videography for Resource Monitoring.* American Society of Photogrammetric Engineering and Remote Sensing, pp. 185–199.
Wiegand, C.L., Leamer, R.W., Weber, D.A., and Gerbermann, A.H. 1971. Multibase and multi-emulsion space photos for crops and soils. Photogram, Eng. XXXVII: 147–156.
Wiegand, C.L., Everitt, J.H., and Richardson, A.J., 1992. Comparison of multispectral video and SPOT-1hrv observations for cotton affected by soil salinity. *International Journal Remote Sensing,* 13:1511–1525.
Wiegand, C.L., Richardson, A., Escobar, D., and Gerbermann, D., 1991. Vegtation indices in crop assessment. *Remote Sensing of Environment,* 35:105–119.
WRI-IIED-UNEP, 1988. World resources 1988–89. An assessment of the resource basethat supports the global economy. New York.
Yun, S., Qingrong, H., Huadong, G., Yuan, L., Qing, D., and Chunming, H., 2003. Effect of dielectric properties of moist salinized soils on backscattering coefficients extracted from RADARSAT images. *IEEE Transactions on Geosciences and Remote Sensing,* 41:1879–1888.
Zhang, Y., Gao, J., and Wang, J., 2007. Detailed mapping of a salt farm from Landsat TM imagery using neural network and maximum likelihood classifiers: a comparison, *International Journal of Remote Sensing,* 28(10):2077–2089.
Zhang, T.T., et al., 2011. Using hyperspectral vegetation indices as a proxy to monitor soil salinity. *Ecological Indicators,* 11(6):1552–1562.

9

Soil Acidification

9.1 Introduction

Agriculture is the main stay of most developing countries. However, agricultural productivity remains critically low in most of these countries. The low agricultural production is largely attributed to low and decreasing soil fertility due to many factors such as soil acidity, soil salinity and/or alkalinity, soil erosion, continuous cropping, and inadequate sustainable soil fertility management (Berga et al., 2001; Van Straaten, 2002; Kiiya et al., 2006; Crawford et al., 2008).

Acidification of agricultural soils is primarily associated with the net removal of base cations like calcium, magnesium, and the direct addition of acidifying inputs (e.g., ammonium-based N fertilizer) to inherently low-pH soils, which have a low buffering capacity. The buffering capacity is the amount of acid or base a buffer can accept until it can no longer maintain a constant pH. A *buffer solution* is one which resists changes in pH when small quantities of an acid or an alkali are added to it. Acidification is most prevalent on ancient, highly weathered soils. For instance, the acidity affects the fertility of soils through deficiencies of P, Ca, and Mg, and the presence of phytotoxic nutrients such as soluble aluminum (Al) and manganese (Mn). Liming is an effective response to control acidity of surface horizons, but rates of lime addition lag behind required levels even in developed countries like Australia and continuing loss of yield occur (FAO and ITPS, 2015). Acidification of the soil due to agronomic practices, the most notable being the application of ammonical fertilizer, has long been identified as a source of crop yield decline, particularly in noncalcareous soils.

9.2 Background

Soil pH is a measure of the number of hydrogen ions in the soil solution; the higher the concentration of hydrogen ions, the more acidic the solution. A thorough understanding of soil pH is essential for the proper soil management and optimum crop productivity. In aqueous (liquid) solutions, an acid is a substance that donates hydrogen ions (H^+) to some other substance (Tisdale et al., 1993). Soil pH is an excellent chemical indicator of soil quality. Theoretically, soil acidity is quantified on the basis of hydrogen (H^+) and aluminum (Al^{3+}) concentrations of soils (Fageria and Baligar, 2008).

Soil acidity occurs when there is a build-up of acid-forming elements in the soil. The production of acid in the soils is a natural process; caused by rainfall and leaching, acidic parent materials and organic matter decay (Havlin et al., 2005). Hence many soils in high-rainfall areas are inherently acidic (McCauley et al., 2009). Acidification is a slow process but it is accelerated by agriculture through use of some fertilizers, like nitrogenous fertilizers, soil structure disturbance, and harvest of high yielding crops (Fageria and Baligar, 2008). According to an estimate, global land has received more than 50 kg/ha accumulated N deposition during 2000–2010 (Penuelas et al., 2013), which has been well documented as the main cause of soil acidification in terrestrial ecosystems (Wright et al., 2001, Guo et al., 2010, Yang et al., 2012). As soils become more acidic, the growth and development of plants susceptible to acidic conditions are adversely affected leading to productivity decline. The N-induced soil acidification (i.e., decrease in pH) has been a significant threat to species diversity and terrestrial ecosystem functioning (Chen et al., 2013). With the unprecedented increasing N deposition in the context of global change (IPCC, 2013), soil acidification is becoming a major problem for global terrestrial ecosystems (Lucas et al., 2011; Yang et al., 2012).

9.3 Global Scenario

Acid soils occupy approximately 3,950 million hectares or 30% of the world's ice free land area and occur mainly in two global belts where they have developed under *udic* or *ustic* moisture regimes. There are two main belts of acid soils:

- In the humid northern temperate zone, which is covered mainly by coniferous forests,
- In the humid tropics, which is covered by savannah and tropical rain forest.

The northern belt (cold and temperate climate) is dominated by Spodosols, Alfisols, Inceptisols, and Histosols whereas the southern tropical belt consists largely of Ultisols and Oxisols. Sixty-seven percent of the acid soils support forests and woodlands and approximately 18% are covered by savanna, prairie, and steppe vegetation. Only 4.5% (179 million hectares) of the acid soil area is used for arable crops. A further 33 million hectares is utilized for perennial tropical crops. The value of the annual production in these areas is approximately US$ 129 billion. Value of products from forests, woodlands, and permanent pastures on acid soils is difficult to evaluate.

Acid sulfate soils are usually left under natural vegetation or used for mangrove forestry. If water is managed well, they can support oil palm and rice. Some other crops grown on acid soils around the world include rice, cassava, mango, cashew, citrus, pineapple, cowpeas, blueberries, and certain grasses.

Acid soils are those that have a pH value of less than 5.5 for most of the year. They are associated with a number of toxicities (aluminum) as well as deficiencies (molybdenum) and other plant restricting conditions. Many of the acid soils belong to Acrisols, Alisols, Podzols, and Dystric subgroups of other soils. An extreme case of an acid soil is the acid sulfate soil (Thionic Fluvisols and Thionic Cambisols). There are two main belts of acid soils:

- In the humid northern temperate zone, which is covered mainly by coniferous forests,
- In the humid tropics, which is covered by savannah and tropical rain forest.

Forests of the tropics and wetlands have an invaluable role in global, regional, and local ecosystem balance and a protective role for flora, fauna, and water resources. While acid soils in the northern belt are increasingly protected and reafforested, the destructive exploitation of timber and abusive modern shifting cultivation have contributed to the loss of >250 million hectares of tropical forest during the second half of this century leaving vast areas of anthropic savannas on heavily eroded and degraded acid soils.

The authors believe that attempts to develop acid soils for agriculture and agroforestry in the tropics should concentrate on these deforested and abandoned areas of degraded acid soils. However, this will be difficult without significant initial investment and adequate technology. A three-step development approach is suggested, which could help prevent or halt the annual destruction of >5 million hectares tropical forests by "untraditional shifting cultivators." It would help to protect the fragile natural ecosystems on tropical acid soils now considered to be indispensable for the future life on earth (von Uexküll and Mutert, 1995). Globally, more than 50 kg/ha accumulated N deposition during 2000–2010 (Penuelas et al., 2013), which has been well documented as the main cause of soil acidification in terrestrial ecosystems (Wright et al., 2001, Guo et al., 2010, Yang et al., 2012).

The N-induced soil acidification (i.e., decrease in pH) has been a significant threat to species diversity and terrestrial ecosystem functioning (Chen et al., 2013). With the unprecedented increasing N deposition in the context of global change (IPCC, 2013), soil acidification is becoming a major problem for global terrestrial ecosystems (Lucas et al., 2011, Yang et al., 2012).

9.4 Development of Soil Acidity

The development of soil acidity essentially involves leaching of bases mainly calcium and magnesium under very high rainfall conditions. With the removal of bases soil becomes acidic which in turn leads to toxicity of aluminum and molybdenum deficiency. The acidification of soil is caused by the following.

9.4.1 Causative Factors of Soil Acidification

9.4.1.1 Acidic Precipitation

"Pure" rain is usually slightly acid, with a pH of between 5 and 5.6 because of the dissolution of carbon dioxide (CO_2) and the dissociation of the resulting carbonic acid (H_2CO_3). A soil exposed to such rain, but no other acidifying inputs and receiving no lime, would attain the same equilibrium pH as that of the rain. There are, however, very strong localized effects because human activity has increased the acidity of precipitation through emissions of acidifying compounds such as SO_2 and nitrogen oxides (NOx) from industry and motor vehicles, and NH_3 volatilized from manures and fertilizers.

9.4.1.2 Acidifying Gases and Particles

From the beginnings of the Industrial Revolution until the 1970s, sulfur emissions increased and SO_2 was the main component of acid deposition (RoTAP, 2012). However, by the 1990s, there has been slight decrease in sulfur deposition because of the decline in heavy industry and the switch from coal to natural gas as an energy source. Data for Woburn Farm, Bedfordshire, showed a decline in total S deposition (sulfate, SO_4^{-4}, in precipitation plus SO_2) from approximately 85 kg/ha/year in 1970 to approximately 15 kg/ha/year in 1995 (McGrath and Zhao, 1995). The current total S deposition at Woburn is <5 kg/ha/year (RoTAP, 2012) and is likely to decline even further toward approximately 2.5 kg/ha/year by 2,030 (S. P. McGrath, personal communication). Nitrogen (N) deposition had become the dominant pollutant in acid rain in 1998, much of which was acidifying. It remains dominant, but at Rothamsted its deposition has decreased from a peak of 40–50 kg/ha/year to arable land in the early 1990s (Goulding et al., 1998) to approximately 25 kg/ha/year today. Much of this decline has been caused by the fitting of catalytic converters to vehicles and a general reduction in emissions from combustion (RoTAP, 2012). Current acid deposition rates are equivalent to no more than 2 kmol H^+/ha/year.

9.4.1.3 Acidifying Fertilizers and Legumes

The most important causes of soil acidification on agricultural land are the application of ammonium-based fertilizers and urea, elemental S fertilizer, and the growth of legumes (Bolan and Hedley, 2003). Ammonium salts strongly acidify soils through the process of nitrification:

$$NH_4^+ + O_2 = NO_3^- + 2H^+ + H_2O \tag{9.1}$$

If the nitrate (NO_3^-) is taken up by the crop, there is no net acidification because the NO_3^- takes up protons with it (Marschner, 2012; 14.4). Acidification only occurs when NH_4^+ is nitrified and the NO_3^- leached. The same is true of urea: there is no net acidification if all the N in the urea is utilized by the crop; acidification only occurs when the urea is converted to NH_4^+, the NH_4^+ nitrified and the leached. Ammonium sulfate applied to some plots of the Park Grass Experiment at Rothamsted has caused a rapid decrease in pH, starting in the surface soil but occurring throughout the profile to at least 1 m. The larger the application rate the more rapid the rate of acidification (Johnston et al., 1986).

Examples of three of the buffering mechanisms in soils are as follows: In carbonate/bicarbonate buffer in the soil under Broadbalk Wilderness, acid deposition has been buffered by very large quantities of chalk applied in the 19th century. The decline in S deposition noted above has resulted in the need for farmers to apply S fertilizer as explained in the Fertilizer Manual (RB209; Defra, 2010). Inputs of S as elemental S or as SO_2 from the atmosphere produce acidity when they are oxidized, but SO_2^{-4} produces no acidity because it is not subject to further oxidation.

$$2S + 3O_2 + 2H_2O = 2H_2SO_4 \tag{9.2}$$

$$2SO_2 + O_2 + 2H_2O = 2H_2SO_4 \tag{9.3}$$

The fixation of atmospheric N_2 by legumes results in the formation of NH_4^+ within the root nodules by nitrogenase, the uptake of an excess of cations, especially K+, and therefore a net release of protons to balance the charge (Marschner, 2012; Ch 14.4). Bolan and Hedley (2003) reported that where legumes had been grown continuously in Australia for >30 years, soil pH had declined by 1 unit.

9.4.1.4 *Nutrient Uptake by Crops and Root Exudates*

Plant growth and nutrient uptake result in some localized acidification around plant roots through the exudation of acids from the roots (Hinsinger et al., 2003). Excluding the particular case of legumes, the contribution of this to bulk soil acidification is small (<10%) when compared with N and S fertilizer inputs (Johnston et al., 1986), but it has an important influence on the bioavailability of plant nutrients in the rhizosphere (Marschner, 2012).

9.4.1.5 *Mineralization*

When microorganisms decompose soil organic matter they produce CO_2, which dissolves in soil water to form H_2CO_3 in the same way as in rain. Thus, soil and root respiration can result in a large concentration of CO_2 in soil air, but because acidic soil solutions hold very little CO_2, the process is unlikely to cause soil pH to decline below 5 (Bolan et al., 2003).

9.4.2 The Impact of Soil Acidification

Soils become acidic for several reasons. The most common source of hydrogen is the reaction of aluminum ions with water. Aluminum toxicity in combination with low pH (Budianta and Vanderdeelen, 1995) is one of the major reasons that render acidic soils unsuitable for the growth of many plants in the humid tropic countries. The forms of aluminum ions present vary with pH. The increased soil acidity causes solubilization of Al, which is the primary source of toxicity to plants at pH below 5.5 (Kariuki et al., 2007).

As observed by Carson and Dixon (1979) under very acidic conditions of pH less than 4.5, the major form of aluminum is Al^{3+}, and pH between 4.5 and 6.5, aluminum-hydroxyl dominates. As the pH increases, exchangeable Al^{3+} precipitates as insoluble Al hydroxyl forms at a rate of 1,000-fold decrease for each unit increase in pH (equation 9.4).

$$Al^{3+} + H_2O \leftrightarrow AlOH_2 + H^+ \tag{9.4}$$

Equation 9.4 explains the reaction of aluminum hydroxyl in very acid soils. However, at pH greater than 6.5, aluminum becomes increasingly soluble as negatively charged aluminates form (Haynes, 1984).

The heavy rainfall can also contribute to the soil acidification by natural causing parent materials to be acidic due to leaching of cations. There are other important causes of soils acidification, such as ammonium fertilizers, release of organic acids in decomposition of crop residues or organic wastes, and continuous cultivation of legumes (Bolan and Hedley, 2003). The acidification caused by the use of ammonium fertilizers is explained by the release of H^+.

$$NH_4 + 2O_2 \leftrightarrow NO_3^- + H_2O + 2H^+ \tag{9.5}$$

The acidification due to legumes is explained by higher absorption of basic cations of legumes and the release of H^+ ions by the root of legume crops to maintain ionic balance, and during N_2 fixation through a function of carbon assimilation.

9.4.3 Soil Acidity and Base Saturation and Buffering Capacity

A relatively high base saturation of CEC (70%–80%) should be maintained for most cropping systems, since the base saturation determines in large measure the availability of bases for plant uptake, and strongly influences soil pH as well. Low base saturation levels result in very acid soils and potentially toxic cations such as Al and Mn in the soil. A high base saturation (>50%) enhances Ca, Mg, and K availability and prevents soil pH decline. Low base saturation (<25%) is indicative of a strongly acidic soils that may maintain Al^{3+} activity high enough to cause phytotoxicity (Soil Survey Division Staff, 1993).

The resistance of soils to changes in pH of the soil solution is termed buffering. In practical terms, buffering capacity for pH increases with increase in the amount of clay and organic matter (Soil Survey Staff, 1993). Thus, soils with high clay and organic matter content (high buffer capacity) will require more lime to increase pH than sandy soils with low amounts of organic matter (low or weak buffering capacity) (Figure 9.1).

9.4.4 Soil Acidity and Crop Responses

Soil pH affects crops in many ways and its effects are mostly indirect, through its influence on chemical factors and biological processes. Chemical factors include aluminum (Al) toxicity, and calcium (Ca), phosphorus (P) and magnesium (Mg) deficiencies (Uchida and Hue, 2000). Optimum nutrient uptake by most crops occurs at a soil pH near 7.0. The nutrients availability such as nitrogen, phosphorus, and potassium is generally reduced as soil pH decreases. Phosphorus is particularly sensitive to pH and can become a limiting nutrient in strongly acid soils. Thus, reduced fertilizer use efficiency and crop performance can be expected when soil acidity is not properly controlled (McFarland et al., 2001). Hardy et al. (1990) reported exchangeable Al to affect crops by shallow rooting, poor use of soil nutrients, and Al toxicity (Figure 9.2).

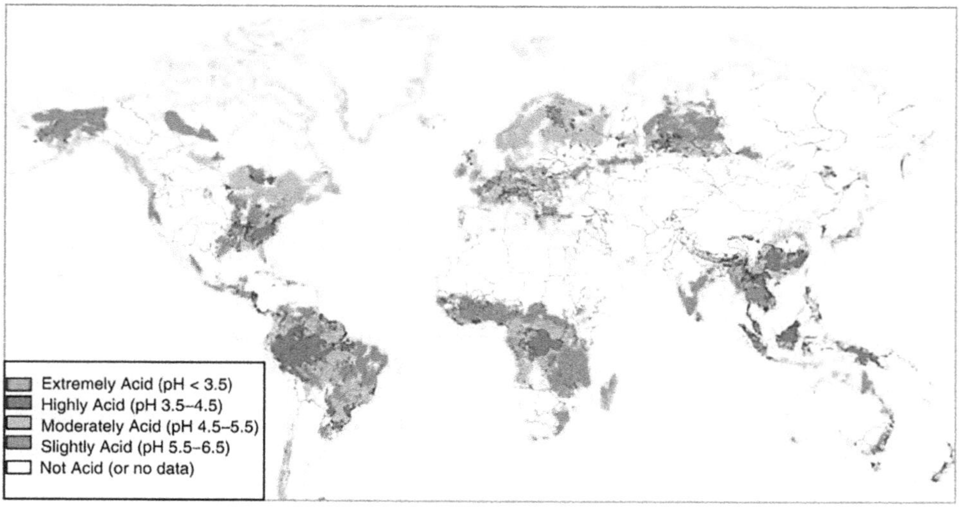

FIGURE 9.1
Soil pH map (Source: Soil Survey Staff. Natural Resources Conservation Service, United States Department of Agriculture).

(a) 0 μM Al (b) 25 μM Al (c) 50 μM Al

FIGURE 9.2
(a–c) Root growth comparison in sweet potato cv. Meriken grown in nutrient solution with aluminum (V. Ila'ava) (Source: Adapted from O'Sullivan et al. (1997)).

9.5 Delineation and Mapping of Acid Soils

The procedure for delineation and mapping of acid soils is similar to that used for mapping other soils. Since acid soils are, generally, associated with tropical climate, savannah vegetation and the acidic parent materials, visual interpretation involving integration of various soil-forming factors (e.g., parent material, relief and vegetation cover) to arrive at a soilscape units seems to be most appropriate approach. Before availability of aerial photograph sand orbital sensor data, namely availability of conventional techniques involving dividing the area to be surveyed into grids of different sizes depending on level of details on soil sought, studying soil profiles and auger bores, collection of soil samples there from, chemical analysis in the laboratory, classification soils according to existing soil classification system, delineation of soilscape boundaries onto base map (e.g., topographic maps of appropriate scale or cadastral maps). The availability of aerial photographs in the early 1930s ushered in a new era in soil resources mapping. Subsequently with the launch of Landsat-1 in 1972 the soil survey and mapping program received a major breakthrough in terms of timeliness and reliability and cost-effectiveness.

9.5.1 Aerial Photographs

By virtue of providing 3D view of the terrain, aerial photographs have been used in soil surveys since the early 1930s. For deriving information on soils from aerial photographs, the dynamic interrelationship between physiography and soils is utilized (Christian and Stewart, 1968). The approach, generally, relies on relating patterns of drainage, relief, land use, tonal features, and cultural features to soil characteristics. The guiding principle has been that soils are the product of the same natural processes and conditions that sculpture the land they dwell in. However, this does not imply that any given physiographic unit will contain a single class of soils; but that the soils within the physiographic unit normally vary within a certain range. Three broad groups, namely *p* aspect/*elemental analysis, physiographic analysis, and morphogenetic analysis.* A brief outline of these approaches is presented here. For details on air-photointerpretation, readers may refer to Buringh (1960), Vink (1964), Goose (1967), and Bennema and Gelens (1969a, b).

9.5.1.1 Aspect/Elemental Analysis

Aerial photointerpretation may be done for many purposes each with its specific interest in the objects or the features on the imagery. The aspect/element refers to the kind of objects or features on which a particular analysis is based. The aspects of interest for soil resources inventory can be divided into (Bennema and Gelens, 1969a, b):

- *Basic aspects* individually visible on the images, e.g., slope and relief, natural vegetation, crops, soils, and rock surface;
- *Compound aspects* visible on the images through a combination of two or more basic aspects, e.g., land types, drainage ways or pattern, faults, and joints;
- *Inferred aspects* not directly visible on the images but deducted from basic or compound aspects, e.g., soil depth, parent material, drainage, and erosion conditions.

Bennema and Gelens (1969a, b) observed the following relationships between the aspects of image interpretation and soil conditions:

- The aspect has direct relation to soil, e.g., the color or gray tone of the top soil, and the drainage condition;
- The aspect indicates certain conditions of soil formation; changes in the aspect mean changing conditions of soil formation and most likely different soils, e.g., differences in slope and relief or in parent material;
- The aspect shows consequences of soil differences, e.g., differences in natural vegetation and in a number of cases in land use.

9.5.1.2 Physiographic Analysis

Physiographic analysis involves the study of external terrain features as manifested in the stereo-photographs (Bennema and Gelens, 1969a, b). Physiographic analysis is based on a thorough knowledge of physiographic processes and their reflection on aerial photographs. The physiographic analysis includes the analysis of basic and compound aspects, such as relief, slope as well as vegetation and land use, and aspects important for the description of the drainage system. The same elements of the aspect analysis are used but in a different way. Those areas which have uniformity in appearance are delineated and characterized by the same symbol on the images. When the physiography changes, a different unit is delineated. The terrain is classified into physiographic units, each of which contains a unique association of soils.

9.5.1.3 Morphogenetic Analysis

Morphogenetic analysis comprises the delineation of units not only on the basis of their appearance, but also on the basis of the processes which have shaped these units. However, most important for morphogenetic analysis is of course the geomorphology of the area, but also hydrology and other factors may play their parts. Specific knowledge is required of the morphogenetic processes and it is unlikely that this analysis is the in the interpretation of air photos of an unknown area. However, in image analysis some geomorphological units may be applied with great certainty, such as river levees, river basins, point bars, and beach ridges. It will be understandable that, in practice, a combination of the physiographic and morphogenetic analysis is generally applied.

9.5.2 Spaceborne Multispectral Measurements

For deriving information on soils two approaches, namely monoscopic visual interpretation and computer-assisted digital analysis, are often employed. The monoscopic visual interpretation involves delineation of soilscape boundaries on the hard copies of satellite image by a human interpreter. Alternatively, digital image of an area of interest could be interpreted through on-screen/heads-up visual interpretation approach using image elements. The computer-assisted digital analysis, on the other hand, involves delineating soilscape boundaries based solely on the spectral response pattern (digital numbers: DN values) of surface soil in different spectral bands of multispectral images, representing the intensity of reflected or emitted radiation. In this approach, the features on the ground are identified/delineated based solely on the spectral response patterns of terrain features as recorded by the sensor aboard air/spaceborne platforms. For further details on applications of remote sensing in soil resources mapping readers may refer to Mulder (1987), Mulder et al. (2011), Wulf et al. (2015), and Dwivedi (2017).

9.5.2.1 Visual Interpretation

Sensors onboard air and spaceborne platforms operating in visible infrared and microwave region capture the reflected/scattered/emitted electromagnetic radiation. With regard to soil resources mapping wherein we are looking at information not only on Earth's surface but also on the third dimension of the earth, three scenarios emerge: first, in the case of bare surfaces reflectance data offer direct information about the soil surface; second, where surfaces are covered with natural vegetation in varying degrees—either partially (with some contribution in reflected/scattered radiant energy from soils) or fully (wherein the spectral response is only from vegetation canopy); and lastly, in cases where soil is covered fully by vegetation, the reflected/scattered/emitted electromagnetic radiation captured by the sensor is essentially from vegetation canopy. Under such conditions, it is possible to derive information on soils if there is a direct correlation of vegetation reflectance and underlying soils. In bare soils partially vegetated soils, the procedures used for deriving information on soils are discussed hereunder, and the schematic diagram of the approach is appended as Figure 9.3.

9.5.2.1.1 Selection of Remote Sensing Data

Depending on the purpose and scale of soil mapping satellite data with spatial resolution matching with the map scale is acquired. With respect to season of satellite data to be used for soil resources mapping it is dry season (May–June for Asian region) data with the maximum soil exposure (with east vegetation cover) is preferred since the presence of vegetation obscures the reflected/scattered/emitted radiation from the soil surface. However, in a spectrally homogenous terrain like alluvial plain or aeolian plain, the presence of vegetation (February–March) may sometimes be helpful in delineation of soilscape boundaries. The presence of vegetation indicates soil moisture availability, moderate to fine texture, and moderately deep to deep soils.

9.5.2.1.2 Preparation of Database

For deriving information on soils the multispectral data from individual sensor corresponding to the mapping scale could be used. In case if it is not so, an alternative is to make conjunctive use of coarser spatial resolution multispectral data and finer resolution panchromatic data. In this process the spectral information comes from multispectral data and the spatial information form panchromatic data wavelet several digital data fusion approaches, namely HIS transformation, Brovey transformation and Principal Component Analysis (PCA), etc. available in the standard digital data analysis software are utilized. This step is relevant to both the approaches, i.e., visual interpretation as well as digital analysis.

The basic objective of the image interpretation or analysis is to develop a soil map with a good geometric fidelity in terms of scale, map projection, and datum. In order to realize the objective standard, satellite data products available from remote sensing data providers which are generally bulk corrected for radiometric and geometric errors are utilized. In case soil mapping is to be undertaken at scales larger than 1:25,000 (1:10,000/1:5,000) precise geometric correction using ground control points (GCPs) collected with Differential Global Positioning System (DGPS) needs to be resorted to. A schematic diagram showing various steps of the interpretation to derive information on acid soils is appended as Figure 9.3.

9.5.2.1.3 Preliminary Visual Interpretation

For the visual interpretation, lithological (rock type) information may be taken from published geological maps. Later on, within each lithological unit, broad physiographic units are delineated based on slope (derived from contour information) or digital elevation model (DEM), and image elements, namely color, texture, pattern, shape, association, etc. These units may be further divided based on land use/land cover, slope, drainage, degree of erosion, wetness, etc. A tentative legend in terms of lithology, physiography, soil erosion status, land use/land cover, and image elements is then developed. Sample strips representing broad lithological and physiographic units are selected for field verification. The units in which soil sampling is to be done are located precisely on the image with the help of published topographic maps.

9.5.2.1.4 Ground Truth Collection

The digital image data contains spectral response pattern of the features on the Earth including exposed soil surface or the surface with partial or full vegetation cover. On the other hand, in the case of hard copies of the image gray levels/different shades of colors representing various terrain features are manifested. Ground truth mission or field visit is undertaken to establish the relationship between image elements/a homogenous cluster of DN values and the features on the ground. Initially, a reconnaissance traverse of the area is made to assess the trafficability of the terrain and to locate the sample points precisely. Subsequently, the points where observations are to be made are marked precisely on the topographical maps and the observation with respect to land use/land cover, soil erosion status, salt encrustation, surface drainage, irrigation source and depth and quality of groundwater, crop condition—vigor and density/and local relief are made. Later, in typical areas, soil profiles are excavated and morphological features studied, and soil samples from typical profiles were collected for analysis in the laboratory. Besides, auger bore and surface observations are made outside the sample strip to account for variations outside the sample areas. Detailed physical and chemical analysis of soil samples collected during field visits was carried out using the standard chemical analysis procedure.

9.5.2.1.5 Soil Chemical Analysis

Soil samples collected during the fieldwork are air-dried, pounded with wooden hammer, and sieved with 2 mm sieve in the laboratory, and were used for analysis. Soil samples are analyzed for physical and chemical properties, namely soil texture, pH, electrical conductivity, organic matter content, calcium carbonate, exchangeable calcium,

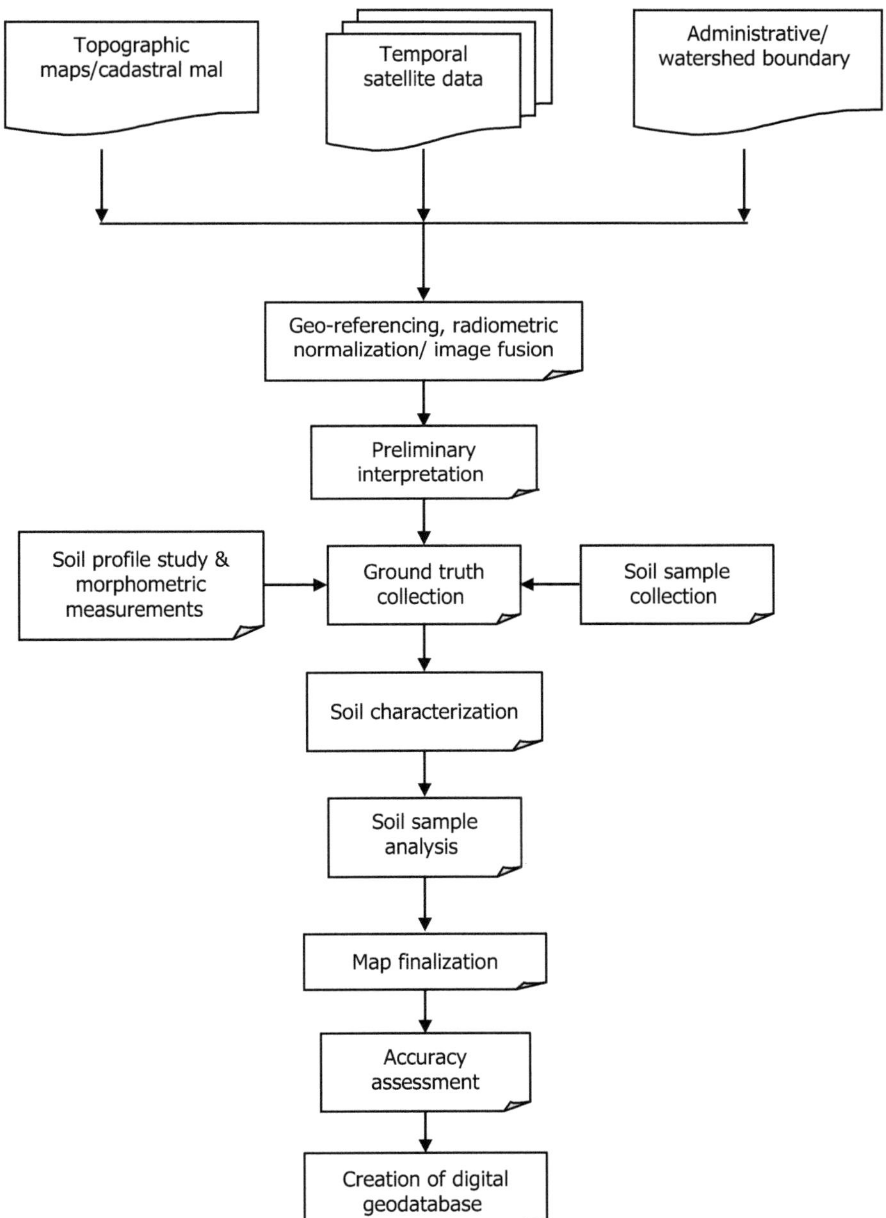

FIGURE 9.3
Schematic diagram of the approach.

magnesium, potassium, and sodium, cation exchange capacity following standard analytical procedures.

9.5.2.1.6 Post-field Interpretation

Based on morphological characteristics and chemical analysis data soils are classified according to Soil Taxonomy (Soil Survey Staff, 1993) or any other national/international soil classification (FAO-UNESCO, 1974; WRB, 2015). The tentative legend, developed earlier, is modified vis-à-vis field observations. A critical analysis of taxonomic unit/units identified in each polygon on the map is done to decide upon whether soil unit mapped is pure (homogenous with respect to its composition) or it is an association of more than one taxonomic unit. The soil association generally consists of a dominant unit occupying 80% of the aerial extent of the polygon in soil map.

After finalization of the soil map, the boundaries of the polygons are transferred onto a base map. A base map is the one which contains important natural and cultural features that enable the map users to locate the soilscape units of interest. The natural features include water bodies: rivers, lakes, tanks. The cultural features include location of settlements, roads, and railways. In the case of soil map generated from hard copies of the image, after finalizing soil boundaries, it could be digitized, and superimposed over digital base map generated from the published digital topographic maps. In the event of visual interpretation of digital image data through heads-up or on-screen visual interpretation approach, the soilscape boundaries/soil polygons are directly overlaid onto digital base map as one of the layers. The area under each soilscape unit is computed using available tool in GIS software.

9.5.2.1.7 Accuracy Estimation

For quantitative estimates of the classification accuracy of soil maps, sample areas representing different types of soils are selected randomly (Congalton, et al., 1983). An adequate number of sample points representing different types of soils are identified on the soil map for accuracy estimation. A one-to-one comparison of the categories of soil mapped and ground truth data, and the information available in the published report is made. Accuracy estimation in terms of overall accuracy, error of omission and error of commission; and Kappa coefficient (K) is subsequently made after generating confusion matrix. The Kappa coefficient (K) is computed as follows (Bishop et al., 1975):

$$K = \frac{N\sum_{i=1}^{r} xii - \sum_{i=1}^{r}(xi+)(x+i)}{N^2 - \sum_{i=1}^{r}(xi+)(x+i)} \tag{9.6}$$

where "r" is the number of rows in the matrix, xii is the number of observations in row i and column i (the ith diagonal elements), xi+ and x + i are the marginal totals of row "r" and column i, respectively; and N is the number of observations.

As an illustration for mapping acid soils in part of southern India using spaceborne multispectral data is given in Figure 9.4. The figure shows Resourcesat-2 LISS-III data for three cropping seasons, namely *kharif* (monsoon season), *rabi* (winter season), and *zaid* (summer season), at 1:50,000 scale. The November 2005 LISS-III data corresponds to post-monsoon season, January 2006 to winter season and May 2006 to summer season. The image covers

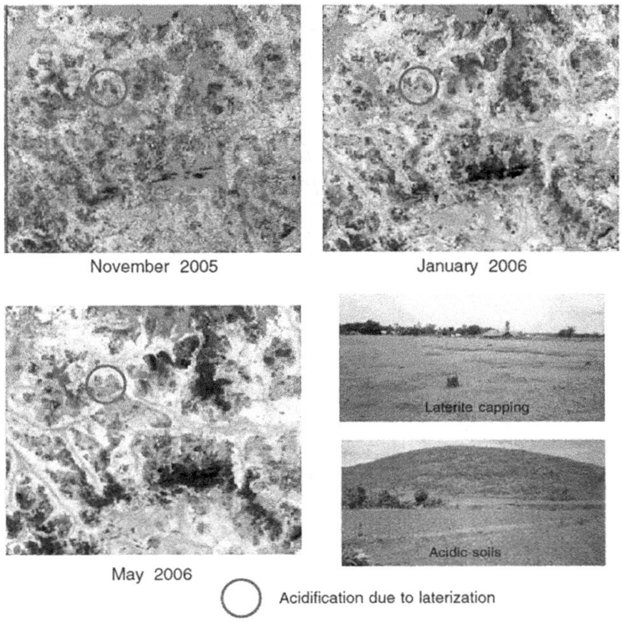

FIGURE 9.4
Acid soils (laterites) as seen in the Resourcesat-2 LISS-III image of various cropping seasons, and on the ground (Courtesy: National Remote Sensing Centre, Indian Space Research Organization, Department of Space, Government of India).

an area in part of southern India with predominantly laterite soils. For illustration, a small pocket of laterite soils (acid soils) are marked with blue circle. It is interesting to note that owing low rainfall and poor fertility of these soils except during monsoon there is no vegetation cover. Vegetation cover is manifested in different hues of red color and sometimes pink color. The ground photograph, adjacent to May 2006 in the figure attests this fact. The laterite soils are seen as light-to-medium cyan color in the images.

9.5.2.1.8 A Case Study

A study was taken up to map acid soils in Mailanhalli village, Bangalore District, Karnataka with coarse-textured ferruginous (red) soils as dominant soils. The Indian Remote Sensing Satellite (IRS-ID) Linear Imaging Self-scanning Sensor (LISS-III) was used to generate soil resources map through monoscopic visual interpretation approach (Figure 9.5). The approach in brief involves delineation of broad lithological units, physiographic units therein, slope, and land use/land cover. Each unit, thus delineated, is termed landscape unit. It is followed by identification of sample points to study for morphological and chemical characterization of soils in each landscape unit; classification of soils based on morphological and physical and chemical characteristics according standard soil classification system, map finalization, accuracy estimation, and creation of digital database. The soil map, thus prepared, is appended as Figure 9.6 and chemical characteristics of a typical soil is appended as Table 9.1. As evident from the table, soils are acidic in nature with pH values (1:2 soil: water suspension) within the pedon ranging from 5.4 to 6.47 (Table 9.1).

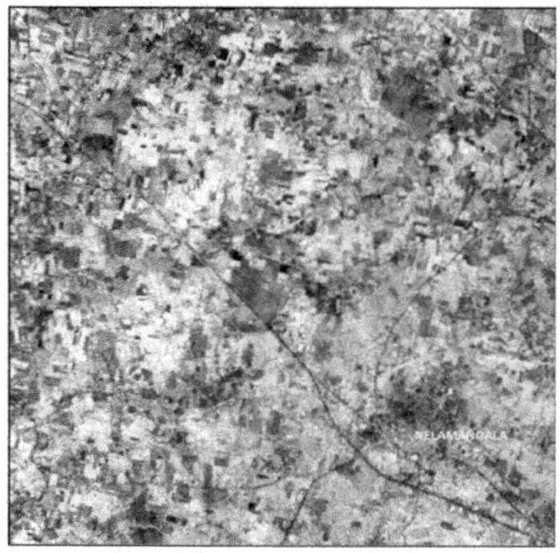

FIGURE 9.5
A standard false color composite of IRS LISS III image of the area (Courtesy: National Remote Sensing Centre, Indian Space Research Organization, Department of Space, Government of India).

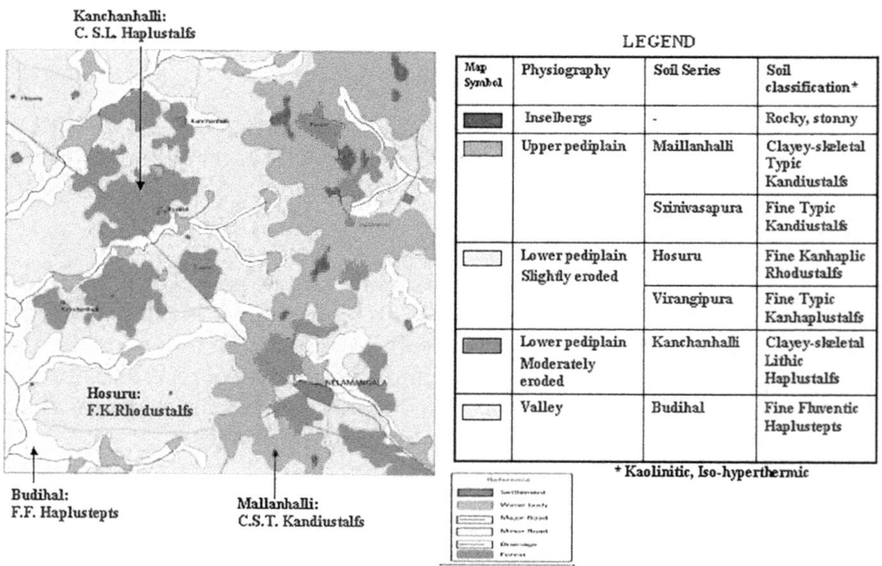

FIGURE 9.6
Soil map of the area (Source: Adapted from National Remote Sensing Agency (2008)).

9.5.2.2 Computer-Assisted Digital Analysis

The digital analysis approach relies solely on the reflected/scattered radiant energy from surface soils as measured by the sensors aboard air/spaceborne platforms. The approach is quite effective in areas which are bare/or have very sparse (<20%) vegetation cover, and the surface reflectance is related well with the type of soils occurring therein. However, multispectral digital data are generally used to a limited extent especially in areas where

TABLE 9.1

Physical and Chemical Properties Soils of Virangipura Series

Horizon	Depth (cm)	pH (1:2)	EC 1:2 (dSm^{-1})	Org. C (%)	Sand (%)	Silt (%)	Clay (%)	Texture
Ap	0–12	5.40	0.0025	0.20	87.44	2.05	10.51	ls
Bt-1	12–38	6.20	0.0041	0.45	50.10	5.34	44.56	sc
Bt2	38–73	6.00	0.0300	0.33	52.75	4.92	42.33	sc
BC1	73–85	6.35	0.0048	0.22	60.04	7.43	32.53	scl
BC2	85–106	6.47	0.0100	0.21	57.48	9.23	33.29	gscl
2Bt	106–120	6.42	0.0500	0.14	44.68	10.33	44.99	c
2BC	120–145	6.00	0.0700	0.07	46.04	30.33	23.63	l

spectral response pattern has direct bearing on the type of soils occurring therein, e.g., red soils, laterite soils peat, or muck without vegetation cover. Image enhancement techniques are also sometimes employed to improve the image contrast. Subsequently either spectral soil maps are generated using supervised or unsupervised approach or the enhanced data product is used for visual interpretation. Several studies have been carried out to utilize the potential of spaceborne multispectral measurements. For instance, the digital analysis of Landsat MSS data enabled Peterson et al. (1975) delineating mollic areas and mollic inclusions in a predominantly Alfisols in Indiana, USA. A good correlation between spectral soil maps and soil series could be achieved when ancillary data, namely physiographic boundaries, were included in the analysis of multispectral data (Weismiller et al., 1977). Furthermore, various image enhancement techniques were evaluated for deriving information on spatial distribution of soils (Dwivedi, 1984; Agbu and Niziyimana, 1991). The steps involved in digital analysis are discussed hereunder.

9.5.2.2.1 Database Preparation

The first step in generating multi-sensor datasets is the geo-referencing of the image to a common map grid. The multispectral data from a sensor if it meets the spatial resolution requirement could be used. Otherwise the multispectral data with coarser resolution could be digitally merged either the finer spatial resolution panchromatic data using appropriate algorithms.

9.5.2.2.2 Preliminary Digital Analysis

For deriving information on soil resources, the digital multispectral data could be displayed onto a color monitor of an image analysis system, and spectrally homogenous clusters are delineated using unsupervised classification algorithm. Based on the analyst's experience, the areas likely to represent a particular type of soils are identified as polygons after displaying the individual sensor's multispectral data or digitally merged multispectral and panchromatic data. These polygons are called training sets. Depending on spectral variations, as observed on a digital image, the training set represents almost all kind of variations. Spectral signatures of these training sets in terms of mean, standard deviation, correlation coefficients, and variance and co-variance matrices are generated. The spectral signature is then used for classifying the entire area of interest, and output in the form of a hard copy is generated. A color is assigned to each soil categories. Sample strips covering likely variation in the salt-affected soils are selected and precisely located onto topographic map at the mapping scale for subsequent ground truth collection/field verification.

9.5.2.2.3 Ground Truth Collection

The procedure adopted for field verification is same as in the case of soil mapping through visual interpretation. The location of sample points is recorded with the Global Positioning System (GPS).

9.5.2.2.4 Final Digital Analysis

To begin with, soils are classified based on morphological and chemical analysis data according the soil classification system prevailing in a particular country. Alternatively, the soil classification system developed by FAO/UNESCO, 1978 or World Resources Base for soil resources (IUSS, 2015) could be used. Thereafter the digital analysis of satellite data is initiated by locating the sample points and by identifying the categories into which soils have been initially classified during preliminary interpretation phase. After incorporating ground truth information the training set polygons are modified, training set signatures in terms of mean, standard deviation, and variance and co-variance matrix are regenerated, and the image is classified. In order to assess the spectral separability of soil classes, the transformed divergence (TD), Bhattacharya distance, and Jeffries–Matusita distances are utilized. Spectral separability indicates overlap or otherwise of the spectral classes. The routine is available in most of the image analysis software. For example, in the case of TD the values range between <0 and 2.0 or a multiple of 1,000 of these values. As a tentative guide

0.0< to <1.0 (poor separability)
1.0< to <1.9 (moderate separability)
1.9< to <2.0 (good separability)

Poor separability (0.0< to <1.0) indicates that the two signatures are statistically very close to each other. There are two ways to improve these parability of spectral signatures. One signature can be arbitrarily discarded when the separability is closer to 0, or the two signatures can be merged using merge option when the separability is closer to 1. Moderate separability (1.0< to <1.9) indicates that the two signatures are separable, to some extent. However, it is desirable to improve separability, if possible, perhaps by adding or modifying training areas. Low signature separability is usually caused by improper combinations of image bands, and/or training sites which have large internal variability within each class.

Further refinements in the training sets, if required, are made based on training set signature separability analysis. Several iterations are made till an optimal separability is achieved. Using these training set signatures as input, entire image is classified using per-pixel Gaussian maximum likelihood algorithm or any other suitable classification algorithm, and category-wise area statistics is generated.

9.5.2.2.5 Accuracy Estimation

For quantitative estimates of the classification accuracy of soil map generated though digital analysis of multispectral data the procedure is given in Section 9.5.2.1.

9.5.3 Mapping Vegetation-Covered Soils

As already mentioned, soils are either bare or covered with vegetation in varying degrees. In the case of bare soils the reflected/scattered/emitted radiant energy is from soil surface, and, in most cases, is related to soil properties and the type of soils occurring therein. In partially vegetation-covered soils the measured spectral response is admixture of soils and vegetation. The more the soil cover, the higher the contribution of reflected radiation from soils resulting, thereby decreasing normalized difference vegetation index (NDVI) values (indicator of vegetation density and vigor) with increasing soil brightness under identical environmental conditions (Huete, 1988; Tucker et al., 1985). According to Bartholomeus et al. (2007), accurate estimation of soil attributes is hampered if the pixels have a vegetation cover over 20%. In order to account for contribution of soils to spectral response of vegetation in partially vegetated terrain several vegetation indices, namely the soil adjusted vegetation index (SAVI) (Huete, 1988; Rondeaux et al., 1996), the transformed soil adjusted vegetation index (TSAVI) (Baret et al., 1989; Rondeaux et al., 1996), the Modified SAVI (MSAVI), and the global environment monitoring index (GEMI) (Rogan and Yool, 2001), have been used. For a more extensive overview of vegetation indices the reader may refer to Dorigo et al. (2007). Various researchers found a relationship between NDVI values and local-level soil properties like root zone soil moisture (Wang et al., 2007), soil color (Singh et al., 2006), soil texture and water-holding capacity (Lozano-Garcia et al., 1991), and soil carbon and nitrogen content (Sumfleth and Duttmann, 2008). Alternatively, NDVI time series have been used to derive soil patterns by analyzing changing NDVI values during a growing season and the onset of senescence during a dry season (Lozano-Garcia et al., 1991).

In fully/densely vegetated regions, the measured reflected/backscattered/emitted by remote sensing sensors is essentially from vegetation canopies. To retrieve soil properties, in addition to spectral response pattern, more detailed information on the vegetation cover is needed. Two proxy indicators: plant functional types (PFT) and Ellenberg indicator values have been used to retrieve soil properties from remote sensing data. PFTs is a system used by climatologists to classify plants according to their physical, phylogenetic and phenological characteristics as part of an overall effort to develop a vegetation model for use in land use studies and climate models. PFTs provide a finer level of modeling than biomes, which represent gross areas such as desert, savannah, and deciduous forest.

9.5.4 Digital Soil Mapping

Soil mapping procedures discussed in Sections 9.5.2.1 and 9.5.2.2 and conventional approaches are basically qualitative in nature. That is, soilscape units are defined in terms taxonomic units having a broad range of soil properties. Such information may be adequate for land valuation for taxation, agronomic planning, and even in military operations and developmental planning. However, in the context of global environmental changes that are currently altering key ecosystem services that soils provide, quantitative information on soil properties is needed to comprehend the role soil plays in the biophysical and biogeochemical functioning of the planet. Quantitative models have been developed, especially within the last three decades, to describe, classify, and study the spatial distribution patterns of soil more objectively enable generation of precise information on soil properties.

Digital soil mapping (DSM), one of the pedometric techniques, can quantitatively predict the spatial distribution of soil taxonomic classes. *Pedometrics* is a neologism derived from the Greek roots, pedos (soil) and metron (measurement), and is formed and used analogously to other terms such as biometrics, psychometrics, econometrics, hemometrics, and the oldest of all geometrics (Webster, 1994). DSM is defined as *"the creation and population of spatial soil information by the use of field and laboratory observational methods coupled with spatial and non-spatial soil inference systems"* (McBratney et al., 2003; Carré et al., 2007; Lagacherie et al., 2007). Key components of DSM are the method and the set of environmental covariates used to predict soil classes. The basis of DSM is the application of pedometric methods that predict the spatial and temporal distribution of soil types and soil properties. DSM relies on quantitative methods to integrate diverse soil observations from field, laboratory, and remote sensing and proximal sensing data (Grunwald, 2010) for inferring spatial patterns of soils across various spatial and temporal scales. Various methods that have been, or could be, used for fitting quantitative relationships between soil properties or classes and their "environment." These include *generalized linear models, classification and regression trees, neural networks, fuzzy systems, and geostatistics.* Terrain attributes derived from DEMs, and spectral reflectance from satellite imagery, have been the most commonly used in digital soil mapping.

The conceptual framework in which the pedometric methods are applied is the State Factor Equation of soil formation, first introduced by Jenny (1941) which states that soils can be described by the main environmental soil-forming factors, namely climate, organisms, relief, parent material, and time (CLORPT). DSM uses this concept to develop empirical models that relate observations of soil properties with environmental variables describing the main soil-forming factors (i.e., CLORPT). Refinements of the modeling framework were made over the years, including the SCORPAN (McBratney et al., 2003) framework, which is spatially explicit, and the STEPAWBH (Grunwald, 2011), which is both spatially and temporally explicit. McBratney et al. (2003) also developed a generic framework, the scorpan-SSPFe soil spatial prediction function with spatially autocorrelated errors method, which is particularly relevant for those places where soil resource information is limited. It is based on the seven predictive SCORPAN factors, a generalization of Jenny's five factors, namely: (1) s: soil, other or previously measured attributes of the soil at a point; (2) c: climate, climatic properties of the environment at a point; (3) o: organisms, including land cover and natural vegetation; (4) r: topography, including terrain attributes and classes; (5) p: parent material, including lithology; (6) a: age, the time factor; and (7) n: space, spatial or geographic position. Interactions (*) between these factors are also considered. It is particularly relevant for those places where soil resource information is limited.

The SCORPAN-SSPFe method essentially involves the following steps: (i) define soil attribute(s) of interest and decide resolution q and block size b, (ii) assemble data layers to represent Q, (iii) spatial decomposition or lagging of data layers, (iv) sampling of assembled data (Q) to obtain sampling sites, (v) GPS field sampling and laboratory analysis to obtain soil class or property data, (vi) fit quantitative relationships (observing Ockham's razor) with auto correlated errors, and (vii) predict digital map (McBratney et al., 2003). Different techniques produced different errors of interpolation. Hybrid methods such as CLORPT with geostatistics offer powerful spatial prediction methods, especially up to the catchment and regional extent (McBratney et al., 2000).

This approach is suitable if legacy soil data are scarce or unavailable and exhaustive REMOTE SENSING or proximal sensing data is available. Alternatively, if legacy soil data are available, soil and terrain attributes, derived from remote sensing or soil proxies, can be used as secondary variables to improve the interpolation of existing soil data (McBratney et al., 2003). In case remote sensing represents the primary data source, spatial interpolation using geostatistical techniques can be employed to map spatial patterns in areas with sparse soil data. In heterogeneous areas, methods like simple kriging and (generalized) linear models with independent variables, such as slope, curvature, wetness index, and soil profile information, have been used to derive soil attribute maps (Moore et al., 1993; Saby et al., 2011).

9.6 Conclusion

In order to ensure food security the productivity of existing cultivated land to be improved, additional land to be brought under plough. Rehabilitation and management of acid soils can contribute significantly in this endeavor. The information on acid soils with regard to nature, extent, magnitude of the problem is a prerequisite for their rehabilitation and management. Geospatial technology offers immense potential in providing such information is a timely and cost-effective manner. Treatment of acid soils with lime or materials rich in calcium and magnesium is the only strategy for their rehabilitation. However, with state-of-the-art technology the detection of sub-surface acidity and the deficiencies of calcium and magnesium, aluminum, being sub-surface phenomenon, does not seem feasible. Integration of remote sensing, proximal sensing, and *in situ* observations may provide some insight into sub-surface soil acidity.

References

Agbu, P.A., Niziyimana, E., 1991. Comparisons between spectral units derived from SPOT image texture and field soil map units. *Photogrammetric Engineering and Remote Sensing* 4:397–405.

Bartholomeus, H., Epema, G., Schaepman, M.E., 2007. Determining iron content in Mediterranean soils in partly vegetated areas, using spectral reflectance and imaging spectroscopy. *International Journal of Applied Earth Observation and Geoinformation* 9(2):194–203.

Baret, F., Guyot, G., Major, D.J., 1989. TSAVI: a vegetation index which minimizes soil brightness effects on LAI and APAR estimation, Geosciences and Remote Sensing Symposium, 1989. IGARSS'89/12th International Canadian Symposium on Remote Sensing. IEEE, New York, Vancouver, pp. 1355–1358.

Bennema, J., Gelens, H.F., 1969a. *Aerial Photo-Interpretation for Soil Surveys.* Lecture Notes ITC, Enschede, pp. 1–87.

Bennema, J., Gelens, H.F., 1969b. *Aerial Photo-Interpretation for Soil Surveys.* International Institute for Aerial Surveys and Earth Science, Delft, p 87.

Berga, L., Siriri, D., Ebanyat, P., 2001. Effect of soil amendments on bacterial wilt incidence and yield of potatoes in southernwestern Uganda. *African Crop Science Journal* 9(1):267–278.

Bishop, Y., Fienberg, S., Holland, P., 1975. *Discrete Multivariate Analysis-Theory and Practices.* MIT Press, Cambridge, MA, 575 p.

Bolan, N.S., Hedley, M.J., 2003. Role of carbon, nitrogen and sulfur cycles in soil acidification. In: Rengel, Z., (ed.) *Handbook of Soil Acidity.* New York, pp. 29–56.

Buringh, P., 1960. *The Application of Aerial Photography in Soil Surveys. Manual of Photographic Interpretation*, American Society of Photogrammetry, pp 631–666
Budianta, D., Vanderdeelen, J., 1995. *Dynamics of Exchangeable Aluminium in Ultisol.* International Conference on Soil Resources and Sustainable Agriculture; Kuala Lumpur, Malaysia 1995.
Carré, F., McBratney, A.B., Mayr, T., Montanarella, L., 2007. Digital soil assessments: beyond DSM. *Geoderma* 142:69–7.
Carson, C.D., Dixon, J.B., 1979. *Acidity. The Encyclopedia of Soil Science.* Hutchinson & Ross Inc., Stroudsburg, PA, pp. 1–3.
Chen, D., Lan, Z., Bai, X., Grace, J.B., Bai, Y., 2013. Evidence that acidification-induced declines in plant diversity and productivityare mediated by changes in below-ground communities and soil properties in a semi-arid steppe. *Journal of Ecology* 101:1322–34.
Christian, C.S., Stewart, G.A., 1968. Methodology of integrated surveys. In: *Aerial Surveys and Integrated Studies. Proceedings Toulouse Conference.* 1964. Natural Resources Service, UNESCO 6.
Congalton, R., Oderwald, R., Mead, R., 1983. A quantitative method to test for consistency and correctness in photo interpretation. *Photogrammetric Engineering & Remote Sensing* 49(1):69–74.
Crawford, T.W., Singh, U., Breman, H., 2008. Solving agricultural problems related to soil acidity in central Africa's great lakes region. CATALIST project report. Alabama: International center for soil fertility and agriculture development.
Defra, 2010. *The Fertiliser Manual (RB209).* TSO, London.
Dorigo, W.A., Zurita-Milla, R., de Wit, A.J.W., Brazile, J., Singh, R., Schaepman, M.E., 2007. A review on reflective remote sensing and data assimilation techniques for enhanced agroecosystem modeling. *International Journal of Applied Earth Observation and Geoinformation* 9:165–193.
Dwivedi, R.S., 1984. Utility of some image enhancement techniques for reconnaissance soil mapping: A case study from Southern India. In: *Proceedings of the International Symposium on Machine Processing of Remotely Sensed Data with Special reference to Thematic Mapper Data and Geographic Information System Laboratory for Applications of Remote Sensing*, Purdue University, Indiana, USA, June 12–14, 1984.
Dwivedi, R.S., 2017. *Remote Sensing of Soils.* Springer-Verlag, Berlin, Heidelberg.
Fageria, N.K., Baligar, V.C., 2008. Ameliorating soil acidity of tropical Oxisols by liming for sustainable crop production. In: Sparks, D.L., (ed.) *Advances in Agronomy.* Academic Press, Brazil, pp. 345–389.
FAO and ITPS, 2015. Status of the World's Soil Resources (SWSR)—Main Report. Food and Agriculture Organization of the United Nations and Intergovernmental Technical Panel on Soils, Rome, Italy.
Food and Agriculture Organization, 1974. FAO-UNESCO soil map of the world: Vol. 1, Legend. UNESCO, Paris. IUSS Working Group WRB, 2015.
Goosen, D., 1967. *Aerial Photo-Interpretation in Soil Survey.* FAO Soils Bulletin No.6: pp.1–55.
Goulding, K.W.T., Bailey, N.J., Bradbury, N.J., Hargreaves, P., Howe, M.T., Murphy, D.V., Poulton, P.R., Willison, T.W., 1998. Nitrogen deposition and its contribution to nitrogen cycling and associated soil processes. *New Phytologist* 139:49–58.
Grunwald, S. 2010. The current state of digital soil mapping and what is next. In: Boetinger, D.W., Howell, A.C., Moore, Hartemink, A.E., Kienst-Brown, S., (eds.) *Digital Soil Mapping: Bridging Research, Production and Environmental Applications.* Springer, Heidelberg, pp. 3–12.
Grunwald, S., 2011. Digital soil mapping and modeling at continental scales: Finding solutions for global issues. *Soil Science Society of America Journal* 75(4):1201–1213.
Guo, J.H., et al., 2010. Significant acidification in major Chinese croplands. *Science* 327:1008–1010.
Hardy, D.H., Raper, C.D., Miner, G.S., 1990. Chemical restrictions of root in Ultisol subsoils lessened by long-term management. *Soil Science Society of America Journal* 54:1657–1660.
Havlin, J.L., Beaton, J.D., Tisdale, S.L., Nelson, W.L., 2005. *Soil Fertility and Fertilizers: An Introduction to Nutrient Management.* Pearson prentice hall, Upper Saddle River, NJ.
Haynes, R.J., 1984. Lime and phosphate in the soil–plant system. *Advances in Agronomy* 37:249–315.
Hinsinger, P., Plassard, C., Tang, C., Jaillard, B., 2003. Origins of root-mediated pH changes in the rhizosphere and their responses to environmental constraints: a review. *Plant and Soil* 248:43–59.

Huete, A.R., 1988. A soil-adjusted vegetation index (SAVI). *Remote Sensing of Environment* 25(3):295–309.

IPCC, 2013. Climate change 2013: The physical science basis. Contribution of working group I to the fifth Assessment Report of the Intergovernmental Panel on Climate Change, Stocker, T.F., et al., (ed.) (Cambridge: Cambridge University Press).

IUSS Working Group WRB, 2015. World reference base for soil resources 2014: International soil classification system for naming soils and creating legends for soil maps-Update 2015. Food and Agriculture Organization of the United Nations Rome, 2015.

Jenny, H., 1941. *Factors of Soil Formation: A System of Quantitative Pedology*, McGraw Hill Book Company, Inc., New York.

Johnston, A.E., Goulding, K.W.T., Poulton, P.R., 1986. Soil acidification during more than 100 years under permanent grassland and woodland at Rothamsted. *Soil Use and Management* 2:3–10.

Kariuki, S.K., Zhang, H., Schroder, J.L., Edwards, J.E., Payton, M., Carver, B.F., et al., 2007. Hard red winter wheat cultivar responses to a pH and aluminium concentration gradient. *Agron Journal* 99:88–98.

Kiiya, W.W., Mwwogam, S.W., Obura, R.K., Musandu, A.O., 2006. Soil acidity amelioration as a method of sheep sorrel (rumex acetosella) weed management in Potato (Solanum tuberosum L.) in cool highlands of the north Rift, Kenya. KARI Biannual scientific conference; Nairobi: KARI; 2006.

Lagacherie, P., McBratney, A.B., Voltz, M., (ed.), 2007. *Digital Soil Mapping: An Introductory Perspective. Developments in Soil Science, 31.* Elsevier, Amsterdam.

Lozano-Garcia, D.F., Fernandez, R.N., Johannsen, C.J., 1991. Assessment of regional biomass–soil relationships using vegetation indexes. *IEEE Transactions on Geoscience and Remote Sensing* 29(2):331–339.

Lucas, R.W., Klaminder, J., Futter, M.N., Bishop, K.H., Egnell, G, Laudon, H., Hogberg, P., 2011. A meta-analysis of the effects of nitrogen additions on base cations: Implications for plants, soils, and streams, *Forest Ecology Manage* 262:95–104.

Marschner, P., (ed.), 2012. *Marschner's Mineral Nutrition of Higher Plants*, 3rd edn. Academic Press, London.

McBratney, A.B., Odeh, I.O.A., Bishop, T.F.A., Dunbar, M.S., Shatar, T.M., 2000. An overview of pedometric techniques for use in soil survey. *Geoderma* 97:293–327.

McBratney, A.B., Mendon Santosb, M.L., Minasny, B., 2003. On digital soil mapping. *Geoderma* 117:3–52.

McCauley, A., Jones, C., Olson-Rutz, K., 2009. *Soil pH and organic matter, nutrients managements module.* Montana State University, Bozeman.

McFarland, M.L., Harby, V.A., Redmon, L.A., Bade, D.H., 2001. *Managing Soil Acidity.* Texas Agricultural Experiment Station, Texas.

McGrath, S.P., Zhao F.J., 1995. A risk assessment of sulphur deficiency in cereals using soil and atmospheric deposition data. *Soil Use and Management* 11:110–114.

Moore, I.D., Gessler, P.E., Nielsen, G.A., Peterson, G.A., 1993. Soil attribute prediction using terrain analysis. *Soil Science Society of America Journal* 57(2):447–452.

Mulder, M.A., 1987. *Remote Sensing in Soil Science.* Elsevier, Oxford.

Mulder, V.L., Bruin, S., de Schaepman, M.E., Mayr, T.R., 2011. The use of remote sensing in soil and terrain mapping-A review. *Geoderma* 162(2011):1–19.

National Remote Sensing Agency, 2008. Development of diagnostic spectral response pattern for some terrain features using field spectroradiometer. Technical report. National Remote Sensing Agency, Indian Space Research Organization, Department of Space, Government of India.

O'Sullivan, J.N., Asher, C.J., Blarney, F.P'.C., 1997. Nutrient disorders of sweet potato. ACIAR Monograph No 48, 136 p.

Peñuelas, J., Poulter, B., Sardans, J., Ciais, P., van der Velde, M., Bopp, L., Boucher, O., Godderis, Y., Hinsinger, P., Llusia, J., Nardin, E., Vicca, S., Obersteiner, M., Janssens, I.A., 2013. Human-induced nitrogen-phosphorus imbalances alter natural and managed ecosystems across the globe. *Nature Communication.* Article no. 42934.

Peterson, J.B., Goodrick, F.E., Melhorn, W.N., 1975. Delineation of the boundaries of a buried preglacial valley with Landsat-1 data. Proc. *1st NASA Earth Resources Survey Symposium, Houston, Texas*, 97–103.

Rondeaux, G., Steven, M., Baret, F., 1996. Optimization of soil-adjusted vegetation indices. *Remote Sensing of Environment* 55(2):95–107.

Rogan, J., Yool, S.R., 2001. Mapping fire-induced vegetation depletion in the Peloncillo Mountains, Arizona and New Mexico. *International Journal of Remote Sensing* 22(16), 3101–3121.

RoTAP. 2012. Review of transboundary air pollution: Acidification, eutrophication, ground level ozone and heavy metals in the UK. Contract Report to the Department for Environment, Food and Rural Affairs. Centre for Ecology & Hydrology. ISBN 978-1-906698-22-5, 294 pp.

Saby, N.P.A., Marchant, B.P., Lark, R.M., Jolivet, C.C., Arrouays, D., 2011. Robust geostatistical prediction of trace elements across France. *Geoderma* 162(3–4):303–311.

Singh, D., et al., 2006. Environmental degradation analysis using NOAA/AVHRR data. *Advances in Space Research* 37(4):720–727.

Soil Survey Division Staff, 1993. *Soil Survey Manual*. US Department of Agriculture Handbook. US Government Printing Office, Washington, DC.

Sumfleth, K., Duttmann, R., 2008. Prediction of soil property distribution in paddy soil landscapes using terrain data and satellite information as indicators. *Ecological Indicators* 8(5):485–501.

Tisdale, S.L., Nelson, W.L., Beaton, J.D., Havlin, J.L., 1993. *Soil Acidity and Basicity*, 5th edn. Macmillan Publishing, New York.

Tucker, C.J., Vanpraet, C.L., Sharman, M.J., van Ittersum, G., 1985. Satellite remote sensing of total herbaceous biomass production in the Senegalese Sahel: 1980–1984. *Remote Sensing of Environment* 17(3):233–249.

Uchida, R., Hue, N.V., 2000. *Soil Acidity and Liming. Plant Nutrient Management in Hawaii soils, Approaches for Tropical and Subtropical Agriculture*. College of tropical agriculture and human resources, University of Hawaii, Manoa, HI.

Van Straaten, P., 2002. *Rocks for Crops: Agrominerals of Sub-Saharan Africa*. ICRAF, Nairobi.

Vink, A.P.A., 1964. Aerial photographs and soil sciences. *Proceedings of the Toulouse Conference on Aerial Surveys and Integrated Studies*: pp. 81–141.

von Uexküll, H.R., Mutert, E., 1995. Global extent, development and economic impact of acid soils. *Plant Soil* 171(1):1–15. DOI:10.1007/BF00009558.

Wang, X., Xie, H., Guan, H., Zhou, X., 2007. Different responses of MODIS-derived NDVI to root-zone soil moisture in semi-arid and humid regions. *Journal of Hydrology* 340(1–2):12–24.

Webster, R., 1994. The development of pedometrics. *Geoderma* 62:1–15.

Weismiller, R.A., Persinger, I.D., Montgomery, O.L., 1977. Soil inventory for digital analysis of satellite scanner and topographic data. *Soil Science Society of America Journal* 41:1166–1170.

World Reference Base for Soil Resources 2014, Update 2015. International soil classification system for naming soils and creating legends for soil maps. World Soil Resources Reports No. 106, Rome: FAO.

Wright, R.F., Alewell, C., Cullen, J.M., Evans, C.D., Marchetto, A., Moldan, F., Prechteland, A., Rogora, M., 2001. Trends in nitrogen deposition and leaching in acid-sensitive streams in Europe, *Hydrology and Earth System Sciences* 5:299–310.

Wulf, H., Mulder, T., Schaepman, M.E., Keller, A., Jörg, P., 2015. Zurich Open Repository and Archive, University of Zurich.

Yang, Y., et al., 2012. Significant soil acidification across northern China's grasslands during 1980s–2000s, *Global Change Biology* 18:2292–2300.

10

Waterlogging

10.1 Introduction

Waterlogging and subsequent soil salinization and/or alkalinization is the major land degradation process operating in arid, in semiarid, and, occasionally, in dry subhumid regions. Waterlogging, mainly caused by human intervention in natural drainage system, is the major form of physical soil degradation in central America but is also identified in South America, and other parts of the world. Globally, waterlogged areas account for some 10 million hectares (Oldeman et al., 1992). In India, the area estimates for waterlogging range from 4.75 (Irrigation Commission, 1972) to 6 million hectares (out of which 2.6 million hectares are on account of rising ground water table) (National Commission on Agriculture, 1976). Furthermore, according to an estimate provided by the Working Group of the Ministry of Water Resources (Anonymous, 1991) waterlogged areas cover 1.6 million hectares which is subject to waterlogging due to rising ground water table.

When water from rainfall or irrigation exceeds the net water movement out of the soil profile, soil pores will become water filled, displacing soil air and becoming "waterlogged" Waterlogging can occur in both irrigated and rain-fed areas, but it is more prevalent in irrigated areas (Cannell, 1977), where there is a greater chance of extra water from rain or additional irrigation on already wet soils. Waterlogging is commonly referred to as a condition where in there is obvious saturated soil and/or water ponded on the soil surface. Waterlogging can occur either where there is a rising water table or from continuous addition of water to the surface (e.g., during irrigation or high-intensity rainfall where soils have slow drainage). Impeded drainage or very slow soil water movement produce waterlogged conditions. This may occur because of clay texture, soil pans, compaction, smearing, the presence of local high water table, or surface ponding (Ross, 1989). Under natural conditions, low-lying areas, local depressions, and lower element of slopes are generally susceptible to waterlogging. Apart from this, fine-textured soils and shallow soils with hard sub-surface pans like *calcic, gypsic, or ferric* layers do experience waterlogging. Human interventions in the form of irrigated agriculture accelerate this process. The application of excess irrigation water and recharge from irrigation network causes gradual rise of ground water table and creates waterlogging condition. The excess moisture (waterlogging conditions) affects crop growth because of deficient aeration.

Excess water in soil can affect the crop growth and yield by way of (i) the direct effect on plant physiological processes from low oxygen concentrations in the soil around roots (rhizosphere), (ii) changes in soil chemistry and reduced nutrient availability resulting in toxicities of some elements and deficiencies of others, and (iii) changes in microbial activity affecting normal rhizosphere processes (Shaw et al., 2013).

Waterlogging may either be due to surface flooding or rise in ground water table, as a result of excess inflow compared to outflow which may be either on account of excess rain or over-irrigation and drainage congestion. Generally, waterlogging is referred to as stagnation of water on the land surface. However, from agriculture production point of view, waterlogging could be described as a situation of adverse air–water proportion in soil system with respect to an acceptable level of crop productivity response. Waterlogging refers to a soil condition whereby excess water in the root zone inhibits gas exchange with the atmosphere. Consequently, there is a significant decrease in the level of oxygen in the root zone with the accompanying increase in carbon dioxide concentration.

10.2 The Effects of Waterlogging

The deleterious effects of waterlogging on soils as well as in plants are briefly described hereunder. For further details readers may refer to Shaw et al. (2013).

10.2.1 The Effects on Soils

As mentioned in the previous section, water logging leads to low oxygen level in soils. Low oxygen conditions of waterlogging bring about changes in soil chemistry and microbial activities. The process of decomposition of organic matter in waterlogged soils is different from and slower than in aerated soils, resulting indifferent end products within the soil matrix. In well-aerated soil, the products from decomposition include carbon dioxide, nitrate, and sulfate, whereas in a waterlogged soil, the decomposition results are carbon dioxide, hydrogen, methane, ammonia, amines, mercaptans, and hydrogen sulfide (Ponnamperuma, 1972) (Academic Press, Inc.: Waltham, MA). In an aerated soil, oxygen acts as an electron acceptor, but when there is limited oxygen available, as with waterlogging, other compounds become electron acceptors; hence, redox potential is reduced. A lowered redox potential affecting O_2, NO_3^-, Mn^{4+}, Fe^{3+}, SO_4^{2-}, CO_2, and H^+ can result in potentially toxic chemicals accumulating within the rhizosphere (Ohlsson, 1979).

In waterlogged soils, oxygen deficiency predisposes to denitrification, with the consequent rapid loss of nitrate. Since nitrate is essential for physiological function, plant growth and appearance are quickly affected.

Nutrient availability is also affected by soil pH, as it is an important factor controlling chemical reactions within the soil matrix. Under low oxygen, the pH of acid soil changes toward neutral, whereas for alkalines oils the pH will decrease (Ponnamperuma, 1972). Hence, in waterlogged anoxic soil, the range and availability of nutrients available to plant roots will also change (Neumann and Romheld, 2012). Plant shoot concentrations of Na, B, Mn, Fe, and Al can increase and they may become toxic (Setter et al., 2009), while concentrations of NO_3^- and SO_4^{2-} may become deficient.

10.2.2 Plant Responses to Waterlogging

Different plant species exhibit different degrees of damage due to a waterlogging event. Sensitivity of plants to waterlogging might be expected to vary. During a waterlogging event, the oxygen concentration in stagnant water will decrease through hypoxic (low soil oxygen concentration) conditions to anoxic (zero soil oxygen concentration) conditions.

Experiments have been conducted on a variety of species to determine the "critical oxygen pressure," the minimum oxygen concentration required before the plant starts to shows signs of stress (Girton, 1979; Saglio et al., 1984), and is generally ~10% at 25°C (Meyer et al., 1985; Drew, 1997). Consequently, many plant species undergo adaptation shortly after waterlogging occurs. Aerobic respiration is affected through a change in the glycolysis process in waterlogged plants (Zabalza et al., 2009). Less ATP is produced, resulting infermentation producing only one-eighteenth of the energy (Bramley et al., 2007), implying that adaptation has a cost that will most likely lead to reduced growth and yield. It has been suggested that the magnitude of the effect of waterlogging on yield is related to the occurrence at particular development stages.

Setter and Waters (2003) provide a thorough review of plant waterlogging tolerances at different growth stages. In their study, wheat and barley were found to be more tolerant to waterlogging during tillering and grain filling than during preemergence, seedling growth, and reproductive stages. Plants that are subject to waterlogging often develop symptoms of leaf yellowing, wilting, root rotting, and root blackening. As discussed in Section 10.1, waterlogging can cause a change in microbial activity within the rhizosphere (Ponnamperuma, 1972), which can lead to root rotting and blackening. Curiously, leaf wilting, usually strongly associated with water deficit, has been observed in several species under waterlogged conditions. Tomato leaves wilt when waterlogged for 4–6 h (Jackson, 2002). Tobacco plants undergo leaf yellowing and wilting when waterlogged for 2 h (Kramer and Jackson, 1954). The aboveground symptoms are often the first indicators that crop yield may be reduced. Almost certainly, the aboveground symptoms indicate conditions below the ground are suboptimal (Shaw et al., 2013).

Experimental investigations on root growth in relation to aeration suggest that the roots may tolerate quite low levels of oxygen, but the accompanying increase in carbon dioxide may inhibit growth, water intake, and nutrient absorption in some species when it exceeds 4–5% in the soil atmosphere (Etherington and Armstrong, 1975). Waterlogging induces the redox (reduction) reactions in soils and influences organic matter decomposition. The main effects of waterlogging include (i) the production of gleying conditions in soils; (ii) changes in pH; (iii) the accumulation of organic matter due to slower anaerobic decomposition or fermentation process; (iv) the production of toxic by-products; and (v) the influence of reducing condition on retention, transformation, and losses of native plant nutrients. The sequence of events when a well-aerated soil becomes waterlogged is generally as follows. Patrick (1978) has recognized four stages. In the first stage NO_3^- reduction begins before complete removal of oxygen. Reduction of Mn^{4+} to Mn^{3+} occurs during the reduction of oxygen and nitrate. The reduction of iron (Fe^{3+} to Fe^{2+}) is usually the second stage but will not occur if oxygen or nitrate is present in the system. The reduction of NO_3 Mn^{4+} and Fe_3 is undertaken by facultative anaerobic bacteria. The third stage is the reduction of SO_4^{2-} to S_2 (usually Fe_2S or H_2S) by true anaerobic bacteria and shows the complete absence of O_2, NO_2, and NO_3^-. Finally, CH_4 is not produced until most of the sulfate is reduced to sulfide. Besides, waterlogging has marked effects on organic matter accumulation and decomposition. Furthermore, nutrient cycling especially nitrification is affected by waterlogging. Inorganic nitrogen in the form of NH_4-N occurs.

In arid, semiarid and dry subhumid regions especially in areas with well-defined wet and dry seasons, and adequate salt reserves in the sub-surface, waterlogging results in the development of soil salinity and/or alkalinity. The incremental rise in ground water table particularly in irrigated commands due to percolation of excess irrigation water and lateral seepage from canal results in the rise in ground water table. When ground water table reaches very close to the surface (<2 m from the land surface), it establishes capillary linkage. During dry period (winter and summer), to meet the evaporative demand, shallow ground water moves to the surface through capillary action. Dissolved salts in the ground water too move to the surface along with the water. Water evaporates leaving the dissolved salts behind. Thus, continued salt accretion on the land surface leads to build up of salts on the land surface, and ultimately results in the formation of salt-affected soils.

10.3 Norms for Categorization

As mentioned earlier, the water table, which affects adversely the yield of agricultural crops, varies depending upon rooting zone of crops, the type of soil, and the quantity of water. The actual depth of ground water table, when it starts affecting the yield of crops adversely, may vary over a wide range from zero for rice to about 1.5 m for other crops. For shallow-rooted crops like rice, potato, cauliflower, cabbage, lettuce, onion, etc., the root zone depth is taken as 60 cm whereas for moderately deep-rooted crops like wheat, groundnut, musk melon, carrot, peas, beans, chilies, etc., it is 90 cm. For deep-rooted crops like maize, cotton, sorghum, pearl millet, soybean, sugar beat, tomato, etc., it is 120 cm, and for very deep-rooted crops like sugarcane, citrus, coffee, apple, grape, wine, sunflower, lucern, etc., it is 180 cm. The water table within the above-mentioned depths may have adverse effects on crop growth and yield. However, various states in India have defined their own norms for water-table depth when it starts affecting or likely to affect crop growth and yield. Taking into account the details of norms followed by different states, rooting depths of different crops under different soil and fringe areas of water table which affect the upper layer of the soil due to capillary action/evaporation, the Working Group of the Ministry of Water Resources (Anonymous, 1991) has adopted the following norm for irrigated lands.

Waterlogged areas	Water table within 2 m of the land surface
Potential areas for	Water table between 2 and 3 m below land surface waterlogging
Safe areas	Water table below 3 m of the land surface

10.4 Role of Remote Sensing

Several groundwater flow modeling studies have focused on assessing the waterlogged areas and to control problems of waterlogging and salinization. These studies include Konikow and Bredehoeft (1974), Ramana et al. (1995), Gates et al. (2002), Dawoud and Allam (2004), etc.

Wildman (1982) found that plant response is a more accurate means of detecting poorly drained soils in California mainly because of a build-up of the water table. He also indicated that due to the accumulation of organic matter, soil color is generally darker in poorly drained areas than well-drained soils. The visible bands in Landsat-MSS and SPOT data can be used to identify this color. Baber (1982) pointed out that color-infrared photography could indicate drainage problems by soil moisture saturation or plant stress. Shallow water tables exhibit an increase in surface moisture, which can be detected from visible reflectance and microwave emissivity.

Remote sensing data acquired in the visible, near-infrared (NIR) and short-wave infrared (SWIR) regions have shown encouraging results in providing information on spatial pattern of waterlogging (Kalubarme et al., 1981; Sharma and Bhargava, 1988; Chaubey and Chaubey, 2000). Such studies have, however, enabled detection of waterlogged areas with either standing water (surface ponding) or a thin film of water at the surface or the land with wet surface, using Landsat-MSS and TM; and Indian Remote Sensing Satellite IRS-1A/-1B/-1C/-1D Linear Imaging Self-Scanning Sensor (LISS-I, -II, and -III) data. In Australia, conjunctive use of spectral measurements made by a Portable Field Spectroradiometer and a 13-channel airborne scanner covering 0.45–12.0 μm region of the electromagnetic spectrum and the Landsat-TM data helped in segregation of crops suffering from waterlogging, from those which are normal (Wallace et al., 1993). The dynamics of waterlogging has also been studied in the Sharada Sahayak and Periyar–Vaigai command areas (National Remote Sensing Agency, 1995; 1997), in parts of Gandak command in Bihar (Dwivedi, 1997) and eastern Uttar Pradesh (Sujatha et al., 2000) using multi-temporal multispectral data.

As pointed out in Section 10.3 waterlogging could be a surface phenomenon and may be confined to local depressions or low-lying areas or lower element of slopes. However, sometimes waterlogging may result from rising ground water table. While waterlogging associated with the surface ponding or standing water/surface wetness/presence of vegetation could be amenable to remote sensing, the detection of sub-surface waterlogging has been a challenging task due to the lack of any visible manifestation on the land surface. Optical, thermal, and microwave sensors data apart from geophysical sensors data have been used for detection of waterlogging.

10.4.1 *In situ* Spectral Reflectance Studies

Water typically looks blue or blue-green due to stronger reflectance at these shorter wavelengths, and darker if viewed at red or NIR wavelengths. In fact, in the NIR region there is almost total absorption of incident radiation making the water look dark in this region of the spectrum (Figure 10.1). If there is suspended sediment present in the upper layers of the water body, then this will allow better reflectivity and a brighter appearance of the water. Generally, the spectral response pattern of suspended sediment(s) and that of shallow (but clear) water is quite similar. Chlorophyll in algae absorbs more of the blue wavelengths and reflects the green, making the water appear greener in color when algae are present. The topography of the water surface (rough, smooth, floating materials, etc.) can also lead to complications for water-related interpretation due to potential problems of specular reflection and other influences on color and brightness.

The presence of moisture in the soil affects its spectral response considerably. Clark (1981) noted a dramatic decrease in the albedo, and other changes related to water and lattice-OH from dry to wet condition of montmorillonite at room temperature. Further, adding water to montmorillonite sample enhanced the water OH features at 0.94, 1.2, 1.4, and 1.9 μm, due to relatively high surface area and a corresponding high content of adsorbed water. While studying the reflectance spectra of a representative soil (*Typic Hapludalf*) over various water tensions, Baumgardner et al. (1985) observed a decrease in the general albedo and the water absorption features at 1.4 and 1.9 μm.

In Western Australia, McFarlane et al. (1992) used portable field spectrometer (PFS) with 256 continuous band sand operating in the 0.4–2.5 μm region for measuring the spectra of

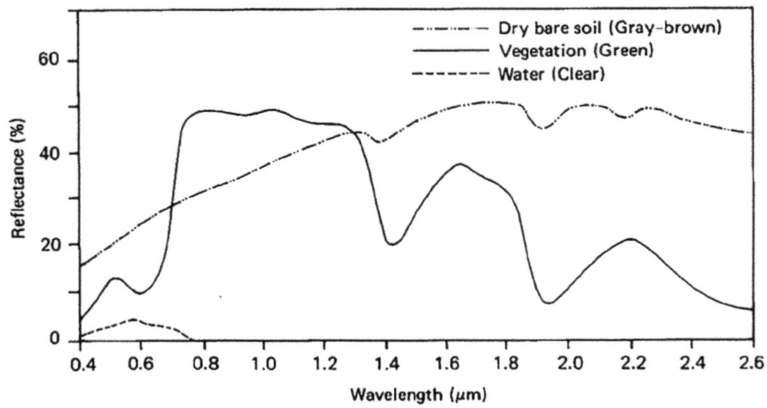

FIGURE 10.1
Spectral reflectance pattern of water and other major terrain features.

crops affected by waterlogging. For wheat crop there was a poor separation in the visible part of the spectrum but a good separation between 0.80 and 1.8 μm in the NIR region of the spectrum. Besides, there were separations between 1.2 and 1.38, between 1.46 and 1.38, and between 1.94 and 2.5 μm. The PFS spectra for waterlogged and non-waterlogged oat canopies showed that there was slightly better separation in the visible part of the spectrum (0.40–0.65 μm) than for wheat. Furthermore, there was an overlap between the spectra of two waterlogged and two non-waterlogged pasture canopies in the month of October. In, general, pasture spectra were more variable than crop spectra.

10.4.2 Aerial Photographs and Airborne Spectral Measurements

Aerial photographs have been used for delineation of the land subject to waterlogging. Wallace et al. (1993) used 1:10,000 scale aerial photographs over part of Western Australia for detection and mapping of waterlogged areas. In India, Sahai and Kalubarme (1985) used black-and-white and color-infrared aerial photographs at 1:30,000 and 1:50,000 scale for delineation of waterlogged areas with water table within 1.5–3.0 m in the Ukai command area in Gujarat, western India.

Wallace et al. (1993) used a 13-channel airborne multispectral scanner (GEOSCAN Mark 1) operating in the visible and NIR region (0.45–0.97 μm), short-wave (1.98–2.40 μm), and thermal-infrared region (8.5–12.5 μm) of the spectrum to detect and map areas in cereal crops where growth has been affected by waterlogging in Western Australia. NIR and thermal channels were found to be important in the discrimination between waterlogged and non-waterlogged areas.

However, the diagnostic absorption bands for uncombined water occur at 1,400, 1,900, and 2,850 nm (Taylor et al., 1994). The soils with the strongest water absorption features occur where the water table is closer to the ground (within 30 cm).

10.4.3 Spaceborne Multispectral Measurements

10.4.3.1 Optical Sensor Data

Not only there is an inter-annual variations in the spatial extent of waterlogged areas but also such variations are observed even intra-annually. Multi-cropping season spaceborne

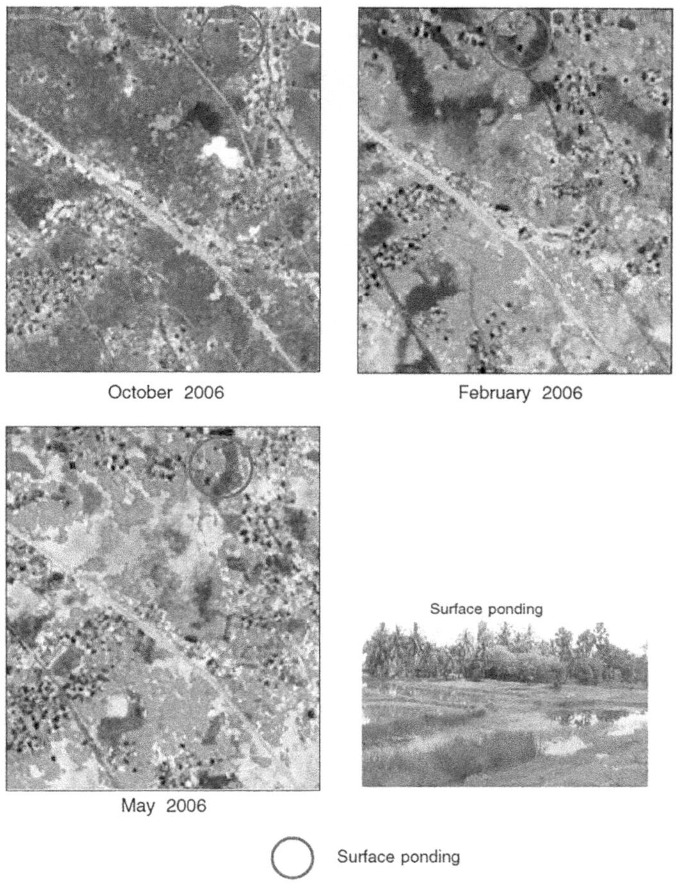

FIGURE 10.2
Waterlogging as manifested on the surface Resourcesat-2 LISS-III images of three cropping seasons, namely *kharif* (monsoon) October 2006, *rabi* (winter season) February 2006, and *zaid* (summer season) May 2006. Cyan color of different hues indicates waterlogged areas. A ground photograph of the waterlogged areas in also appended (Credit: National Remote Sensing Centre, Indian Space Research Organization, Department of Space, Government of India).

optical sensor data offer immense potential in delineating waterlogged areas (Figure 10.2). Besides, multi-temporal spaceborne multispectral data have been found very useful in studying the dynamics of waterlogging. Figure 10.3 vividly demonstrates such a potential of bi-temporal multispectral Landsat MSS and Resourcesat-2 LISS-III data in studying the dynamics of waterlogging in part of Rajasthan, western India.

Optical remote sensing data acquired in the visible, NIR and SWIR regions have shown encouraging results in providing information on spatial pattern of waterlogging (Kalubarme et al., 1981; Sharmaand Bhargawa, 1988; Chaubey and Chaubey, 2000). Landsat-TM data helped in segregation of crops suffering from waterlogging, from those which are normal (Wallace et al., 1993). Using spaceborne multispectral measurements made in the reflective portion of the spectrum waterlogged areas with surface ponding or a thin film of water on the surface or wetness of the surface layer could be delineated (Sahai and Kalubarme, 1985). The potential of the Indian Remote Sensing satellite (IRS-1A)

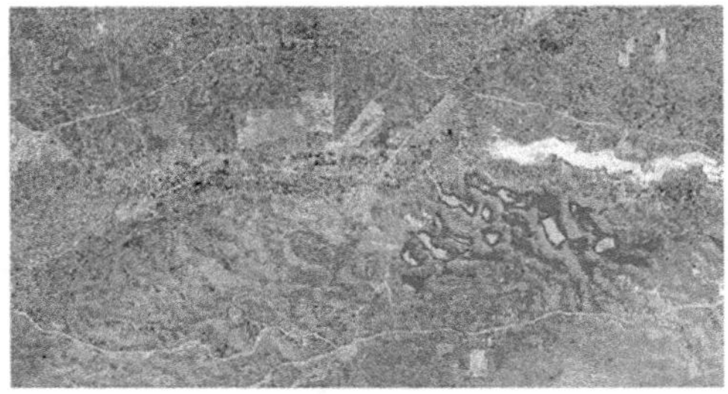

FIGURE 10.3
The principal component transformation of multi-temporal multispectral Landsat MSS and Resourcesat-2 LISS-III data. Within water bodies shallow water with cyan color and deep and clear water with dark blue color are very clearly seen. Waterlogged areas are manifested light yellow to yellow color. In addition canal network also has come out well.

Linear Imaging Self-Scanning (LISS-II) has been evaluated over part of Mahanadi stage-1 command area, Odisha (Figure 10.4). Two categories of waterlogging, namely seasonal and perennial, could be delineated. Shown in the figure are also the changes that have taken place in spatial extent and distribution of waterlogged areas during the period 1980 and 1998 using Indian Remote Sensing satellite (IRS-1D) Linear Imaging Self-Scanning (LISS-III) data of 1999. Such a capability has been operationally used for delineating and monitoring water-logged areas for entire Sharda canal system covering an estimated 6.7 million hectares in Uttar Pradesh, northern India (National Remote Sensing Agency, 1995). Similar studies have been carried out in other irrigation command areas, namely Periyar–Vaigai (Tamil Nadu) command areas (National Remote Sensing Agency, 1997); parts of Gandak command in Bihar (Dwivedi, 1997); Nagarjuna Sagar command area, part of Andhra Pradesh, southern India (Dwivedi et al., 1999).

The potential of the Indian Remote Sensing satellite (IRS-1A) Linear Imaging Self-Scanning (LISS-II) and Landsat-TM data has been evaluated over part of Mahanadi stage-1

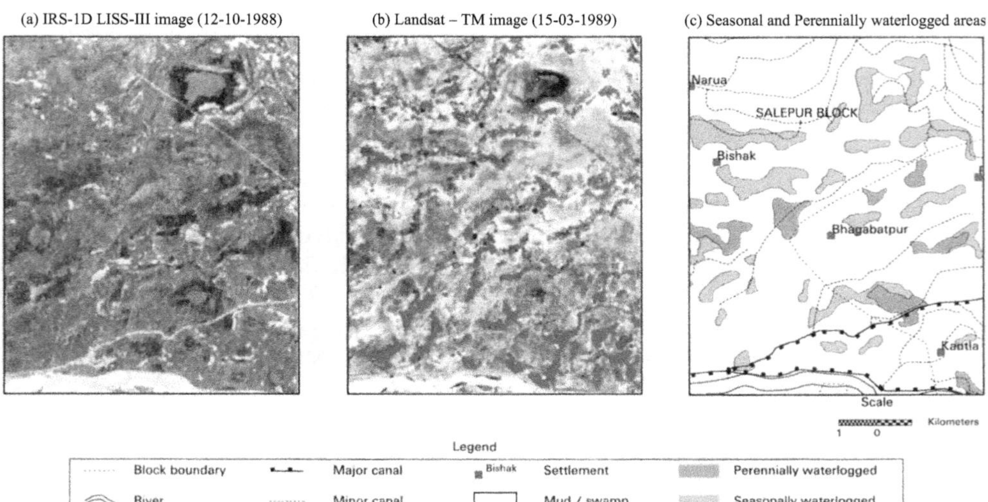

FIGURE 10.4
Waterlogging in part of Mahanadi stage-1 canal command area, Odisha, eastern India.

command area, eastern India for delineation of waterlogged areas (Figure 10.4). As evident from the figure, two categories of waterlogging, namely seasonal and perennial, could be delineated. While permanently waterlogged areas have either standing water or exhibit surface wetness during post-monsoon and summer season, seasonally waterlogged areas have such features post-monsoon season only.

A Case Study: To demonstrate the potential of spaceborne multispectral and multi-temporal data a study that was carried out in part of Indira Gandhi irrigation command area, part of Ganganagar district of Rajasthan, western India is presented here. The Landsat Multi-Spectral Scanner (MSS) and Thematic Mapper (TM), and Resourcesat-2 Linear Imaging Self-Scanning Sensor (LISS-III) data for the years 1975, 1985, 1995, and 2005 were used to study the dynamics of waterlogging. The study revealed that (i) the areas which were non-vegetated during 1975 were found supporting more and more vegetation especially crops in successive years due to introduction of canal irrigation; and (ii) the palaeochannel (old river course) which supported very good crops during 1975 gradually became waterlogged during the 30-year period thereby resulting in the extinction of croplands and consequently abandoning of settlements therein. As evident from Figure 10.5, there was no indication of waterlogging in 1975, with the passage of time surplus irrigation water from agricultural fields and from lateral seepage from canals, the ground water table began building up. Consequently in 1985 waterlogging began

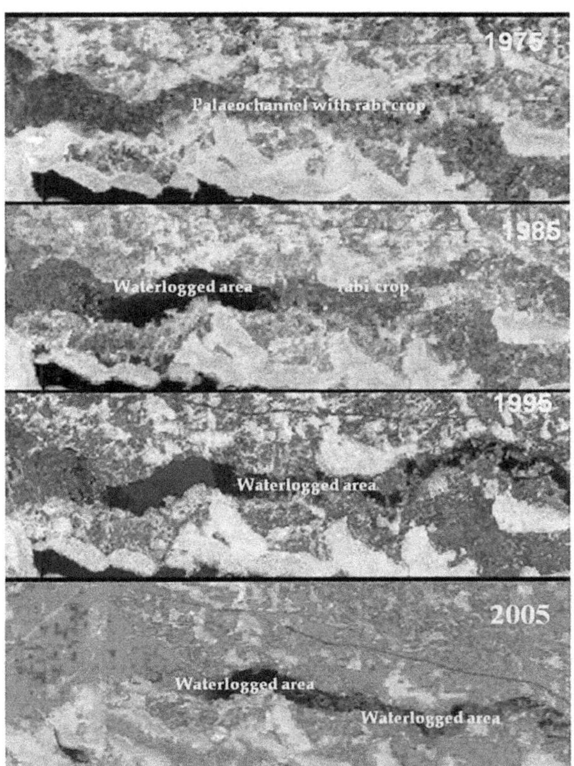

FIGURE 10.5
The dynamics of waterlogging in part of Indira Gandhi canal command area, Ganganagar, Rajasthan, western India as manifested in multispectral temporal images during the period 1975–2005.

manifesting on the surface (Figure 10.5). During the subsequent periods, i.e., 1990 and 1995 additional area began experiencing waterlogging.

The digital analysis of these dataset reveals an increase in waterlogged areas to the tune of 4,172 hectares during 30 years period.

10.4.3.2 Thermal Sensor Data

The effect of groundwater on soil temperature was noted as early as the 1930s (van den Bouwhuysen, 1934). In one of the pioneering investigations, Kappelmeyer (1957) successfully used near surface soil temperatures (1.5 m depth) to locate fissures carrying hot water. Since then, studies have been using soil temperature at shallow depths (0.5–2 m) to locate aquifers and delineate their flow systems. Cartwright (1968) was among the earliest to advocate the use of land surface temperature (LST) measurement to locate shallow ground water. Soil temperature measurements showed that water tables within the depth of annual soil temperature variation constitute a heat sink in the summer and a heat source in the winter. Subsequently, Chase (1969) noted that cool areas on thermal-infrared imagery seemed to correspond with shallow ground water. Heilman and Moore (1982) used data from the Heat Capacity Mapping Mission satellite to examine shallow ground water in this same region, and found good agreement between cool areas in night time thermal images and areas of known shallow ground water. The radiation temperature measurements made by passive microwave sensors have shown encouraging results in areas experiencing waterlogging due to rising ground water table. Similarly, in another study conducted in part of Mahanadi stage-I command, Odisha State, India day and night thermal infrared images from Landsat-TM were used to explore its potential in the detection of sub-surface waterlogging (Figure 10.6). Owing to high thermal inertia, waterlogged areas show up as dull white to white patches within dark-toned background in the night-time thermal image (National Remote Sensing Agency, 2001). On day-time thermal image, waterlogged areas are seen as medium to dark gray tone. The normalized difference temperature index (NDTI) from day-and night-time thermal data was generated as follows:

$$\text{NDTI} = \frac{(\text{D. Thermal DN values} - \text{N. Thermal DN values})}{(\text{D. Thermal DN values} + \text{N. Thermal DN values})} \tag{10.1}$$

where D = day-time, N = night-time, and DN = digital number.

The sub-surface water logged areas appear as dark gray tone of irregular shape in the NDTI image.

In another study utilizing post-monsoon season Terra-1 Moderate Resolution Imaging Spectrometer (MODIS) acquired during day- and night-time over part of Uttar Pradesh Sreenivas et al. (2010) evaluated the potential of spectral indices derived from NIR, SWIR (SWIR), and thermal-infrared spectral channels for detection of sub-surface waterlogging. Among various spectral indices evaluated for their potential in the detection of sub-surface waterlogging, day-and night-time land surface temperature difference (DLST) and night-time LST have been found to correlate well with the incidence of waterlogging (water-table depth <2 m), followed by normalized difference of band-2 (841–876 nm) and band-7 (2,105–2,155 nm). For the reliability of remote sensors in detecting shallow groundwater effect, Alkhaier et al. (2012) concluded from their study that the effect can be sufficiently clear to be detected if at least one of the following conditions met: high-potential evaporation and high contrast between day and night temperatures. Furthermore, under these

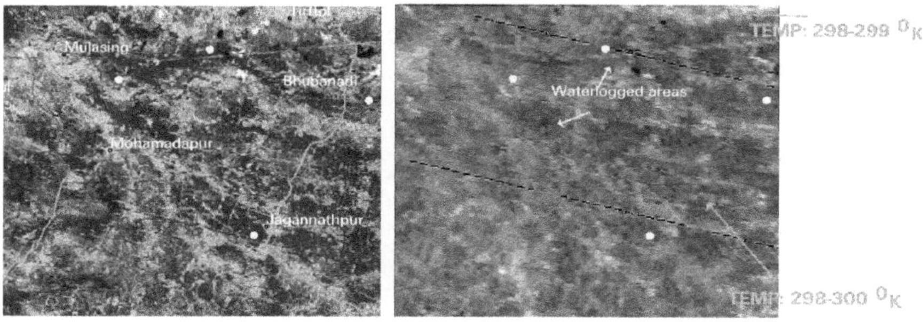

Landsat TM standard FCCof November 2,2000. Day-time thermal image of November2, 2000.

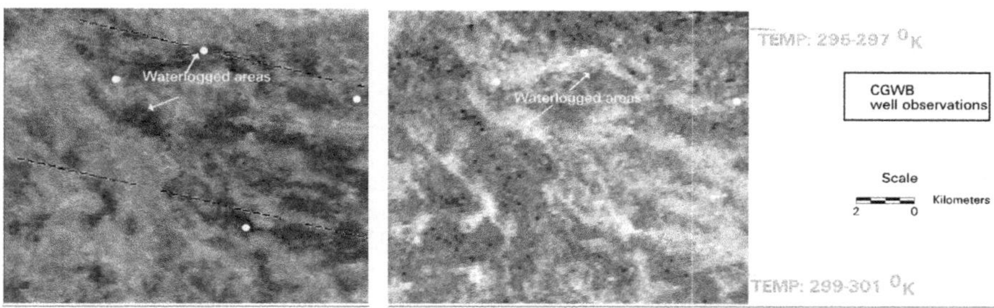

Normalized difference temperature Indeximage. Night-time thermal image November2,2000.

FIGURE 10.6
Detection of sub-surface waterlogging using Landsat TM thermal band (10.5–12.5 µm) day- and night-time data over part Mahanadi Stage-I command area, Odisha, eastern India. Yellow full circles represent ground observation points (Source: Adapted from National Remote Sensing Agency (2001)).

conditions, most day and night hours are suitable for shallow groundwater depth detection using thermal sensor data.

10.4.4 Geophysical Techniques

Ground-penetrating radar (GPR) and electromagnetic induction (EMI) techniques have been used especially for detection of sub-surface waterlogging due to rising ground water table.

10.4.4.1 Ground-Penetrating Radar (GPR)

GPR is a noninvasive geophysical tool that is specifically designed to penetrate earthen materials and provide images of the shallow (0–30 m) subsurface. GPR operates by transmitting pulses of radio-frequency electromagnetic energy into the subsurface. These waves of electromagnetic energy propagate through the soil until they impinge upon a layer or object with contrasting dielectric properties. Contrasts in dielectric properties cause a portion of the transmitted electromagnetic energy to be reflected back to an antenna. The GPR records the strength of reflections for a set duration after each pulse. A plot of these data is called a "trace." The trace signal is usually plotted against a time scale measured in nanoseconds GPR systems generate a rapid succession of traces, which are displayed as a "radar gram." A radar gram is simply a display of one trace after another with the intensity of the reflected signal represented by different colors or shades of gray.

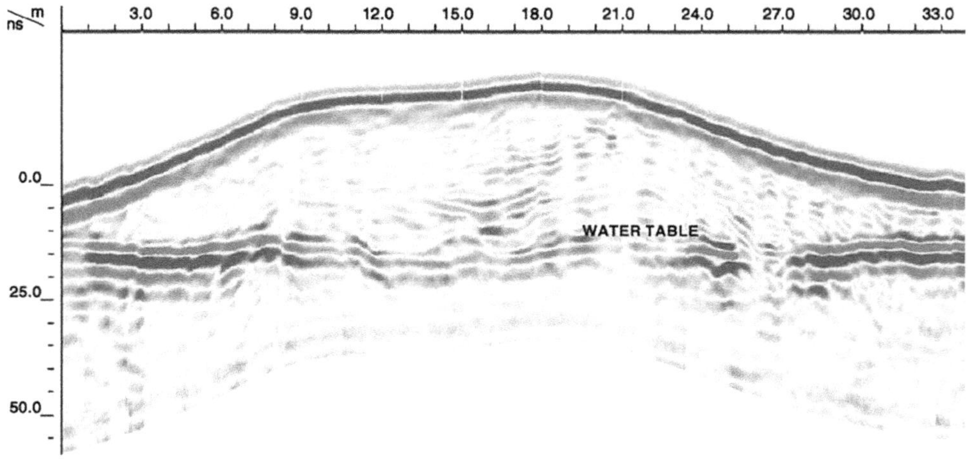

FIGURE 10.7
A radar record obtained with a 200 MHz antenna on a low aeolian dune in North Dakota. The water table provides high-amplitude linear reflections. The depth of penetration is limited by the conductivity of the alkaline ground water (Source: Adapted from Doolittle et al. (2000)).

As pointed out earlier, the major limitation of optical sensors data in detecting the waterlogging is their inability to provide information on areas affected by waterlogging due to rising ground water table. GPR is well known as one of the best tools for charting groundwater table depths (Sellmann et al., 1983; Smith et al., 1992; Doolittle et al., 2000;). In Figure 10.7, the GPR record over part of an aeolian landscape in northwestern Indian is shown. The average depth to the water table ranges from about 0.9 to 2.8 m on this radar record. The water table provides a continuous, high-amplitude reflection that is easily identified and traced across the radar record. Recently, Mahmoudzadeh et al. (2012) used a pulse radar with a single 200 MHz bowtie antenna combined with a differential GPS, and Electrical Resistivity Tomography (ERT) instrument to investigate the water-table depth in weathered granites in a semiarid catchment (Sardon, Salamanca, Spain), and had observed that the shallow water table was sparsely detected in the GPR profiles ranging from ~1 to ~3 m the entire catchment. The results showed a good agreement between ERT and GPR profiles.

10.4.4.2 Electromagnetic Induction (EMI) Sensors

In EMI system, apparent conductivity ECa is measured. The measured conductivity depends on the salt concentration in soil, type, and amount of clay and soil moisture and soil porosity. This technique has been used in electromagnetic conductivity meters such as EM31 and EM38 manufactured by Geonics. The instrument has a data logger incorporated into the control console. Ground conductivity and in-phase measurements are read directly from the data logger screen, or logged for later analysis. The effective depth of exploration for this unit is about 6 m, making it ideal for many geotechnical and groundwater surveys. Sherlock and McDonnell (2003) used simple linear regression analyses to compare terrain electrical conductivity measurements (EC) derived from multiple EMI frequencies to a distributed grid of water-table depth and soil-moisture measurements in the hill slope of Putnam County, New York. Regression analysis indicates that EC measurements from the low-frequency EM31 meter with a vertical dipole orientation could explain over 80% of the variation in water-table depth across the test hill slope.

10.5 Using Models to Simulate Plant Responses to Waterlogging

There are several models used for predicting crop growth and yield, such as CropSyst (Stockle et al., 2013), CROPGRO (Boote et al., 1998), Decision Support System for Agrotechnology Transfer (DSSAT) (Jones et al., 1998), Crop Estimation through Resource and Environment Synthesis (CERES) (Ritchie et al., 1998), System Approach to Land Use and Sustainability (SALUS) (Basso et al., 2006), SWAGMAN® Destiny (Meyer et al., 1996), DRAINMOD (Skaggs, 2008), and the Agricultural Production Systems Simulator (APSIM) (Asseng et al., 1998). However, only three of the listed models predict yield reduction from waterlogging, SWAGMAN Destiny, DRAINMOD, and APSIM.

SWAGMAN (Salt Water And Groundwater MANagement) Destiny is a management tool developed by CSIRO Land and Water designed to estimate growth and yield of a range of crops that are affected by salinization, shallow groundwater, water deficit, and waterlogging. SWAGMAN Destiny inputs weather data as well as soil, crop, and agronomic information that characterize the crop system of interest. It assigns the potential yield that could be expected for the crop of interest, and then calculates the likely effects on crop growth of daily edaphic and climatic stresses. The accumulation of the various abiotic stresses (water deficit, aeration, nitrogen insufficiency, and salinity) is used to reduce the potential yield and hence estimate an actual yield.

DRAINMOD simulates the effectiveness of drainage based on the water balance in the soil profile. DRAINMOD was developed by Dr. Wayne Skaggs at the North Carolina State University in 1980. The model estimates water-table depths, soil water conditions, and drainage rates by input of soil properties, crop parameters, drainage design parameters, weather parameters, and irrigation data. It calculates infiltration, evapotranspiration, drainage, surface runoff, subirrigation, deep seepage, lateral seepage, water-table depth, and soil water status (Skaggs, 2008). Crop yields have been analyzed using DRAINMOD, and model variations have been documented to include crop parameters (Hardjoamidjojo et al., 1982; Wang et al., 2006; Hassanpour et al., 2011; Kale, 2011). As with SWAGMAN, DRAINMOD calculates the relative yield using the assigned potential yield and the reduced crop yield due to combined stresses. The relative yield depends on factors specific to individual crops and the water table.

APSIM was developed by the Agricultural Production Systems Research Unit, a collaborative group of CSIRO and Queensland State Government agencies. APSIM uses a collection of various modules—such as crop growth, soil water (SOILWAT), surface residue (RESIDUE), soil organic matter and soil nitrogen supply (SOILN), and soil pH (SOILpH)—to take account of interactions between soil water, nitrogen, and plant residues on crop growth, development, and yield. Weather, soil, crop, and management data are inputs. The module SOILWAT estimates soil water balance and has evolved from modifications to the models CERES and PERFECT (Productivity Erosion Runoff Functions to Evaluate Conservation Techniques) (Littleboy et al., 1989). The water characteristics of lower limit, drained upper limit, saturated, and volumetric water contents and soil layer thickness are specified. Processes of runoff, evaporation, saturated flow, unsaturated flow, and movement of solutes are represented (Keating et al., 2003).

Like SWAGMAN Destiny, the model uses a "tipping bucket" approach to redistribute water between layers, while a perched water table is calculated as the proportion of water content between the drained upper limit and saturation in a layer when the immediate layer below is saturated (Asseng et al., 1997).

10.6 Conclusions

Waterlogging affects the soil properties and crop growth and yields of agricultural crops. However, there is a large variation in the sensitivity of a particular crop to waterlogging. Waterlogging could either be due to prolonged surface ponding or because of rising groundwater table. Spaceborne multispectral optical sensor data hold very good promise for detection and monitoring of surface-waterlogged areas with surface ponding, the presence of a thin film of water and surface wetness. However, waterlogging due to rising ground water table, which in general, does not have any surface manifestation, could not be resolved using multispectral measurements made in the reflective region of the spectrum. The geophysical techniques, namely GPR operating in low frequencies (10–500 MHz), EMI sensors, and temperature measurements made by thermal scanners and microwave radiometers have shown promising results in the detection of sub-surface waterlogging. Since the geophysical instruments used for charting ground water table capture backscattered electromagnetic radiation from a sub-surface materials with contrasting dielectric properties, namely salts and water/soil moisture/metallic objects, the measurements from these sensors need to validated with the piezometric observations and ground water table data. Further research to identify system parameters for geophysical instruments, namely GPR and EMI best suited for detection of sub-surface waterlogging need to be taken up.

References

Alkhaier, G., Flerchinger, N., Su, Z., 2012. Shallow groundwater effect on land surface temperature and surface energy balance under bare soil conditions: modeling and description. *Hydrology and Earth System Sciences* 16, 1817–1831.

Anonymous, 1991. Waterlogging, soil salinity and alkalinity. Report of the Working Group on Problem Identification in Irrigated Areas with Suggested Remedial Measures, Ministry of Water Resources, Government of India.

Asseng, S., Keating, B.A., Huth, N.I., Eastham, J., 1997. Simulation of perched watertables in a duplex soil. In: *Proceedings of the International Congress on Modelling and Simulation – MODSIM97*, Hobart, Tasmania, Modelling and Simulation Society of Australia, Canberra, pp. 538–543.

Asseng, S., Keating, B.A., Fillery, I.P.R., Gregory, P.J., Bowden, J.W., Turner, N.C., Palta, J.A., Abrecht, D.G., 1998. Performance of the APSIM-wheat module in Western Australia. *Field Crops Research* 57, 163–179.

Baber, J.J. Jr., 1982. Detection of crop conditions with low-altitude aerial photography. In: *Remote Sensing for Resource Management*, C.J. Johannsen, J.L. Sanders, eds., Soil Conservation Society of America, 407–412.

Basso, B., Ritchie, J.T., Grace, P.R., Sartori, L., 2006. Simulation of tillage systems impact on soil biophysical properties using the SALUS model. *Italian Journal of Agronomy* 4, 677–688.
Baumgardner, M.F., Silva, L.F., Biehl, L.L., Stoner, E.R, 1985. Reflectance properties of soils. *Advances in Agronomy* 38, 1–44.
Boote, K.J., Jones, J.W., Hoogenboom, G., Pickering, N.B., 1998. The CROPGRO model for grain legumes. In: *Understanding Options for Agricultural Production*, G.Y. Tsuji, G. Hoogenboom, P.K. Thornton, eds., Kluwer Academic Publishers, Wageningen, the Netherlands, 99–128.
Bramley, H., Turner, D.W., Tyerman, S.D., Turner, N.C., 2007. Waterflow in the roots of crop species: the influence of root structure, aquaporin activity and waterlogging. *Advances in Agronomy* 96, 133–196.
Cannell, R., 1977. Soil aeration and compaction in relation to rootgrowth and soil management. *Applied Biology* 2, 1–86.
Cartwright, K., 1968. Temperature prospecting for shallow glacial and alluvial aquifers in Illinois. Illinois State Geological Survey Circular 433, Urbana, Illinois.
Chase, M.E., 1969. Airborne remote sensing for ground water studies in a prairie environment. *Canadian Journal of Earth Sciences* 6, 737–741.
Chaubey, V.K., Chaubey, R., 2000. Remote sensing of waterlogged areas in Sri Ram Sagar command. In: *Proceedings of the Eighth ICID International Drainage Workshop*, New Delhi, India, 31st January to 4th February, 2000, Vol. II, pp. 375–385.
Clark, R.N., 1981. The reflectance of water-mineral mixture at low temperatures. *Journal of Geophysical Research* 86, 3074–3086.
Dawoud, M.A., Allam, A.R., 2004. Effect of new Nag Hammadi Barrage on ground water and drainage conditions and suggestion of mitigation measures. *Water Resources Management* 18(4), 321–337.
Doolittle, J.A., Jenkinson, B.J., Franzmeier, D.P., Lynn, W., 2000. Improved radar interpretations of water-table depths and ground-water flow patterns using predictive equations. In: Noon, D. *Proceedings Eight International Conference on Ground-Penetrating Radar*, Goldcoast, Queensland, Australia, May 23–26, 2000, The University of Queensland, pp. 488–495.
Drew, M.C., 1997. Oxygen deficiency and root metabolism: injury and acclimation under hypoxia and anoxia. *Annual Review of Plant Physiology and Plant Molecular Biology* 48, 223–250.
Dwivedi, R.S., 1997. Mapping waterlogged areas in part of Indo-Gangetic plains using remote sensing data. *Geocarto International* 12(3), 65–70.
Dwivedi, R.S., Sreenivas, K., Ramana, K.V., 1999. Inventory of salt-affected soils and waterlogged areas: A remote sensing approach. *International Journal of Remote Sensing* 20(8):1589–1599.
Etherington, J.R., Armstrong, W., 1975. *Environment and Plant Ecology*, Willey Eastern Ltd., New Delhi, p. 126.
Gates, T.K., Burkhalter, J.P., Labadie, J.W., Valliant, J.C., Broner, I., 2002. Monitoring and modeling flow and salt transport in a salinity threatened irrigated valley. *Journal of Irrigation and Drainage Engineering* 128(2), 87–99.
Girton, R.E., 1979. Effect of oxygen concentration on the respiration of excised root tip segments of maize and rice, and germinating grains of rice and buckwheat. *Physiologia Plantarum* 46, 58–62. doi:10.1111/j.1399-3054.1979.tb03186.x.
Hardjoamidjojo, S., Skaggs, R.W., Schwab, G.O., 1982. Corn yield response to excessive soil water conditions. *Transactions of the American Society of Agricultural Engineers* 25, 922–927.
Hassanpour, B., Parsineejad, M., Yazdani, M.R., Salahshour Dalivand, F., Kossari, H., 2011. Evaluation of modified DRAINMOD in predicting groundwater table fluctuations and yield of canola in paddy fields under snowy conditions (case study: Rasht, Iran). *Irrigation and Drainage Systems* 60, 660–667. doi:10.1002/ird.614.
Heilman, J.L., Moore, D.G., 1982. Evaluating depth to shallow ground water table using Heat Capacity Mapping Mission (HCMM) data. *Photogrammetric Engineering and Remote Sensing* 48(12), 1903–1906.
Irrigation Commission, 1972. *Report of the Irrigation Commission*, Vol. I, Ministry of Irrigation and Power, Government of India, New Delhi.

Jackson, M.B., 2002. Long distance signalling from roots to shoots assessed: the flooding story. *Journal of Experimental Botany* 53, 175–181.

Jones, J.W., Tsuji, G.Y., Hoogenboom, G., Hunt, L.A., Thornton, P.K., Wilkens, P.W., Imamura, D.T., Bowen, W.T., Singh, U., 1998. Decision support system for agrotechnology transfer: DSSAT v3. In: Tsuji, G.Y., Hoogenboom, G., and Thornton, P.K., (eds) *Understanding Options for Agricultural Production*. Systems Approaches for Sustainable Agricultural Development, vol. 7, Springer, Dordrecht, pp. 157–177.

Kale, S., 2011. Estimating effects of drainage design parameters on crop yields under irrigated lands using DRAINMOD. *Scientific Research and Essays* 14, 2955–2963.

Kalubarme, M.H., Sahai, B., Bapat, M.V., 1981. Remote sensing of waterlogged and salt-affected soils in the Mahi command area. Presented at the National Symposium on Development and Management of Water Resources, 1981.

Kappelmeyer, O., 1957. The use of near surface temperature measurements for discovering anomalies due to causes at depths. *Geophysical Prospecting* 5, 239–258.

Keating, B.A., Carberry, P.S., Hammer, G.L., Probert, M.E., Robertson, M.J., Holzworth, D., Huth, N.I., Hargreaves, J.N.G., Meinke, H., Hochman, Z., McLean, G., Verburg, K., Snow, V., Dimes, J.P., Silburn, M., Wang, E., Brown, S., Bristow, K.L., Asseng, S., Chapman, S., McCown, R.L., Freebairn, D.M., Smith, C.J., 2003. An overview of APSIM, a model designed for farming systems simulation. *European Journal of Agronomy* 18, 267–288. doi:10.1016/S1161–0301(02)00108-9SOILWAT.

Konikow, L.F., Bredehoeft, J.D., 1974. Modeling flow and chemical quality changes in an irrigated stream-aquifer system. *Water Resources Research* 10(3), 546–562.

Kramer, P.J., Jackson, W.T., 1954. Causes of injury to flooded tobacco plants. *Plant Physiology* 29, 241–245.

Littleboy, M., Silburn, D.M., Freebairn, D.M., Woodruff, D.R., Hammer, G.L., 1989. PERFECT—A computer simulation model of Productivity, Erosion, Runoff Functions to Evaluate Conservation Techniques, Queensland Department of Primary Industries, Brisbane.

Mahmoudzadeh, M.R., Francés, A.P., Lubczynski, M., Lambot, S., 2012. Using ground penetrating radar to investigate the water table depth in weathered granites —Sardon case study, Spain. *Journal of Applied Geophysics* 79, 17–26.

McFarlane, D.J., Wheaton, G.A., Negus, T.R., Wallace, J.F., 1992. Effects of waterlogging on crop and pasture production in the Upper Great Southern, Western Australia. Technical Bulletin No. 86, Department of Agriculture, and Western Australia.

Meyer, W.S., Barrs, H.D., Smith, R.C.G., White, N.S., Heritage, A.G., Short, D.L., 1985. Effect of irrigation on soil oxygen status and root shoot growth of wheat in a clay soil. *Australian Journal of Agricultural Research* 36:171–185.

Meyer, W.S., Godwin, D.C., White, R.J.G., 1996. SWAGMAN® Destiny: A tool to predict productivity change due to salinity, waterlogging and irrigation management. In: *Proceedings of the 8th Australian Agronomy Conference*, Toowomba, Queensland, pp. 425–428.

National Commission on Agriculture, 1976. Report of the National Commission on Agriculture, Parts V, IX and Abridged Report. Ministry of Agriculture and Irrigation, Government of India, New Delhi.

National Remote Sensing Agency, 1995. Study of land degradation problems in Sharada Sahayak command area for sustainable development. Technical report, National Remote Sensing Agency, Hyderabad.

National Remote Sensing Agency, 1997. Satellite assessment of waterlogging and soil salinity/alkalinity in Periyar-Vaigai command area, Technical Report, National Remote Sensing Agency, Hyderabad.

National Remote Sensing Agency, 2001. Assessment and monitoring of waterlogging and soil salinity and alkalinity in Mahanadi Stage-1 command area, Orissa. Project report, National Remote Sensing Agency Hyderabad, Department of Space, Government of India.

Neumann, G., Romheld, V., 2012. Rhizosphere chemistry in relation to plant nutrition. In: *Marschner's Mineral Nutrition of Higher Plants*, 3rd edn, P. Marschner, ed., Elsevier Limited, UK, pp. 347–368.

Ohlsson, T., 1979. Redox reactions in soil sequence of redox reactions in a waterlogged soil. *Nordic Hydrology* 10, 89–98.

Oldeman, L.R., Hakkeling, R.T.A., and Sombrock, W.G., 1992. *World Map of the Status of Human-induced Soil Degradation: An Explanatory Note. Global Assessment of Soil Degradation (GLASOD).* ISRIC, Wageningen.

Patrick, W.H. Jr., 1978. Critique of "Measurement and Prediction of Anaerobics in soils". In: *Nitrogen in the Environment*, Vol. 1, D.R. Neilson, J.G. McDonald, eds., *Nitrogen Behaviour in Field Soil*, Academic Press, London, pp. 449–457.

Ponnamperuma, F.N., 1972. The chemistry of submerged soils. *Advanced agronomy* 24, 29–95.

Ramana, D.V., Rai, S.N., Singh, R.N., 1995. Water table fluctuations due to transient recharge in a 2D aquifer system with inclined base. *Water Resources Management* 9(2), 127–138.

Ritchie, J.T., Singh, U., Godwin, D., Bowen, W.T., 1998. Cereal growth, development and yield. In: *Understanding Options for Agricultural Production*, G.Y. Tsuji, G. Hoogenboom, P.K. Thornton, eds., pp. 79–98.

Ross, S., 1989. *Soil Processes: A Systematic Approach*, Routledge, London.

Saglio, P.H., Rancillac, F., Bruzen, F., Prader, A., 1984. Critical oxygen for growth and respiration of excised and intact roots. *Plant Physiology* 76, 151–154. doi:10.1104/pp.76.1.151.

Sahai, B., Kalubarme, M.H., 1985. Ecological studies in the Ukai command area. *International Journal of Remote Sensing* 6(3, 4), 401–409.

Sellmann, P.V., Arcone, S.A., Delaney, A.J., 1983. Radar profiling of buried reflectors and the ground water table. USA Cold Region Research and Engineering Laboratory Report 83-11, Hanover, New Hampshire.

Setter, T.L., Waters, I., 2003. Review of prospects for germ plasm improvement for waterlogging tolerance in wheat, barley and oats. *Plant and Soil* 253, 1–34.

Setter, T.L., Waters, I., Sharma, S.K., Singh, K.N., Kulshreshtha, N., Yaduvanshi, N.P.S., Ram, P.C., Singh, B.N., Rane, J., McDonald, G., Khabaz-Saberi, H., Biddulph, T.B., Wilson, R., Barclay, I., McLean, R., Cakir, M., 2009. Review of wheat improvement for waterlogging tolerance in Australia and India: the importance of anaerobiosis and element toxicities associated with different soils. Annals of Botany 103(2):221–235.

Sharma, R.C., Bhargava, G.P., 1988. Landsat imageries for mapping saline soils and wetlands in north-west India. *International journal of Remote Sensing*, 39–44.

Shaw, R.E., Meyer, W.S., McNeill, A., Tyerman, S.D., 2013. Waterlogging in Australian agricultural landscapes: a review of plant responses and crop models. *Crop and Pasture Science* 64, 549–562. doi:10.1071/CP13080.

Sherlock, M.D., McDonnell, J.J., 2003. A new tool for hill slope hydrologists: spatially distributed groundwater level and soil water content measured using electromagnetic induction. *Hydrology Process* 17, 1965–1977.

Skaggs, R.W., 2008. DRAINMOD: a simulation model for shallow water table soils. In: South Carolina Water Resources Conference, Clemson University Restoration Institute, Clemson, SC.

Smith, M.C., Vellidis, G., Thomas, D.L., Breve, M.A., 1992. Measurement of water table fluctuations in a sandy soil using ground penetrating radar. *Transactions of the American Society of Agricultural Engineers* 35, 1161–1166.

Sreenivas, K., Dwivedi, R.S., Singh, A.N., Raviprakash, S., 2010. Detection of sub-surface waterlogging using Terra-1 MODIS data. *Journal of the Indian Society of Remote Sensing* 38(1), 119–132.

Sujatha, G., Dwivedi, R.S., Sreenivas, K., Venkataratnam, L., 2000. Mapping and monitoring of degraded lands in part of Jaunpur district of Uttar Pradesh using temporal spaceborne multispectral data. *International Journal of Remote Sensing* 21(3), 519–531.

Stockle, C.O., Nelson, R., Kemanian, A., Donatelli, M., Bechini, L., Ferrer, F., Van Evert, F., Campbell, G.S., McCool, D., Debaeke, P., 2013. CropSyst Abstract, Washington State University.

Taylor, G.R., Bennett, B.A., Mah, A.H., Hewson, R.D., 1994. Spectral properties of salinated land and implications for interpretation of 24 channel imaging spectrometer data. Presented at the International Airborne Remote Sensing Conference and Exhibition, Strasburg, France, September 11–15, 1994.

van den Bouwhuysen, J.N.A., 1934. The thermocouple proves useful on a geophysical survey. *Engineering and Mining Journal* 135, 342–344.
Wallace, J.F., Campbell, N.A., Wheaton, G.A., McFarlane, D.J., 1993. Spectral discrimination and mapping of cereal crops in Western Australia. *International Journal of Remote Sensing* 14(14), 2731–2743.
Wang, X., Mosley, C.T., Frankenberger, J.R., Kladivko, E.J., 2006. Subsurfaced rain flow and crop yield predictions for different drain spacings using DRAINMOD. *Agricultural Water Management* 79, 113–136. doi:10.1016/j.agwat.2005.02.002.
Wildman, W.E., 1982. Detection and management of soil, irrigation and drainage problems. In: *Remote Sensing for Resource Management*, C.J. Johannsen, J.L. Sanders, eds., Soil Conservation Society of America, 387–401.
Zabalza, A., van Dongen, J.T., Froehlich, A., Oliver, S.N., Faix, B., Gupta, K.J., Schmalzlin, E., Igal, M., Orcaray, L., Royuela, M., Geigenberger, P., 2009. Regulation of respiration and fermentation to control the plant internal oxygen concentration. *Plant Physiology* 149, 1087–1098.

11

Land Degradation due to Mining, Aquaculture, and Shifting Cultivation

11.1 Introduction

Mining, in general, and opencast mining, in particular, may lead to severe environmental degradation. Paradoxically, from an environmental point of view, coal mining is a major habitat transforming activity which has a number of detrimental environmental consequences, namely soil erosion by water, acid-mine drainage, and increased sediment load as a result of abandoned and unreclaimed mined lands (Parks et al., 1987). Besides, considerable amount of solid wastes piled in the form of huge overburden dumps, destruction and degradation of forest and agricultural lands, and discharge of effluents from coal washeries into nearby water-bodies are some of the other associated problems that have adverse environmental impacts. Continuous monitoring of these lands is, therefore, essential for their effective reclamation and environmental management.

Aquaculture is an essential activity for world's welfare. It is considered a double-edged sword, because it has not only tremendous benefits to the humanity, but also great impact on environment. Considering the benefits, seafood produced by fisheries and aquaculture contributes with 15%–20% of average animal protein consumption to 2.9 billion people worldwide (Smith et al., 2010) without considering the contribution of freshwater or brackish water species. The aquaculture affects the ecosystem and environment in several ways. Important among them are discussed here under: For further details, readers may refer to Martinez-Porchas and Martinez-Cordova (2012) and Fisheries and Oceans Canada (2006).

Destruction of Natural Ecosystem: For laying out aquaculture farms/ponds forests especially mangrove forests are cleared (Stickney and McVey, 2002). The mangrove forest cover has dwindling worldwide: from 19.8 million hectares in 1980 to less than 15 million in 2000. The annual deforestation rate was 1.7% from 1980 to 1990 and 1.0% from 1990 to 2000, and the problem continues even today (FAO, 2006).

Soil Salinization/Acidification: Chemical properties of water and feed used in aquaculture have significant impact on soil fertility. In the event of abandoning aquaculture farms due to multiple problems (operative, economic, sanitary, etc.) soils become hyperacidic/saline rending these soils unfit for cultivation (Rodríguez-Valencia et al., 2010). Consequently, these soils cannot be used for agricultural purposes for long periods. In addition, the application of lime and other chemicals used in aquaculture to treat the soil can also modify its physicochemical characteristics regime, which could aggravate the problem.

Water Pollution: Inland aquaculture has been reported to be responsible for the deterioration the quality of surface as well as ground water for human consumption (Paez-Osuna, 2001). For instance, preliminary assessments revealed that an intensive aquaculture system farming three tons of freshwater fish can be compared, in respect to waste generation, to a community of around 240 inhabitants (Avnimelech, 2009).

Eutrophication and Nitrification of Effluent Receiving Ecosystems: The eutrophication or organic enrichment of water column results mainly from unconsumed feed especially due to overfeeding, lixiviation of aquaculture feedstuffs (Focardi et al., 2005; Crab et al., 2007), decomposition of dead organisms, and overfertilization. In fact, from the total nitrogen supplemented to the cultured organisms, only 20%–50% is retained as biomass by the farmed organisms, while the rest is incorporated into the water column or sediment and eventually discharged in the effluents toward the receiving ecosystems, causing diverse impacts such as phytoplankton blooms, sometimes of toxic microalgaes, such as red tides, burring, and death of benthic organisms, as well as undesirable odors and the presence of pathogens in the discharge sites (Martinez-Cordova and Enriquez-Ocana, 2007).

Shifting Cultivation: Shifting cultivation is the most complex and multifaceted form of agriculture in the world. It is highly diverse land use systems that have been evolving since as early as 10,000 BC in a wide range of distinct socioeconomic and ecological conditions, from montane to lowland ecosystems, and from tropical forests to grasslands (Spencer, 1966). Shifting cultivation encompasses cropping systems such as horticulture and annual cropping, perennial tree crops, animal husbandry, and management of forests and fallows in sequential or rotational cycles. Shifting cultivation is currently practiced in a wide variety forms by 500 million to 1 billion people around the world. Shifting cultivation generally refers to any temporally and spatially cyclical agricultural system that involves clearing of land—usually with the assistance of fire—followed by phases of cultivation and fallow periods (Thrupp et al., 1997). According to Brookfield and Padoch (1994), "Shifting cultivation is not one system but hundreds or thousands of systems." The forms assumed by shifting cultivation are more varied than in practically any other land-use system (Ruthenberg, 1980). Processes of expansion and intensification in shifting cultivation have been widely reported for Asia (van Vliet et al., 2013) but less so for South America and especially for Amazonia (Vosti et al., 2002; Brondizio, 2009).

11.2 Global Distribution

Shifting Cultivation: Shifting cultivation, also known as swidden agriculture or slash-and-burn farming, is an age-old and prevailing subsistence farming practice in the tropical regions. There are 40–50 countries globally (Mertz, 2009) with almost 300–500 million people directly or indirectly carrying out this traditional swiddening system (Kleinman et al., 1996). Evidence from a recent meta-analysis published by van Vliet et al., (2012) suggests that these swidden cultivators are mostly located in the mountainous and hilly parts of Latin America, Central Africa, and Southeast Asia (SEA). Based on the Institute of Scientific Information (ISI) Web of Science database (October 8, 2013), swidden agriculture is mainly practiced by smallholder farmers in a conservative estimate of 64 developing countries from Africa, Latin America, and South/Southeast Asia (Figure 11.1).

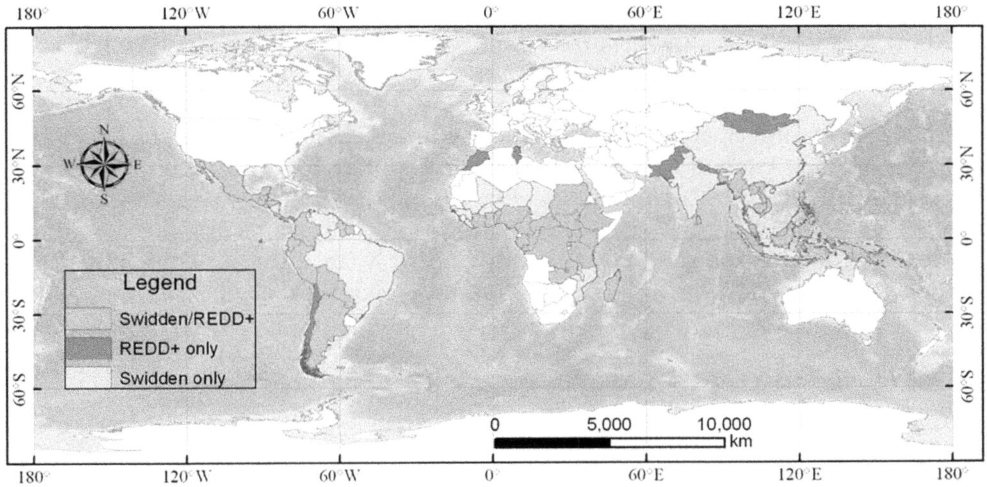

FIGURE 11.1
Spatial distribution of swidden practice (including shifting cultivation and slash-and-burn agriculture) in pantropical developing countries (Source: Adapted from Li et al. (2014)).

11.3 Role of Remote Sensing

Both aerial photographs and spaceborne multispectral measurements have been utilized for assessment and monitoring of areas subject to mining, aquaculture, and shifting cultivation.

11.3.1 Aerial Photographs

11.3.1.1 Mining

Remote sensing technique is quite useful in the detection, inventory, monitoring, and quantifying active surface mining activities or abandoned mining problems (Anderson, 1980). Aerial photographs have been used during the late seventies and early eighties for assessment and monitoring of some activities related to mining which lead to deterioration of soil health by way of developing the toxicity of certain elements. Carrel et al., (1978) used both visual interpretation and computer-aided techniques to interpret aerial photographs for monitoring reclamation activities at a strip mining site in Missouri, USA. In another study, Mroczynski and Wiesmiller (1982) used color infrared aerial photographs for evaluating reclaimed and unreclaimed land in southwestern Indiana, USA.

11.3.1.2 Aquaculture

Initially, aerial photographs have been used for detection of wetlands. Carter et al., (1979) concluded that seasonal color infrared photographs provide sufficiently detailed information to delineate wetland areas as small as 0.5 hectare in size and 20 m in width. Carter and

Stewart (1977) and Moore and North (1974) have discussed the use of high altitude color infrared photographs to document surface water boundaries. In Bangladesh both black-and-white and infrared color aerial photographs have been used for delineation and monitoring the spatial extent of shrimp farming (aquaculture) areas (Quader et al., 1986; Shahid et al., 1992). NASA Stennis Space Center's Lear Jet-mounted Calibrated Airborne Multispectral Scanner (CAMS) has been utilized to monitor turbid finfish aquaculture impoundments in Mississippi (Millie et al., 1992). Unmanned radar installations have also been used in Canada to support an "integrated maritime surveillance system" for protecting the Canadian 200 nm exclusive economic zone and for monitoring not only illicit fish poaching but also drug trafficking, smuggling, piracy and illegal immigration (Ponsford et al., 2001).

11.3.1.3 Shifting Cultivation

Inventory and monitoring of shifting cultivation areas using remote sensing especially aerial photographs essentially aims at delineating the areas within the forest where there has been degradation in forest cover or the forests have been cleared. Aerial photographs have been found ideal for mapping small ecosystems and fine-scale landscape features, such as riparian areas or individual trees (Fensham and Fairfax, 2002; Tuominen and Pekkarinen, 2005), because they often possess a high level of spatial and radiometric (tonal) detail. Besides, aerial photographs have been used to assess and map the changes in landscape. Aerial photographs also provide the longest-available, temporally continuous, and spatially complete record of landscape change, dating from the early 1930s in some cases. As a result, aerial photographs are a source of valuable historical information on vegetation cover and condition (Cohen et al., 1996; Fensham and Fairfax, 2002).

Aerial photographs can be cost-effective in mapping, inventorying, and planning (Paine and Kiser, 2003) and, as such, are used for applications ranging from forest inventories, disturbance mapping, productivity estimates, and wildlife management (Avery and Berlin, 1992). Thus, many important management decisions are routinely made on the basis of maps derived from aerial photographs (Cohen et al., 1996; Paine and Kiser, 2003).

11.3.2 Sapaceborne Multispectral Measurements

11.3.2.1 Mining

Spaceborne multispectral measurements have been used for deriving information on extent, spatial distribution, temporal dynamics as well as assessment of sediment transport.

11.3.2.1.1 Detection and Delineation

By virtue of repetitive coverage at regular intervals, spaceborne multispectral measurements offer immense potential in deriving information on various features associated with mining activities in a timely and cost-effective manner. An illustration of the approach for detection and delineation of various features related to mining in part of Goa, southern India using Indian Remote Sensing satellite (IRS-1D)-Linear Imaging Self-Scanning (LISS-III) and panchromatic (PAN) sensor data is presented hereunder (National Remote Sensing Agency, 2002).

Preparation of Database: Preparation of database involves geo-referencing and fusion of multispectral and panchromatic digital data (Figure 11.2). Initially, the IRS-1C PAN data digitally co-registered to Survey of India (SOI) topographic maps at 1:25,000 scale and resampled to 6 m spatial resolution using first-order polynomial transform on a Silicon

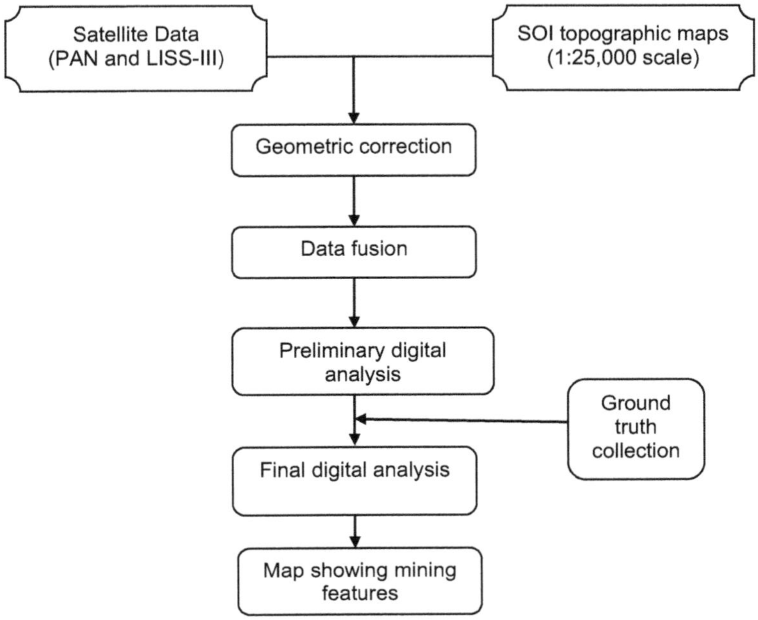

FIGURE 11.2
Schematic diagram of the approach for delineation of mining features.

Graphics work station using ERDAS/IMAGINE version 8.4 software. The digital data, thus generated, were used as a reference for geo-referencing IRC-1C LISS-III multispectral data, which were resampled to 6 m spatial resolution.

Preliminary Visual Interpretation: For objective assessment of the lands affected by mining activities, quantification of sediment from these sites is a prerequisite. To predict the sediment detachment, information on the land cover, slope, management practices, and soil erodibility is essential. Hence, preliminary interpretation of satellite digital data has been carried out to delineate various land use/land cover categories in conjunction with the SOI topographical maps. Simultaneously, preliminary interpretation was also done for mapping soil properties by exploiting the inherent relationship among lithology, physiography, and land cover. The iron-rich overburden material dumps are clearly seen in the image as various shades of green and yellow colors depending on iron and gravel contents. The mining areas could be broadly delineated on the LISS-III data by their characteristic spectral response pattern. However, after merging the LISS-III with PAN data, various stages of mining and other associated disturbed areas could also be discerned.

Ground Truth Collection: Ground truth mission was subsequently planned to establish the relationship among image elements, namely color, texture, shape, size, shadow, pattern, association, etc. and various features associated with the mining in random sample strips. Initially, a reconnaissance traverse of the area was made to assess the traffic ability and to precisely locate sample areas. Intensive field observations were then made in sample areas on various land use/land cover categories and soil properties. Precise location of the sampling sites was recorded with the help of a Global Positioning System (GPS) receiver model Magellan ProMark-X.

Final Digital Analysis: Various categories of land use/land cover verified during ground truth collection phase were marked in the geometrically corrected digital satellite data

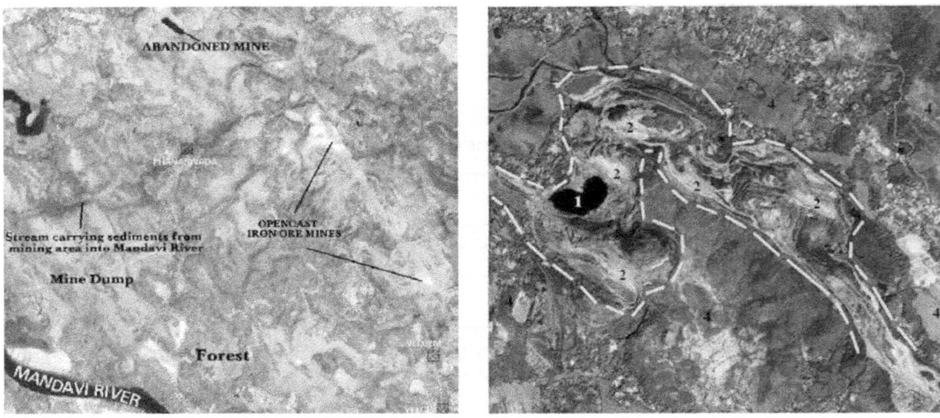

IRS-1C LISS-III image (29-05-1997) Resourcesat-2 LISS-IV image

FIGURE 11.3
Mining features as seen in IRS-1C LISS-III image and Resourcesat-2 LISS-IV image around Mugaon, west Madhban, Goa, southern India. The numeral '1' in LISS-IV image indicates mining pond, '2' mine dump, and '4' agricultural land. White patches within red-colored background are opencast iron ore mining, and dark red color indicates the forests (Courtesy: National Remote Sensing Centre, Indian Space Research Organization, Department of Space, Government of India).

as training sets on a Silicon Graphics workstation using ERDAS/IMAGINE version 8.4 software. Training sets were defined and the spectral response patterns of training sets were generated for various land use/cover categories using signature editor module. Subsequently, the separability analysis of various classes in terms of transformed divergence (TD) was carried out. Further refinement in the training sets was made to improve the separability of various land use/cover categories. The entire image was then classified using Gaussian maximum likelihood per-pixel classifier.

Final Digital Analysis: The spatial distribution of soil properties delineated during preliminary visual interpretation has been modified vis-à-vis ground truth and soil chemical analysis data by displaying the PAN-merged LISS-III data on the color monitor of the Silicon Graphics workstation using ERDAS/IMAGINE version 8.4 software. And the vector coverage was generated and its topology built. The physical and chemical properties of soils were attached as attribute information in a Polygon Attribute Table (PAT).

The false color composite of IRS-1C LISS-III image is appended as Figure 11.3. It is evident from the image that various features associated with the mine, namely areas with opencast iron ore mining, abandoned mines, mine dump and mining ponds are very clear.

Another illustration of the potential of orbital sensor multispectral and multi-temporal data is shown in Figure 11.4. Here, of the three cropping season multispectral data, the February month data suit well for delineation of active mining areas. Aground photo of the mining area is also appended adjacent to April 2006 LISS-III image.

11.3.2.1.2 State-of-the-Art

An extensive review of the work carried out on mining using satellite data has been made by Rathore and Wright (1993). The potentials of remote sensing data include

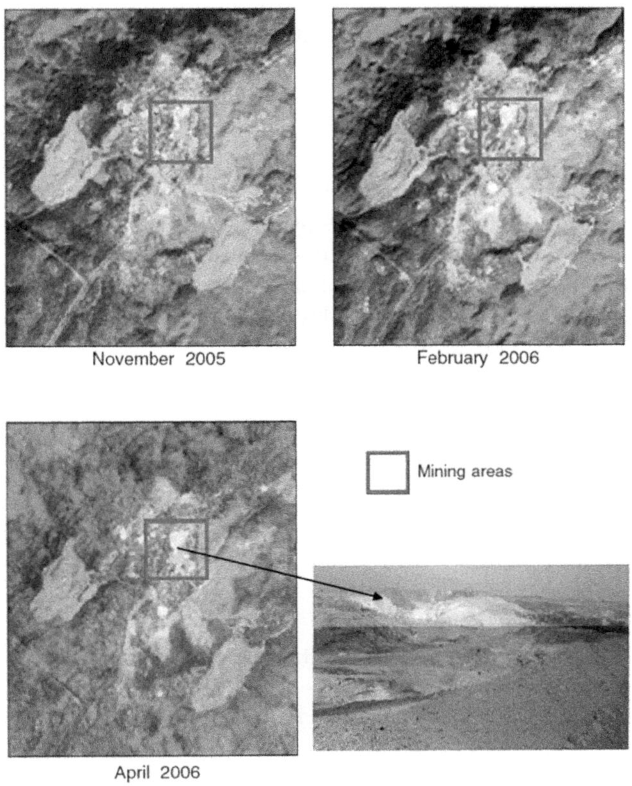

FIGURE 11.4
Mining areas as captured by Resourcesat-2 LISS-III in multi-temporal images ranging from post-rainy season to summer. Accompanying ground photograph provides a glimpse of various features associated with mining (Courtesy: National Remote Sensing Centre, Indian Space Research Organization, Department of Space, Government of India).

monitoring land disruption due to mining (Wier et al., 1973; Irons et al., 1980; Parks et al., 1987), detection of mine fires (Richards, 1983; NRSA, 1992), mining revegetation and the monitoring of reclaimed lands (Legg, 1986), water pollution assessment (Repic et al., 1991), monitoring (Lathrop and Lillesand, 1986; Lillesand et al., 1987), and detection of land subsidence (Mechaffie and Seargent, 1985; Volk et al., 1990). In India, Landsat MSS/TM and IRS-1A/-1B LISS I/-II data have been used for detection of mine dumps and the lands affected by them in the down slope (Thayalan and Nagabhushana, 1987; Dakhole et al., 1997).

Satellite data have been used in monitoring the impacts of coal mining in the Jharia coal fields of Bihar, India (Prakash and Gupta, 1998). While working in the Eastern Ghats covering parts of Vishakhapatnam district of Andhra Pradesh, southern India, using IRS-1B LISS-II data, Murthy et al., (1997) predicted a loss of forest cover within 3 km radius from proposed mining site. Digital classification of remotely sensed data of mining areas can present special problems as a considerable amount of spectral confusion can exist between mining-induced land cover types such as quarries, mine pits, coal dumps, overburden dumps, and surrounding areas like scrubs, barren lands, agricultural land,

and built-up areas. The use of conventional classification techniques and algorithms has not been found suitable in such cases, and alternative classification strategies have been adopted (Solomon et al., 1979).

11.3.2.2 Aquaculture

Acoustic techniques for estimating fish abundance initially served as precursors for satellite remote sensing and have been operational for at least the past thirty-five years (Ehrenberg et al., 1972). Moreover, "sonarecho statistics" have been deployed for estimating fish school density and classification of fish and rough sea floor "by analyzing the nature of echo fluctuations" (Stanton and Clay, 1986). Spaceborne multispectral data from multispectral scanner system (MSS), Thematic Mapper (TM) aboard Landsat series of satellites, and Multispectral Linear Array (MLA) aboard SPOT-1 covering parts of Bangladesh have also been used to identify the clusters of shrimp farming. It was concluded that the area under shrimp farming registered an increasing trend. While inventorying and monitoring shrimp cultivation using aerial photos, Landsat MSS and SPOT data, Shahid et al., (1992) observed an increase in the spatial extent of shrimp cultivation areas at the expense of paddy lands. Li et al., (1995) showed that wetland identification is possible using either Landsat TM.

11.3.2.2.1 Delineation of Aquaculture Farm Ponds

A case study illustrating the potential of orbital sensor multispectral and multi-temporal data over an area around east coast of India is presented hereunder.

Database: The Landsat-TM, SPOT-MLA, and IRS-1C LISS-III digital data were used to monitor the temporal changes in the spatial extent and distribution of aquaculture. Besides, IRS-1C LISS-III, and PAN-fused digital data were used to study the details within the areas where aquaculture is practiced (National Remote Sensing Agency, 2002).

Approach: The approach essentially involves database preparation and systematic on-the–screen visual interpretation of both concurrent and historical spaceborne multispectral and multi-temporal digital data (Figure 11.5). Various steps involved are discussed hereunder.

Database Preparation: Preparation of database involves geo-referencing, radiometric normalization, and data fusion. The details of the steps involved are given hereunder.

Geo-referencing: To begin with, the IRS-1C LISS-III data of the test site acquired during February, 1997 were digitally co-registered to SOI topographic maps at 1:50,000 scale and resampled to 24 m pixel dimension using first-order polynomial transform on a Silicon Graphics work station using ERDAS/IMAGINE version 8.4 software. The digital data, thus generated, were used as a reference for geo-referencing other satellite datasets of 1986, 1988, and 2001. The Landsat-TM datasets were resampled to 30 m × 30 m, SPOT-MLA to 20 m × 20 m, and LISS-III to 24 m × 24 m pixel dimensions. Besides, for detailed study, as mentioned earlier, high-resolution PAN data in conjunction with multispectral LISS-III data were used. For digital fusion of the two datasets, the LISS-III data of 1997 were resampled to 6 m × 6 m so as to generate a hybrid image by merging IRS-1C LISS-III multispectral data with the PAN data acquired on the same day.

Radiometric Normalization: Since multi-temporal satellite data were used, the radiometric normalization forms an important element to bring down the variations in spectral reflectance introduced due to variations in sun elevation and radiometric gain settings. Initially, the digital number (DN) values were converted into radiance values using the gain settings and saturation radiance values provided in the satellite data header/supplied by the NRSA Data Centre (NDC) for IRS-1C LISS-III and SPOT-MLA sensors. For Landsat-TM

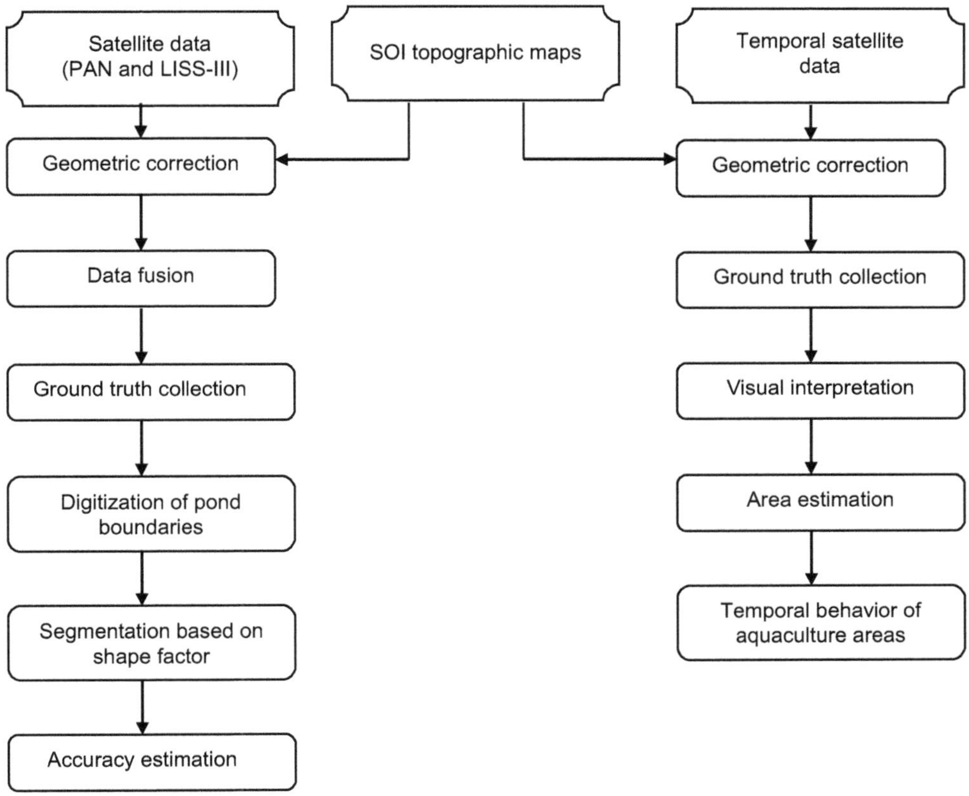

FIGURE 11.5
Schematic diagram of approach for monitoring aquaculture farm ponds.

images, the values provided by Markham and Barker (1986) were used. Later, corrections for sun elevation angle variations using cosθ correction (where θ is the sun elevation angle) were made. Subsequently, all the radiance values were scaled using a common linear scaling factor.

$$L\lambda = LMIN\lambda + \left(\frac{LMAX\lambda}{QCALMAX}\right)QCAL \tag{11.1}$$

where

QCAL = calibrated and quantified scaled radiance in units of DN
LMINλ = spectral radiance at QCAL = 0
LMAXλ = spectral radiance at QCAL = QCALMAX
QCALMAX = range of rescaled radiance in DN
Lλ = spectral radiance.

Data Fusion: Data fusion is one of the important steps in bringing out the spatial information available in PAN data and spectral details with the multispectral data. As mentioned earlier, the LISS-III and PAN data were resampled to 6 m spatial resolution and were fused using principle component data fusion technique provided in the ERDAS/IMAGINE 8.4 software. The algorithm, initially, converts the raw satellite data into principle components. Subsequently the first component is replaced with the high-resolution panchromatic data and an inverse principle component transform is subsequently performed.

Preliminary Visual Interpretation: The detection and delineation of aquaculture areas is based solely on their spectral response pattern as seen in the image or as portrayed in the changes in spectral radiance values of spaceborne multispectral data. Since not only image elements but also associations in terms of terrain conditions are taken into consideration while delineating aquaculture areas, on-the-screen visual interpretation approach was employed. In general, aquaculture, both prawn and fish cultivation, involve stocking of water, such areas are clearly discernible on the image with a characteristic spectral response of water girdled by well-laid regular-shaped field bunds, and are associated with field channels and drains. In the IRS-1C LISS-III and PAN-fused images, the aquaculture sites are more clear and the boundaries are distinct.

Ground Truth Collection: The ground truth mission was planned to establish the relationship between spectral response pattern, as seen in the multispectral data, and the existence of aquaculture ponds apart from establishing the spectral signatures of prawn and fish ponds. Information on the type of aquaculture practiced, package of practices, yield, etc. was noted. Soil samples were collected from aquaculture ponds (both fish and prawn ponds) and from normal cultivated (paddy lands) fields to study the impact of the practices on soil properties. Soil samples were analyzed in the laboratory for pH (1:2 soil: water suspension), electrical conductivity (1:2 soil: water suspension), cation exchange capacity (CEC), organic matter content, etc. following the standard procedure.

Map Finalization: Initially, the spaceborne multispectral data of 1986 were displayed onto color of monitor Silicon Graphic workstation and the areas where aquaculture is practiced were delineated manually as a vector coverage vis-à-vis ground truth. Subsequently, the coverage has been duplicated and overlaid onto the 1988 data and the changes in the spatial extent and spatial distribution pattern of aquaculture areas, if observed, were noted and necessary modification in the boundaries were made. Similar exercise was carried out for other period datasets too. These vector coverages were then suitably annotated and topologically corrected. Since both prawn and fish culture involve surface ponding with water, spectrally they look alike. There seems to be some differences with respect to their perimeter. Based on ground truth information, an attempt was made to segregate prawn ponds from those of fish ponds using a shape factor, the ratio of area to perimeter of ponds, as the lone parameter. The manifestation of aquaculture ponds in different databases is appended as Figures 11.6–11.8. Aquaculture ponds could be identified by their regular shape, distinct boundaries which stand out very in water background. In fact, aquaculture ponds are quite clear even at 23.5m resolution LISS-III data.

Further improvement in discernibility of aquaculture pond features especially clarity of water within the ponds, presence of vegetation within, and water channel with some vegetation (red) running diagonally from Southwest to Northeast) and standing water (blue) could be achieved with improvement in spatial resolution of remote sensing data. A look at digitally-merged IRS-1C LISS-III data with 23.5 m spatial resolution panchromatic (PAN) image with 5.8 m resolution attests this fact (Figure 11.7).

Having very high commercial value there has been a spurt in the expansion of the area under aquaculture in areas with mangroves and paddy cultivation. A comparison of the area of aquaculture during the period 1986–2001 reveals a more than fivefold increase in its spatial extent during 15 years period. The area under aquaculture has increased fivefold during the period 1986–2001, i.e., 3,250 hectares in 1986 versus 17,674 hectares in 2001.

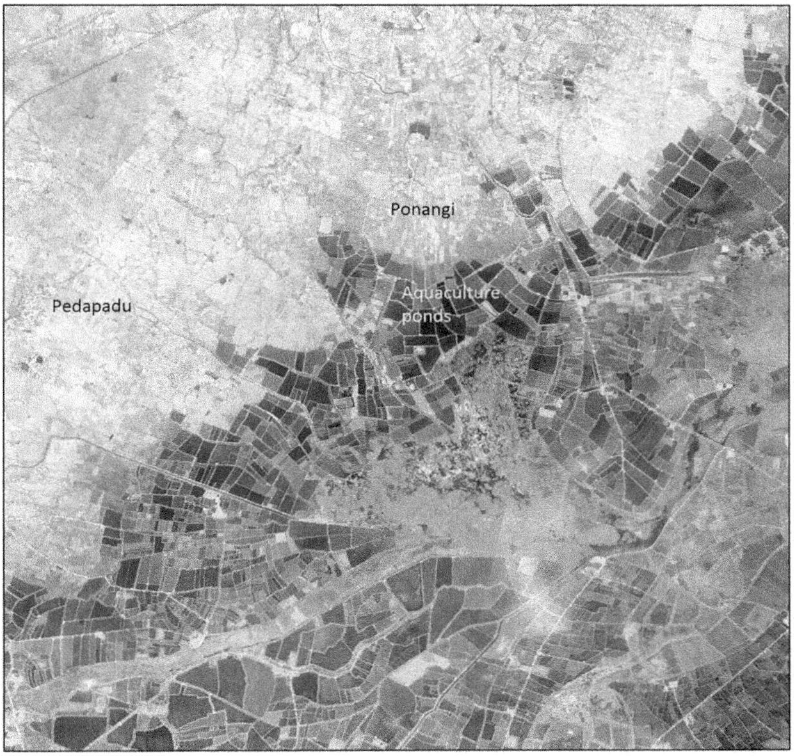

FIGURE 11.6
A regional view of aquaculture ponds in coastal Andhra Pradesh as captured by Resouecsat-2 LISS-III sensor with 23.5 m spatial resolution.

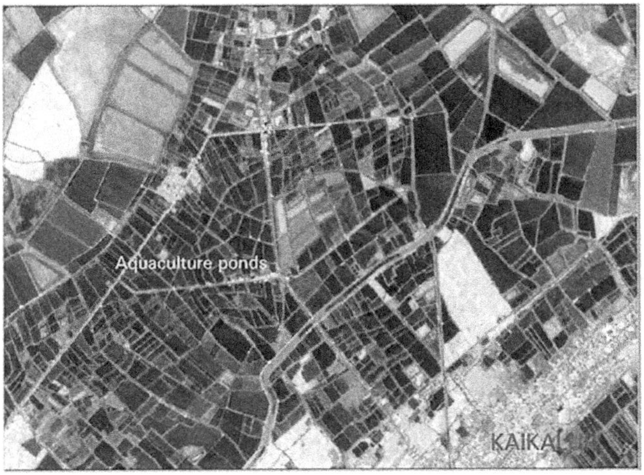

FIGURE 11.7
Aquaculture farm ponds as captured by IRS-1C LISS-III and PAN sensors.

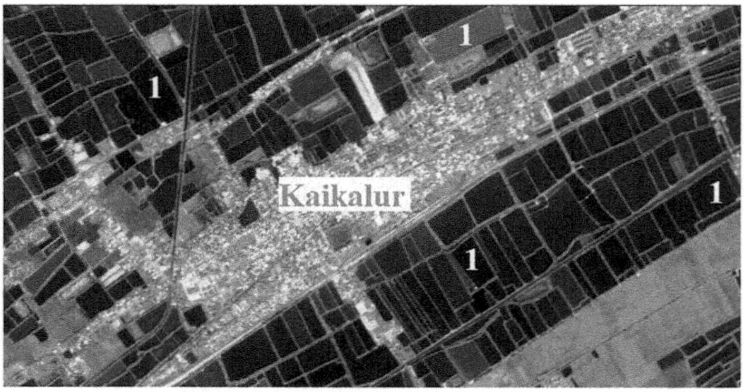

FIGURE 11.8
Aquaculture ponds as seen in Resourcesat-2 LISS-IV (5.8 m spatial resolution) and Cartosat-1 PAN-merged image. The numeral '1' indicates aquaculture farm ponds (National Remote Sensing Centre, Indian Space Research Organization, Department of Space, Government of India).

The area under aquaculture farm ponds in the years 1988 and 1997 was 4,789 and 15,531 hectares, respectively. The increase, as pointed out earlier, has been at the cost of paddy lands. Such a phenomenon is highly deleterious to the environment especially surface and ground water apart from soils.

11.3.2.2.2 State-of-the-Art

Spaceborne multispectral data from MSS, TM aboard Landsat series of satellites, and MLA aboard SPOT-1 covering parts of Bangladesh have also been used to identify the clusters of shrimp farming. It was concluded that the area under shrimp farming registered an increasing trend. While inventorying and monitoring shrimp cultivation using aerial photos, Landsat MSS, and SPOT data, Shahid et al., (1992) observed an increase in the spatial extent of shrimp cultivation areas which is extending into paddy lands.

Honhoulis and Michener (2000) delineated wetlands through a rule-based approach using ancillary information, such as National Water Inventory (NWI) and SPOT-XS data. Tripathi et al., (2000) found the Landsat-TM data useful for delineation of active and abandoned shrimp farming in Thailand.

Using temporal Landsat data over part of Thailand, Vibulsresth et al., (1993) reported a decreasing trend on the order of 1.49% in the area under mangroves during the period 1991 and 1992 due to sudden spurt in shrimp farming. In India, spaceborne multispectral and multi-temporal data over coastal Andhra Pradesh, southern India have been used to study the temporal behavior of aquaculture and an increasing trend in the area under prawn cultivation was noted (Venkataratnam et al., 1997). In another study the spatio-temporal behavior of aquaculture was studied using satellite images acquired in 1986, 1989, 1996, and 2001 satellite image covering an area around Kaikalur (Figure 11.9). As evident from the figure there has been a marked expansion of the area under aquaculture in coastal Andhra Pradesh, India during the period 1986–2001.

Using temporal Landsat data over part of Thailand, Vibulsresth et al. (1993) reported a decreasing trend on the order of 1.49% in the area under mangroves during the years 1991 and 1992 due to sudden spurt in shrimp farming. In India, spaceborne multispectral and multi-temporal data over coastal Andhra Pradesh, southern India have been used to study the temporal behavior of aquaculture and an increasing trend in the area under prawn cultivation was noted (Venkataratnam et al., 1997).

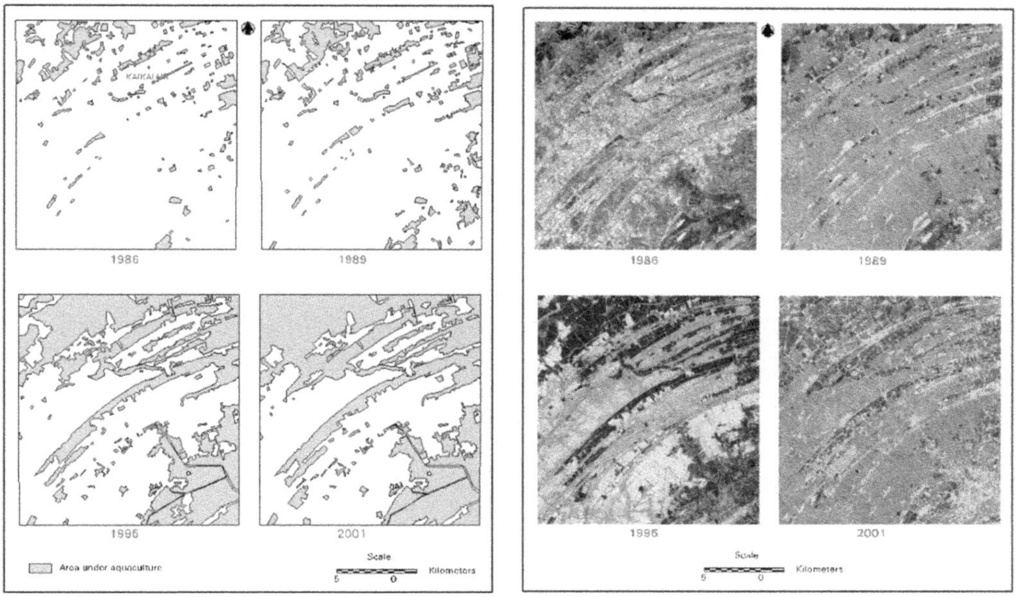

FIGURE 11.9
Monitoring aquaculture using temporal satellite data over surroundings of Kaikalur, Krishna district, Andhra Pradesh, southern India.

11.3.2.3 Shifting Cultivation

Since the 1970s, remote sensing techniques have been providing persistent impetus to environmental changes monitoring. Shifting cultivation or swidden agriculture slash-and-burn practice is widely perceived as one of the primary causes for deforestation throughout the pan-tropical regions (Geis and Lambin, 2001). Because of the diverse, complex, and dynamic features of swidden system, so far, accurate assessment with remote sensing imagery is very challenging (Schmidt-Vogt et al., 2009) due mainly to difficulties in detecting swidden practice with remote sensing methods (van Vliet et al., 2013). Landsat-MSS and TM, and IRS LISS-I and -II data have been used for mapping shifting cultivation or *jhum* cultivation areas (Dwivedi and Ravi Sankar, 1991). In addition, multi-temporal nature of satellite data has been utilized for mapping shifting cultivation areas and studying their dynamics (Kushwaha, 1991). The area that have been freshly cleared for shifting cultivation in part of North Tripura district, Tripura could be clearly identified in Landsat-TM data acquired during April 2018 (Figure 11.10). The shifting cultivation areas appear in different hues of yellow color amidst forest cover (red color) background.

A close look at figure 11.11 showing shifting cultivation reveals that practice of clearing forest for shifting cultivation is in the form of isolated small pockets and shrinkage in size of land under shifting cultivation has been shrinking.

Remote sensing data can provide valuable information on fire events and burned areas due to their synoptic, multi-temporal, multispectral and repetitive coverage capabilities. However, studies of detecting burnt area or fire scar or wildland fire with satellite imagery were widely reported. For example, Mallinis et al. (2012) compared ten methods for mapping burned area in a Mediterranean setting using Landsat TM imagery including

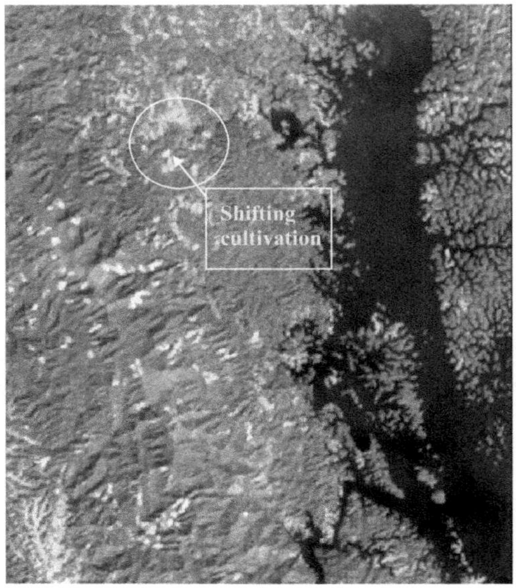

FIGURE 11.10
Shifting cultivation areas (irregular-shaped clearings in light yellow to light green color) within dense vegetation (dark to very dark colored background) around Gumti reservoir, part of North Tripura district, Tripura state, north-eastern region, India as seen in Landsat-8 Operational Land Imager (OLI) collected on April 6, 2018 image.

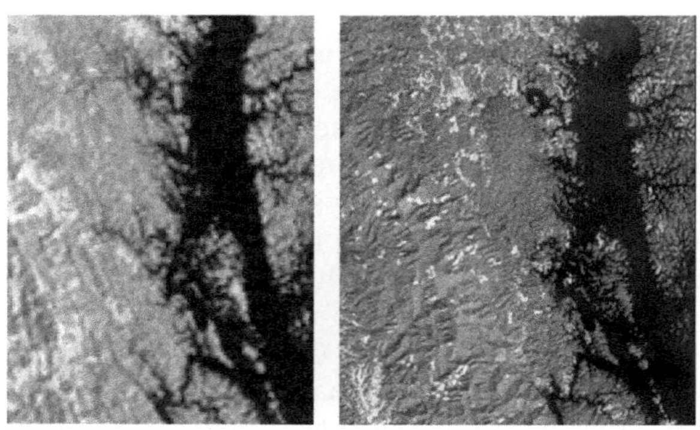

FIGURE 11.11
Shifting cultivation areas (irregular-shaped clearings in light yellow to light green color) within dense vegetation (dark to very dark colored background) around Gumti reservoir, part of North Tripura district, Tripura state, north-eastern region, India as seen in Landsat-3MSS data of April1 5, 1978 (left) and Landsat-8 OLI, April 6, 2018 (right) images (Source: National Remote Sensing Centre, Indian Space Research Organization, Department of Space, Government of India).

spectral unmixing, artificial neutral networks (ANN), logistic regression, maximum likelihood, thresholding of spectral vegetation indices, hue-lightness-saturation (HIS) transformation, principal component analysis, classification and regression trees, object-based image analysis, and support vector machines (SVM).

11.4 Conclusions

Mining activities, aquaculture, and shifting cultivation significantly contribute to soil degradation, vegetation degradation, and water degradation in varying degrees. Their impact is manifested in the form of toxicity of certain elements, salinization and/or alkalinization/acidification in case industrial effluents, contamination of surface and ground water, and/vegetation degradation and water erosion in areas where shifting cultivation is practiced. For rehabilitation of such land timely and reliable information on their nature, extent, and spatial distribution and its dynamics is a prerequisite. Geospatial technologies play a crucial role in providing such information in a timely and cost-effective manner. However, in order to be detected/delineated, the impact of these activities/causative factors of degradation needs to have adequate surface manifestation. An integrated approach involving geospatial technology, proximal sensing and ancillary or legacy information in a GIS environ seems to be a plausible solution for improving the level of information that can be derived on these lands.

References

Anderson, A.T., 1980. Evaluating the environmental effects of past and present surface mining: a remote sensing applied research review. In: *Proceedings of 14th International Symposium on Remote Sensing of Environment held at San Jose,* Costa Rica in April 1980 (Ann Arbor, Michigan: Environmental Research Institute, Michigan), pp. 23–30.

Avery, T.A., and Berlin, G.L., 1992. *Fundamentals of Remote Sensing and Air Photo Interpretation*. Prentice Hall, Upper Saddle River, NJ.

Avnimelech, Y., 2009. *Biofloc Technology. A Practical Guide Book,* The World Aquaculture Society, Baton Rouge, LA.

Brookfield, H., and Padoch, C., 1994. Appreciating Agrodiversity: a look at the dynamism and diversity of indigenous farming practices. *Environment: Science and Policy for Sustainable Development* 36(5): 6–21.

Brondizio, E., 2009. Agriculture intensification, economic identity, and shared invisibility in Amazonian peasantry: Caboclos and colonists in comparative perspective. In: Adams, C., Harris, M., and Murrieta, R., (eds.) *Amazon Peasant Societies in a Changing Environment: Political Ecology, Invisibility and Modernity in the Rain Forest*. Springer, São Paulo, p. 181–214.

Carrel, J.E., Johannsen, T.W., Barney, T.W., and McFarland, W., 1978. Remote measurement of vegetative cover in surface mines. In: *Proceedings of the 14th International Symposium on Remote Sensing of the Environment,* April 20–26th, 1978. Centre for Remote Sensing Information and Analysis., Environmental Research Institute of Michigan (Ann Arbor Michigan, E. R. I. M), III, pp. 1653–1664.

Carter, V., Malone, D.L., and Burbank, J.H., 1979. Wetland classification and mapping in western Tennessee. *Photogrammetric Engineering and Remote Sensing* 45:273–284.

Carter, V., and Stewart, W.R., 1977. Seasonal color infrared photographs for mapping inland wetlands on U.S. Geological Survey 7.5-minute quadrangles, in 5th Bienn. Color Aerial Photography Workshop, Sioux Falls, S. 0, Aug. 19–21, 1975, *Proceedings of the American Society of Photogrammetry,* p. 143–161.

Cohen, W.B., Kushla, J.D., Ripple, W.J., and Garman, S.L., 1996. An introduction to digital methods in remote sensing of forested ecosystems: Focus on the Pacific Northwest, USA. *Environmental Management* 20:421–435.

Crab, R., Avnimelech, Y., Defoirdt, T., Bossier, P., and Verstraete, W., 2007. Nitrogen removal techniques in aquaculture for a sustainable production. *Aquaculture* 270(1–4):1–14.

Dakhole, W.M., Narula, K.K., and Shrivastava, S.I., 1997. Land degradation mapping using IRS-1C LISS-III data: A case study of Jhabua district of Madhya Pradesh. Proc (Abstract) National Conference on Soil and Water Conservation for sustainable Farming Systems. February 6–8, 1997, Nagpur.

Dwivedi, R.S., and Ravi Sankar, T., 1991. Monitoring shifting cultivation areas using spaceborne multispectral and multitemporal data. *International Journal Remote Sensing* 12(3):427–433.

Ehrenberg, J.E., and Lytle, D.W., 1972. Acoustic techniques for estimating fish abundance. *Geoscience Electronics, IEEE Transactions* 10(3):138–145.

FAO, 2006. *El Estado Mundial de la Pesca y Acuicultura*. FAO, Rome.

Focardi, S., Corsi, I., and Franchi, E., 2005. Safety issues and sustainable development of European aquaculture: new tools for environmentally sound aquaculture. *Aquaculture International* 13(1–2):3–17.

Fisheries and Oceans Canada, 2006. A Scientific Review of the Potential Environmental Effects of Aquaculture in Aquatic Ecosystems. Volume V. Behavioural Interactions Between Farm and Wild Salmon: Potential for Effects on Wild Populations (Laura K. Weir and Ian A. Fleming); Overview of the Environmental Impacts of Canadian Freshwater Aquaculture (C.L. Podemskiand P.J. Blanchfield); A Scientific Review of Bivalve Aquaculture: Interaction Between Wild and Cultured Species (T. Landry, M. Skinner, A. LeBlanc, D. Bourque, C. McKindsey, R. Tremblay, P. Archambault, L. Comeau, S. Courtenay, F. Hartog, M. Ouellette and J.M. Sevigny).

Fensham, R.J., and Fairfax, R.J., 2002. Aerial photography for assessing vegetation change: A review of applications and the relevance of findings for Australian vegetation history. *Australian Journal of Botany* 50:415–429.

Geist, H.J., and Lambin, E.F., 2001. What drives tropical deforestation? A meta-analysis of proximate and underlying causes of deforestation based on subnational case study evidence; LUCC International Project Office: Louvain-la-Neuve, Belgium, p. 116.

Photogrammetric Engineering and Remote Sensing Irons, J.R., Lachhowski, H., and Peterson, C., 1980. Remote sensing of surface mines: A comparative study of sensor systems. In: *Proceedings of the 14th International Symposium on Remote Sensing of the Environment*. San Jose, Costa Rica, April, 1990, Ann Arbor Michigan: Environmental Research Institute of Michigan. pp. 1041–1053.

Kleinman, P., Bryant, R.B., and Pimentel, D., 1996. Assessing ecological sustainability of slash-and-burn agriculture through soil fertility indicators. *Agron Journal* 88:122–127.

Kushwaha, S.P.S., 1991. Remote sensing of shining cultivation in North-East India. In: Murai, S.,(ed.) *Applications of Remote Sensing in Asia and Oceania Environmental Change Monitoring*. Asian Association of Remote Sensing, pp. 60–77.

Lathrop, R.G., and Lillesand, T.M., 1986. Utility of thematic mapper data to assess water quality in Southern Green Bay and West-Central Lake, Michigan. *Photogrammetric Engineering and Remote Sensing* 52:671–680.

Legg, C.A., 1986. Monitoring of open cast coal mining and reclamation works in the United Kingdom using MSS and TM imagery. In: *Proceedings of the 20th International Symposium on Remote Sensing of the Environment*. San Jose, Costa Rica April, 1990. (Ann Arbor Michigan: Environmental Research Institute of Michigan) 2, pp. 931–941.

Lillesand, T.M., Lathrop, R.G., and Vandre Castle, J.R., 1987. Towards an integrated system for satellite remote sensing of water quality in the Great Lakes. In: *Proceedings of the Falls ASPRS Meeting*, Reno N. V. (Virginia: American Society for Photogrammetry and Remote Sensing) pp. 342–347.

Li, P., Feng, Z, Jiang, L, Liao, C., and Zhang, J., 2014. A Review of Swidden Agriculture in Southeast Asia. *Remote Sensing* 6:1654–1683.

Mallinis, G., and Koutsias, N., 2012. Comparing ten classification methods for burned area mapping in a Mediterranean environment using Landsat TM satellite data. *International Journal of Remote Sensing* 33:4408–4433.

Markham, B.L., & Barker, J.L. (1986). *Landsat MSS and TM post-calibration dynamic ranges, exoatmospheric reflectances and at-satellite temperatures.* Earth Observation Satellite Co., Lanham, MD, Landsat Tech. Note 1.

Martinez-Cordova, L.R., and Enriquez-Ocana, F., 2007. Study of the benthic fauna in a discharge lagoon of a shrimp faro with special emphasis on polychaeta. *Journal of Biological Sciences* 7:12–17.

Martinez-Porchas, M., and Martinez-Cordova, L.R., 2012. World aquaculture: Environmental impacts and troubleshooting alternatives. *The Scientific World Journal* 9, DOI:10.1100/2012/389623.

Mechaffie, P.H., and Seargent, R.E., 1985. Detection and delineation of mine related subsidence in Western Kentucky coalfields. ASCM-ASPRS Fall Meet Proceedings, Indianapolis, September 8–13, 1985 (Falls Church Virginia: American Society for Photogrammetry and Remote Sensing and American Congress on Surveying and Mapping), pp. 538–554.

Mertz, O., 2009. Trends in shifting cultivation and the REDD mechanism. *Current Opinion in Environmental Sustainability* 1:156–160.

Millie, D.F., Baker, M.C., Tucker, C.S., Vinyard, B.T., and Dionigi, C.P., 1992. High-Resolution Airborne Remote Sensing of Bloom-Forming Phytoplankton. *Journal of Phycology JPYLAJ* 28(3):281–290.

Moore, G.K., and North, G.W., 1974. Flood inundation in the southeastern United States from aircraft and satellite imagery. *Water Resources Bulletin* 10(5):1082–1097.

Mroczynski, T.P., and Weismiller, R.A., 1982. Aerial photography: A tool for strip mine reclamation. In: Johannsen, C.J., and Sanders, J.L., (eds.) *Remote Sensing for Resources Management*. Soil Conservation Society of America, Ankeny, IA, pp. 331–337.

Murthy, M.S.R., Subramanian, S.K., and Dutt, C.B.S., 1997. Environmental impact analysis of bauxite mining, Eastern Ghats, Andhra Pradesh using remote sensing data. NNRMS Bulletin. Department of Space, Government of India, Bangalore. (B) 21:30–35.

National Remote Sensing Agency, 1992. Airborne scanner survey and data analysis for underground and surface coal mine fire detection in Jharia coalfield, Bihar. Technical report no. NRSA-AG-GD-TR-2/92.

National Remote Sensing Agency, 2002. Assessment of land degradation due to mining and aquaculture, Project report National Remote Sensing Agency, Department of Space, Government of India. Hyderabad.

Paez-Osuna, F., 2001. The environmental impact of shrimp aquaculture: causes, effects, and mitigating alternatives. *Environmental Management* 28(1):131–140.

Paine, D.P., and Kiser, J.D., 2003. *Aerial Photography and Image Interpretation*, 2nd edn. Wiley, Hoboken, NJ.

Parks, N.F., Peterson, G.W., and Baumer, G.M., 1987. High resolution remote sensing of spatially and spectrally complex coal surface mines of Central Pennsylvania: A comparison between SPOT, MSS and Landsat-TM. *Photogrammetric Engineering and Remote Sensing* 53:415–420.

Ponsford, A.M., Sevgi, L., and Chan, H.C., 2001. An integrated maritime surveillance system based on high-frequency surface-wave radars. 2. Operational status and system performance. Raytheon Syst. Canada Ltd., Waterloo, Ont., Canada; This paper appears in: Antennas and Propagation Magazine, IEEE Publication Date: Oct. 2001 Volume: 43, Issue: 5 On page(s): 52–63.

Prakash, A., and Gupta, R.P., 1998. Land use mapping and change detection in a coal mining area: A Case study in the Jharia coat field, India. *International Journal of Remote Sensing* 19(3):391–410.

Quader, M.O., Pramanik, M.A.H., Khan, F.A., and Polcyn, F.C., 1986. *Mangrove Ecosystem Study of Chakoria Sunderban of Chittagong with Special Emphasis on Shrimp Ponds by Remote Sensing Techniques*. Marine interfaces Ecohydrodynamics, Elsevier Oceanography series, 42, Amsterdam/Tokyo, pp. 646–653.

Rathore, C.S., and Wright, R., 1993. Monitoring environmental impacts of surface coal mining. *International Journal of Remote Sensing* 14(6):1021–1042.

Repic, R.L., Lee, J.K., and Mausel, P.W., 1991. An analysis of selected water parameters in surface coal mines using multispectral videography. *Photogrammetric Engineering and Remote Sensing*. 57:1589–1596.

Richards, Jr., W., 1983. Geological applications. In: Colwell. R N., (ed.) *Manual of Remote Sensing*. American Society of Photogrammetry and Remote Sensing, Falls Church, VA.

Rodríguez-Valencia, J.A., Crespo, D., and L'opez-Camacho, M., 2010, La camaroniculturay la sustentabilidad del Golfo de California, 2010.

Ruthenberg, H., 1980. *Farming Systems in the Tropics*. Clarendon Press, Oxford.

Schmidt-Vogt, D., Leisz, S.J., Mertz, O., Heinimann, A., Thiha, T., Messerli, P., Epprecht, M., Cu, P.V., Chi, V.K., and Hardiono, M., et al., 2009. An assessment of trends in the extent of swidden in southeast Asia. *Human Ecology* 37:269–280.

Shahid, M.A., Pramanic, M.A.H., Jabar, M.A., and Ali, S., 1992. Remote sensing application to study the coastal shrimp farming areas in Bangladesh. *Geocarto International* 7(2):5–14.

Smith, M.D., Roheim, C.A., Crowder, L.B., Halpern, B.S., Turnipseed, M., Anderson, J.L., Asche, F., Bourillon, L., Guttorm-sen, A.G., Khan, A., Liguori, L.A., McNevin, A., O'Connor, M.I., Squires, D., Tyedmers, P., Brownstein, C., Carden, K., Klinger, D.H., Sagarin, R., and Selkoe, K.A., 2010. Sustainability and global seafood. *Science* 327(5967):84–786.

Solomon, J.L., Miller, W.F., and Quattrochi, D.A., 1979. Development of a tree classifier for discrimination of surface mine activity from Landsat digital data. Proceedings of the American Society of photogrammetry 45th Annual meeting held at Washington, DC on 18–24 march 1979 (Falls Church, Virginia: American society of Photogrammetry), II, pp. 607–613.

Spencer, J.E., 1966. *Shifting Cultivation in Southeastern Asia, Vol. 19*. University of California Press, Berkeley, CA.

Stanton, T., and Clay, C., 1986. Sonar echo statistics as a remote-sensing tool: Volume and sea floor, *IEEE Journal of Oceanic Engineering* 11(1):79–96.

Stickney, R.R., and McVey, J.P., 2002. Responsible Marine Aquaculture, World Aquaculture Society, New York.

Thayalan, S., and Nagabhushana, S.R., 1987. Impact of mining and land cover in Kudremukh area, Karnataka. Proceedings (Abstract) National Symposium on Land Transformation and Management. Dec 21–23, 1987, Hyderabad. pp. 58–59.

Thrupp, L.A., Hecht, S., Browder, J.O., Lynch, O.J., Megateh, N., and O'Bnen, W., 1997. *The Diversity and Dynamics of Shifting Cultivation: Myths, Realities and Policy Implications*. World Resources Institute, Washington, DC.

Tripathi, N.K., Annachhatre, A., and Patil, A.A., 2000. On the role of remote sensing in environmental impact analysis of shrimp farming. Map India 2000.

Tuominen, S., and Pekkarinen, A., 2005. Performance of different spectral and textural aerial photograph features in multi-source forest inventory. *Remote Sensing of Environment* 94:256–268.

van Vliet, N., Adams, C., Vieira, ICG., and Mertz, O., 2013. Slash and burn and shifting cultivation systems in forest agriculture frontiers from the Brazilian amazon. *Society & Natural Resources*; 26(12):1454–67.

Van Vliet, N., Mertz, O., Heinimann, A., Langanke, T., Pascual, U., Schmook, B., Adams, C., Schmidt-Vogt, D., Messerli, P., Leisz, S.J., Castella, J.C., Jørgensen, L., Birch-Thomsen, T., Hett, C., Bech-Bruun, T., Ickowitz, A., Vu, K.C., Yasuyuki, K., Fox, J., Padoch, C., Dressler, W., and Ziegler, A.D., 2012. Trends, drivers and impacts of changes in swidden cultivation in tropical forest-agriculture frontiers: A global assessment. *Global Environmental Change* 22:418–429.

Venkataratnam, L., Thammappa, S.S., and Ravishankar, T., 1997. Mapping and monitoring prawn farming areas through remote sensing techniques. *Geocarto International* 12(2):23–29.

Vibulsresth, S., Ratanasermpong, S., Dowreang, D., and Silapathong, C., 1993, Mangrove monitoring using satellite data. *TRSC Newsletter* 10(1):1–2.

Volk, P., Hayden, R., and Saradeth, S., 1990. Landsat and SPOT data—important tools for environmental assessment and hydrogeological applications. *Proceedings of the Conference on Remote Sensing: An operational Technology for Mining and Petroleum Industries*. London 29–31 October, 1990.

Vosti, S.A., Witcover, J., and Carpentier, C.L., 2002. *Agricultural Intensification by Smallholders in the Western Brazilian Amazon: from Deforestation to Sustainable Land Use*. International Food Policy Research Institute, Washington, DC.

Wier, C.W., Wobber, F.J., Orville, R.R., and Amato, R.V., 1973. Fracture mapping and strip mine inventory in the mid-west by using ERTS-1 imagery. In: *Proceedings of the Symposium on Significant Results Obtained from the EARTH Resources Technology Satellite-1*. Maryland, USA, March 5–9, 1973 (Washington, DC: National Aeronautics and Space Administration), pp. 553–560.

12

Drought

12.1 Introduction

Drought is a creeping disaster and a silent killer. It rolls back development gains and exacerbates poverty, especially in least developed countries. Due to climate change, the frequency, severity, and duration of droughts is likely to be more serious in the future. Droughts are, in fact, the world's costliest natural disasters, causing an average $6–$8 billion in global damages annually and collectively affecting more people than any other form of natural disaster (Wilhite, 2000a). The socioeconomic and environmental impacts of droughts have increased significantly, particularly among the most vulnerable groups in Asia and the Pacific.

Droughts occur in virtually all climatic zones, such as high as well as low rainfall areas, and are mostly related to the reduction in the amount of precipitation received over an extended period of time, such as a season or a year. Temperatures, high winds, low relative humidity, timing and characteristics of rains, including distribution of rainy days during crop-growing seasons, intensity and duration of rain, and onset and termination, play a significant role in the occurrence of droughts. In contrast to aridity, which is a permanent feature of climate and is restricted to low rainfall areas (Wilhite, 1992), drought is a temporary aberration.

12.2 Background

Drought, in general, refers to extreme precipitation shortage. However, objectively characterizing it for planning and management is a challenging issue. Often there is confusion between a heat wave and a drought. While a typical timescale associated with a heat wave is on the order of a week, a drought may persist for months or even years (Chang and Wallace, 1987). The combination of a heat wave and a drought has dire socioeconomic consequences Mishra and Singh (2010). No universally acceptable definition of drought is available because there area wide variety of sectors affected by drought, its diverse spatial and temporal distribution and the demand placed on water supply by human-used systems.

Drought can generally be defined as the extreme persistence of precipitation deficit over a specific region for a specific period of time (González and Valdés, 2006). Besides, the elements of "persistence" of "substantial precipitation deficit," "bounded by time and space," definitions have expanded to include impacts on environment and society

(Tsakiris and Vangelis, 2004). Sheffield and Wood (2007) have defined drought as a continuous period where soil moisture remains below the 20th percentile at monthly scale. It is, however, widely accepted that the major cause of drought is lower than the average rainfall. The rainfall deficit will have different impacts depending on other factors, including meteorological conditions ecosystem type and socioeconomic circumstances.

Drought impacts are functions of both the enormity of the water shortage and susceptibility on ground conditions. Being affected by drought is thus a context-dependent matter. Part of the complexity in drought definition stems from such subjectivity of extent of drought impacts. This challenge is reflected in the conceptual development of non-meteorological drought indices; although more than 91 drought impacts can be identified (NDMC, 2006), drought indices make use of a handful of impact indicators including vegetation health, evapotranspiration (ET), or water resources levels.

Three main physical drought types have been established: meteorological, agricultural, and hydrological droughts. These droughts occur in a particular order: precipitation deficiency induces meteorological drought, which subsequently impacts soil moisture content (i.e., agricultural drought) (Figure 12.1). Low recharge from the soil to water features such as streams and lakes cause a delayed hydrological drought. However, two more drought types, namely socioeconomic and ground water droughts, have also been established. The definitions of different drought types are given hereunder (Wilhite et al., 2007; FAO, 2008):

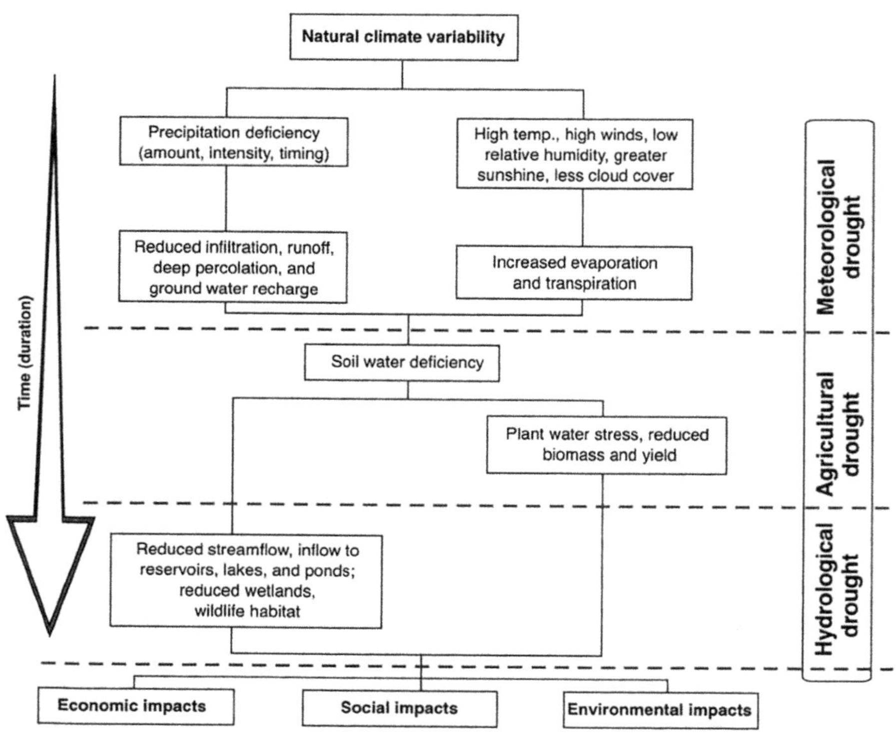

FIGURE 12.1
Relationship between various types of drought and duration of drought events (Source: Adapted from Wilhite (2000b)).

- *Meteorological or climatological drought* which is generally regarded as being lower than average precipitation for some time period; in some cases air temperature and precipitation anomalies could be combined.
- *Agro-meteorological or agricultural drought* occurs when plant available water from precipitation and stored in the soils falls short of below that required during critical growth stage. This leads to below-average yield in both pasture and grain-producing crops.
- *Hydrological drought* is generally defined by one or a combination of factors such as stream flow, reservoir storage, and ground water.
- *Socioeconomic drought* is defined in terms of loss from an average or expected return and can be measured by both social and economic indicators (McVicar and Jupp, 1998).
- *Groundwater drought* is defined by the decrease in groundwater level (Chang and Teoh, 1995; Eltahir and Yeh, 1999). However, groundwater storage, or groundwater recharge or discharge can be and has also been used to define or quantify a groundwater drought.

In addition to type, droughts are fundamentally characterized in three dimensions: severity, duration, and spatial distribution. Additional characteristics include: frequency, magnitude (cumulated deficit), predictability, rate of onset, and timing. However, usage of the terms *severity*, *intensity*, and *magnitude* is not universal, and sometimes they are interpreted uniformly. *Duration:* Depending on the region, drought's duration can vary from a week up to a few years. Because of drought's dynamic nature, a region can experience wet and dry spells simultaneously when considering various timescales. As such, in shorter durations the region experiences dryness or wetness, while in longer term, it experiences otherwise.

Magnitude: The cumulative deficit of water (e.g., precipitation, soil moisture, or runoff) below some threshold during a drought event.

Intensity refers to the ratio of drought magnitude to its duration.

Severity: Severity is used in two ways: (i) the degree of the precipitation deficit (i.e., magnitude), or the degree of impacts resultant from the deficit (Wilhite, 2004); (ii) geographic extent: the areal coverage of the drought which is variable during the event; and (iii) frequency: the frequency or return period of a drought is defined as the average time drought events that have a severity that is equal to or greater than a threshold.

12.2.1 Drought Indicators

Variables such as ET and stream flow are used in conjunction with precipitation deficit to more comprehensively characterize drought. Using different models (e.g., water balance/hydrological models), such variables or indicators are used in combination to derive a drought index. A drought index provides a comprehensive picture for drought analysis and decision-making that is more readily useable compared with raw data from indicators (Hayes, 2006). Drought indices are derived by assimilating drought indicators into a single numerical value. Such indicators can be meteorological, hydrological, or water supply-and-demand in nature. Meteorological indicators include precipitation and cloud cover; hydrological indicators include stream flow and groundwater level; water supply-and-demand indicators include reservoir storage. In practice, however, some indicators such as precipitation, potential evapotranspiration (PET), and soil- and vegetation-cover characteristics have had wider applications.

Generally, drought indices are categorized based on the type of impacts they relate to. The categorization can also be based on the variables they relate to or use of disciplinary data. Three popular categories are meteorological, agricultural, and hydrological drought indices. Niemeyer (2008) adds three more categories to this list: comprehensive, combined, and remote sensing-based drought indices. Comprehensive drought indices use a variety of meteorological, agricultural, and hydrological variables to draw a comprehensive picture of drought. The Palmer drought severity index (PDSI) is an example of this approach.

12.3 Global Scenario

Drought is a widespread natural hazard. It is evident from Figure 12.2 that a fairly large area across the globe experiences moisture stress which is further acerbated by incidences of droughts. Generally, arid and semiarid regions across the globe very often experience drought events (Figure 12.2). Apart from impacting food and water security, droughts also have profound effect on global economy. For example, recently, large-scale intensive droughts events have occurred and affected large areas in Europe, Africa, Asia, Australia, South America, Central America, and North America. For example, during the growing seasons over the period 1980–2003 in USA, drought accounted for $144 billion (41.2%) of the estimated $349 billion total cost of all weather-related disasters (Ross and Lott, 2003).

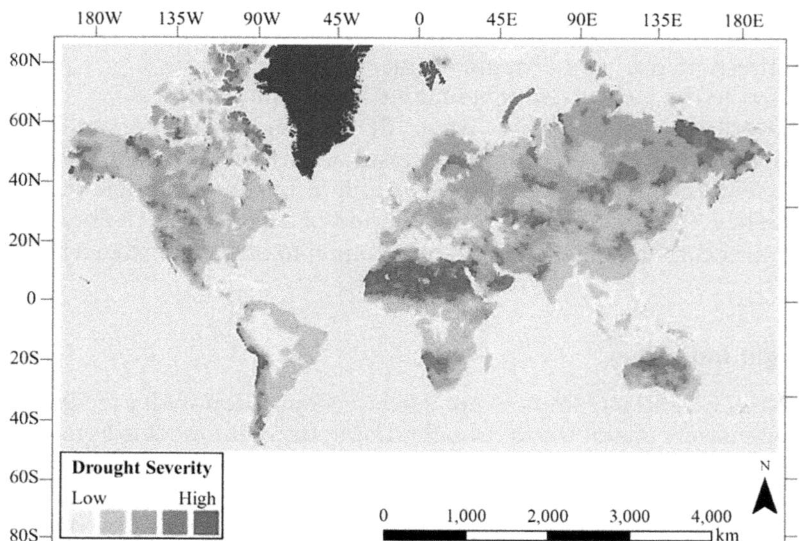

FIGURE 12.2
World drought severity distribution map computed over the 1901–2008 period (modified after World Resources Institute, 2015). Drought is defined as a continuous period where soil moisture remains below the 20th percentile at monthly scale (Sheffield and Wood, 2007).

The Canadian Prairies are the most drought-susceptible area in Canada due to high variability of precipitation. For instance, the drought event during 2001 and 2002 growing seasons resulted in an estimated loss of \$3.6 billion in agricultural production (Wheaton et al., 2005).

In Australia, the winter cereal crop was reduced by 36% and costed around AUD\$3.5 billion during the 2006 drought event (Wong et al., 2010). During the past 30 years, several major drought events occurred in Europe. The most severe drought event is the one that caused 10% reduction in the overall European cereal yields in the Iberian Peninsula in 2005. Since 1991, the European Union has estimated a yearly average economic impact of drought by €5.3 billion (European Communities, 2007). In Asia, the Intergovernmental Panel on Climate Change reported that most of rice, maize, and wheat production has declined in many Asian countries in the last few decades (Bates et al., 2009). For examples, during 1999–2000 crop-growing seasons, around 60 million people in Central and Southwest Asia were affected by drought and around 40 million hectares of agricultural land were affected in China alone. In India, drought has been reported at least once in every three years in the last five decades (World Bank, 2003).

North African countries such as, Morocco, Algeria, and Tunisia faced similar situation during that drought event as the acreage of cereal crop was shrunken by 31% comparing to the previous year's harvest (De Pauw, 2005). The West Asia—including Jordan—and North Africa region experienced several drought events in the last three decades resulting in reduced food production. For example, in 1999, in Iraq, rainfall was 30% below average resulting in 70% failure in rain-fed agriculture crops.

12.4 Drought Assessment and Monitoring

For taking up any relief measure to combat drought, a systematic assessment of drought with respect to its severity and spatial distribution is a prerequisite. In the event of drought there is a change in meteorological variables, namely precipitation, soil moisture, etc. Consequently it impacts ecosystem in terms of vegetation health and growth. Traditionally, reservoir storage, stream flows, soil moisture, and ground water table have been used for assessment and monitoring of drought. Since the incidence of drought is intimately related to quantum, frequency, and spatial distribution of rainfall and due to its effects on several facets of terrain in terms of change in density and vigor of vegetation including crops, shrinkage in the extent of water spread and the number of water bodies, and soil moisture stress, a single parameter may not adequately reflect its impact.

Several drought indices have, therefore, been developed for drought assessment. Drought indicators are variables that describe features of drought (Steinemann et al., 2005). Drought indices integrate large amounts of data, such as precipitation, snowpack, stream flow, and other water supply indicators, to monitor drought severity in a comprehensive framework and to measure how much the climate in a given period has deviated from historically established normal conditions (Narasimhan and Srinivasan, 2005). Several methodologies for drought characterization exist; however, using drought indices is prevalent. More than 150 drought indices have been developed (Niemeyer, 2008) and additional indices have recently been proposed (Zargar et al., 2011; Vicente-Serrano et al., 2010).

12.4.1 Meteorological Indicators

Rainfall is the most important single factor influencing the incidence of drought. Many studies have utilized the relationship between actual and average rainfall for assessment of the nature and frequency of droughts Gibbs and Meher (1967). In fact, rainfall deciles are generally used as drought indicator. To calculate deciles the rainfall data are arranged in terms of rainfall totals, in ascending order (from lowest to highest). The ranked dataset is then divided into ten equal parts (i.e., ten blocks of 10%). Some of the commonly used drought indices are described hereunder.

12.4.1.1 Deciles

The method of deciles or 10% iles is based on dividing the distribution of monthly record precipitation into 10% parts (Gibbs and Maher, 1967). Extended lengths of precipitation data record are required for accurate estimation. Deciles may be computed for any chosen period or window.

12.4.1.2 Percent of Normal Precipitation

The percent of normal precipitation is one of the simplest measurements of rainfall for a location. Analyses using percent of normal are very effective when used for a single region or a single season. It is computed by dividing actual precipitation by normal precipitation—typically considered to be a 30-year mean—and multiplying by 100%. This can be calculated for a variety of timescales, including monthly, seasonal, annual, or water year. Normal precipitation for a specific location is considered to be 100%.

12.4.1.3 Palmer Drought Severity Index

PDSI provides a standardized measurement of soil moisture conditions to compare between locations and over time (Palmer, 1965). The index estimates duration and intensity of drought events by measuring departure of the moisture supply based on a supply-and-demand concept of the water balance equation. It incorporates precipitation and temperature data, and local available water content of the soil from an unspecified period that best corresponds to past 9–12 months. Past conditions are incorporated because long-term drought is cumulative, so the intensity of drought at a particular time is dependent on the current conditions and the cumulative patterns of previous months.

All the basic terms of the water balance equation are determined, including ET, soil recharge, runoff, and moisture loss from the surface layer. By accounting for moisture the equations are described in Palmer's original study (Palmer, 1965). The PDSI estimates when a drought (or wet spell) begins, ends, and the duration of the event (Palmer, 1965; Alley, 1984). The PDSI generally ranges from −4.0 to +4.0 and it is designed so that an extreme drought (−4.0) in one climate division has the same meaning in terms of the moisture deficit as an extreme drought in any other climate division (Alley, 1984) (Table 12.1).

The Palmer hydrological drought index (PHDI) is a derivative of the PDSI which is based on daily inflow (precipitation) and soil moisture storage (Karl and Knight, 1985).

TABLE 12.1

Drought Categories Based on PDSI

PDSI	Weather Condition
4.0 or more	• Extremely wet
• 0.5 to 0.99	• Incipient wet spell
• 0.49 to –0.49	• Near normal
• –0.5 to –0.99	• Incipient dry spell
• –1.0 to –1.99	• Mild drought
• –2.0 to –2.99	• Moderate drought
• –3.0 to –3.99	• Severe drought
• –4.0 or less	• Extreme drought

Source: Palmer (1965).

12.4.1.4 Standardized Precipitation Index

The standardized precipitation index (SPI) assigns a single numerical value to the precipitation that can be compared across the regions with markedly different climatic conditions (McKee et al., 1993).

$$P(x) = \frac{x^{\gamma-1}\exp(-x/\beta)}{\beta^{\gamma}\Gamma(\gamma)} \quad \gamma > 0 \tag{12.1}$$

$$SPI = \frac{X_{ij} - X_{im}}{\sigma} \tag{12.2}$$

These timescales reflect the impact of drought on availability of different water resources. While ground water, stream flow, and reservoir reflect the long-term precipitation anomalies, soil moisture conditions respond to precipitation anomaly for a relatively short scale. Essentially, the SPI is the number of standard deviations that an observed value of precipitate would deviate from the long-term mean, for a normally distributed random variable. Since precipitation is not normally distributed a transformation is applied first so that transformed precipitation values follow a normal distribution.

Calculation of SPI for a specific time period at any location requires a long-term precipitation database of 30 years or more. Cumulative distribution is then transformed using equal probability to a normal distribution with a mean zero and standard deviation one, so that the values of SPI are really in standard deviations. SPI values generally range from +2 to –2, representing extremely wet to extremely dry conditions, respectively. The drought categories based on SPI values are shown in Table 12.2.

12.4.1.5 Crop Moisture Index

The crop moisture index (CMI)—a Palmer derivative reflects moisture supply in the short term across major crop-producing regions. The index uses a meteorological approach developed by Palmer (1968) to monitor week-to-week crop conditions. The CMI has been designed to evaluate short-term moisture conditions across major crop-producing regions while PDSI monitors long-term meteorologically wet or dry spells. The index is based on weekly mean temperature and total precipitation within a climate division,

TABLE 12.2

SPI Drought Categories

SPI Value	Drought Class
2.0+	• Extremely wet
• 1.55 to 1.99	• Very wet
• 1.0 to 1.49	• Moderately wet
• –0.99 to 0.99	• Nearly normal
• –1.0 to 1.49	• Moderately dry
• –1.5 to –1.99	• Severely dry
• –2.0 and less	• Extremely dry

Source: Shafer and Dezman (1982).

and incorporates the CMI value from the previous week. The CMI responds rapidly to changing conditions, and is weighted by location and time.

12.4.1.6 Standardized Precipitation Evapotranspiration Index

The standardized precipitation evapotranspiration index (SPEI) is an extension of the widely used SPI. The SPEI is designed to take into account both precipitation and PET in determining drought (Vicente-Serrano et al., 2010).

$$\mathrm{SPEI} = \mathrm{W} - \frac{\mathrm{C_0} + \mathrm{C_1 W} + \mathrm{C_2 W^2}}{1 + \mathrm{d_1 W} + \mathrm{d_2 W^2} + \mathrm{d_3 W^3}}, \tag{12.3}$$

where

$$\mathrm{W} = \sqrt{-2\ln(\mathrm{P})} \quad \text{for } \mathrm{P} \le 0.5 \tag{12.4}$$

and P is the probability of exceeding a determined D value, $P = 1-F(x)$. If $P > 0.5$, then P is replaced by $1-P$ and the sign of the resultant SPEI is reversed. The constants are $C_0 = 2.515517$, $C_1 = 0.802853$, $C_2 = 0.010328$, $d_1 = 1.432788$, $d_2 = 0.189269$, and $d_3 = 0.001308$. The average value of SPEI is 0, and the standard deviation is 1. The SPEI is a standardized variable, and it can therefore be compared with other SPEI values over time and space. An SPEI of 0 indicates a value corresponding to 50% of the cumulative probability of D, according to a log-logistic distribution.

12.4.1.7 Soil Moisture Deficit Index

Since drought indices like the PDSI and SPI have coarse spatial (7000–100,000 km^2) and temporal resolution (monthly), in order to improve spatial and temporal resolutions Narasimhan and Srinivasan (2005) developed two new drought indices, namely soil moisture deficit index (SMDI) and evapotranspiration deficit index (ETDI) based on weekly soil moisture deficit and ET deficit, respectively. The distributed hydrologic model was used to simulate soil moisture and ET from daily weather data at a high spatial resolution (16 km^2) using GIS. Using this simulated data the drought indices SMDI and ETDI were developed. The weekly soil moisture deficit model takes into account the full profile as well as different depths (at 0.61, 1.23, and 1.83 m) which makes it adaptable to different crop types.

$$SD_{i,j} = \frac{SW_{i,j} - MSW_j}{MSW_j - \min SW_j} \times 100, \quad \text{if } SW_{i,j} = MSW_j$$
$$SD_{i,j} = \frac{SW_{i,j} - MSW_j}{\max SW_j - MSW_j} \times 100, \quad \text{if } SW_{i,j} > MSW_j \tag{12.5}$$

$$\sum_{t=1}^{j} Z_t = -100t - 100 \tag{12.6}$$

$$SMDI_j = \frac{\sum_{t=1}^{j} SD_t}{25t + 25} \tag{12.7}$$

12.4.1.8 Surface Water Supply Index

The surface water supply index (SWSI) was designed to complement the Palmer drought indices for moisture conditions (Shafer and Dezman, 1982). The Palmer indices are not designed for large topographic variations across a region, and do not account for snow accumulation and subsequent runoff. On the contrary, SWSI incorporates mountain snowpack levels and was designed specifically to assess surface water conditions. Thus, the SWSI integrates both hydrological and climatological features into a single index value for each major river basin (Shafer and Dezman, 1982). These values are standardized to allow comparisons between basins. Four inputs, snowpack, stream flow, precipitation, and reservoir storage, are used to calculate SWSI.

12.4.2 Remote Sensing-Based Methods

In situ-based drought indices usually provide very accurate estimates of agricultural drought conditions at the points of observations. However, the uneven spatial distribution of these points (hydrometeorological stations) across the landscape often imposes uncertainty in delineating spatial context. In order to address this, geographic information system (GIS)-based interpolation techniques (e.g., inverse distance, krigging, nearest neighbor) are usually employed. However, these techniques often generate different outcomes despite using the same set of input variables (Li and Heap, 2014). By virtue of providing synoptic view in discrete bands of electromagnetic spectrum at regular interval spaceborne spectral measurements offer immense potential in drought vulnerability analysis, damage assessment, and warning.

Spaceborne observations have been used to monitor drought-related variables from a meteorological viewpoint and also to assess and quantify drought impacts from an ecosystem perspective. In the former, spaceborne multispectral, thermal infrared, or microwave (MW) data have been used to retrieve a drought-related variable including precipitation soil moisture or ET. The specific hydrologic variable is then converted into a drought indicator by calculating the extent of an anomaly or departure from the longer-term environmental baseline. Those data are used to quantitatively assess and categorize drought severity.

Spaceborne spectral measurements have also been used to assess the impact of drought on ecosystem including vegetation health and growth—by assessing the photosynthetic capacity of plants (Tucker and Choudhury, 1987). Precipitation deficits can lead to reduced photosynthetic capacity and changes in absorption of solar radiation in photosynthetically active wavelengths by plants (Asrar et al., 1984). The major applications of remote sensing in deriving information on drought are provided hereunder. For further details the reader may refer to McVicar and Jupp, 1998, AghaKouchak et al., 2015, Murthy and Seshasai (2010) and Hazaymeh, K. and Quazi K., 2016.

12.4.2.1 Estimation of Meteorological Parameters

Precipitation: Precipitation is generally measured *in situ* using rain gauges. Besides, precipitation is commonly assessed using a precipitation-based drought index, i.e., SPI, which has been recommended by the World Meteorological Organization (WMO) as a global measure of meteorological drought (WMO, 2009). Deriving SPI involves describing the frequency distribution of precipitation using either a parametric distribution function or a nonparametric approach for different precipitation accumulation periods (e.g., 1-, 3-, or 6-month period). The SPI is computed by transforming the cumulative probability of precipitation into the standard normal distribution.

Spaceborne visible and infrared images of cloud top temperature can be converted into a precipitation rate using empirical statistical relationships (Turk et al., 1999). Passive MW sensors offer a more physically based approach to instantaneous precipitation estimation (Kummerow et al., 2001). While MW sensors provide more accurate precipitation information, they are limited by their infrequent overpasses (≈2 observations per day for any location). Geostationary Earth Observation visible and infrared data, on the other hand, provide more frequent precipitation information (≈15–30 minutes) although with higher uncertainty.

Studies suggest that a combination of spaceborne spectral measurements in MW and infrared regions leads to better precipitation estimates, especially of diurnal patterns (Joyce et al., 2004). Currently, several satellite precipitation data products are available including the Climate Predicting Center (CPC), Morphing Technique (CMORPH) (Joyce et al., 2004), Tropical Rainfall Measuring Mission (TRMM), Multi-satellite Precipitation Analysis (TMPA) (Huffman et al., 2007), Precipitation Estimation from Remotely Sensed Information using Artificial Neural Networks (PERSIANN) (Hong et al., 2004), and the Global Precipitation Climatology Project (GPCP) (Adler et al., 2003). For a comprehensive review of precipitation algorithms refer to Levizzani et al. (2007) and Kidd (2001).

Soil Moisture: Soil moisture is often used as a measure of agricultural drought since it affects plant growth and productivity. A number of soil moisture-based indices have been developed and used for drought monitoring, including the standardized soil moisture index (SSI) (Hao and AghaKouchak, 2013) and the soil moisture percentile (Wang et al., 2009). The soil moisture input to these drought indicators can be obtained from land surface model simulations or from satellite estimates. Most satellite soil moisture algorithms are based on passive MW, active MW or a blend of data from multiple sensors MW-based soil moisture retrieval relies on the relationship between soil permittivity and liquid water content. There are empirical relationships that link passive MW brightness temperature and active MW backscattering to volumetric water content of soil. MW soil moisture observations typically represent the top 2–5 cm of soil depth (Entekhabi et al., 2010). For root-zone soil surface moisture estimates, MW soil moisture observations can be coupled to an appropriate land surface model (Reichle et al., 2004). For a comprehensive review of optical, thermal, passive MW, and active MW soil moisture monitoring approaches readers may refer to Hazaymeh and Hassan (2016).

Groundwater and Terrestrial Water Storage: Estimates of drought impacts on terrestrial water storage at regional to global scales can be obtained using the Gravity Recovery and Climate Experiment (GRACE) mission. GRACE responds to all factors that change the gravity field of an area, including the terrestrial water storage. Launched in 2002, the GRACE mission tracks global variations in gravity fields that can be converted into estimates of terrestrial water storage (Rodell and Famiglietti, 2002). The principle of gravimetry and terrestrial water storage estimation is that bulk surface and subsurface water mass has a gravitational potential that alters the Earth's gravity field, allowing changes in the Earth's gravitational field to be used as an indicator of changes in the total amount of water stored in the vertical continuum. GRACE consists of two identical satellites separated by a distance of about 220 km that orbit the Earth at an altitude of approximately 500 km. Changes in the Earth's gravity fields alter the distance between the two spacecraft. Having accurate measures of the distance between the two spacecraft, one can quantify temporal terrestrial water storage anomalies, which include the sum of surface water, groundwater, soil moisture, snow/ice, and moisture stored in vegetation (Rodell, 2012).

Having terrestrial water storage, changes to groundwater ΔG can be approximated as $\Delta G = \Delta TWS - \Delta SM - \Delta SWE$, where SM and SWE represent soil moisture and snow water equivalent, respectively (Rodell et al., 2007).

Evapotranspiration: While ground-based measurement of ET at large spatial extents is challenging, remotely sensed data offer a unique opportunity to provide large-scale estimates of ET. A unique feature of ET for drought monitoring is that it describes both water/moisture availability and the rate at which it is consumed. Broadly, remote sensing-based ET estimation methods can be categorized into the following groups based on (a) principles of water balance, (b) principles of surface energy balance, (c) vegetation indices, and (d) hybrid approaches based on vegetation indices and surface temperature information (Yang and Shang, 2013). Water balance models track changes in moisture in the soil for ET estimation, whereas energy balance models use land surface temperature (LST) and meteorological data to estimate sensible heat flux and moisture/water fluxes (Morillas et al., 2013).

Snow: Snow is considered a natural reservoir of water resources, and in some regions snow melt constitutes a substantial fraction of the annual runoff. A deficit in the winter snowpack could potentially lead to a summer hydrological drought resulting in reduced stream flows and/or groundwater levels or agricultural drought (e.g., depleted soil moisture reserves to support plant functions). Hence, monitoring snow is fundamental to drought assessment in many regions. Snow water equivalent, snow-covered area, snow depth, and snow albedo (Painter et al., 2013) are the snow parameters that need to be regularly monitored. Spaceborne spectral measurements in both optical and MW regions hold great promise in deriving above-mentioned snow parameters. Optical-based products provide estimates of only snow-covered area, whereas MW alone or in conjunction with optical sensors data provides information on snow-covered area, snow depth, and snow water equivalent. Snow exhibits a strong spectral gradient in reflectance, from high albedo in visible wavelengths to low reflectance in middle infrared region (Dozier et al., 2009; Kongoli et al., 2012). Thus, snow can be delineated using the ratio of the reflectance in visible reflectance and the middle infrared reflectance region (Romanov et al., 2000). Furthermore, snow can also be detected using the normalized difference snow index (NDSI) defined as follows (Hall et al., 2002):

$$NDSI = \frac{\rho_{Green} - \rho_{SWIR}}{\rho_{Green} + \rho_{SWIR}} \tag{12.8}$$

By virtue of the cloud penetration, and day-and-night operation capability, microwave sensors provide a unique opportunity for continuous snow monitoring (Kongoli et al., 2012).

More importantly, MWs can penetrate into snow, allowing estimation of snow water equivalent and snow depth that cannot be obtained from optical-based methods (Durand et al., 2008). The empirical relationship between variations in observed snow water equivalent and snow depth and the difference in brightness temperature in two low-frequency channels forms the basis for delineation and monitoring snow cover using MW data (Kongoli et al., 2007). A number of algorithms have been developed for estimation of snow covered area using MW datasets (Grody and Basist, 1996). There are static empirical algorithms in which one set of regression parameters are used (Goodison, 1989), as well as dynamic empirical algorithms in which different regression coefficients are used in various regions and for different seasons (Foster et al., 2005).

Unlike other drought-related variables, snow-based indicators of drought are still in their infancy, primarily because there is a lag between snow occurrence and change in surface water and soil moisture availability that varies in space and time. Runoff from snowmelt could affect water availability on timescales from a few weeks in case of low-elevation snow that occurs at lower latitudes) to a few months at high-elevation snow prevalent at higher latitudes. This lag time is a significant strength for drought early warning. However, even over one particular location, depending on seasonal temperatures and the timing of snow accumulation, the lag time from snowfall, snowmelt, and runoff varies substantially. This highly variable lag is the main challenge in deriving snow-based indicators for drought monitoring.

12.4.2.2 Drought Impacts

The impact of drought can be assessed based on observed changes in vegetation health and land cover from remotely sensed data. Quantitative assessment of vegetation condition is generally based on the spectral response pattern of vegetation greenness expressed in the red (R) and near-infrared (NIR) portions of the electromagnetic spectrum. In order to address the spatial distribution and dynamics of agricultural drought several remote sensing-based indices have been developed. These indices are based on unique spectral response patterns of soil surface and canopy characteristics, particularly in the red, NIR, shortwave infrared (SWIR), and thermal regions of the electromagnetic spectrum.

In fact, the applications of remote sensing in agricultural drought assessment and monitoring relies on the fact that drought might affect the biophysical and chemical properties of soil and vegetation, namely soil moisture, organic matter, vegetation biomass, chlorophyll, and canopy and soil temperature (Anjum et al., 2011). Consequently, it may change their spectral and thermal responses, which in turn can be used as indicators of drought occurrence. The utility of remote sensing-based methods depends on different factors including satellite data availability, cost, data quality, preprocessing, and post-processing requirements including the expertise available to use remote sensing data.

Vegetation Indices in Optical Region: The normalized difference vegetation index (NDVI) is the most frequently used vegetation index and the first remote sensing-based measure used to monitor agricultural drought (Rouse et al., 1974). NDVI is the difference between reflected R and NIR radiation divided by their sum (Rouse et al., 1974):

$$\mathrm{NDVI} = \frac{(\rho\mathrm{NIR} - \rho\mathrm{R})}{(\rho\mathrm{NIR} + \rho\mathrm{R})} \tag{12.9}$$

Since the soil spectrum and non-vegetated surfaces do not exhibit distinct differences in spectral absorption between the red and NIR bands, NDVI can be used to separate vegetation from the soil background (Karnieli et al., 2010) and provide a measure of general vegetation health. Time series decomposition provides the means to assess persistent and recurrent vegetation conditions, which may be related to deep- or shallow-rooted vegetation types in ecosystems.

A significant relationship has been reported between NDVI and precipitation and soil moisture (Adegoke and Carleton, 2002), and thus, NDVI (or its derivatives) has been widely used for drought assessment and vegetation health monitoring. The general vegetation health can be temporally decomposed to monitor changes in recurrent (shallow-rooted) and persistent (deeper-rooted) vegetation for many landscapes across the globe (Donohue et al., 2009). This is particularly important when monitoring drought in savannahs or areas with mixed trees and ecosystems across semiarid regions.

Building on the original definition of NDVI, a number of other vegetation indicators, such as the transformed vegetation index (TVI) (Rouse et al., 1974):

$$\mathrm{TVI} = (\mathrm{NDVI} + 0.5)^{1/2} \tag{12.10}$$

perpendicular vegetation index (PVI) (Jackson et al., 1980):

$$\mathrm{PVI} = \frac{(\mathrm{NIR} - \mathrm{aR} - \mathrm{b})}{\sqrt{\mathrm{a}^2 + 1}} \tag{12.11}$$

and Thiam's transformed vegetation index (TTVI) (Thiam, 1997):

$$\mathrm{TTVI} = \mathrm{ABS}(\mathrm{NDVI} + 0.5) \tag{12.12}$$

where ABS indicates absolute, have been developed. These indices describe the vegetation condition by combining spectral information from different parts of the electromagnetic spectrum that are sensitive to biophysical characteristics of vegetation, such as chlorophyll content, water content, and internal leaf structure.

The vegetation condition index (VCI), for example, scales NDVI values between its minimum,

$$\mathrm{VCI_j} = \frac{(\mathrm{NDVI_j} - \mathrm{NDVI_{min}}) * 100}{\mathrm{NDVI_{max}} - \mathrm{NDVI_{min}}} \tag{12.13}$$

and maximum values to separate the short-term weather signal from the long-term ecological signal for drought monitoring (Kogan and Sullivan, 1993). Use of the monthly VCI is more appropriate in areas with a large land management signal (e.g., cropping) and hence is suitable for monitoring agricultural drought (McVicar and Jupp, 1998). A standardized form of NDVI, known as the standardized vegetation index (SVI), is based on the z score of NDVI values (Park et al., 2008) (Figure 12.3).

For comprehensive list of vegetation indices readers may refer to Silleos et al. (2006) and Payero et al. (2004).

Vegetation Indices in SWIR Region: A number of vegetation stress and drought indicators have been developed using SWIR data such as the normalized difference water index (NDWI) (Gao, 1996). The NDWI is defined as the difference between two NIR bands. These two channels sense similar depth through the vegetation canopy and are less sensitive to atmospheric scattering effects than NDVI.

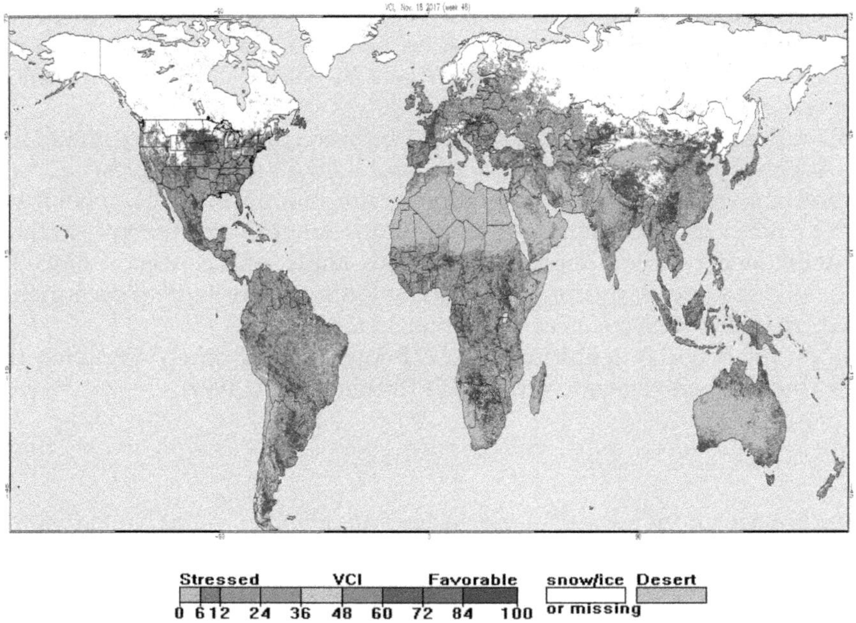

FIGURE 12.3
VCI for the last week of November 2017 (Source: NOAA/NESDIS Global Vegetation Health Products).

$$\text{NDWI} = \frac{\rho_{\text{NIR}} - \rho_{\text{SWIR}}}{\rho_{\text{NIR}} + \rho_{\text{SWIR}}} \tag{12.14}$$

To combine information from NDVI and NDWI, the normalized difference drought index (NDDI) has been proposed (Gu et al., 2007) as follows:

$$\text{NDDI} = \frac{(\text{NDVI} - \text{NDWI})}{\text{NDVI} + \text{NDWI}} \tag{12.15}$$

It should be noted that SWIR bands respond to soil moisture and leaf water content differently and, thus, combining multiple SWIR bands (rather than one SWIR band) with a NIR band may improve sensitivity for drought monitoring (Wang et al., 2008). To address this issue, the normalized multi-band drought index (NMDI) has been proposed for monitoring soil and vegetation moisture condition using moderate resolution imaging spectroradiometer (MODIS) data (Wang and Qu, 2007):

$$\text{NMDI} = \frac{\left(0.86 - (\rho 1.64 - \rho 2.13)\right)\mu\text{m}}{\left(\rho 0.86 + (\rho 1.64 - \rho 2.13)\right)\mu\text{m}} \tag{12.16}$$

In NMDI, the 0.86 μm band is NIR spectral band, whereas the 1.64 and 2.13 μm are SWIR bands. By combining information from different channels, the NMDI enhances the sensitivity to drought severity (Wang and Qu, 2007). Similar efforts have been focused on combining visible data with SWIR information which led to the development of the *visible and shortwave infrared drought index* (Zhang et al., 2013) which combines MODIS blue (band 3), red (band 1), and SWIR (band 6) information. Furthermore, US Drought Monitor (USDM) is another index which the National Oceanic and Atmospheric Administration (NOAA) and the US Department of Agriculture (USDA) Foreign Agricultural Service (FAS) is using for assessment and monitoring of drought in United States (Svoboda et al., 2002; http://droughtmonitor.unl.edu/). The USDM assimilates several widely used climatic drought indices, including the PDSI (Palmer, 1965), SPI percent of normal precipitation (PNP; Werick et al., 1994), land soil moisture and stream flow, satellite NDVI, and many supplementary indicators for regional drought detection (Mu et al., 2013).

Vegetation Indices in Thermal Region: Drought stress can also be quantified using remotely sensed surface brightness temperature derived from thermal channels from multiple satellite instruments (e.g., AVHRR, MODIS, ETM+, and TIRS). The LST computed from thermal infrared (TIR) bands has been found to provide valuable information on surface moisture conditions (Gutman, 1990). The temperature condition index (TCI) is based on TIR observations to determine temperature-related vegetation stress. The TCI is defined as

$$TCI = \frac{(T_{max} - T_i)}{(T_{max} - T_{min})} \tag{12.17}$$

where T, T_{max}, and T_{min} denote the smoothed weekly temperature and its multiyear maximum and minimum, respectively (Kogan, 1995). Using data from NOAA-AVHRR and Terra MODIS, TCI may be computed at weekly timescales.

One limitation of TCI is that it does not account for day of year and/or time of day, since it only relies on smoothed weekly temperatures and their multiyear maxima and minima (Figure 12.4). This issue has been addressed in the development of the normalized difference temperature index (NDTI) that can be considered a specific time-of-day version of the crop water stress index (CWSI) (McVicar and Jupp, 2002).

Composite Vegetation Indices: Studies show that TCI coupled with VCI provides a powerful tool for monitoring vegetation stress and drought condition (Singh et al., 2003), and the two indices have been widely used over different regions (Jain et al., 2009). The reflective- and thermal-based information have been combined for effective and integrated (vegetation temperature) drought information using a combination of the reflective and thermal channels (e.g., combination of NDVI and LST). The *vegetation temperature condition index (VTCI)* (Wan et al., 2004) integrates NDVI, LST and thermal properties and provides one index that reflects drought information based on both temperature and vegetation. The *vegetation health index (VHI)* is among the commonly used reflective-thermal indicators that integrates the VCI and TCI (Kogan, 1995):

$$VHI = \alpha VCI + (1 - \alpha)TCI \tag{12.18}$$

where α refers to the relative contribution of the VCI and TCI (Figure 12.5). Similar to the VCI and TCI, the VHI is typically computed on a weekly timescale and has been used for both drought detection and early warning in different regions (Kogan, 2001).

Vegetation Indices in MW Region: MW remote sensing provides unique information of water content through detecting the change in the dielectric constants between water, soil, and vegetation. Passive and active MW remote sensing-based models/indices showed promising results for water content estimation and agricultural drought studies (Moran, 2004;

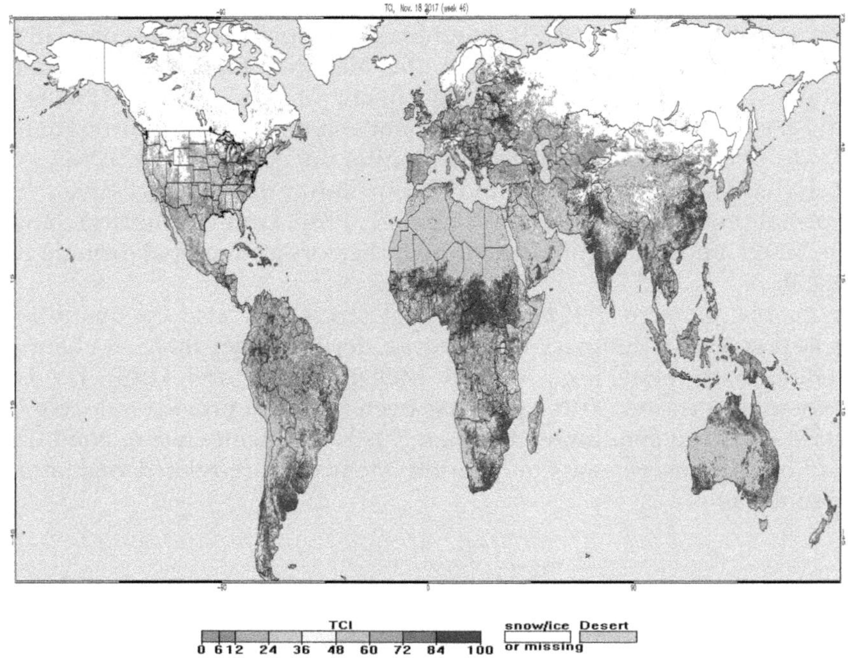

FIGURE 12.4
TCI for the last week of November 2017 (Source: NOAA/NESDIS Global Vegetation Health Products).

Anderson et al., 2011). Passive MW remote sensors, e.g., Scanning Multichannel Microwave Radiometer (SMMR), Special Sensor Microwave/Imager (SSM/I), Soil Moisture and Ocean Salinity (SMOS), and Soil Moisture Active and Passive (SMAP), have been used for surface water content monitoring through measuring the intensity of MW emission from soil and vegetation which is related to water content (Njoku et al., 2003; Moran, 2004).

Surface Emission Models can be grouped into three categories such as (i) bare soil emission models which are basically a function of surface roughness and dielectric properties (e.g., Q/H model and its modifications (Mo and Schmugge, 1987), (ii) vegetative areas emission models which is based on the optical depth and albedo (e.g., t~w model) (Wigneron et al., 2003), and (iii) soil and vegetation model (e.g., *Microwave Polarization Difference Index (MPDI)*) (Meesters et al., 2005).

While working in Iberian Peninsula, Sanchez et al. (2016) developed soil moisture agricultural drought index (SMADI) based on synergistic fusion of the surface soil moisture (SSM) from the Soil Moisture and Ocean Salinity (SMOS) mission, with the MODIS derived LST, and water/vegetation indices for agricultural drought monitoring.

Soil Moisture Retrieval Methods: These methods include statistical and forward model inversion techniques (Jackson et al., 1982). It is interesting to note that though passive MW has solid physical basis for water content retrieval and high temporal resolution, it has different major challenges including spatial resolution (i.e., 10–20 km), the available wavelength does not provide adequate water content sensitivity over different levels of vegetation covers, and technical and engineering challenges.

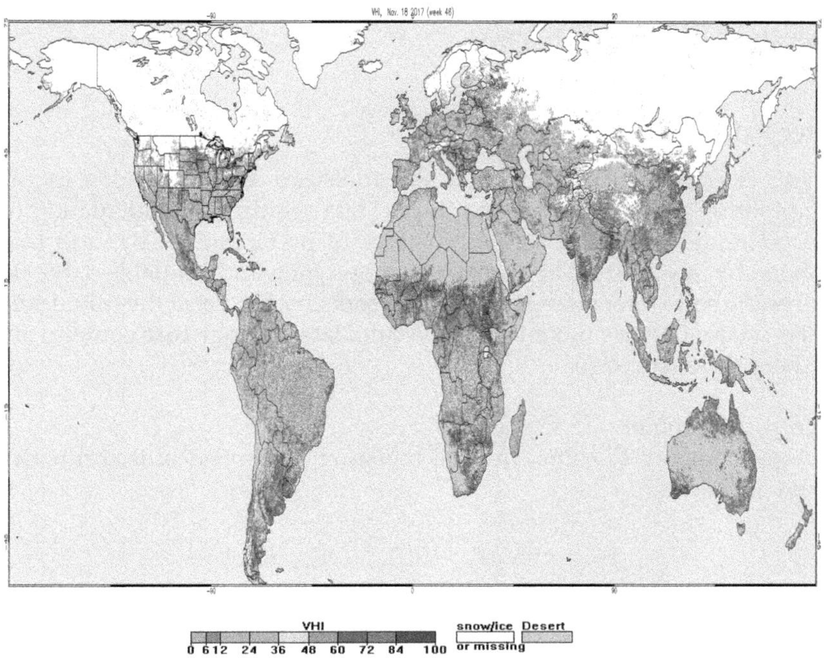

FIGURE 12.5
VHI for the last week of November 2017 (Source: NOAA/NESDIS Global Vegetation Health Products).

12.4.3 Process-Based Indicators

By modeling the energy and matter transfer between atmosphere and surface, a process-based indicator known as evaporative fraction (EF) is derived. EF is defined as the fraction of available energy used for ET. EF is derived as follows:

$$R_n = G + H + Le \qquad (12.19)$$

where

R = net radiation, G = soil heat flux, H = sensible heat flux, Le = latent heat flux

$Le = (R_n - G) - c_p^{*}\gamma_{air} * 1/r_a * (T_s - T_x)$

c_p = Sp. heat of air

$\gamma_{air}{}^{*}$ = air density

r_a = surface roughness, T_s = surface temperature

T_x = maximum temperature

EF = Le/(R– G)

EF = evaporative fraction.

As long as adequate soil moisture is available, energy will be used for its ET and EF will be close to one (no water stress). In the absence of soil moisture, all available energy will be directed into warming up surface and the ambient air and EF will approach zero indicating serious water stress (Bastiaanssen, 1998).

12.4.4 Water Balance Approach

Using the soil characteristics and initial soil moisture values, weekly/monthly water balance computations can be carried out. Input data required for calculation of monthly water balances model consist of monthly values of precipitation, P_m, and potential ET, PET_m (calculated by any method appropriate to the region and available data). Besides, the model requires the soil-water storage maximum capacity, S_{max}, and the initial soil moisture content, S_0. The water balance model will then calculate soil moisture content, m, actual ET and runoff (here water surplus).

i. Soil Moisture Content
   If for a given month P > PET, the soil moisture at the end of that month is then obtained as follows:

$$S_m = \min\{(P_m - PET_m) + S'_{m-1}, S_{max}\} \tag{12.20}$$

Otherwise, if $P_m < PET_m$, the soil moisture is given as follows:

$$S_m = S_{m-1} * \exp\frac{(P_m - PET_m)}{S_{max}} \tag{12.21}$$

ii. Actual ET
   For the first of the cases,

$$P_m > PET_m \tag{12.22}$$

$ET_m = PET_m$, otherwise

$$ET_m = P_m + S_{m-1} - S_m \tag{12.23}$$

iii. Water Surplus
   It is assumed that soils will only produce water surplus whenever the soil moisture content at the end of a month will equal its maximum water storage capacity. In that case water surplus, T_m, will be calculated as follows:

$$T_m = P_m - PET_m + S'_{m-1} - S_m \tag{12.24}$$

Rainfall departures and SMI values are useful for contingency crop planning based on the climatic conditions. Spatial distribution of weekly SMI (actual/available moisture) maps reveals moisture status spatially and temporally in a given area. Time-to-time contingencies can be suggested to the farmers, by closely monitoring the weather conditions during the season (Murthy and Seshasai, 2010).

12.5 Drought Forecasting

Drought forecasting plays a crucial role in risk management, drought preparedness, and mitigation. Several models, namely regression analysis, time series analysis, probability models, artificial neural network (ANN) model, and hybrid models, have been developed to address various aspects of drought, such as identification and prediction of its duration and severity (Mishra and Singh, 2011). The variables include (i) precipitation for meteorological drought analysis as precipitation deficits lead to a drought; (ii) stream flow, reservoir, and lake level data for hydrologic drought analysis; (iii) groundwater levels for ground water drought; and (iv) soil moisture and crop yield for agricultural drought. The input variables required for drought forecasting depend upon the type of drought being addressed (Mishra and Singh, 2010). Different approaches of drought modeling are discussed hereunder:

12.5.1 Regression Analysis

The relationship among two or more quantitative variables, a dependent variable whose value is to be predicted, and independent variables about which information is available can be explored using regression analysis. For example, a multiple regression predicts one variable from two or more independent variables:

$$Y = a + bX_1 + cX_2 + dX_3 \qquad (12.25)$$

where Y is the dependent variable; X_1, X_2, and X_3 are the independent variables; and a, b, c, and d are constants. Here the dependent variable is a drought quantifying parameter, e.g., a drought index, whereas the independent variables are explanatory variables for the drought quantifying parameter, i.e., precipitation, stream flow, and soil moisture.

12.5.2 Time Series Analysis

Time series models have been commonly used in a broad range of scientific applications, including hydrology. Their application for drought forecasting, however, have been limited. Some of the major strengths of time series models include their systematic search capability for identification, estimation, and diagnostic check for model development (Mishra and Desai, 2005). Time series models, like autoregressive integrated moving average (ARIMA) model effectively consider serial linear correlation among observations; whereas seasonal autoregressive integrated moving average (SARIMA) models can describe time series that exhibits non-stationarity both within and across seasons. Two types of stochastic models are commonly used (Box et al., 2005) which are briefly discussed here.

i. General nonseasonal ARIMA model:
   It may be written as follows:

$$\phi(B)\nabla^d z_t = \theta(B)a_t \qquad (12.26)$$

   In short, a seasonal ARIMA (SARIMA) model is described as follows: ARIMA (p, d, q) (P, D, Q)s, where (p, d, q) is the nonseasonal part of the model; and (P, D, Q)s is the seasonal part of the model.

In ARIMA model approach, the past observations are analyzed to formulate a model describing the inner correlation among them. The time series is then extrapolated into the future according to the model. The development of ARIMA model includes three stages: identification, estimation, and diagnostic checking.

ii. General seasonal ARIMA model may be written as follows:

$$\phi_P(B)\Phi_P(B^S)\nabla^d\nabla_S^D Z_t = \theta_q(B)\Theta_Q(B^S)a_t \tag{12.27}$$

where $\phi(B)$ and $\theta(B)$ are polynomials of order p and q, respectively; $\Phi(B^s)$ and $\Theta(B^s)$ are polynomial in B^s of degrees P and Q, respectively; p is the order of nonseasonal autoregression; d is the number of regular differencing; q is the order of nonseasonal moving average; P is the order of seasonal auto regression; D is the number of seasonal differencing; Q is the order of seasonal moving average; and s is the length of season. Using time series models a time series of a drought quantifying parameter can be used for drought forecasting, depending upon the previous observations.

12.5.3 Probability Models

By virtue of their complex nature as well as to quantify uncertainties associated with hydrometeorological variables causing droughts, probability models are useful for drought forecasting. Markov chain models have been commonly used for drought forecasting. A Markov chain is a stochastic process having the property that the value of the process at time t, X_t, depends only on its value at time $t-1$, X_{t-1}, and not on the sequence of values X_{t-2}, X_{t-3} ... X_0 that the process passed through in arriving at X_{t-1}. This can be written as follows (Haan, 2002):

$$\begin{aligned} &\text{Prob}(X_t = a_j \mid X_{t-1} = a_i, X_{t-2} = a_k, X_{t-3} = a_1, \ldots X_0 = a_q) \\ &= \text{Prob}(X_t = a_j \mid X_{t-1} = a_i) \end{aligned} \tag{12.28}$$

The conditional probability, $\text{Prob}(X_t = a_j \mid X_{t-1} = a_i)$, gives the probability that the process at time t will be in "state j" given that at time $t-1$ the process was in "state i."

The term $\text{Prob}(X_t = a_j \mid X_{t-1} = a_i)$ is commonly called one-step transition probability meaning thereby it is the probability that the process makes the transition from state a_i to state a_j in one time period or one step, which is commonly denoted by p_{ij}. Transition probabilities are basic to the structure of Markov chains.

12.5.4 ANN Model

Neural networks are a class of flexible nonlinear models that can adaptively recognize patterns from the data. Theoretically, it has been shown that given an appropriate number of nonlinear processing units, neural networks can learn from experience and estimates any complex functional relationship with high accuracy. For a detailed review of ANN models and their application in hydrology the reader may refer to ASCE (2000) and Govindaraju and Rao (2000). Generally, a three-layer feed-forward model consisting of input layer, hidden layer, and output layer is used for forecasting. The input nodes can be (i) suitable previously lagged observations of drought quantifying time series, (ii) explanatory variables for quantifying drought, or (iii) a combination of both (i) and (ii). The hidden nodes with appropriate nonlinear transfer functions are used to process the information received by the input nodes, while the output layer is used for forecasting for different lead times.

12.5.5 Hybrid Models

Hybrid models are useful in deriving advantages of individual models for predicting droughts with better accuracy as well as higher lead time in comparison to individual models. For example, wavelet transform methods can capture useful information at various resolution levels, whereas a neural network model can forecast the decomposed sub-signals at various resolution levels obtained from wavelets and reconstruct forecasted sub-signals to the original series.

12.6 Long-Lead Drought Forecasting

Long-lead drought forecasts are made using climate indices as well as the periodic nature of hydrometeorological variables. Drought indices, derived over decades, have rainfall as a major parameter causing droughts (Mishra and Singh, 2010). Therefore, long-lead drought forecasting is possible, if precipitation is predicted at long lead time. Precipitation has been shown to be related globally to broad-scale atmospheric phenomena, such as El Nino-Southern Oscillation [ENSO], Sea Surface Temperature (SST), and Geopotential Height (GpH) (Ropelewski and Halpert, 1996). A relatively strong statistical relationship among ENSO, SSTs, and rainfall has aroused considerable interest in long-range rainfall forecasting (Palmer and Anderson, 1994; Goddard et al., 2001). However, the relation of large climate indices to rainfall varies from one region to another. For instance, ENSO is a good indicator of droughts in Australia (Chiew and McMahon, 2002), but not necessarily in central and northern parts of Asia (Morid et al., 2006).

Several studies demonstrated that ocean–atmosphere forcing by persistent SST influenced the timing of drought events, and their duration and magnitude over continental regions were largely governed by land–atmosphere feedbacks (Ferguson et al., 2010). Improvements in climate forecasts have created potential to improve seasonal to inter-annual drought forecasting for efficient water resources management. This potential seems to remain largely untapped. Barriers to the broad use of drought forecasts include difficulties in understanding, applying, evaluating, and trusting forecasts (Steinemann, 2006).

12.7 Drought Monitoring Systems: Global Scenario

In recent years, several research and operational drought monitoring models have been developed. However, drought warning and prediction systems are still the least developed systems among other natural disasters primarily because of the complex processes involved in drought onset and development. On a regional scale, there are a number of drought monitoring tools tailored for local to continental scale applications such as the US Agency for International Development (USAID), Famine Early Warning System Network (FEWS Net) (Funk, 2009), United States Drought Monitor (Svoboda et al., 2002), African

Drought Monitor (Sheffield et al., 2008), and the University of Washington Experimental Surface Water Monitor (Shukla et al., 2011). In an interconnected world where a drought in the USA, Russia, or Australia could affect global food prices a regional perspective to drought monitoring and prediction may not be sufficient (Wegren, 2011).

Currently, few global drought models are available, including the Global Information and Early Warning System on Food and Agriculture (GIEWS), Humanitarian Early Warning Service (HEWS) by the World Food Programme (WFP), Global Drought Monitor by the University College London (UCL), the Standardized Precipitation-Evapotranspiration Index Global Drought Monitor (Vicente-Serrano et al., 2010), and the Global Drought Portal (GDP) by the United States National Climatic Data Centre (Pozzi et al., 2013).

The aforementioned models, however, do not provide probabilistic forecasts of drought condition for risk assessment. For a comprehensive review of drought forecasting readers may refer to Grasso and Singh (2011). For example, the FAO-GIEWS provides monthly briefings on countries under drought that are facing food crisis. The HEWS gathers drought information from various sources and summarizes the information into brief descriptions of the current conditions. The UCL and GDP provide interactive maps of the latest drought condition. While these outputs are valuable, probabilistic seasonal drought forecasts can substantially improve drought early warning capability.

12.7.1 Global Integrated Drought Monitoring and Prediction System

United Nations Environment Programme (UNEP) report calls for a drought prediction system based on a comprehensive and integrated approach that would include multiple drought indicators (Grasso and Singh, 2007). According to the United Nations (UN) Strategy for Disaster Reduction (ISDR) an early warning system should have the following features (United Nations, 2006): (a) monitoring and predicting components, (b) risk knowledge, (c) disseminating information, and (d) response. The Global Integrated Drought Monitoring and Prediction System (GIDMaPS) was developed in pursuance of there commendation to provide drought information based on multiple drought indicators and input datasets. The system provides meteorological and agricultural drought information based on multiple satellite-, and model-based precipitation and soil moisture datasets. It includes a near real-time monitoring component and a seasonal probabilistic prediction module. The datasets include historical drought severity data from the monitoring component, and probabilistic seasonal forecasts from the prediction module. The probabilistic forecasts provide essential information for early warning, taking preventive measures, and planning mitigation strategies. GIDMaPS datasets are a significant extension to current capabilities and datasets for global drought assessment and early warning system.

12.7.1.1 Approach

The algorithm of the GIDMaPS is schematically illustrated in Figure 12.6. GIDMaPS integrates precipitation and soil moisture data from model simulations and remote sensing observations including the Modern-Era Retrospective analysis for Research and Applications (MERRA-Land) (Bosilovich et al., 2011), North American Land Data Assimilation System (NLDAS), Global Land Data Assimilation System (GLDAS), and the Global Drought Climate Data Record (GDCDR). GDCDR combines real-time precipitation estimation from remotely sensed information using Artificial Neural Networks (PERSIANN) satellite data with long-term GPCP observations using a Bayesian algorithm.

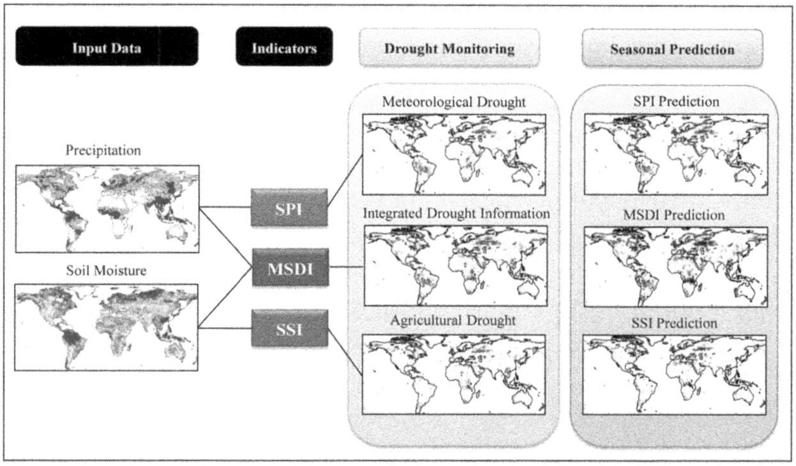

FIGURE 12.6
Schematic view of the GIDMaPS algorithm. SPI, standardized precipitation index; SSI, standardized soil moisture index; MSDI, multivariate standardized drought index (Source: Adapted from Hao et al. (2014)).

GIDMaPS uses three drought indicators for monitoring and prediction: the SPI, SSI, and multivariate standardized drought index (MSDI). In this method, the marginal probability distribution of precipitation and soil moisture are computed using the empirical Gringorten plotting position. In other words, instead of fitting a parametric distribution function, the probabilities of observed precipitation and soil moisture are computed empirically. The empirical probabilities are then standardized as SPI = $\phi-1(pp)$ and SSI = $\phi- 1(ps)$ where ϕ is the standard normal distribution (probability density function), and pp and ps denote the empirical probabilities of precipitation and soil moisture, respectively.

The joint empirical probabilities of P and S can be standardized to derive the MSDI (MSDI = $\phi-1(p)$).

As shown in Figure 12.5, GIDMaPS includes a monitoring component that provides drought information based on historical observations, and a prediction component for seasonal drought forecasting. The prediction component of GIDMaPS is based on the concept of the Ensemble Stream flow Prediction (ESP) method that has been used in numerous hydrology and climate studies (Day, 1985).

This approach assumes that historical data from any location are possible scenarios for the future. The GIDMaPS's prediction component extends persistence-based drought prediction to a multi-index framework based on multiple variables, providing short-term forecasts based on both univariate (SPI, SSI) and multivariate (MSDI) drought indicators (Hao et al., 2014).

12.7.2 European Drought Monitoring System

Fostering European Drought Research and Science-Policy Interfacing (DROUGHT-R&SPI) is a collaborative project funded by the European Commission (Acácio et al., 2013). The aim of the project is to improve drought preparedness in Europe through increased knowledge of drought-generating processes, drought early warning and management plans, and improved science-policy interfacing. The European Drought Reference database (EDR) has been developed during the project and was used to disseminate a detailed summary of for major European drought events. The database was made publicly available through the website of the European Drought. It contains a tool to display drought conditions on any day in the historical period 1958–2009, including the recommended meteorological indicators, namely SPI, SPEI, and hydrological drought indicators.

12.7.3 Drought Monitoring System for South Asia

The South Asian Climate Outlook Forum (SASCOF) was established at a meeting of the Permanent Representatives of the South Asian countries with WMO in August 2009 Global Water Partnership, South Asia with support from the WMO/GWP Integrated Drought Management Programme (IDMP) 2014. The countries participating in SASCOF include Afghanistan, Bangladesh, Bhutan, India, Maldives, Myanmar, Nepal, Pakistan, and Sri Lanka. The main objectives of SASCOF are (1) to review the progress made in understanding and long-range prediction of summer monsoon both regionally and globally; (2) to make available detailed information on climate variability in South Asia for dissemination along with the seasonal outlook; (3) to provide a platform for the stakeholders of SASCOF to share and exchange experience and knowledge on summer monsoon and its prediction; (4) to initiate capacity building/human resource development activities for the South Asian region, particularly in seasonal prediction; (5) to build collaboration and partnerships among the members of SASCOF for mutual benefit; and (6) to identify needs of user sectors through a dialog among different groups.

The forum brings together the experts from a climatologically homogeneous region and provides consensus-based climate prediction and information usually for the season having critical socioeconomic significance. This information has been applied to reducing climate-related risks and supporting sustainable development. In the field of drought management, there has not been much cross sharing of knowledge and experience among the countries in the region, although every country of South Asia has decades of experience of dealing with droughts.

12.7.4 Indian National Agricultural Drought Assessment and Monitoring System

National Remote Sensing Agency, Department of Space, Government of India, in collaboration with India Meteorological Department (IMD) and various state agriculture departments had developed an operational national program *"National Agricultural Drought Assessment and Monitoring System (NADAMS)"* in 1990 for providing information on agricultural drought in terms of prevalence, severity, and persistence at state, district, and subdistrict levels (administrative units). The program covers 13 agriculturally important and drought-vulnerable states of the country. Monitoring of agricultural drought is restricted to *Kharif* season (June–October/November), since this season is agriculturally more important and rainfall dependent.

The state government declares the agricultural drought based on a large number of drought indicators, many of which are not amenable to remote sensing. The emphasis of NADAMS is, therefore, on assessment of agricultural crop health. The agricultural crop health is monitored at state/district/subdistrict (administrative units) level using daily observed coarse resolution (1.1 km) NOAA-AVHRR data for nine states. Moderate resolution data from Advanced Wide Field Sensor (AWiFS) on board Resourcesat-1/-2/2a with 56 m spatial resolution are being used for detailed assessment of agricultural drought at district and subdistrict level.

In NADAMS, compositing of NDVI with maximum value approach for the period of 15 days and 1 month, is being adopted. Drought assessment is carried out on fortnightly basis and reporting of information to the concerned user departments is done on monthly basis. The assessment of agricultural drought situation in each district/block/taluk is based on (1) seasonal NDVI/NDWI progression, i.e., transformation of NDVI/NDWI from the beginning of the season; (2) comparison of NDVI/NDWI profile with previous normal years—relative deviation and VCI; (3) weekly rainfall status compared to normal; and (4) weekly progression of sown. The relative deviation of NDVI/NDWI from that of normal and the rate of progression of NDVI/NDWI from month to month provides an indication crop health in the district which is then complemented by *in situ* observations as evident from rainfall and the crop sown area.

The methodology involves integration of satellite-derived crop condition/surface wetness indices and their anomalies with *in situ* rainfall measurement and crop area progression to evolve decision rules on the prevalence, intensity, and persistence of agricultural drought. The indices used in agricultural drought assessment include shortwave angle slope index (SASI), normalized difference wetness index (NDWI), NDVI, and soil moisture index (SMI) derived from soil water balance approach. The rainfall data—rainfall in terms of deviation from normal, number of dry weeks—is taken from India Meteorological Department. The coarse resolution NDVI from Oceansat-2 Ocean Colour Monitor (OCM), NOAA-AVHRR, and Terra MODIS data are being used for assessment of crop condition and wetness anomalies, respectively, which in turn are utilized for detecting drought intensity at district/state level (Figure 12.7).

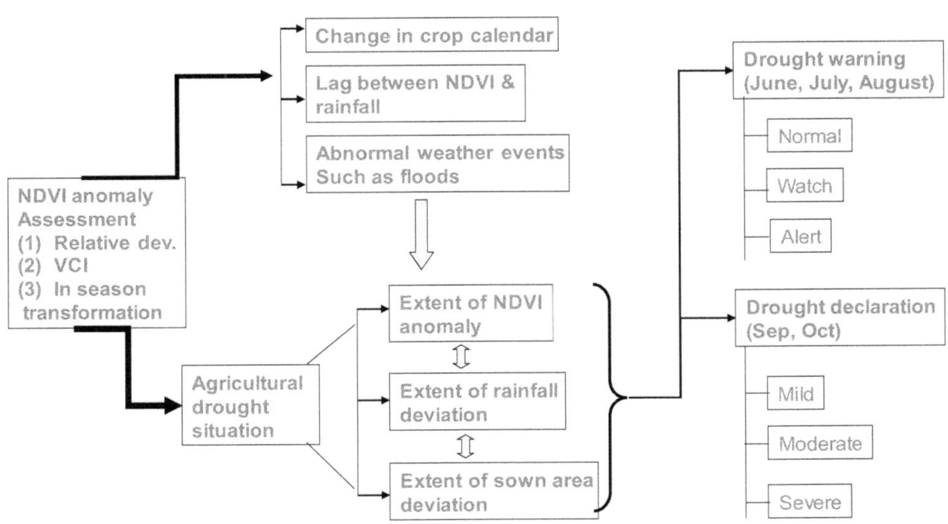

Workshop on Harmonization of data for land use and evaluation, 28-29 July 2006, NRSA - Drought

FIGURE 12.7
Methodology for agricultural drought assessment (Source: Adapted from Murthy and Seshasai (2010)).

During June to August, drought warning information is issued in terms of "Watch," "Alert," and "Normal" categories. In the case of "Watch," external intervention is required if similar drought like conditions persist during the successive month, while "Alert" calls for immediate external intervention, in terms of crop contingency plans. During September and October, based on NDVI anomalies corroborated by *in situ* observation on crop health, drought declaration is done in terms of mild, moderate and severe drought is made (Sesha Sai et al., 2016). The operational component of NADAMS project has been carried out by the Mahalanobis National Crop Forecast Centre (MNCCF), Ministry of Agriculture and Farmers Welfare, Government of India since June 2012.

12.8 Conclusion

Quite a good amount of research has been carried out toward comprehending the nature, causative factors, assessments and monitoring, and management of drought using traditional as well as advanced techniques and tools. Although five types of drought, namely meteorological, agricultural, hydrological, socioeconomic drought, and ground water drought have been recognized, it is agricultural that has drawn the attention of researchers more because of the fact that the agriculture sector is affected the most due to drought event. Geospatial technology has been utilized extensively for assessment, monitoring, and modeling agricultural drought. It is, in fact, the impact of drought on agricultural crops that is captured by orbital sensors which in turn has been used to develop various types of indices across the electromagnetic spectrum ranging from visible to MW through SWIR and thermal infrared regions. However, no single index either meteorological, hydrological or vegetation/soil indices has been found to represent the magnitude of drought. Integrated index/indices have, therefore, been very often been developed for drought investigation.

Substantial effort has gone into developing remote sensing-based techniques and tools for assessment and monitoring and impact assessment of agricultural drought. Despite, there are quite a few challenges, which demand further research. Those include (i) monitoring agricultural drought progression in relatively small area, (ii) filling gaps in the data, (iii) developing consistent historical database, and (vi) developing standard validation schema (Hazaymeh and Hassan, 2016). With respect to management of drought, it is the preparedness phase which seems to be very crucial and calls for forewarning of drought event with a reasonable lead time ranging from a few weeks to 1 year. Attempts in this direction have also been made and several agricultural drought forecasting models have been developed. However, most of them are agro-climatic region-specific. Therefore, the need is to develop a universal model for agricultural drought forecasting with adequate lead time.

Although some drought indices like SPI and NDVI are popularity adopted, the variety of drought indices reflect a fundamental lack of universal definition, and concepts, and different operational requirement. In addition to the variability in the types and applications of droughts (e.g., meteorological versus hydrological), the dissociation of drought indices with drought impacts calls for development of aggregate drought indices to cover more aspects and applications.

References

Acácio, V., Andreu, J., Assimacopoulos, D., Bifulco, C., di Carli, A., Dias, S., Kampragou, E., Monteagudo, D.H., Rego, F., Seidl, I., Vasiliou, E., Wolters, W., 2013. Review of current drought monitoring systems and identification of (further) monitoring requirements. Technical report No.6.

Adegoke, J.O., Carleton, A.M., 2002. Relations between soil moisture and satellite vegetation indices in the US corn belt. *J. Hydrometeorol.*, 3(4): 395–405, doi:10.1175/1525-7541(2002)003<0395:RBSMAS>2.0.CO;2.

Adler, R.F., Huffman, G.J., Chang, A., Ferraro, R., Xie, P.P., Janowiak, J., Rudolf, B., Schneider, U., Curtis, S., Bolvin, D., Gruber, A., Susskind, J., Arkin, P., Nelkin, E., 2003. The version-2 Global Precipitation Climatology Project (GPCP) monthly precipitation analysis (1979–present), *J. Hydrometeorol.*, 4(6): 1147–1167.

AghaKouchak, A., Farahmand, A., Melton, F.S., Teixeira, J., Anderson, M.C., Wardlow, B.D. and Hain, C.R., 2015. Remote sensing of drought: Progress, challenges and opportunities. *Reviews of Geophysics.*, 53(2): 452–480.

Alley, W.M., 1984. The Palmer Drought Severity Index: limitations and assumptions. *J. Clim. Appl. Meteor.*, 23: 1100–1109.

Anderson, M.C., Kustas, W.P., Norman, J.M., Hain, C.R., Mecikalski, J.R., Schultz, L., Gonzalez-Dugo, M.P., Cammalleri, C., d'Urso, G., Pimstein, A., Gao, F., 2011. Mapping daily evapotranspiration at field to continental scales using geostationary and polar orbiting satellite imagery. *Hydrol. Earth Syst. Sci.*, 15: 223–239.

Anjum, S.A., Xie, X., Wang, L., Salem, M.F., ManCand Lei, W., 2011. Morphological, physiological and biochemical responses of plants to drought stress. *Afr. J. Agric. Res.*, 6(9): 2026–2032.

ASCE Task Committee on Application of Artificial Neural Networks in Hydrology, 2000. Artificial neural networks in hydrology. I. Preliminary concepts. *J. Hydrol. Eng.*, 5(2): 124–137.

Asrar, G., Fuchs, M., Kanemasu, E., Hatfield, J., 1984. Estimating absorbed photosynthetic radiation and leaf area index from spectral reflectance in wheat. *Agron. J.*, 76(2): 300–306, doi:10.2134/agronj1984.00021962007600020029x.

Bastiaanssen, W.G.M., 1998. *Remote Sensing in Water Resources Management: The state of the art*, IWMI publication, Colombo, p. 118.

Bates, B.C., Kundzewicz, Z.W., Wu, S., et al., 2009. Technical Paper, International Panel on Climate Change (IPCC) Secretariat, Geneva, 210 pp. Available from: www.ipcc.ch/pdf/technical-papers/climate-change-water-en.pdf.

Bosilovich, M.G., Robertson, F.R., Chen, J., 2011. Global energy and water budgets in MERRA. *J. Climate*, 24: 5721–5739.

Box, G.E.P., Jenkins, G.M., Reinsel, G.C., 2005. *Time Series Analysis: Forecasting and Control*, 3rd ed., Post & Telecommunications Press, Beijing, 327 pp.

Chang, F.C., Wallace, J.M., 1987. Meteorological conditions during heat waves and droughts in the United States great plains. *Mon. Weather Rev.*, 115(7): 1253–1269.

Chang, T.J., Teoh, C.B., 1995. Use of the Kriging method for studying characteristics of ground water droughts. *J. Am. Water Resour. Assoc.*, 257: 1001–1007.

Chiew, F.H.S., McMahon, T.A., 2002. Global ENSO-stream flow teleconnection, stream flow forecasting and interannual variability. *Hydrol. Sci. J.*, 47: 505–522.

Day, G.N., 1985. Extended streamflow forecasting using NWSRFS. *J. Water Resour. Plann. Manag.* 111: 157–170.

De Pauw, E., 2005. Monitoring agricultural drought in the Near East. Monitoring and Predicting Agricultural Drought: a global study, 208–226.

Donohue, R.J., McVicar, T., Roderick, M.L., 2009. Climate-related trends in Australian vegetation cover as inferred from satellite observations, 1981–2006, *Global Change Biol.*, 15(4): 1025–1039, doi:10.1111/j.1365-2486.2008.01746.x.

Dozier, J., Green, R.O., Nolin, A.W., Painter, T.H., 2009. Interpretation of snow properties from imaging spectrometry. *Remote Sens. Environ.*, 113(1): S25–S37.

Durand, M., Kim, E.J., Margulis, S.A., 2008. Quantifying uncertainty in modeling snow microwave radiance for a mountain snowpack at the point-scale, including stratigraphic effects. *IEEE Trans. Geosci. Remote Sens.*, 46(6): 1753–1767, doi:10.1109/TGRS.2008.916221.

Eltahir, E.A.B., Yeh, P.J.F., 1999. On the asymmetric response of aquifer water level to floods and droughts in Illinois. *Water Resour. Res.*, 35(4): 1199–1217.

Entekhabi, D., et al., 2010. The soil moisture active passive (SMAP) mission. *Proc. IEEE*, 98(5): 704–716.

European Communities, 2007. Addressing the challenge of water scarcity and droughts in the European Union. Commission of the European communities 2007, 414 Final, Brussels.

FAO, 2008. A Review of Drought Occurrence and Monitoring and Planning Activities in the Near East Region. Cairo, Egypt. Available from: www.ais.unwater.org/ais/pluginfile.php/516/course/section/175/Drought%20Occurrence%20and%20Activities%20in%20the%20Near%20East.pdf.

Ferguson, I.M., Dracup, J.A., Duffy, P.B., Philip, P., Siegfried, S., 2010. Influence of SST Forcing on Stochastic Characteristics of Simulated Precipitation and Drought. *J. Hydrometeorol.*, 11: 754–769.

Foster, J.L., Sun, C., Walker, J.P., Kelly, R., Chang, A., Dong, J., Powell, H., 2005. Quantifying the uncertainty in passive microwave snow water equivalent observations. *Remote Sens. Environ.*, 94(2): 187–203.

Funk, C., 2009. New satellite observations and rainfall forecasts help provide earlier warning of African drought. *The Earth Obs.*, 21: 23–27.

Gao, B.C., 1996. NDWI a normalized difference water index for remote sensing of vegetation liquid water from space. *Remote Sens. Environ.*, 58(3): 257–266.

Gibbs, W.J., Meher, J.V., 1967. *Rainfall deciles* as drought indicators, Bureau of Meteorology, Melbourne, Australia, Bulletin No. 48, 117.

Goddard, L., Mason, S.J., Zebiak, S.E., Ropelewski, C.F., Basher, R., Cane, M.A., 2001. Current approaches to seasonal to interannual climate predictions. *Int. J. Climatol.*, 21: 1111–1152.

González, J., Valdés, J., 2006. New drought frequency index: Definition and comparative performance analysis. *Water Resour. Res.*, 42(11): W11421.

Goodison, B.E., 1989. Determination of areal snow water equivalent on the Canadian prairies using passive microwave satellite data, in Geoscience and Remote Sensing Symposium, 1989. IGARSS'89. 12th Canadian Symposium on Remote Sensing, vol. 3, pp. 1243–1246, IEEE.

Govindaraju, R.S., Rao, A.R., 2000. *Artificial Neural Networks in Hydrology*, Kluwer Academic Publishers, Amsterdam.

Grasso, V.F., Singh, A., 2011. *Early warning systems: State-of-art analysis and future directions*, United Nations Environment Programme.

Grody, N.C., Basist, A.N., 1996. Global identification of snow cover using SSM/I measurements. *IEEE Trans. Geosci. Remote Sens.*, 34(1): 237–249, doi:10.1109/36.481908.

Gu, Y., Brown, J.F., Verdin, J.P., Wardlow, B., 2007. A five-year analysis of MODIS NDVI and NDWI for grassland drought assessment over the central Great Plains of the United States. *Geophys. Res. Lett.*, 34: L06407.

Gutman, G.G., 1990. Towards monitoring droughts from space. *J. Clim.*, 3(2): 282–295.

Haan, C.T., 2002. *Statistical Methods in Hydrology*, The Iowa State University Press, Ames, IA.

Hall, D.K., Riggs, G.A., Salomonson, V.V., DiGirolamo, N.E., Bayr, K.J., 2002. MODIS snow-cover products. *Remote Sens. Environ.*, 83(1): 181–194.

Hao, Z., AghaKouchak, A., 2013. Multivariate standardized drought index: a parametric multi-index model. *Adv. Water Res.*, 57: 12–18.

Hao, Z., AghaKouchak, A., Nakhjiri, N., Farahmand, A., 2014. Global integrated drought monitoring and prediction system. *Scientific Data*, 1: 140001. Published online 2014 Mar 11.

Hayes, M.J., 2006. *Drought indices. Van Nostrand's Scientific Encyclopedia*, John Wiley and Sons Inc., Hoboken, NJ.

Hazaymeh, K., Hassan, Q.K., 2016. Remote sensing of agricultural drought monitoring: a state of art review. *AIMS Environ. Sci.*, 3(4): 604–630.

Hong, Y., Hsu, K., Gao, X., Sorooshian, S., 2004. Precipitation estimation from remotely sensed imagery using an artificial neural network cloud classification system. *J. Appl. Meteorol. Climatol.*, 43(12): 1834–1853, doi:10.1175/JAM2173.1.

Huffman, G., Adler, R., Bolvin, D., Gu, G., Nelkin, E., Bowman, K., Stocker, E., Wolff, D., 2007. The TRMM Multi-satellite Precipitation Analysis: Quasi-global, multiyear, combined-sensor precipitation estimates at fine scale. *J. Hydrometeorol.*, 8(1): 38–55.

Jackson, R. D., Pinter, P. J. Jr., Reginato, R. J., and Idso, S. B., 1980. Hand-held radiometry, USDA-SEA-AR, Agricultural Reviews and Manuals, W-19, pp. 66.

Jackson, T.J., Schmugge, T.J., Wang, J.R., 1982. Passive microwave sensing of soil moisture under vegetation canopies. *Water Resour. Res.*, 18: 1137–1142.

Jain, S.K., Keshri, R., Goswami, A., Sarkar, A., Chaudhry, A., 2009. Identification of drought-vulnerable areas using NOAA AVHRR data. *Int. J. Remote Sens.*, 30(10): 2653–2668.

Joyce, R., Janowiak, J., Arkin, P., Xie, P., 2004. CMORPH: A method that produces global precipitation estimates from passive microwave and infrared data at high spatial and temporal resolution. *J. Hydrometeorol.*, 5(3): 487–503.

Karl, T.R., Knight, R.W., 1985. *Atlas of Monthly Palmer Hydrological Drought Indices (1931–1983) for the Contiguous United States*, Historical Climatology Series 3–7, National Climatic Data Center, Asheville, NC.

Karnieli, A., Agam, N., Pinker, R.T., Anderson, M., Imhoff, M.L., Gutman, G.G., Panov, N., Goldberg, A., 2010. Use of NDVI and land surface temperature for drought assessment: Merits and limitations. *J. Clim.*, 23(3): 618–633.

Kidd, C., 2001. Satellite rainfall climatology: A review. *Int. J. Climatol.*, 21(9): 1041–1066, doi:10.1002/joc.635.

Kogan, F., 1995. Application of vegetation index and brightness temperature for drought detection. *Adv. Space Res.*, 15(11): 91–100, doi:10.1016/0273-1177(95)00079-T.

Kogan, F.N., 2001. Operational space technology for global vegetation assessment. *Bull. Am. Meteorol. Soc.*, 82(9): 1949–1964.

Kogan, F., Sullivan, J., 1993. Development of global drought-watch system using NOAA/AVHRR data. *Adv. Space Res.*, 13(5): 219–222.

Kongoli, C., Dean, C.A., Helfrich, S.R., Ferraro, R.R., 2007. Evaluating the potential of a blended passive microwave-interactive multi-sensor product for improved mapping of snow cover and estimations of snow water equivalent. *Hydrol. Processes*, 21(12): 1597–1607, doi:10.1002/hyp.6722.

Kongoli, C., Romanov, P., Ferraro, R., 2012. Snow cover monitoring from remote sensing satellites, in *Remote Sensing of Drought: Innovative Monitoring Approaches*, pp. 359–386, CRC Press.

Kummerow, C., Hong, Y., Olson, W., Yang, S., Adler, R., McCollum, J., Ferraro, R., Petty, G., Shin, D.B., Wilheit, T., 2001. The evolution of the Goddard profiling algorithm (GPROF) for rainfall estimation from passive microwave sensors. *J. Appl. Meteorol.*, 40(11): 1801–1820.

Levizzani, V., Bauer, P, Turk, F.J. (eds.), 2007. *Measuring Precipitation From Space: EURAINSAT and the Future*, Springer, Dordrecht, Netherlands.

Li, J., Heap, A.D., 2014. Spatial interpolation methods applied in the environmental sciences: a review. *Environ. Model Softw.*, 53: 173–189.

McKee, T.B., Doesken, N.J., Kleist, J., 1993. The relationship of drought frequency and duration to time scales. Preprints, AMS 8th Conference on Applied Climatology, pp. 179–184. January 17–22, Anaheim, CA.

McVicar, T.R., Jupp, D.L., 1998. The current and potential operational uses of remote sensing to aid decisions on drought exceptional circumstances in Australia: a review. *Agric. Syst.*, 57(3): 399–468, doi:10.1016/S0308-521X(98)00026-2.

McVicar, T.R., Jupp, D.L., 2002. Using covariates to spatially interpolate moisture availability in the Murray–Darling Basin: A novel use of remotely sensed data. *Remote Sens. Environ.*, 79(2): 199–212.

Meesters, A.G.C.A., De Jeu, R.A.M., Owe, M., 2005. Analytical derivation of the vegetation optical depth from the microwave polarization difference index. *IEEE Geosci. Remote. Sens. Lett.*, 2: 121–123.

Mishra, A.K., Desai, V.R., 2005. Drought forecasting using stochastic models. *J. Stoch. Environ. Res. Risk Assess.*, 19: 326–339.

Mishra, A.K., Singh, V.P., 2010. A review of drought concepts. *J. Hydrol*, 391(2010): 202–216.

Mishra, A.K., Singh, V.P., 2011. Drought modelling—a review. *J. Hydrol.*, 403(1–2): 157–175.

Mo, T., Schmugge, T., 1987. A parameterization of the effect of surface roughness on microwave emission. *IEEE Trans. Geosci. Remote Sens.*, 25: 481–486.

Moran, M., 2004. Thermal infrared measurement as an indicator of planet ecosystem health. *Therm. Remote Sens. L. Surf. Process*: 257–282.

Morid, S., Smakhtin, V., Moghaddasi, M., 2006. Comparison of seven meteorological indices for drought monitoring in Iran. *Int. J. Climatol.*, 26: 971–985.

Morillas, L., Leuning, R., Villagarcía, L., García, M., Serrano-Ortiz, P., Domingo, F., 2013. Improving evapotranspiration estimates in Mediterranean drylands: The role of soil evaporation. *Water Resour. Res.*, 49: 6572–6586.

Mu, Q., Zhao, M., Kimball, J.S., McDowell, N.G., Running, S.W., 2013. A remotely sensed global terrestrial drought severity index. *J. Am. Meteorol. Soc.*, doi:10.1175/BAMS-D-11-00213.1. URL: http://droughtmonitor.unl.edu/.

Murthy, C.S., Seshasai, M.V.R., 2010. Agricultural drought monitoring and assessment. In: P.S. Roy, R.S. Dwivedi, D. Vijayan (eds.), *Remote Sensing Applications*, National Remote Sensing Centre, Indian Space Research Organization, India.

Narasimhan, B., Srinivasan, R., 2005. Development and evaluation of Soil Moisture Deficit Index (SMDI) and Evapotranspiration Deficit Index (ETDI) for agricultural drought monitoring. *Agric. For. Meteorol.*, 133(1): 69–88.

NDMC, 2006. *Impacts of Drought*. National Climatic Data Center, Asheville, NC.

Niemeyer, S., 2008. New drought indices. Options Méditerranéennes. *Série A: Séminaires Méditerranéens*, 80: 267–274.

Njoku, E.G., Jackson, T.J., Lakshmi, V., et al., 2003. Soil moisture retrieval from AMSR-E. *IEEE Trans. Geosci. Remote Sens.*, 41: 215–228.

Painter, T.H., Seidel, F.C., Bryant, A.C., McKenzie Skiles, S., Rittger, K., 2013. Imaging spectroscopy of albedo and radiative forcing by light-absorbing impurities in mountain snow. *J. Geophys. Res. Atmos.*, 118: 9511–9523, doi:10.1002/jgrd.50520.

Palmer, W.C., 1965. Meteorological drought. Weather Bureau Research Paper No. 45, US Department of Commerce, Washington, DC, 58 pp.

Palmer, W.C., 1968. Keeping track of crop moisture conditions, nationwide: The new Crop Moisture Index. *Weatherwise*, 21(4): 156–161.

Palmer, T.N., Anderson, D.L.T., 1994. The prospects for seasonal forecasting. *Quart. J. Roy. Meteorol. Soc.*, 120: 755–793.

Park, J.S., Kim, K.T., Choi, Y.S., 2008. Application of vegetation condition index and standardized vegetation index for assessment of spring drought in South Korea, in Geoscience and Remote Sensing Symposium, 2008. IGARSS 2008. IEEE International, vol. 3, pp. 774–777, IEEE, Boston, Mass.

Payero, J., Neale, C., Wright, J., 2004. Comparison of eleven vegetation indices for estimating plant height of alfalfa and grass. *Appl. Eng. Agric.*, 20(3): 385–393.

Pozzi, W., et al., 2013. Toward global drought early warning capability. *Bull. Amer. Meteor. Soc.*, 94.

Reichle, R.H., Koster, R.D., Dong, J., Berg, A.A., 2004. Global soil moisture from satellite observations, land surface models, and ground data: Implications for data assimilation. *J. Hydrometeorol.*, 5(3): 430–442.

Rodell, M., Famiglietti, J., 2002. The potential for satellite-based monitoring of groundwater storage changes using GRACE: The high plains aquifer, central US. *J. Hydrol.*, 263(1): 245–256.

Rodell, M., Chen, J., Kato, H., Famiglietti, J.S., Nigro, J., Wilson, C.R., 2007. Estimating groundwater storage changes in the Mississippi River Basin (USA) using GRACE. *Hydrogeol. J.*, 15(1): 159–166.

Rodell, M., 2012. Satellite gravimetry applied to drought monitoring, in B. D. Wardlow, M. C. Anderson, and J. P. Verdin (eds.). *Remote Sensing of Drought: Innovative Monitoring Approaches*, CRC Press/Taylor & Francis. Also available at http://digitalcommons.unl.edu/nasapub/218.

Romanov, P., Gutman, G., Csiszar, I., 2000. Automated monitoring of snow cover over North America with multispectral satellite data. *J. Appl. Meteorol.*, 39(11): 1866–1880.

Ropelewski, C.F., Halpert, M.S., 1996. Quantifying southern oscillation–precipitation relationships. *J. Climate*, 9: 1043–1059.

Ross, T., Lott, N., 2003. A climatology of 1980–2003 extreme weather and climate events. National Climatic Data Center Technical Report No. 2003-01. NOAA/NESDIS. National Climatic Data Center, Asheville, NC, 2003.

Rouse, J., Haas, R., Schell, J., Deering, D., Harlan, J., 1974. Monitoring the Vernal Advancement and Retrogradation (Greenwave Effect) of Natural Vegetation, 362 pp., Texas A & M Univ., Remote Sens. Cent., College Station, Tex.

Sánchez, N., González-Zamora, Á., Piles, M., Martínez-Fernández, J., 2016. A new soil moisture agricultural drought index (SMADI) integrating MODIS and SMOS products: a case of study over the Iberian Peninsula. *Remote Sens.*, 8: 287.

Sesha Sai, M.V.R., Murthy, C.S., Chandrasekar, K., Jeyaseelan, A.T., Diwakar, P.G., Dadhwal, V.K., 2016. Agricultural drought: assessment & monitoring. *Mausam*, 67(1): 131–142.

Shafer, B.A., Dezman, L.E., 1982. Development of a Surface Water Supply Index (SWSI) to Assess the Severity of Drought Conditions in Snowpack Runoff Areas. *Proceedings of the Western Snow Conference*, Colorado State University, Fort Collins, CO, 164–175.

Sheffield, J., Wood, E.F., 2007. Characteristics of global and regional drought, 1950–2000: analysis of soil moisture data from off-line simulation of the terrestrial hydrologic cycle. *J. Geophys. Res.*, 112: D17115.

Sheffield, J., Wood, E., Lettenmaier, D., Lipponen, A., 2008. Experimental drought monitoring for Africa. *GEWEX News*, 18: 4–6.

Shukla, S., Steinemann, A.C., Lettenmaier, D.P., 2011. Drought monitoring for Washington State: indicators and applications. *J. Hydrometeorol.*, 12: 66–83.

Silleos, N.G., Alexandridis, T.K., Gitas, I.Z., Perakis, K., 2006. Vegetation indices: advances made in biomass estimation and vegetation monitoring in the last 30 years. *Geocarto. Int.*, 21(4): 21–28.

Singh, R.P., Roy, S., Kogan, F., 2003. Vegetation and temperature condition indices from NOAA AVHRR data for drought monitoring over India. *Int. J. Remote Sens.*, 24(22): 4393–4402.

Steinemann, A., 2006. Using climate forecasts for drought management. *J. Appl. Meteorol. Climatol.*, 75: 1353–1361.

Steinemann, A.C., Hayes, M.J., Cavalcanti, L., 2005. *Drought indicators and triggers. Drought and water crises: Science, technology, and management issues.* CRC Press, Boca Raton, FL, p. 71.

Svoboda, M., LeComte, D., Hayes, M., Heim, R., Gleason, K., Angel, J., Rippey, B., Tinker, R., Palecki, M., Stooksbury, D., Miskus, D., Stephens, S., 2002. The drought monitor. *Bull. Am. Meteorol. Soc.*, 83(8): 1181–1190.

Thiam, A.K., 1997. Geographic Information Systems and Remote Sensing Methods for Assessing and Monitoring Land Degradation in the Sahel: The Case of Southern Mauritania. Doctoral Dissertation, Clark University, Worcester Massachusetts.

Tsakiris, G., Vangelis, H., 2004. Towards a drought watch system based on spatial SPI. *Water Res. Manag.*, 18(1): 1–12.

Tucker, C.J., Choudhury, B.J., 1987. Satellite remote sensing of drought conditions. *Remote Sens. Environ.*, 23(2): 243–251.

Turk, F.J., Rohaly, G.D., Hawkins, J., Smith, E.A., Marzano, F.S., Mugnai, A., Levizzani, V., 1999. Meteorological applications of precipitation estimation from combined SSM/I, TRMM and infrared geostationary satellite data. In: *Microwave Radiometry and Remote Sensing of the Earth's Surface and Atmosphere*, P. Pampaloni, S. Paloscia (eds.), VSP Int. Sci. Publisher, Utrecht, Netherlands, Passive, pp. 353–363.

United Nations, 2006. Global survey of early warning systems. (United Nations Report, 2006).

Vicente-Serrano, S.M., Beguería, S., López-Moreno, J.I., 2010. A multiscalar drought index sensitive to global warming: the standardized precipitation evapotranspiration index. *J. Clim.*, 23(7): 1696–1718.

Wan, Z., Wang, P., Li, X., 2004. Using MODIS land surface temperature and normalized difference vegetation index products for monitoring drought in the southern Great Plains, USA. *Int. J. Remote Sens.*, 25(1): 61–72.

Wang, L., Qu, J.J., 2007. NMDI: A Normalized Multi-band Drought Index for monitoring soil and vegetation moisture with satellite remote sensing. *Geophys. Res. Lett.*, 34: L20405.

Wang, L., Qu, J.J., Hao, X., Zhu, Q., 2008. Sensitivity studies of the moisture effects on MODIS SWIR reflectance and vegetation water indices. *Int. J. Remote Sens.*, 29(24): 7065–7075.

Wang, A., Bohn, T.J., Mahanama, S.P., Koster, R.D., Lettenmaier, D.P., 2009. Multimodel ensemble reconstruction of drought over the continental United States. *J. Clim.*, 22(10): 2694–2712.

Wegren, S.K., 2011. Food security and Russia's 2010 drought. *Eurasian Geogr. Econ.*, 2: 140–156.

Werick, W., Willeke, G., Guttman, N., Hosking, J., Wallis, J., 1994. National drought atlas developed. *EOS Trans. AGU*, 75(8): 89.

Wheaton, E.E., Wittrock, V., Kulshreshtha, S., et al., 2005. Lessons learned from the Canadian drought years of 2001 and 2002: synthesis report. Saskatchewan Research Council, Publication No. 11602-46E03.

Wigneron, J.P., Calvet, J.C., Pellarin, T., Wigneron, J.P., Calvet, J.C., Pellarin, T., Van de Griend, A.A., Berger, M., Ferrazzol, P., 2003. Retrieving near-surface soil moisture from microwave radiometric observations: Current status and future plans. *Remote Sens. Environ.*, 85: 489–506.

Wilhite, D.A., 1992. *Preparing for Drought: A Guidebook for Developing Countries*, Climate Unit, United Nations Environment Program, Nairobi, Kenya.

Wilhite, D.A., 2000a. Drought as a natural hazard: Concepts and definitions. In: *Drought: A Global Assessment*, D.A. Wilhite (ed.), Routledge, Abingdon, 3–18.

Wilhite, D.A., 2000b. Chapter 1 *Drought as a Natural Hazard: Concepts and Definitions*. Drought Mitigation Center Faculty Publications. 69.

Wilhite, D.A., 2004. Drought as a natural hazard. In: *International Perspectives on Natural Disasters: Occurrence, Mitigation, and Consequences*, J.P. Stoltman, J. Lidstone, L.M. Dechano (eds.), Kluwer Academic Publishers, Dordrecht, Netherlands, pp. 147–162.

Wilhite, D.A., Svoboda, M.D., Hayes, M.J., 2007. Understanding the complex impacts of drought: A key to enhancing drought mitigation and preparedness. *Water Resour. Manag.*, 21: 763–774.

WMO, 2009. Inter-Regional Workshop on Indices and Early Warning Systems for Drought, *World Meteorological Organization*, Lincoln, Nebraska, 8–11 December 2009.

Wong, G., Lambert, M.F., Leonard, M., Metcalfe, A.V., 2010. Drought analysis using trivariate copulas conditional on climatic states. *J. Hydrol. Eng.*, 15: 129–141.

World Bank, 2003. Report on financing rapid onset natural disaster losses in India: A risk management approach. Report No. 26844-IN, Washington, DC. Available from: www.gfdrr.org/sites/gfdrr/files/publication/India%20Financing%20Rapid%20Onset%20Natural%20Disaster%20Losses%20in%20India-A%20Risk%20Management%20.

World Resources Institute, 2015. Available from: www.wri.org/applications/maps/aqueduct-atlas/#x=39.92&y=18.14&s=ws!20!28!c&t=waterrisk&w=def&g=0&i=BWS-16!WSV-16!SV-2!HFO-4!DRO-4!STOR2&tr=ind-1!prj-1&l=3&b=ter (accessed on 23-01-2018).

Remote sensing of agricultural drought monitoring: A state of art review (PDF Download Available). Available from: www.researchgate.net/publication/308776690_Remote_sensing_of_agricultural_drought_monitoring_A_state_of_art_review (accessed on 23-01-2018).

Yang, Y., Shang, S., 2013. A hybrid dual-source scheme and trapezoid framework-based evapotranspiration model (HTEM) using satellite images: Algorithm and model test. *J. Geophys. Res. Atmos.*, 118: 2284–2300, doi:10.1002/jgrd.50259.

Zargar, A., Sadiq, R., Naser, B., Khan, F.I., 2011. A review of drought indices. *Environ. Rev.*, 19: 333–349.

Zhang, N., Hong, Y., Qin, Q., Liu, L., 2013. VSDI: A Visible and Shortwave infrared Drought Index for monitoring soil and vegetation moisture based on optical remote sensing. *Int. J. Remote Sens.*, 34(13): 4585–4609.

URL

www.unescap.org/our-work/ict-disaster-risk-reduction/regional-space-cooperation mechanisms-drought-monitoring-and-sustainable-development/about (accessed on 14-07-2017).

www.drought.unl.edu/risk/impacts.htm (accessed on 22-04-2011).

http://swat.tamu.edu/software/swat-executables.

http://droughtmonitor.unl.edu.

http://cordis.europa.eu/result/rcn/176011_en.html.

www.ais.unwater.org/ais/pluginfile.php/516/course/section/175/Drought%20Occurrence%20and%20Activities%20in%20the%20Near%20East.pdf.

www.agr.gc.ca/eng/programs-and-services/list-of-programs-and-services/drought-watch/managing-agroclimate-risk/lessons-learned-from-the-canadian-drought-years-2001-and-2002/?d=1463593613430.

www.wri.org/applications/maps/aqueductatlas/#x=39.92&y=18.14&s=ws!20!28!c&t=waterrisk&w=def&g=0&i=BWS-16!WSV-16!SV-2!HFO-4!DRO-4!STOR-8!GW-8!WRI-4!ECOS-2!MC-4!WCG-8!ECOV-2&tr=ind-1!prj-1&l=3&b=terr.

www.eea.europa.eu/policy-documents/addressing-the-challenge-of-water.

www.ipcc.ch/pdf/technical-papers/climate-change-water-en.pdf.

www.gfdrr.org/sites/gfdrr/files/publication/India%20Financing%20Rapid%20Onset%20Natural%20Disaster%20Losses%20in%20India-A%20Risk%20Management%20.

http://wwa.colorado.edu/climate/iwcs/archive/IWCS_2007_July_feature.pdf (accessed on 21-07-2017).

http://drought.unl.edu/ranchplan/DroughtBasics/WeatherDrought/MeasuringDrought.asp (accessed on 21-07-2017).

www.geo.uio.no/edc.

http://cordis.europa.eu/result/rcn/176011_en.html (accessed on 25-09-2017).

www.aimspress.com/journal/environmental.

www.drought.gov/drought/data-maps-tools/soil-moisture (accessed on 21-07-2017).

13

Land Degradation Information Systems

13.1 Introduction

Timely and reliable information on nature, extent, and magnitude of the problem and spatial distribution of degraded lands is a prerequisite for planning and implementation of appropriate preventive and curative measures, and to assess the success and progress of various land reclamation programs. Having discussed the role of geospatial technologies in detection, mapping, and monitoring of various types of degraded lands in previous chapter, the next logical step is to organize available information on degraded lands in the form of digital database with standard projection and datum, and to develop the information system for easy access by the perspective users.

The inventories on degraded lands have been carried out world over since last several decades at different scales depending on the purpose for which the survey is to be carried out. However, for identification of magnitude of limitations of degraded lands, and their impact on global change, generation of maps on degraded lands at global level following universal standard procedures for classifying, describing, and mapping is a prerequisite. Once the information on land degradation is generated using a common database and approach for its inventorying and monitoring, the next step would be to develop a standard, uniform, and usable database. For developing a degraded land database two types of information, namely the inventory of degraded land sand mapping the changes therein, are generally considered.

Historically, inventory on degraded lands has been carried out using a traditional approach [(Global Assessment of Human-induced Soil Degradation (GLASOD)] as well as those using air/spaceborne remote sensing data as hard copies or transparent sheets. Such maps have been used for various purposes including planning and rehabilitation of degraded lands, and monitoring the progress and success of rehabilitation interventions made. These maps, however, suffer from two major limitations, namely (i) tremendous amount of efforts involved in case soil data are to be integrated with other themes for deriving meaningful information for reclamation and subsequent management, and (ii) very limited durability and longevity. There is, therefore, a need for alternative and efficient approaches for storage and archival information on degraded lands. In fact, efficiency in storing and processing of these primary data, and in displaying the derived soil information, is considerably increased when using computer-assisted procedures operating on appropriately structured databases. The Geographic Information System (GIS) with a capacity to store, retrieve, manipulate, and analyze large quantities of spatial and attribute data offer immense potential in generating soil resources database.

13.2 Background

Let us familiarize ourselves with certain terms which are very often used while developing an information system. System is a concept that is useful to study objects, especially when they are complex. A **system** is a regularly interacting or interdependent group of items forming a unified whole. Every **system** is delineated by its spatial and temporal boundaries, surrounded and influenced by its environment, described by its structure and purpose and expressed in its functioning. A system is the result of viewing the active world from a certain point of view (Carvalho, 1999).

In the context of natural resources and degraded lands we deal with the data on their nature, extent, spatial distribution, potential, and limitations. Data are representation of facts, concepts and/or instructions, suitable for interpretation or processing by humans or automated systems (Paresi, 1998) but only data placed in context can produce information. Information is an answer to specific question, in terms of problem solving within managerial/decision-making activities.

An information system (IS) is an organized system for the collection, organization, storage, and communication of information. More specifically, it is the study of complementary networks that people and organizations use to collect, filter, process, create, and distribute data. There are several definitions available for IS. Two of them are quoted here. *"Information systems are combinations of hardware, software, and telecommunications networks that people build and use to collect, create, and distribute useful data, typically in organizational settings"* (Valacich and Schneider, 2010). Another definition goes as *"Information systems are interrelated components working together to collect, process, store, and disseminate information to support decision making, coordination, control, analysis, and visualization in an organization"* (Kenneth et al., 2013). In essence, technically an IS is a set of interrelated components that collect (or retrieve), process, store, and distribute information to support decision making and control in an organization.

As it is evident from foregoing, these definitions focus on two perspectives: one relating to its *function*; the other relating to its *structure*. From a functional perspective; an IS is a technologically implemented medium for the purpose of recording, storing, and disseminating linguistic expressions as well as for the supporting of inference making. From a structural perspective, an IS consists of a collection of processes, data, models, technology, and partly formalized language people, forming a cohesive structure which serves some organizational purpose or function.

An IS encompasses three activities, namely input, processing, and output. Input captures or collects relevant raw data from different sources, processing converts the raw input into a more meaningful form, whereas output transfers the processed information to the users for end use. Additionally, ISs also require feedback, which helps the IS managers to carry out corrections and/or making necessary improvements in the quality of the information and its utility, and efficiency of the system.

13.2.1 Components of an IS

The IS consists of (1) hardware (physical computer equipment and associated device, machines, and media), (2) software (programs and procedures), (3) data (data and knowledge base), (4) networks (communications media and network support), and (5) human resources (end users and IS specialists, system analyst, programmers, data administrators,

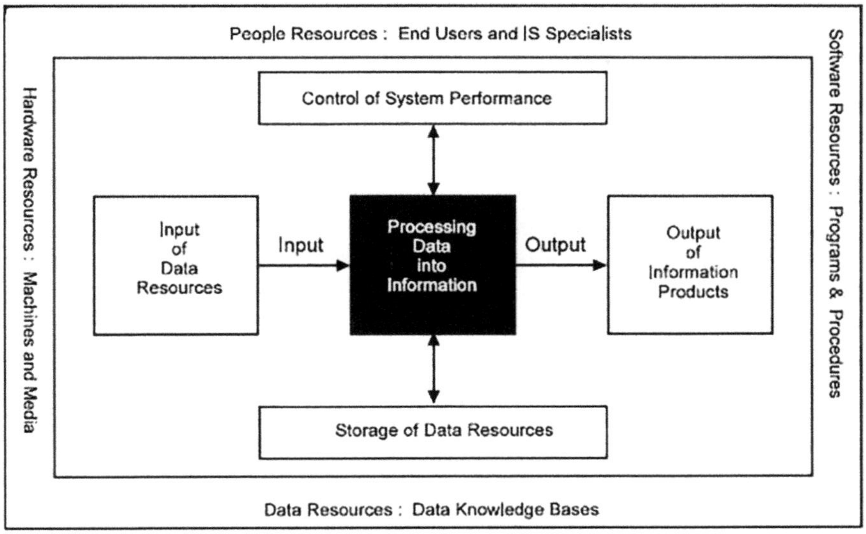

FIGURE 13.1
Components of an information system.

etc.) (Figure 13.1). Hardware of an IS includes computers, keyboards, disk drives, iPads, and flash drives. Software is a set of instructions that guide the hardware to perform the function/s. There are several categories of software, with the two main categories being operating system software, which makes the hardware usable, and application software, which does something useful. Examples of operating systems include Microsoft Windows on a personal computer and Google's Android on a mobile phone. Examples of application software are ERDAS IMAGINE, ILWIS, ENVI, and ArcGIS.

The third component of IS is data. In the context of land degradation, data may consist of location, spatial extent, magnitude of the problem, physical and/or chemical properties of degraded lands wherever applicable. For instance, physicochemical properties of acid soils and salt-affected soils required for rehabilitation of acid soil and salt-affected soils. Apart from the components of hardware, software, and data, which have long been considered the core technology of ISs, communication has been added recently. Technically, the networking communication component is made up of hardware and software, but it is such a core feature of today's ISs that it has become its own category.

Among these components data resource or database is most important from users' perspective. Data resources include data (which is raw material/input of ISs) and database. Database refers to a set of related data and the way it is organized. Data can take many forms, including traditional alphanumeric data, composed of numbers and alphabetical and other characters that describe business transactions and other events and entities. Text data, consisting of sentences and paragraphs used in written communications; image data, such as graphic shapes and figures; and audio data, the human voice and other sounds, are also important forms of data.

The data resources of IS are typically organized into processed and organized database, and knowledge in a variety of forms such as facts, rules, and case studies about successful usage of information on degraded lands.

13.3 Database

Designing database involves careful decisions about the kind of database structure to be used and the number of layers of thematic data, namely base maps, parent material, landform, Digital Elevation Models (DEMs), soil resources maps, and attribute data constituting the database. The database design is subsequently conceptualized and the kind of database structure to be used for creating database is selected. Additionally, a provision in the database design is also made to establish appropriate linkages between spatial and attribute data for easy input, retrieval, and subsequent analysis/integration with the databases on other natural resources data.

13.3.1 Database Model

A database model is a type of data model that determines the logical structure of a database and fundamentally determines in which manner data can be stored, organized, and manipulated. The most popular example of a database model is the relational model, which uses a table-based format. A data model is a collection of concepts and rules for the description of the structure of the database. Structure of the database means the data types, the constraints, and the relationships for the description or storage of data, respectively. Generally, relational database model is used for storage of digital soil resources data. However, very often other database models, namely hierarchical, network- and object-oriented, and a combination thereof are also utilized. A brief description of various database models is given hereunder.

13.3.1.1 Hierarchical Model

In a hierarchical database management system (DBMS) one data item is subordinate to another one. This is called a parent–child relationship. The hierarchical data model organizes data in a tree-like structure. One of the rules of a hierarchical database is that a parent can have multiple children, but a child can only have one parent. For example, think of an online store that sells many different products. The entire product catalog would be the parent, and the various types of products, such as books and electronics, would be the children. Each type of product can have its own children categories. For example, books could be broken up into fiction and nonfiction. Each of these categories can be broken up into subcategories. You can continue like this by listing individual authors and then the individual book titles. This is a rather simple way to represent data, but it is very efficient. This model works best for data that are inherently hierarchical in nature. Many datasets cannot easily be organized in this manner and require a more complex approach (Figure 13.2).

13.3.1.2 Network Model

In a network DBMS, every data item can be related to many other ones. The database structure is like a graph. This is similar to the hierarchical model and also provides a tree-like structure. However, a child is allowed to have more than one parent. In the example of the product catalog, a book could fall into more than one category. The structure of a network database becomes more like a cobweb of connected elements (Figure 13.3).

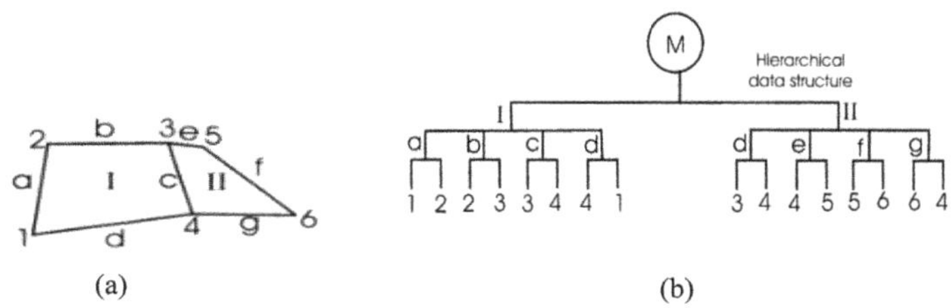

FIGURE 13.2
A sample polygon map (a) and its representation in a hierarchical database model (b) (Source: Adapted from Burrough and McDonell (2005)).

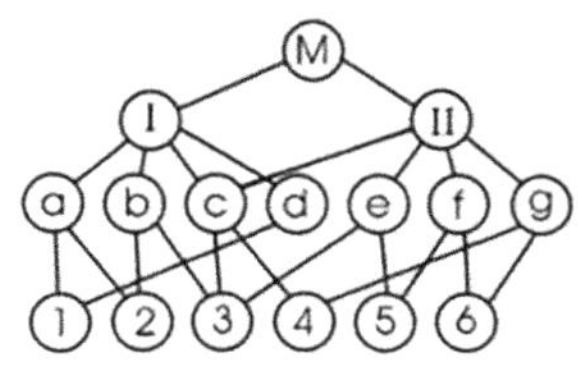

FIGURE 13.3
Network database model (Source: Adapted from Batjes et al. (2013)).

For example, consider an organization with an employee database. For each employee, there are different pieces of data, such as their name, address, telephone number, social security number, and job function. Different units in the organization need different levels of access. For example, the human resources department needs to have access to the social security information for each employee so they can take care of tax deductions and set up benefits. This is somewhat sensitive information, so other departments do not need access to this part of the database. All the pieces of data are connected in a network that implements these rules. While conceptually relatively simple, this database structure can quickly become very complicated.

13.3.1.3 Relational Model

In a relational DBMS, all data are organized in the form of tables. This DBMS model emerged in the 1970s and has become by far the most widely used type of DBMS. Most of the DBMS software developed over the past few decades uses this model. In a table, each row represents a record, also referred to as an entity. Each column represents a field, also referred to as an attribute of the entity. In relational database model the terms *relations*, *attributes*, and *tuples* are used whereas most of us use the more common terms *tables*, *columns*, and *rows*, respectively.

A relational DBMS uses multiple tables to organize the data. Relationships are used to link the various tables together. Relationships are created using a field that uniquely identifies each record. For example, for a table of books, you could use the ISBN number since there are no two books with the same ISBN. For a table of authors, you would create a unique Author ID to identify each individual author (Figure 13.4).

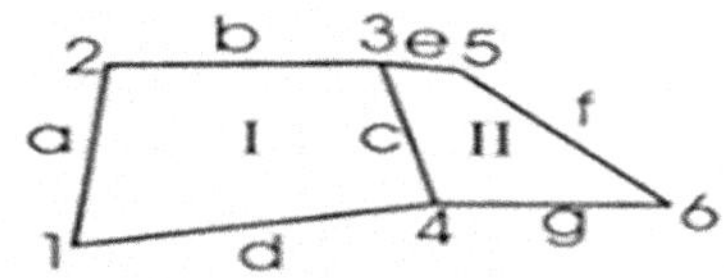

Map

	p1	p2
M	I	II

Polygons

	L1	L2	L3	L4
I	a	b	c	d
II	c	e	f	g

Lines

	Pnt1	pnt2
a	1	2
b	2	3
c	3	4
d	4	1
e	3	5
f	5	6
g	6	4

FIGURE 13.4
A relational database model for a map polygon (Source: Adapted from Batjes et al. (2013)).

13.3.1.4 Object-Oriented Model

The previous DBMS models work primarily with text and numbers. Object-oriented databases are able to handle many newer data types, such as images, audio, and video. These data items are the objects stored in the database. An object-oriented database management system (OODBMS), also sometimes just called object database (ODBMS), is a DBMS that stores data in a logical model that is closely aligned with an application program's object model. Of course, an OODBMS will have a physical data model optimized for the kinds of logical data model it expects. Object-oriented database models have been around since the seventies when the concept of object-oriented programming was first explored. It is only in the last 10 or 15 years that companies are utilizing object-oriented DBMSs (OODBMS). The major problem for OODBMSs was that relational DBMSs (RDBMS) were already implemented industry wide. OODBMS should be used when there is a business need, high performance required, and complex data are being used. Due to the object-oriented nature of the database model, it is much simpler to approach a problem with these needs in terms of objects (Figure 13.5).

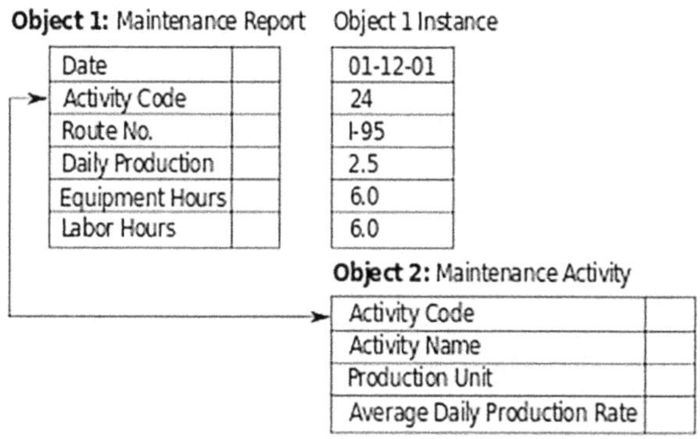

FIGURE 13.5
Object-oriented database model.

The result can be a performance increase of 10–1,000 times while writing as little as 40% of the code (this is because it requires no intermediate language such as SQL; everything is programmed in the OO language of choice). This code can be directly applied to a database, and thus saves time and money in development and maintenance. An object-oriented database interface standard is being developed by an industry group, the Object Data Management Group (ODMG). The Object Management Group (OMG) has already standardized an object-oriented data brokering interface between systems in a network.

13.4 Land Degradation ISs

The land degradation IS consists of appropriately organized database, data retrieval and processing and/or integration system. Typically, land degradation is a derivative information generated by integrating information on soils, biota, and overhead climatic parameters. Furthermore, in the context of new concept of land degradation which refers to *"the reduction in the capacity of the land to provide ecosystem goods and services over a period of time for its beneficiaries"* (Kellner et al., 2011) the scope of land degradation further broadens. In fact, ecosystem goods refer to absolute quantities of land produce having an economic or social value for human beings. They include animal and vegetal production, land availability, and water quality and quantity. Ecosystem services concern more qualitative characteristics and their impact on the beneficiaries and the environment. An overview of ecosystem goods and services is given in Table 13.1. These should first be transferred into tangible, measurable entities because terms such as *regulating climate* are not readily measured or estimated as such. Therefore, the goods and services have been further segregated into more tangible components such as biomass, soil, water, biodiversity, economics, and social (Table 13.1).

Since information on soils is an integral part of land degradation IS, familiarity with the procedure of the development of soil ISs and currently available soil ISs at global level seems to be a logical step toward development of land degradation ISs. A detailed account of existing soil ISs is beyond the scope of the chapter. A brief procedure for developing soil database and an overview of soil ISs are provided hereunder.

TABLE 13.1

Elements of Ecosystem Goods and Services

Ecosystem "processes" include:
Maintenance of energy flux, dissipation, climate modulation (**vegetation/water/soil**)
Maintenance of hydrologic flux, hydrologic cycle, water quality (**water**)
Biological productivity, plant pollination (**vegetation**)
Maintenance of bio-geochemical cycling, storage, mineral-gaseous cycles, water-air quality (**soil/water**)
Decomposition, weathering, soil development-stability, soil quality (**soil**)
Maintenance of biological diversity (**biodiversity**)
Absorbing, buffering, diluting, detoxifying pollutants-xenobiotics (**soil/water**)
Ecosystem "goods" include:
Food (**economic/vegetation/social**)
Construction materials (**economic**)
Medicinal plants (**economic/biodiversity**)
Wild genes for domestic plants and animals (**biodiversity**)
Water (**water/economic**)
Ecosystem "services" include:
Maintaining hydrological cycles (**water**)
Regulating climate (**water/soil/vegetation**)
Cleansing water and air (**water**)
Maintaining the gaseous composition of the atmosphere (**vegetation**)
Pollinating crops and other important plants (**biodiversity**)
Generating and maintaining soils (**soil**)
Storing and cycling essential nutrients (**soil**)
Absorbing and detoxifying pollutants (**soil**)
Providing beauty, inspiration, and recreation (**social**)

Source: Modified from Ehrlich and Ehrlich (1991), Lubchenco et al. (1993), and Richardson (1994).

13.4.1 Soil Database

A soil database is a tool for the capture, storage, processing, and display of soil data and information generated by soil surveys. The Data Base Management System (DBMS) provides various functions that allow entry, storage, and retrieval of large quantities of information and provides ways to manage the data. The creation of digital geospatial database involves the following steps.

13.4.1.1 Data Acquisition

The existing digital spatial data on soil resources could be procured from concerned national soil survey organizations. If soil resources maps are available as hard copies (analog data), it could be digitized; and if not available, it could be generated through a fresh inventory of soil resources using suitable database and approaches or interpolation of point observations to continuous surfaces.

13.4.1.2 Geo-referencing and Creation of Digital Data

It involves digital image-to-map (standard topographic maps) registration of soil resources maps. In case desired topographic maps are not available, ground control points identified using Global Navigation Satellite Positioning System (GNSS) could be used for geo-referencing.

The digital soil resources maps, if available, could be directly used as an input. Soil maps derived through interpretation of satellite images/aerial photographs could be digitized and subsequently vectorized in a GIS environment using an appropriate software tool.

13.4.1.3 Data Verification and Editing

Errors in the input soil resources data—both thematic and positional errors—if any, need to be checked thoroughly to ensure data quality. The digital input, thus generated, may be edited for continuity of soil polygons and administrative boundaries and annotation, and extension of soil boundaries into the adjoining soil map sheets.

13.4.1.4 Data Updation

Sufficient provision needs to be made in the database to incorporate additional spatial and attribute data, and updation of existing database in the event of the availability of larger-scale or updated information.

13.4.1.5 Soil Degradation Data

The data on kinds, causes, and severity of soil degradation are stored for every soil mapping unit (SMU) or broader geographic unit (e.g., landform units). Information on soil erosion, both water and wind erosion, erosion susceptibility, waterlogging, salinization, and/or alkalinization, acidification, soil degradation due mining, aquaculture, shifting cultivation, industrial effluents, fertilizers, insecticides, pesticides, hardpan formation, surface crusting, fertility depletion, etc. is highlighted to characterize environmental fragility status or ongoing deterioration processes. Identifying the location and mapping the dynamics of such features can be assisted through the digital image processing techniques. When interfaced with the other files of the soil database and the other modules of the ILWIS or any other suitable database, this component can contribute substantially to planning and implementation of land reclamation program apart from serving as database for environmental impact assessments. The information on soils is the basic data which are used to derived land degradation categories in conjunction with terrain's relief, vegetation cover, and climate.

13.4.1.6 Soil ISs

The importance of rehabilitation and sustainable development of degraded lands in addressing major global issues, such as food, energy, water securities, climate change, conservation, and preserving biodiversity, is increasingly recognized. Consequently, there is a growing demand for quality-assessed soil data and information, of various types and scales, by other scientific disciplines, practitioners, policy, and society. Realizing the importance of standardized information on soils including assessment and monitoring of land degradation for various developmental and research purposes, several countries including USA, Canada, Latin America, Australia, Ireland, and India have developed their own soil ISs. Such ISs address two basic issues: (i) developing soil ISs for some areas for which data are not available and (ii) standardization of soil database with respect to information content, scale, projection, datum, and data structure since it is still in the process of development by different countries, and lacks uniformity. Hence there is a need to bring all the databases onto a common platform. In order to have a harmonized soil IS and to bridge the gap with soil information at global level the International Soil Reference and Information Centre (ISRIC) is in the process of the development of global soil IS.

Several countries including USA (National Soils Information Systems—NASIS), Canada (Canadian Soil Information Service—Can SIS), Europe (The European Soil Information System—EUSIS), Latin America (Sistema de Información de Suelos de Latinoamérica), Australia (The Australian Soil Resource Information System—ASRIS), The Irish Soils Information System (ISIS), and India (Indian Integrated National Agricultural Resource Information System (INARIS)) (Maji et al., 2004). Globally, ISRIC has been playing a lead role through its initiative titled "ISRIC Soil Information System (ISIS)" in developing a global-level soil IS. Since the development of soil IS is a continuing process, the progress made so far is briefed hereunder.

13.4.1.6.1 ISRIC Soil Information System

ISIS is a project to bring the ISRIC Soil Reference collection on-line. ISIS characterizes a collection of monoliths with morphological, analytical data that represent the main soil reference groups of the World Reference Base for Soil Resources (WRB). It holds data on the World Soil Reference Collection, a large part of which have been collected during the national soil reference collections (NASREC) program. This new global soil map obviously have advanced interpretation and functionality options, which aim to assist better decisions in a range of global issues like food production, climate change, and environmental degradation (www.globalsoilmap.net).

Using ISIS, users may access:

Site Data: 60 attributes on location, geology, landform, soil surface properties, hydrology, land use, vegetation, and climate.

Soil Data: Soil profile description has been made according to the FAO guidelines. One hundred physical, chemical, and mineralogical attributes have been considered. Attributes for classification have been followed as given in the FAO-UNESCO Legend (1974), FAO, UNESCO, and ISRIC (1988), the revised FAO-UNESCO Legend (1988), World Reference Base for Soil Resources (2006), USDA Soil Taxonomy (Soil Survey Staff, 1975), and in a national classification system. The World Soil Reference Collection at ISRIC includes some 950 monoliths from over 70 countries with detailed soil profile and environmental data.

13.4.1.6.2 World Soil Information System

In pursuance of its mandate to serve the international community as custodian of global soil information and to increase awareness and understanding of the role of soils in major global issues, ISRIC–World Soil Information is strengthening its capabilities in collecting, storing, processing, and disseminating global soil and terrain information for the research and development of sustainable and productive land use. The world Data Centre was created in 1966, following a request by the 1964 UNESCO General Council, and obtained the status of *ICSU World Data Centre* in 1989. Initially named as *WDC for Soil Geography and Classification* and was later rechristened as *WDC for Soils*. In August 2011, ISRIC was accredited as regular member of the new *ICSU World Data System* (WDS). In that capacity, it deals directly with data curation and data analysis services.

13.4.1.6.3 Data Curation

Currently, new soil data are collected less and less while the same legacy (old) data and information derived there from are repeatedly used. Hence, it is very important to secure, maintain, and expand data that support current information, including soil reference collections, soil maps, and reports, and this ideally for public, free access. Various collections are being maintained and regularly expanded by ISRIC. Since its establishment, a primary responsibility has been to create and maintain a world soil reference collection. The collection comprises over 1,000 soil monoliths sampled to be representative of the soil units of the FAO-UNESCO *Soil Map of the World* (FAO, 1971–1981; FAO et al., 2009), forming a unique educational, cultural, and scientific resource. A selection of the monoliths is on display in the World Soil Museum in Wageningen, and the whole collection: monoliths, reference samples, soil morphological and analytical data, thin sections, and images, has been documented in a relational database, queryable on-line. This information is supported by a large collection of country reports, books, and maps that originate largely from developing countries. The map collection comprises mainly small-scale (1:250,000 or smaller) maps, part of which have been digitized in the framework of the EuDASM project (Panagoset et al., 2011).

This wealth of soil geographic and attribute information has been used to develop a range of broad-scale GIS databases (e.g., Batjes, 2009; FAO et al., 2009; Nachtergaele et al., 2011) that were generally distributed on-line as stand-alone, compressed files through both the ISRIC website and Global Change Master Directory (GCMD).

13.4.1.6.4 Major Components

Since June 2011, ISRIC's website has provided access to a range of new facilities that are in various stages of development including the following.

13.4.1.6.5 ISRIC Enterprise Database

All soil-related data managed or maintained by ISRIC will be made available online from one central database environment in one uniform format (Tempel and van Kraalingen, 2011). The spatially enabled relational database management system (RDBMS) will be used to hold and manipulate a variety of data types, scales, and sources. Version 1.0 is being populated using a range of disparate, stand-alone soil databases developed by ISRIC and its partners (e.g., Batjes, 2009; Van Engelen, 2011); for the future, other data (sets) will be added using both on-line and off-line facilities for data entry (www.isric.org/data/wosis). In conjunction with this, a range of web-based tools are being developed to extract user-defined data from the system using agreed data-exchange formats (e.g., SoilML, GeoSciML). The system is implemented using open standards (e.g., OGC, PostGresQL).

13.4.1.6.6 Vocabulary Service

This multilingual, soil-environmental thesaurus is being developed as a reference indexing and retrieval tool. It will provide ontologies of interest in the Soil Science domain and was created at the request of the Working Group on Soil Information Standardisation of the International Union of Soil Sciences (IUSS). Currently, the service is based on selected databases (FAO—Agrovoc; EEA—Gemet). Over the next 2–4 years, the service will be developed further in collaboration with colleagues from Europe (GS-Soil project), Australia (CSIRO SEEGrid), and similar projects around the world. Currently, it uses SEEGrid Vocabulary Service implementation as a frontend while editing is based on FAO's VocBench which is open source.

13.4.1.6.7 Metadata Service (Operational)

Metadata service has been implemented as a multilingual catalog application to enter, search, and retrieve metadata records and manage spatially referenced resources.

Currently, the service harvests soil-related metadata from various sources worldwide, including ISRIC's Library, FAO, and the Consultative Group on International Agricultural Research (CGIAR). The service is based on *GeoNetwork open source* and follows the principles of Free and Open Source Software (FOSS) and International and Open Standards for services and protocols (e.g., ISO/TC211 and OGC, Open Geospatial Consortium).

13.4.1.6.8 One Soil Map Viewer

It provides visualizations of qualitative and quantitative soil properties worldwide. Maps are accessed from various data sources and providers in a distributed way around the world. The viewer will create a dynamic soil map for different soil properties with standardized legends and data exchange formats.

13.4.1.6.9 Web Mapping Service

Web mapping service (WMS) provides OGC-conform maps for different layers and allows users to visualize datasets hosted at the WDC for soils.

13.4.1.6.10 Web Processing Service

Web processing service (WPS) provides the capability to compute soil functions or parameters of interest in the Soil Science domain. Three different services are currently available: (a) interpolation over depth of quantitative soil profile properties using the spline function, (b) extraction of point information of soil information or auxiliary data sources for a given location on the planet and soil information or auxiliary data sources for a given area on the earth specified by the corner coordinates, and (c) overlay of two raster data sources to provide statistical information for a predefined zone-grid (e.g., watershed, countries) for soil information or auxiliary data sources. The WPS delivers OGC conforming XML files.

13.4.1.6.11 Worldgrids

Worldgrids provides a repository of gridded predictors with global or at least partial global coverage. As WDC for Soils, ISRIC facilitates the collation and use of multi-thematic gridded repositories of Digital Soil Mapping covariates. Worldgrids already contains over 100 grid maps at a 5.6 km resolution and several 1 km resolution maps. Functionalities include querying, extraction, and creation of an overlay of a user-specified zone grid and a defined covariate map.

13.4.1.6.12 Global Soil Information Facility

Global Soil Information Facility (GSIF) will provide an overarching methodological framework for the production of consistent global soil information products of various types and scales/resolutions (Hengl et al., 2011). The key principles which are the pillars of GSI comprise (Figure 13.6):

Crowd sourcing: Everyone collecting soil data or working with soil information is welcome to contribute point data, spatial data, and/or modeling algorithms to the GSIF effort. All point datasets are maintained "as is" in GISF data repository together with their metadata, in particular the license.

Data license: The soil data provided by an individual remain property of the original contributors, unless otherwise indicated by the contributor. The contributor is required to indicate the way data may be used, processed, and distributed. Datasets with a CC BY or CC BY-NC Creative Commons license or similar will be standardized using the regular WoSIS workflow prior to their distribution to the users with the original license.

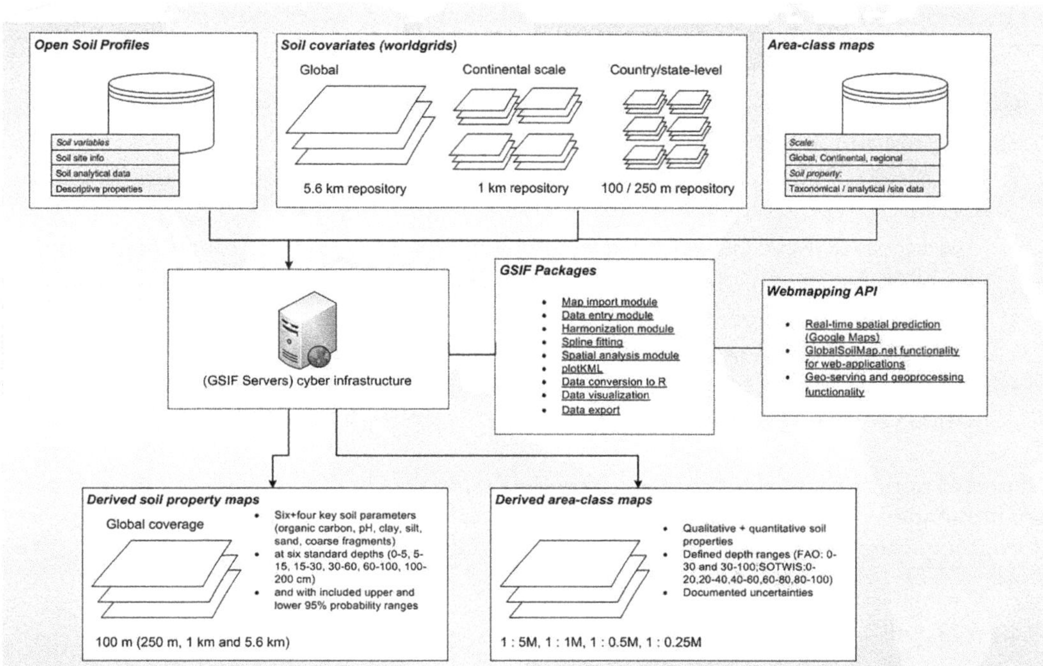

FIGURE 13.6
Simplified representation of ISRIC's GSIF framework (Source: Adapted from Batjes et al. (2013)).

Alternatively, some datasets may only be used to make derivative predictions and visualizations, such as SoilGrids. These derivative products will be made available with an open data license in accordance with the ISRIC Data Policy.

Open Source Software: GSIF products are mainly based on free and open source software, such as Linux, PHP, LaTeX, R and contributed R packages, RStudio, GDAL, GRASS, SAGA GIS, PostgreSQL, Geoserver, PostGIS, Python, Google Earth, and similar. This feature makes GSIF independent of commercial software packages.

International standards: GSIF has been designed to serve global soil mapping initiatives and supports Findable, Accessible, Interoperable, and Reusable (FAIR) guidelines for data sharing. The IS closely follows internationally accepted standards, e.g., International System of Units, international soil classifications systems, FAO soil field description guides, SoilML guidelines, Open Geospatial Consortium standards, World Geodetic System 1984, and the like.

Reproducible research: The SoilGrids component of GSIF is based on automated procedures for mapping, pattern recognition and report/plots generation. The derivative maps thus generated can be updated by rerunning the scripts with limited human intervention once new datasets become available.

User network: ISRIC runs an annual Spring School to build a user network to stimulate use of the facilities, obtain feedback for improvements and further development of GSIF components through collaboration.

The components GSIF: GSIF consists of the following independent and interconnected components:

- *World Soil Information Service (WoSIS):* a web service for serving a selection of standardized soil profile data.
- *WorldGrids:* a web service providing access to all environmental covariates used in GSIF.
- *R-packages:* GSIF packages for automated analysis and mapping functionality and plot KML for visualization of soil field data and output maps.
- *SoilGrids:* predicted global soil property and class maps (currently at resolution of 250 m) produced using automated mapping procedure.
- *SoilInfo App:* a smart phone app (for Android and Apple) to access global soil information on individual's mobile device.

Wherever appropriate, algorithms and model states will be documented and preserved. Recommended procedures are being developed and tested to align with the frameworks of ongoing, large international projects such as GlobalSoilMap.net (Sanchez et al., 2009) and e-SOTER (Van Engelen, 2011).

13.4.1.6.13 SoilGrids 1 km—Global 3D Soil IS

Hengl et al. (2014) developed SoilGrids 1 km—a global 3D soil IS at 1 km resolution. The IS contains spatial predictions for a selection of soil properties (at six standard depths): soil organic carbon (g/kg), soil pH, sand, silt, and clay fractions (%), bulk density (kg/m^3), cation-exchange capacity (cmol+/kg), coarse fragments (%), soil organic carbon stock (t/ha), depth to bedrock (cm), World Reference Base soil groups, and USDA Soil Taxonomy suborders. Their predictions are based on global spatial prediction models which they fitted, per soil variable, using a compilation of major international soil profile databases (ca. 110,000 soil profiles), and a selection of ca. 75 global environmental covariates representing soil forming factors. Some of the major limitations of the current version of SoilGrids 1 km include (i) weak relationships between soil properties/classes and explanatory variables due to scale mismatches, (ii) difficulty to obtain covariates that capture soil forming factors, and (iii) low sampling density and spatial clustering of soil profile locations. However, as the SoilGrids system is highly automated and flexible, increasingly accurate predictions can be generated as new input data become available. SoilGrids 1 km is available for download via http://soilgrids.org under a Creative Commons Non Commercial license.

13.4.1.6.14 Soil Degradation Database for Russia

The development of the Soil Geographic Database (SGDB) of Russia aims at providing a scientific basis for formulating the strategy for sustainable and rational land use, monitoring soil cover, and soil conservation (Shoba et al., 2010). The major objectives of the database are as follows:

1. Inventory and formalization of data on soil resources,
2. To cater to the information needs of research projects and academic programs,
3. Inclusion of Russia into the unified soil-information space of European Community and world, and participation in global and regional soil programs.

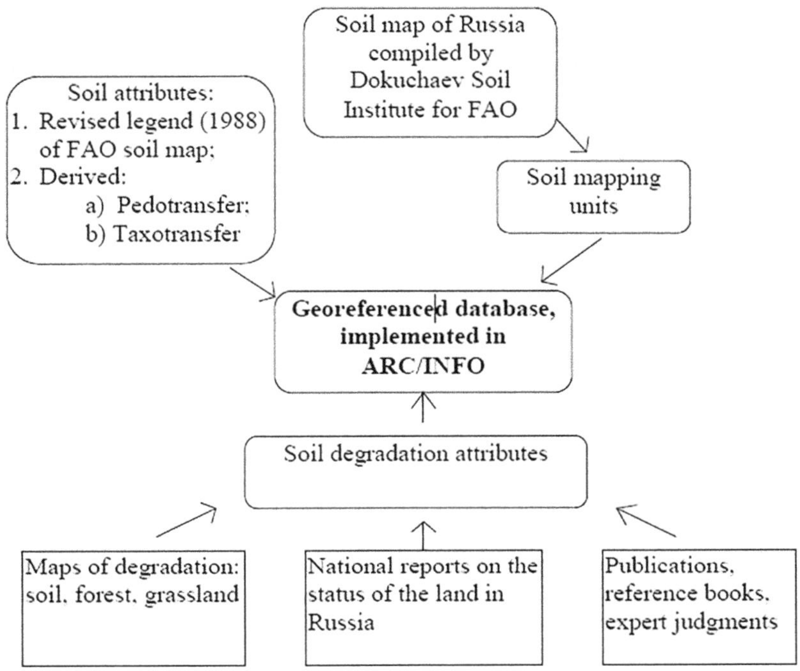

FIGURE 13.7
Principal elements and basic sources of the database (Stolbovoiand Fischer, 1997).

The major components of the SGDB include the Geographic Database and the Specialized Attribute Database. RDBMS is used for data storage and processing (Figure 13.7). The Geographic Information Soil Database forms the cartographic basis of the State Soil-Geographic Database. The soil profile (attribute) database is based on the concept of representative soil profiles. The representative profiles should have an exact geographical location and be provided with a morphological description and a complete set of analytical data. A series of the profiles typical of a given soil type can be used for calculating averaged characteristics for this soil type. For filling the profile attributive database, a special software—Soil-DB (version 1)—was developed. This program allows a provider of information to login to the site, to create and fill the soil description card, and to send it through Internet to the central server of the System.

13.5 Gladis/GIS System

Following up on the LADA/GLADA initiative, a GLADIS/GISA Global Land Degradation Information System has been set up by FAO and partners (Nachtergaele et al., 2011) is at the heart of the GLADIS, and with it the definition of land degradation|| as the reduction in the capacity of the land to provide ecosystem goods and services over a period of time for its beneficiaries. In this context, ecosystem goods refer to absolute quantities of land produce having an economic or social value for human beings, e.g., animal/crop production or water quantity/quality. Ecosystem services on the other hand concerns more qualitative characteristics and their impact on the beneficiaries and the environment. Whereas GLADA had solely focused on the production function, GLADIS set out to determine both the status of ecosystems' capacity to deliver goods and services and the change in this capacity.

The aim of GLADIS/GIS system is to ensure the diffusion of the global data collected and prepared during the LADA project and to promote the dissemination of the obtained results, a Web-GIS IS has been implemented, which is easily accessible through the simple use of any web-provider (Anderson and Moreno-Sanchez, 2003). The Web-GIS system has been designed integrating in a single interface several tools. In particular, the use of open-source software, that have shown multiple advantages in terms of a high reliability—in comparison to the commercial ones—an international spreading (by both private and public subjects), and, finally, a continuous updating by the numerous communities—of both users and developers—that allows to reduce the costs linked to the licenses acquisition.

The used applications follow as well the international standard for the geographical data processing; hence, they may be easily adapted and integrated with other available technological platforms, simplifying the trouble of compatibility with the owned standards. In particular the following open-source sources have been used: Map server (Common Gateway Interface application), Apache (web server), Open Layers (user interface), JpGraph (graph creating library), and FAO GeoNetwork (catalog). The mentioned applications have been integrated, allowing managing, through a simple web interface, all the operations to access the data (e.g., querying, visualization, graph creation, metadata, download).

MapServer is an open source platform for publishing spatial data and interactive mapping applications to the web originally developed in the mid-1990s at the University of Minnesota. OpenLayers is a pure java script library for displaying map data in web browsers, with no server-side dependencies. OpenLayers implements a (still-developing) JavaScript API for building web-based geographic applications similar to the Google Maps and MSN Virtual Earth APIs. JpGraph is an object-oriented graph creating library for PHP that can be used to create numerous types of graphs either on-line or written to a file. GeoNetwork open source is a standards-based geospatial catalog application which allows organizing and publishing geospatial data on the web (Figure 13.8).

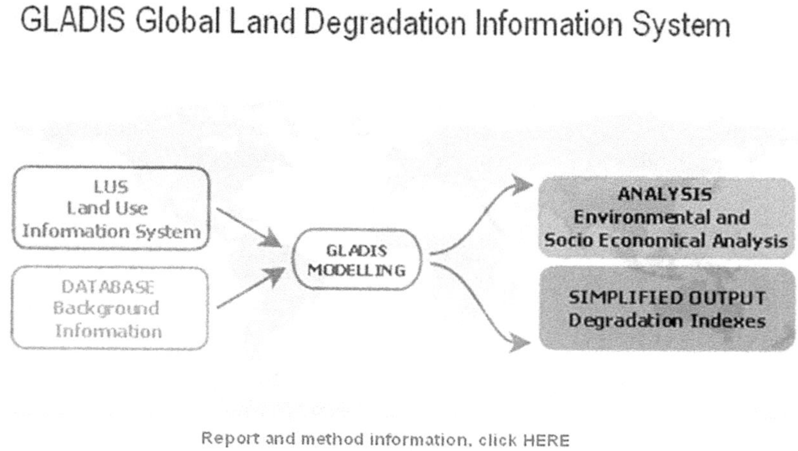

FIGURE 13.8
Snapshot of the GLADIS/GIS system's portal.

Within the ISs, data have been divided into four sections: (1) the first containing all the information required to visualize land status and land degradation process thought the use of a radar chart graphical interface that work at pixel, country, and LUS within country levels; (2) the second containing simplified global outputs; (3) the third dedicated to store the global data resulting by the LADA project that have been used in the definition of the radar graphs inputs; and (4) the fourth committed to store land use systems of the world, the basin cartographical unit of land degradation assessment (Nachtergaele and Petri, 2008) common through all project scales of evaluation (global, national, local). All section allows navigation (zoom, pan, etc.), layer interchange (Layer Switcher), and querying of GIS raster data. The tool will be available offline as pen drive application and is accessible at the address: www.fao.org/nr/lada/index.php?option=com_content&view=article&id=161&Itemid=113&lang=en (Table 13.2).

TABLE 13.2

GLADIS Global Databases: Inputs and Outputs

Data Group	Resource Base Data	Spatial Resolution	Publication Year	Data Collection Year	Units	Original Source Download	Source
General data	Land use systemsa	5 arc minutes	2008	Several		Y	http://lprapp11.fao.org:8080/lus/
	GAUL	1:1,000,000	2009			Y	www.fao.org/nr/lada/index.php?option=com_content&view=article&id=154&Itemid=184&lang=en
				GLADIS Inputs			
Axis 1 biomass	Carbon above ground	5 arc minutes	This work		t/ha	Paper	Nelson and Robertson (2008)
	Carbon below ground	5 arc minutes	2008	Several	t/ha	Y	www.iiasa.ac.at/Research/LUC/External-World-soil- database/HTML/
	Greenes	5 arc minutes	2008	1981–2003	% sum NDVI/y	Y	www.fao.org/geonetwork/srv/en/main.search?any=glada
	Greenes trend significance	5 arc minutes	2008	1981–2003	Classes	Y	www.fao.org/geonetwork/srv/en/main.search?any=glada
	Deforestation trend	Country statistics	2005	1990–2005	%	Paper	FRA (2005)
Axis 2 soil	K factor	5 arc minutes	2010	1980–2000	t.ha.h/MJ.mm	Y	Lada technical report 14
	LS factor	5 arc minutes	2010	1980–2000	Rate	Y	Lada technical report 14
	C factor	5 arc minutes	2010	1980–2000	Rate	Y	Lada technical report 14
	P factor	5 arc minutes	2010	1980–2000	Rate	Y	Lada technical report 14
	R factor	5 arc minutes	2010	1980–2000	MJ.mm/ha.h.y	Y	Lada technical report 14

(Continued)

TABLE 13.2 (*Continued*)

GLADIS Global Databases. Inputs and outputs

Data Group	Resource Base Data	Spatial Resolution	Publication Year	Data Collection Year	Units	Original Source Download	Source
	Water erosion	5 arc minutes	This work	Several	t/ha/y	Y	This work
	Rural compaction	5 arc minutes	This work + FAOSTAT	Several	Rate	Y	This work
	Physical stability	5 arc minutes	This work	Several	Rate	Y	This work
	Management factor	5 arc minutes	2010	Several	Rate	Y	This work
	Nutrient availability	5 arc minutes	2002	Several	Rate	Y	www.iiasa.ac.at/Research/LUC/SAEZ/index.html
	Soil mining and pollution	5 arc minutes	This work	Several	Rate	Y	This work
	Irrigation salinity	Country statistics	2010	Several	Rate	Y	www.fao.org/nr/water/aquastat/dbase/index.stm
	Chemical stability	5 arc minutes	This work	Several	Rate	Y	This work
	Soil resilience	5 arc minutes	This work	Several	Rate	Y	This work
	Soil Improvement	5 arc minutes	This work	Several	Rate	Y	This work
Axis 3 water	Renewable water	Country statistics	2010	Several	bil m3/y	Y	www.fao.org/nr/water/aquastat/dbase/index.stm
	Water withdrawn	Country statistics	2010	Several	Rate	Y	www.fao.org/nr/water/aquastat/dbase/index.stm
	Aridity trend	5 arc minutes	2010	1980–2000	Rate	Y	Lada technical report 14
Axis 4 biodiversity	Ecosystem biodiversity resilience	Refer to source	2008	Several	Rate	Paper	Lee and Jetz (2008)
	Endangered areas	Refer to source	Several	Several	Classes	N	Olson and Dinerstein (2002)
Axis 5 economic	Agricultural value	5 arc minutes	2010	Several	GK$/ha	Y	This work
	Livestock value	Country statistics	This work	Several	Rate	Y	This work
	Forest value	5 arc minutes	This work	1990–2005	Rate	Y	Lebedys (2008)
	Agricultural productivity trend	Country statistics	This work	1990–2005	I$	Y	FAOSTST
	Forest productivity trend	Country statistics	This work	1990–2005	$	Y	Lebedys (2008)

TABLE 13.2 (*Continued*)

GLADIS Global Databases. Inputs and outputs

Data Group	Resource Base Data	Spatial Resolution	Publication Year	Data Collection Year	Units	Original Source Download	Source
Axis 6 social	Accessibility	30 arc seconds	2008	Several	Classes	Y	Nelson (2008)
	Tourism	Country statistics	2009	2005	Rate	Y	UNWTO (2009)
	Protected areas	1: 1 000 000	2009	Several	Classes	Y	WDPA (2009)
			GLADIS Analysis				
Ecosystem status	Biomass	5 arc minutes	2010	Several	Rate	Y	This work
	Soil	5 arc minutes	2010	Several	Rate	Y	This work
	Water resources	5 arc minutes	2010	Several	Rate	Y	This work
	Biodiversity	5 arc minutes	2010	Several	Rate	Y	This work
	Economic prod.	5 arc minutes	2010	Several	Rate	Y	This work
	Social and cultural	5 arc minutes	2010	Several	Rate	Y	This work
Ecosystem processes	Biomass	5 arc minutes	2010	Several	Rate	Y	This work
	Soil	5 arc minutes	2010	Several	Rate	Y	This work
	Water resources	5 arc minutes	2010	Several	Rate	Y	This work
	Biodiversity	5 arc minutes	2010	Several	Rate	Y	This work
	Economic prod.	5 arc minutes	2010	Several	Rate	Y	This work
	Social and cultural	5 arc minutes	2010	Several	Rate	Y	This work
			GLADIS Outputs				
Ecosystem status	Biophysical Status Index	5 arc minutes	2010	Several	Rate	Y	This work
Ecosystem processes	Biophysical land degradation Index	5 arc minutes	2010	Several	Rate	Y	This work
	Land degradation Impact index	5 arc minutes	2010	Several	Rate	Y	This work

[a] For details in inputs and layers of the land use system (LUS) map refer to the specific report (Nachtergaele and Petri, 2008).

The input database included six axes: (1) biomass, (2) soil, (3) water, (4) biodiversity, (5) economic, and (6) social. Single layers and tables identifying categorical and continuous attributes are provided in Table 13.2.

13.5.1 Panning Method

A known application issue is due to the different panning methods of MapServer (acting as a WMS server) and OpenLayers. Due to performance issues, MapServer executes math calculation with integer numbers, and panning only applies to pixel's multiples. OpenLayers, on the other hand, does smooth panning. This means that the map shown by OpenLayers is not in the exact position provided by the WMS server, and offsets by an average of 1/2 pixel. Therefore, the GetFeatureInfo command, allowing retrieving information about features displayed in a map and based on MapServer, may present some slight imprecision (i.e., on radar graph preparation). The issue was not tacked as is only noticeable by user at maximum available zoom scale, and only when querying isolated pixels. Mitigating the issue would have imply to artificially augment the raster layer resolution, and consequently size of maps.

13.5.1.1 Metadata, Formats and Resolution Information, Layers

13.5.1.1.1 Metadata

The metadata are in ISO/DIS 19115 standard format, the FAO standards for metadata information. Data distribution applications such as GeoNetwork implement its metadata according to the scheme and specifications provided by this document.

13.5.1.1.2 Resolution, Projection, and Nomenclature

The datasets are produced in Geographic Coordinates and WGS 84 datum at a 5 arc minute resolution. Each pixel corresponds approximately to 9 km by 9 km at the equator. Most original GIS data were in all cases at 5 arc minutes. In a few cases data sources were national statistics. In other cases original data had a polygon format (.shp). To be consistent, the different databases used are in Geographic Coordinates, and have (un)projected depending in data sources projection. Alphanumeric, not-case-sensitive file-names have been used.

13.5.1.1.3 Formats

Data are produced and provided in ESRI GRID. GRID is a raster data storage format native to ESRI. There are two types of grids: integer and floating point. The use of integer GRIDs is common in representing raster data. Attributes for an integer grid are stored in a value attribute table (.VAT). A VAT has one record for each unique value in the grid. The record stores the unique value (VALUE is an integer that represents a particular class or grouping of cells) and the number of cells (COUNT) in the grid represented by that value. A raster attribute table is generated with three default fields created in the table: OID, VALUE, and COUNT. The ObjectID (OID) is a unique, system-defined, object identifier number for each row in the table. VALUE is a list of each unique cell value in the raster datasets. COUNT represents the number of cells in the raster dataset with the cell value in the VALUE column.

Data are also distributed in TIF (Tagged Image File Format) a format of widespread use in the desktop publishing world. It serves as an interface to several scanners and graphic arts packages. TIFF supports black-and-white, grayscale, pseudo-color, and true-color images, all of which can be stored in a compressed or decompressed format. GeoTIFF is a public domain metadata standard which allows geo-referencing information to be embedded within a TIFF file. Additional information that can be included is map projection, coordinate systems, ellipsoids, data, and everything else necessary to establish the exact spatial reference for the file. Download will be available in version 1 and is not included in beta version.

13.5.1.1.4 Attributes of the Database

For the GLADIS radar chart application and for the GLADIS database, data are provided as separate layers and they are not inserted into a database. The Land use systems version 1.1 is provided for download as detailed in Nachtergaele and Petri (2008).

13.5.1.1.5 Legends

Legends are provided for ESRI GRID format, both in.lyr and in.avl format. Legends will be available in version 1 together with downloadable data.

13.5.1.1.6 Quality of the Database

Data quality is and remains a major concern. Putting together by simple overlay global data layers of variable quality and resolution is a risky exercise that is bound to result in erroneous conclusions on the land use system practiced. Major problems with the individual databases used are well known (FAO, 2005) and are discussed in the report.

13.5.1.1.7 Position Accuracy

Most databases used came from FAO internal sources, so the position accuracy was already corrected. In other cases some consideration need to be done. In FAO/IIASA data the boundaries were buffered and a simple clipping was realistic. Soil map from the beta version of the Harmonized World Soil Database was corrected to match with GAUL by IIASA.

Expansion of outer boundaries and clipping was realized using a reference layer at a 5 arc minute resolution for data downloaded from sources external than FAO. Data were in all cases first reclassified, second resampled, and finally compared to the reference layer. All manipulating technique was applied with a snap to the reference grid.

13.5.1.1.8 Completeness

The work previously described allowed to create a quite consistent database in term of completeness. Few pixels in islands or far northern areas have most of layers containing no data.

Coastlines and water areas have been checked for consistency, the latter been compared with the LUS database (as the LUS layer is used as cartographic base for cartographic assessment). In a few cases the ESRI NIBBLE generalization command was used to solve issues of missing pixels. The command NIBBLE (ESRI) replaces cells of a raster with the values of the nearest neighbors.

13.5.1.1.9 Correctness

The correctness of this database is linked to the correctness of each single input layer.

13.5.1.1.10 Timeliness

Mixing up this amount of different data create issues on time precision. Data sources were from the early nineties (some statistical data) to the present. In some cases the year of data collection is unknown.

13.6 Conclusion

Land degradation is more than an environmental problem alone and should be considered in a holistic way, taking into account all ecosystem goods and services, biophysical as socioeconomic. The need for precise, up-to-date, and geo-referenced spatial digital data on degraded land for various purposes including assessment, monitoring, and modeling land degradation and biophysical and biogeochemical processes calls for development of an IS with standardized database and common scale, projection and datum, and format at global, regional, and local levels. At global level, the information on degraded land is needed to identify the regions experiencing degradation and the impact of land degradation on climate change. At regional level, such information serves as the base for planning the rehabilitation program for degraded lands. The information on land degradation derived from medium-resolution orbital sensor data (e.g., Landsat Operational Linear Imager-OLI, Resources at Linear Imaging Self-scanning Sensor-LISS-III, and the like) may be ideal database for generating information on degraded land useful for planning rehabilitation of degraded land.

GLADIS/GIS system—a global land degradation information system—is a timely initiative in this direction. For addressing the issue of implementation of rehabilitation program for degraded lands, detailed level information in terms of nature, extent, magnitude, and spatial distribution is a prerequisite. The database of GLADIS/GIS system, which is global in nature needs to be strengthened with further detailed level information coming in from various scientific research organizations and academia across the world using medium- and high-resolution earth observation systems such as Worldview-3 ans-4 and Cartosat-2. Furthermore, the emphasis also needs to be placed on eliminating the subjectivity in the existing IS, and on generating quantitative information on terrain and soils that could go into various ecological and other models including land degradation models.

References

Anderson, G., and Moreno-Sanchez. R., 2003. Building web-based spatial information solutions around open specifications and open source software. *Transactions in GIS*, 7(4), 447–466.

Batjes, N.H., 2009. Harmonized soil profile data for applications at global and continental scales: updates to the WISE database. *Soil Use and Management*, 25, 124–127.

Batjes, N.H., Reuter, H.I., Tempel, P., Hengl, T., Leenaars, J.G.B., and Bindraban, P.S., 2013. Development of global soil information facilities. *Data Science Journal*, 12, WDS70–WDS74.

Burrough, P.A., and McDonnell, R.A., 2005. *Principles of Geographic Information Systems*. Oxford University Press, Oxford.

Carvalho, J.A., 1999. Information System? Which One Do You Mean? In: Falkenberg, E., Lyytinen, K., and Verrijn-Stuart, A., (eds.) *Information Systems Concepts: An Integrated Discipline Emerging (Proceedings of the ISCO 4 Conference, Leiden, Holanda, September 20–22, 1999)*, Kluwer. Academic Publishers, 2000, pp. 259–280.

Ehrlich, P.R., and Ehrlich, A.H., 1991. *Healing the Planet: Strategies for Resolving the Environmental Crisis. Reading, Mass.* Addison-Wesley Publishing Co.

FAO, 1971–1981. FAO-UNESCO Soil Map of the World, 1:5,000,000 (Vol. 1 to 10). United Nations Educational, Scientific, and Cultural Organization, Paris; FAO, IIASA, ISRIC, ISSCAS, & Organization of the United Nations (FAO), International Institute for Applied Systems Analysis (IIASA), ISRIC—World Soil Information, Institute of Soil Science—Chinese Academy of Sciences (ISSCAS), Joint Research Centre of the European Commission (JRC).

FAO, IIASA, ISRIC, ISSCAS, and JRC, 2009. Harmonized World Soil Database (version 1.1). Food and Agriculture Organization of the United Nations (FAO), International Institute for Applied

Systems Analysis (IIASA), ISRIC—World Soil Information, Institute of Soil Science—Chinese Academy of Sciences (ISSCAS), Joint Research Centre of the European Commission (JRC), Laxenburg, Austria.

FAO-UNESCO, 1974. *FAO-UNESCO Soil Map of the World: Vol. 1, Legend*. UNESCO, Paris.

FAO, 2005. An inventory and comparison of globally consistent geospatial databases and libraries. Environmental and Natural Resources Working Paper 19.

FAO, 2009. Coping with water scarcity: the role of agriculture. A framework for action. A report submitted in preparation for the Expert Consultation—Coping with water scarcity—The role of agriculture, organized at FAO Headquarters, Rome, December 14–16, 2009.

FRA, 2005. *Global Forest Resources Assessment 2005, Food and Agricultural Organization*. Available at http://www.fao.org/forest-resources-assessment/past-assessments/fra-2005/en/.

Hengl, T., MacMillan, R.A., Walsh, M.G., and Reuter, H.I., 2011. *Global Soil Information Facilities: A Methodological Framework for Open Soil Information*. ISRIC—World Soil Information, Wageningen, p. 231

Hengl, T., de Jesus, J.M., MacMillan, R.A., Batjes, N.H., and Heuvelink, G.B.M., et al., 2014 SoilGrids 1 km—Global soil information based on automated mapping. *PLoS One*, 9(8), e105992. DOI:10.1371/journal.pone.0105992.

IUSS Working Group WRB , 2006. World reference base for soil resources: A framework for international classification, correlation and communication. World Soil Resources Report 103. Food and Agriculture Organization of The United Nations, Rome, 2006.

Kellner, K., Risoli, C., and Metz, M., 2011. Land Degradation Assessment in Drylands (LADA), Terminal Evaluation of the UNEP/FAO/GEF Project Evaluation Office, United Nations Environment Programme May 2011.

Kenneth, C., Laudon, K.C., Laudon, J.P., and Elragal, A.A., 2013. *Management Information Systems: Managing the Digital Firm*. Prentice Hall, Upper Saddle River, NJ.

Lebedys, A., 2008. Contribution of the forestry sector to national economies, 1990-2006. Forest Finance Working Papers Working Paper FSFM/ACC/08.

Lee, T.M., and Jetz, W., 2008. Future battlegrounds for conservation under global change. *Proceedings of the Royal Society B Biological Sciences* 275, 1261–1270.

Lubchenco, J., Risser, P.G., Janetos, A.C., Gosz, J.R., Gold, B.D., and Holland, M.M., 1993. Priorities for an environmental science agenda for the Clinton-Gore Administration: Recommendations for transition planning. *Bulletin of the Ecological Society of America*, 74, 4–8.

Maji, A.K., Reddy, G.P.O., CCPI-Regional Centres, 2004. NATP-(MM) Project on Integrated National Agricultural Resource Information System (INARIS Sub-Project- Soil Resource Database, Annual Report, 2004, pp. 47–51.

Nachtergaele, F.O., Petri, M., 2008. Mapping Land Use Systems at global and regional scales for land degradation assessment. LADA Technical Report #8, FAO, Rome.

Nachtergaele, F.O., Petri, M., Biancalani, R., van Lynden, G., van Velthuizen, H., and Bloise, M., 2011. Global Land Degradation Information System (GLADIS). Version 1.0. An Information database for Land Degradation Assessment at Global Level. LADA Technical Report No.17.

Nelson, A., 2008. *Estimated Travel Time to the Nearest City of 50,000 or More People in Year 2000. Global Environment Monitoring Unit - Joint Research Centre of the European Commission, Ispra, Italy.* Available at http://bioval.jrc.ec.europa.eu/products/gam/.

Nelson, G.C., Robertson, R.D., 2008. Green Gold or Green Wash: Environmental Consequences of Biofuels in the Developing World. Review of Agricultural Economics, 30(3), 517–529

Olson, D.M., and Dinerstein, E., 2002. The Global 200: Priority ecoregions for global conservation. (PDF file). *Annals of the Missouri Botanical Garden* 89, 125–126.

Panagos, P., Jones, A., Bosco, C., and Kumar, P.S.S., 2011. European digital archive on soil maps (EuDASM): Preserving important soil data for public free access. *International Journal of Digital Earth*, 4, 434–443.

Paresi, C.M., 1998. Study guide, Information Systems Development, Introduction to Information Systems; Development and to Information Systems; Development methodologies; ITC.

Richardson, C.J., 1994. Ecological functions and human values in wetlands: A framework for assessing forestry impacts. *Wetlands,* 14, 1–9.

Sanchez, P.A., Ahamed, S., Carré, F., Hartemink, A.E., Hempel, J., Huising, J., et al., 2009. Digital soil map of the world. *Science* 325:680–681. doi:10.1126/science.1175084

Shoba, S., Alyabina, I., and Kolesnikova, V., 2010. Soil-Geographic Database of Russia: Database Management System Soil-DB. 19th World Congress of Soil Science, Soil Solutions for a Changing World 1–6 August 2010, Brisbane, Australia. Published on DVD.

Stolbovoi, V., and Fischer, G., 1997. A New Digital Georeferenced Database of Soil Degradation in Russia. IIASA Interim Report. IIASA, Laxenburg, Austria: IR–97–084.

Soil Survey Staff, 1975. *Soil Taxonomy: A Basic System of Soil Classification for Making and Interpreting Soil Surveys.* Agricultural Handbook 436. US Govt, Printing Office, Washington D.C.

Tempel, P., and van Kraalingen, D., 2011. Global Soil Information Facility: Database component. ISRIC—World Soil Information, ISRIC Report 2011/03.

UNWTO, 2009. *Tourism Highlights 2009 Edition. United Nations World Tourism Organization.* Available at https://www.e-unwto.org/doi/book/10.18111/9789284413591.

Valacich, J., and Schneider, C., 2010. *Information Systems Today: Managing in the Digital World,* 4th edn. Prentice Hall, Upper Saddle River, NJ.

Van Engelen, V.W.P., (2011) Standardizing soil data (e-SOTER regional pilot platform as EU contribution to a Global Soil Information System). International Innovation June, pp. 48–49.

WDPA, 2009. *The World Dataset on Protected Areas.* Available at https://www.iucn.org/theme/protected-areas/our-work/world-database-protected-areas.

Index

J

K

L

S

T

For Product Safety Concerns and Information please contact our EU representative GPSR@taylorandfrancis.com
Taylor & Francis Verlag GmbH, Kaufingerstraße 24, 80331 München, Germany

www.ingramcontent.com/pod-product-compliance
Ingram Content Group UK Ltd.
Pitfield, Milton Keynes, MK11 3LW, UK
UKHW050756180926
475701UK00007B/1089